ELEMENTARY ALGEBRA
CONCEPTS AND APPLICATIONS

FIFTH EDITION

BITTINGER • ELLENBOGEN

STUDENT'S SOLUTIONS MANUAL

JUDITH A. PENNA

Indiana University—Purdue University at Indianapolis

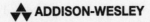

ADDISON-WESLEY

An imprint of Addison Wesley Longman, Inc.

Reading, Massachusetts • Menlo Park, California • New York • Harlow, England
Don Mills, Ontario • Sydney • Mexico City • Madrid • Amsterdam

ISBN 0-201-30497-X

2 3 4 5 6 7 8 9 10 CRS 00 99 98

Table of Contents

Chapter 1

Introduction to Algebraic Expressions

Exercise Set 1.1

1. Substitute 9 for a and multiply.
   $5a = 5 \cdot 9 = 45$

2. 63

3. Substitute 2 for t and add.
   $t + 6 = 2 + 6 = 8$

4. 4

5. $\dfrac{x+y}{2} = \dfrac{2+14}{2} = \dfrac{16}{2} = 8$

6. 7

7. $\dfrac{m-n}{7} = \dfrac{20-6}{7} = \dfrac{14}{7} = 2$

8. 3

9. $\dfrac{a}{b} = \dfrac{45}{9} = 5$

10. 6

11. $\dfrac{9m}{q} = \dfrac{9 \cdot 6}{18} = \dfrac{54}{18} = 3$

12. 3

13. $rt = (55 \text{ mph})(3 \text{ hr}) = 165$ mi

14. 24 hr

15. $bh = (6.5 \text{ cm})(15.4 \text{ cm})$
    $= (6.5)(15.4)(\text{cm})(\text{cm})$
    $= 100.1 \text{ cm}^2$, or 100.1 square centimeters

16. $\dfrac{3}{11}$, or about 0.273

17. $A = \dfrac{1}{2}bh$
    $= \dfrac{1}{2}(5 \text{ cm})(6 \text{ cm})$
    $= \dfrac{1}{2}(5)(6)(\text{cm})(\text{cm})$
    $= \dfrac{5}{2} \cdot 6 \text{ cm}^2$
    $= 15 \text{ cm}^2$, or 15 square centimeters

18. 150 sec, 450 sec, 10 min

19. $7x$, or $x7$

20. $4a$

21. $b + 6$, or $6 + b$

22. $t + 8$, or $8 + t$

23. $c - 9$

24. $d - 4$

25. $q + 6$, or $6 + q$

26. $z + 11$, or $11 + z$

27. Let n represent "a number." Then we have $n + 9$, or $9 + n$.

28. $d + c$, or $c + d$

29. $y - x$

30. Let x represent "a number;" $x - 2$

31. $x \div w$, or $\dfrac{x}{w}$

32. Let s and t represent the numbers; $s \div t$, or $\dfrac{s}{t}$

33. $n - m$

34. $q - p$

35. Let a and b represent the numbers. Then we have $a + b$, or $b + a$.

36. $d + f$, or $f + d$

37. $9 \cdot 2m$

38. $t - 2r$

39. Let y represent "some number." Then we have $\dfrac{1}{4}y$, or $\dfrac{y}{4}$.

40. Let m and n represent the numbers; $\dfrac{1}{3}mn$, or $\dfrac{mn}{3}$

41. Let x represent "some number." Then we have 64% of x, or $0.64x$.

42. Let y represent "a number;" 38% of y, or $0.38y$.

43. $\$50 - x$

44. $65t$ mi

45. $\underline{\quad x + 17 = 32 \quad}$ Writing the equation
    $15 + 17 \;?\; 32$ Substituting 15 for x
    $\quad\quad 32 \mid 32$ $32 = 32$ is TRUE.

    Since the left-hand and right-hand sides are the same, 15 is a solution.

46. No

47. $\underline{a - 28 = 75}$　　　Writing the equation

$\quad 93 - 28 \ ? \ 75$　　Substituting 93 for a

$\quad \quad 65 \ | \ 75$　　　$65 = 75$ is FALSE.

Since the left-hand and right-hand sides are not the same, 93 is not a solution.

48. Yes

49. $\dfrac{t}{7} = 9$

$\quad \dfrac{63}{7} \ ? \ 9$

$\quad \quad 9 \ | \ 9$　　$9 = 9$ is TRUE.

Since the left-hand and right-hand sides are the same, 63 is a solution.

50. No

51. $\dfrac{108}{x} = 36$

$\quad \dfrac{108}{3} \ ? \ 36$

$\quad \quad 36 \ | \ 36$　　$36 = 36$ is TRUE.

Since the left-hand and right-hand sides are the same, 3 is a solution.

52. No

53. Let x represent the number.

$$\underbrace{\text{What number}}\ \underbrace{\text{added to}}\ 73\ \text{is}\ 201?$$
$$\downarrow \qquad\quad \downarrow \qquad\ \downarrow\ \downarrow\ \downarrow$$

Translating:　　　　x　　　　$+$　　　$73 = 201$

$x + 73 = 201$

54. Let w represent the number; $7w = 2303$

55. Let y represent the number.

Rewording:　　42 times $\underbrace{\text{what number}}$ is 2352?

$$\downarrow\ \downarrow \qquad\quad \downarrow \qquad\quad \downarrow\ \downarrow$$

Translating:　42　$\cdot$　　　y　　　$= 2352$

$42y = 2352$

56. Let x represent the number; $x + 345 = 987$

57. Let s represent the number of squares your opponent gets.

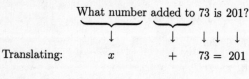

Rewording:　$\underbrace{\begin{array}{c}\text{The number}\\ \text{of squares}\\ \text{your}\\ \text{opponent}\\ \text{gets}\end{array}}$　$\underbrace{\text{added to}}$ 35 is 64.

$$\downarrow \qquad\qquad \downarrow \quad \downarrow\ \downarrow\ \downarrow$$

Translating:　　　s　　　　　$+$　　$35 = 64$

$s + 35 = 64$

58. Let y represent the number of hours the carpenter worked; $25y = 53,400$

59. Let m represent the length of the average commute in the West, in minutes.

Rewording:　$\underbrace{\begin{array}{c}\text{The}\\ \text{length}\\ \text{of the}\\ \text{average}\\ \text{commute}\\ \text{in the}\\ \text{West}\end{array}}$　is　$\underbrace{\begin{array}{c}24.5\\ \text{minutes}\end{array}}$　less　$\underbrace{\begin{array}{c}1.8\\ \text{minutes.}\end{array}}$

$$\downarrow \qquad \downarrow \qquad \downarrow \qquad \downarrow \qquad \downarrow$$

Translating:　m　$=$　24.5　$-$　1.8

$m = 24.5 - 1.8$

60. Let f represent the amount of fuel used by trucks in the U.S. in 1995, in billions of gallons; $f = 2(31.4)$

61. ◈

62. ◈

63. ◈

64. ◈

65. Let n represent "a number." Then we have $3n - 5$.

66. Let a and b represent the numbers; $\dfrac{1}{3} \cdot \dfrac{1}{2}ab$, or $\dfrac{1}{6}ab$, or $\dfrac{ab}{6}$

67. $l + w + l + w$, or $2l + 2w$

68. $s + s + s + s$, or $4s$

69. $a + 2 + 7$, or $a + 9$

70. \$337.50

71. The shaded area is the area of a rectangle with dimensions 20 cm by 10 cm less the area of a triangle with base 20 cm $- 4$ cm $- 5$ cm, or 11 cm, and height 7.5 cm. We perform the computation

$$(20 \text{ cm})(10 \text{ cm}) - \frac{1}{2}(11 \text{ cm})(7.5)$$

$$= 200 \text{ cm}^2 - 41.25 \text{ cm}^2$$

$$= 158.75 \text{ cm}^2, \text{ or } 158.75 \text{ square centimeters}$$

72. 6

73. $y = 35$, $x = 2y = 2 \cdot 35 = 70$;
$$\frac{x-y}{7} = \frac{70-35}{7} = \frac{35}{7} = 5$$

74. 6

75. $\dfrac{y+x}{2} + \dfrac{3 \cdot y}{x} = \dfrac{4+2}{2} + \dfrac{3 \cdot 4}{2} = \dfrac{6}{2} + \dfrac{12}{2} = 3 + 6 = 9$

76. $w + 4$

77. The preceding odd number is 2 less than $d + 2$:
$$d + 2 - 2 = d$$

78. d

Exercise Set 1.2

1. $x + 3$ Changing the order

2. $2 + a$

3. $bc + a$

4. $3y + x$

5. $3y + 9x$

6. $7b + 3a$

7. $2(3 + a)$

8. $9(5 + x)$

9. $a \cdot 8$ Changing the order

10. yx

11. ts

12. $b7$

13. $5 + ba$

14. $x + y3$

15. $(a + 3)2$

16. $(x + 5)9$

17. $a + (3 + b)$

18. $5 + (m + r)$

19. $(r + t) + 9$

20. $(x + 2) + y$

21. $ab + (c + d)$

22. $m + (np + r)$

23. $3(xy)$

24. $9(ab)$

25. $(4x)y$

26. $(9r)p$

27. $(3 \cdot 2)(a + b)$

28. $(5x)(2 + y)$

29. a) $r + (t + 6) = (t + 6) + r$ Using the commutative law
$$= (6 + t) + r \quad \text{Using the commutative law again}$$

   b) $r + (t + 6) = (t + 6) + r$ Using the commutative law
$$= t + (6 + r) \quad \text{Using the associative law}$$
Answers may vary.

30. (a) $v + (w + 5)$; (b) $(v + 5) + w$; answers may vary

31. a) $(5a)b = b(5a)$ Using the commutative law
$$= b(a5) \quad \text{Using the commutative law again}$$

   b) $(5a)b = (a5)b$ Using the commutative law
$$= a(5b) \quad \text{Using the associative law}$$
Answers may vary.

32. (a) $3(yx)$; (b) $(3x)y$; answers may vary

33. $(7 + x) + 2$
$$= (x + 7) + 2 \quad \text{Commutative law}$$
$$= x + (7 + 2) \quad \text{Associative law}$$
$$= x + 9 \quad \text{Simplifying}$$

34. $(2a)4 = 4(2a)$ Commutative law
$$= (4 \cdot 2)a \quad \text{Associative law}$$
$$= 8a \quad \text{Simplifying}$$

35. $(m3)7 = m(3 \cdot 7)$ Associative law
$$= (3 \cdot 7)m \quad \text{Commutative law}$$
$$= 21m \quad \text{Simplifying}$$

36. $4 + (9 + x)$
$$= (4 + 9) + x \quad \text{Associative law}$$
$$= x + (4 + 9) \quad \text{Commutative law}$$
$$= x + 13 \quad \text{Simplifying}$$

37. $3(a + 4) = 3 \cdot a + 3 \cdot 4 = 3a + 12$

38. $4x + 12$

39. $6(1 + x) = 6 \cdot 1 + 6 \cdot x = 6 + 6x$

40. $6v + 24$

41. $3(x + 1) = 3 \cdot x + 3 \cdot 1 = 3x + 3$

42. $9x + 27$

43. $4(1 + y) = 4 \cdot 1 + 4 \cdot y = 4 + 4y$

44. $7s + 35$

45. $9(2x + 6) = 9 \cdot 2x + 9 \cdot 6 = 18x + 54$

46. $54m + 63$

47. $5(r + 2 + 3t) = 5 \cdot r + 5 \cdot 2 + 5 \cdot 3t = 5r + 10 + 15t$

48. $20x + 32 + 12p$

49. $(a + b)2 = a(2) + b(2) = 2a + 2b$

50. $7x + 14$

51. $(x + y + 2)5 = x(5) + y(5) + 2(5) = 5x + 5y + 10$

52. $12 + 6a + 6b$

53. $7x + 7z = 7(x + z)$ The common factor is 7.
   Check: $7(x + z) = 7 \cdot x + 7 \cdot z = 7x + 7z$

54. $5(y + z)$

55. $5 + 5y = 5 \cdot 1 + 5 \cdot y$ The common factor is 5.
       $= 5(1 + y)$ Using the distributive
                                       law
   Check: $5(1 + y) = 5 \cdot 1 + 5 \cdot y = 5 + 5y$

56. $13(1 + x)$

57. $18x + 3y = 3 \cdot 6x + 3 \cdot y = 3(6x + y)$
   Check: $3(6x + y) = 3 \cdot 6x + 3 \cdot y = 18x + 3y$

58. $5(x + 4y)$

59. $5x + 10 + 15y = 5 \cdot x + 5 \cdot 2 + 5 \cdot 3y = 5(x + 2 + 3y)$
   Check: $5(x+2+3y) = 5 \cdot x + 5 \cdot 2 + 5 \cdot 3y = 5x + 10 + 15y$

60. $3(1 + 9b + 2c)$

61. $12x + 9 = 3 \cdot 4x + 3 \cdot 3 = 3(4x + 3)$
   Check: $3(4x + 3) = 3 \cdot 4x + 3 \cdot 3 = 12x + 9$

62. $6(x + 1)$

63. $9x + 3y = 3 \cdot 3x + 3 \cdot y = 3(3x + y)$
   Check: $3(3x + y) = 3 \cdot 3x + 3 \cdot y = 9x + 3y$

64. $5(3x + y)$

65. $2a + 16b + 64 = 2 \cdot a + 2 \cdot 8b + 2 \cdot 32 = 2(a + 8b + 32)$
   Check: $2(a+8b+32) = 2 \cdot a + 2 \cdot 8b + 2 \cdot 32 = 2a + 16b + 64$

66. $5(1 + 4x + 7y)$

67. $11x + 44y + 121 = 11 \cdot x + 11 \cdot 4y + 11 \cdot 11 = 11(x + 4y + 11)$
   Check: $11(x + 4y + 11) = 11 \cdot x + 11 \cdot 4y + 11 \cdot 11 = 11x + 44y + 121$

68. $7(1 + 2b + 8w)$

69. $t - 9$

70. $\frac{1}{2}m$, or $\frac{m}{2}$

71. ◈

72. ◈

73. ◈

74. ◈

75. The expressions are equivalent.
   $8 + 4(a + b) = 8 + 4a + 4b = 4(2 + a + b)$

76. No; for example, let $m = 1$. Then $7 \div 3 \cdot 1 = \frac{7}{3}$, but
   $1 \cdot 3 \div 7 = \frac{3}{7}$.

77. The expressions are equivalent.
   $(rt + st)5 = 5(rt + st) = 5 \cdot t(r + s) = 5t(r + s)$

78. Yes; $yax + ax = (y + 1)ax = (1 + y)ax = ax(1 + y) = xa(1 + y)$.

79. The expressions are not equivalent.
   Let $x = 1$ and $y = 0$. Then we have:
       $30 \cdot 0 + 1 \cdot 15 = 0 + 15 = 15$, but
       $5[2(1 + 3 \cdot 0)] = 5[2(1)] = 5 \cdot 2 = 10$.

80. Yes; $[c(2 + 3b)]5 = 5[c(2 + 3b)] = 5c(2 + 3b) = 10c + 15bc$.

81. ◈

82. ◈

Exercise Set 1.3

1. We write two factorizations of 30. There are other
   factorizations as well.
       $2 \cdot 15, 3 \cdot 10$
   List all of the factors of 30:
   1, 2, 3, 5, 6, 10, 15, 30

2. $2 \cdot 35, 5 \cdot 14$; 1, 2, 5, 7, 10, 14, 35, 70

3. We write two factorizations of 42. There are other
   factorizations as well.
       $2 \cdot 21, 6 \cdot 7$
   List all of the factors of 42:
   1, 2, 3, 6, 7, 14, 21, 42

4. $2 \cdot 30, 5 \cdot 12$; 1, 2, 3, 4, 5, 6, 10, 12, 15, 20, 30, 60

5. $22 = 2 \cdot 11$

6. $3 \cdot 5$

7. We begin factoring 30 in any way that we can and continue factoring until each factor is prime.
$$30 = 2 \cdot 15 = 2 \cdot 3 \cdot 5$$

8. $5 \cdot 11$

9. We begin by factoring 50 in any way that we can and continue factoring until each factor is prime.
$$50 = 2 \cdot 25 = 2 \cdot 5 \cdot 5$$

10. $2 \cdot 2 \cdot 5$

11. We begin by factoring 27 in any way that we can and continue factoring until each factor is prime.
$$27 = 3 \cdot 9 = 3 \cdot 3 \cdot 3$$

12. $2 \cdot 7 \cdot 7$

13. We begin by factoring 18 in any way that we can and continue factoring until each factor is prime.
$$18 = 2 \cdot 9 = 2 \cdot 3 \cdot 3$$

14. $2 \cdot 2 \cdot 2 \cdot 3$

15. We begin by factoring 40 in any way that we can and continue factoring until each factor is prime.
$$40 = 4 \cdot 10 = 2 \cdot 2 \cdot 2 \cdot 5$$

16. $2 \cdot 2 \cdot 2 \cdot 7$

17. 43 has exactly two different factors, 43 and 1. Thus, 43 is prime.

18. $2 \cdot 2 \cdot 2 \cdot 3 \cdot 5$

19. $210 = 2 \cdot 105 = 2 \cdot 3 \cdot 35 = 2 \cdot 3 \cdot 5 \cdot 7$

20. Prime

21. $115 = 5 \cdot 23$

22. $11 \cdot 13$

23. $\dfrac{10}{14} = \dfrac{2 \cdot 5}{2 \cdot 7}$ Factoring numerator and denominator

$\quad = \dfrac{2}{2} \cdot \dfrac{5}{7}$ Rewriting as a product of two fractions

$\quad = 1 \cdot \dfrac{5}{7}$ $\quad \dfrac{2}{2} = 1$

$\quad = \dfrac{5}{7}$ Using the identity property of 1

24. $\dfrac{2}{7}$

25. $\dfrac{49}{14} = \dfrac{7 \cdot 7}{2 \cdot 7} = \dfrac{7}{2} \cdot \dfrac{7}{7} = \dfrac{7}{2} \cdot 1 = \dfrac{7}{2}$

26. $\dfrac{8}{3}$

27. $\dfrac{6}{48} = \dfrac{1 \cdot 6}{8 \cdot 6}$ Factoring and using the identity property of 1 to write 6 as $1 \cdot 6$

$\quad = \dfrac{1}{8} \cdot \dfrac{6}{6}$

$\quad = \dfrac{1}{8} \cdot 1 = \dfrac{1}{8}$

28. $\dfrac{6}{35}$

29. $\dfrac{56}{7} = \dfrac{8 \cdot 7}{1 \cdot 7} = \dfrac{8}{1} \cdot \dfrac{7}{7} = \dfrac{8}{1} \cdot 1 = 8$

30. 12

31. $\dfrac{19}{76} = \dfrac{1 \cdot 19}{4 \cdot 19}$ Factoring and using the identity property of 1 to write 19 as $1 \cdot 19$

$\quad = \dfrac{1 \cdot \cancel{19}}{4 \cdot \cancel{19}}$ Removing a factor equal to 1: $\dfrac{19}{19} = 1$

$\quad = \dfrac{1}{4}$

32. $\dfrac{1}{3}$

33. $\dfrac{100}{20} = \dfrac{5 \cdot 20}{1 \cdot 20}$ Factoring and using the identity property of 1 to write 20 as $1 \cdot 20$

$\quad = \dfrac{5 \cdot \cancel{20}}{1 \cdot \cancel{20}}$ Removing a factor equal to 1: $\dfrac{20}{20} = 1$

$\quad = \dfrac{5}{1}$

$\quad = 5$ Simplifying

34. 6

35. $\dfrac{75}{80} = \dfrac{5 \cdot 15}{5 \cdot 16}$ Factoring the numerator and the denominator

$\quad = \dfrac{\cancel{5} \cdot 15}{\cancel{5} \cdot 16}$ Removing a factor equal to 1: $\dfrac{5}{5} = 1$

$\quad = \dfrac{15}{16}$

36. $\dfrac{21}{25}$

37. $\dfrac{120}{82} = \dfrac{2 \cdot 60}{2 \cdot 41}$ Factoring

$\quad = \dfrac{\cancel{2} \cdot 60}{\cancel{2} \cdot 41}$ Removing a factor equal to 1: $\dfrac{2}{2} = 1$

$\quad = \dfrac{60}{41}$

38. $\dfrac{5}{3}$

39. $\dfrac{210}{98} = \dfrac{2 \cdot 7 \cdot 15}{2 \cdot 7 \cdot 7}$ Factoring

$\phantom{\dfrac{210}{98}} = \dfrac{\cancel{2} \cdot \cancel{7} \cdot 15}{\cancel{2} \cdot \cancel{7} \cdot 7}$ Removing a factor equal to 1: $\dfrac{2 \cdot 7}{2 \cdot 7} = 1$

$\phantom{\dfrac{210}{98}} = \dfrac{15}{7}$

40. $\dfrac{2}{5}$

41. $\dfrac{3}{5} \cdot \dfrac{1}{2} = \dfrac{3 \cdot 1}{5 \cdot 2}$ Multiplying numerators and denominators

$\phantom{\dfrac{3}{5} \cdot \dfrac{1}{2}} = \dfrac{3}{10}$

42. $\dfrac{44}{25}$

43. $\dfrac{17}{2} \cdot \dfrac{3}{4} = \dfrac{17 \cdot 3}{2 \cdot 4} = \dfrac{51}{8}$

44. 1

45. $\dfrac{1}{8} + \dfrac{3}{8} = \dfrac{1+3}{8}$ Adding numerators; keeping the common denominator

$\phantom{\dfrac{1}{8} + \dfrac{3}{8}} = \dfrac{4}{8}$

$\phantom{\dfrac{1}{8} + \dfrac{3}{8}} = \dfrac{1 \cdot \cancel{4}}{2 \cdot \cancel{4}} = \dfrac{1}{2}$ Simplifying

46. $\dfrac{5}{8}$

47. $\dfrac{4}{9} + \dfrac{13}{18} = \dfrac{4}{9} \cdot \dfrac{2}{2} + \dfrac{13}{18}$ Using 18 as the common denominator

$\phantom{\dfrac{4}{9} + \dfrac{13}{18}} = \dfrac{8}{18} + \dfrac{13}{18}$

$\phantom{\dfrac{4}{9} + \dfrac{13}{18}} = \dfrac{21}{18}$

$\phantom{\dfrac{4}{9} + \dfrac{13}{18}} = \dfrac{7 \cdot \cancel{3}}{6 \cdot \cancel{3}} = \dfrac{7}{6}$ Simplifying

48. $\dfrac{4}{3}$

49. $\dfrac{3}{a} \cdot \dfrac{b}{7} = \dfrac{3b}{7a}$ Multiplying numerators and denominators

50. $\dfrac{xy}{5z}$

51. $\dfrac{4}{a} + \dfrac{3}{a} = \dfrac{7}{a}$ Adding numerators; keeping the common denominator

52. $\dfrac{2}{a}$

53. $\dfrac{3}{10} + \dfrac{8}{15} = \dfrac{3}{10} \cdot \dfrac{3}{3} + \dfrac{8}{15} \cdot \dfrac{2}{2}$ Using 30 as the common denominator

$\phantom{\dfrac{3}{10} + \dfrac{8}{15}} = \dfrac{9}{30} + \dfrac{16}{30}$

$\phantom{\dfrac{3}{10} + \dfrac{8}{15}} = \dfrac{25}{30}$

$\phantom{\dfrac{3}{10} + \dfrac{8}{15}} = \dfrac{5 \cdot \cancel{5}}{6 \cdot \cancel{5}} = \dfrac{5}{6}$ Simplifying

54. $\dfrac{41}{24}$

55. $\dfrac{9}{7} - \dfrac{2}{7} = \dfrac{7}{7} = 1$

56. 2

57. $\dfrac{13}{18} - \dfrac{4}{9} = \dfrac{13}{18} - \dfrac{4}{9} \cdot \dfrac{2}{2}$ Using 18 as the common denominator

$\phantom{\dfrac{13}{18} - \dfrac{4}{9}} = \dfrac{13}{18} - \dfrac{8}{18}$

$\phantom{\dfrac{13}{18} - \dfrac{4}{9}} = \dfrac{5}{18}$

58. $\dfrac{31}{45}$

59. $\dfrac{5}{6} - \dfrac{5}{18} = \dfrac{5}{6} \cdot \dfrac{3}{3} - \dfrac{5}{18}$ Using 18 as the common denominator

$\phantom{\dfrac{5}{6} - \dfrac{5}{18}} = \dfrac{15}{18} - \dfrac{5}{18}$

$\phantom{\dfrac{5}{6} - \dfrac{5}{18}} = \dfrac{10}{18}$

$\phantom{\dfrac{5}{6} - \dfrac{5}{18}} = \dfrac{\cancel{2} \cdot 5}{\cancel{2} \cdot 9} = \dfrac{5}{9}$ Simplifying

60. $\dfrac{13}{48}$

61. $\dfrac{7}{6} \div \dfrac{3}{5} = \dfrac{7}{6} \cdot \dfrac{5}{3}$ Multiplying by the reciprocal of the divisor

$\phantom{\dfrac{7}{6} \div \dfrac{3}{5}} = \dfrac{35}{18}$

62. $\dfrac{28}{15}$

63. $\dfrac{8}{9} \div \dfrac{4}{15} = \dfrac{8}{9} \cdot \dfrac{15}{4} = \dfrac{2 \cdot \cancel{4} \cdot \cancel{3} \cdot 5}{\cancel{3} \cdot 3 \cdot \cancel{4}} = \dfrac{10}{3}$

64. $\dfrac{1}{4}$

65. $6 \div \dfrac{3}{7} = \dfrac{6}{1} \cdot \dfrac{7}{3} = \dfrac{2 \cdot \cancel{3} \cdot 7}{1 \cdot \cancel{3}} = 14$

66. $\dfrac{1}{2}$

67. $\dfrac{13}{12} \div \dfrac{39}{5} = \dfrac{13}{12} \cdot \dfrac{5}{39} = \dfrac{\cancel{13} \cdot 5}{12 \cdot 3 \cdot \cancel{13}} = \dfrac{5}{36}$

68. $\dfrac{68}{9}$

69. $24 \div \dfrac{2}{5} = \dfrac{24}{1} \cdot \dfrac{5}{2} = \dfrac{\cancel{2} \cdot 12 \cdot 5}{1 \cdot \cancel{2}} = 60$

70. 468

71. $\dfrac{3x}{4} \div 6 = \dfrac{3x}{4} \cdot \dfrac{1}{6} = \dfrac{\cancel{3} \cdot x}{4 \cdot 2 \cdot \cancel{3}} = \dfrac{x}{8}$

72. $\dfrac{1}{18a}$

73. $\dfrac{5}{3} \div \dfrac{a}{b} = \dfrac{5}{3} \cdot \dfrac{b}{a} = \dfrac{5b}{3a}$

74. $\dfrac{xy}{28}$

75. $\dfrac{x}{6} - \dfrac{1}{3} = \dfrac{x}{6} - \dfrac{1}{3} \cdot \dfrac{2}{2}$ Using 6 as the common denominator

$= \dfrac{x}{6} - \dfrac{2}{6}$

$= \dfrac{x - 2}{6}$

76. $\dfrac{9 + 5x}{10}$

77. $5(x + 3) = 5(3 + x)$ Commutative law of addition

Answers may vary.

78. $(a + b) + 7$; answers may vary.

79. ◈

80. ◈

81. ◈

82. ◈

83. $\dfrac{256}{192} = \dfrac{4 \cdot \cancel{64}}{3 \cdot \cancel{64}} = \dfrac{4}{3}$

84. $\dfrac{p}{t}$

85. $\dfrac{9 \cdot 4 \cdot 16}{8 \cdot 15 \cdot 12} = \dfrac{\cancel{3} \cdot 3 \cdot \cancel{2} \cdot 2 \cdot \cancel{4} \cdot 4}{\cancel{2} \cdot \cancel{4} \cdot \cancel{3} \cdot 5 \cdot \cancel{3} \cdot \cancel{4}} = \dfrac{2}{5}$

86. 1

87. $\dfrac{15 \cdot 4xy \cdot 9}{6 \cdot 25x \cdot 15y} = \dfrac{\cancel{15} \cdot \cancel{2} \cdot 2 \cdot \cancel{x} \cdot \cancel{y} \cdot \cancel{3} \cdot 3}{\cancel{2} \cdot \cancel{3} \cdot 25 \cdot \cancel{x} \cdot \cancel{15} \cdot \cancel{y}} = \dfrac{6}{25}$

88. $\dfrac{5}{2}$

89. We need to find the smallest number that has both 6 and 8 as factors. Starting with 6 we list some numbers with a factor of 6, and starting with 8 we also list some numbers with a factor of 8. Then we find the first number that is on both lists.

6, 12, 18, 24, 30, 36, ...

8, 16, 24, 32, 40, 48, ...

Since 24 is the smallest number that is on both lists, the carton should be 24 in. long.

90.

Product	56	63	36	72	140	96
Factor	7	7	2	36	14	8
Factor	8	9	18	2	10	12
Sum	15	16	20	38	24	20

Product	48	168	110	90	432	63
Factor	6	21	11	9	24	3
Factor	8	8	10	10	18	21
Sum	14	29	21	19	42	24

91. $A = lw = \left(\dfrac{4}{5}\,\text{m}\right)\left(\dfrac{7}{9}\,\text{m}\right)$

$= \left(\dfrac{4}{5}\right)\left(\dfrac{7}{9}\right)(\text{m})(\text{m})$

$= \dfrac{28}{45}\,\text{m}^2$, or $\dfrac{28}{45}$ square meters

92. $\dfrac{25}{28}\,\text{m}^2$

93. $P = 4s = 4\left(\dfrac{5}{9}\text{m}\right) = \dfrac{20}{9}\,\text{m}$

94. $\dfrac{142}{45}\,\text{m}$

95. $x + y = (x + y) \cdot 1$ Identity property of 1

$= (x + y) \cdot \dfrac{2}{2}$ $\dfrac{2}{2} = 1$

$= \dfrac{2}{2} \cdot (x + y)$ Commutative law

$= \dfrac{2x + 2y}{2}$ Distributive law

$= \dfrac{2y + 2x}{2}$ Commutative law

Exercise Set 1.4

1. The real number -3 corresponds to 3 under par, and the real number 7 corresponds to 7 over par.

2. $18, -2$

3. The real number -508 corresponds to a drop of 508 points, and the real number 186.84 corresponds to an increase of 186.84 points.

4. $1200, -560$

5. The real number -1286 corresponds to 1286 ft below sea level. The real number 29,029 corresponds to 29,029 ft above sea level.

6. Jets: -34, Strikers: 34

7. The real number 750 corresponds to a \$750 deposit, and the real number -125 corresponds to a \$125 withdrawal.

8. $27, -9.7$

9. The real numbers 20, −150, and 300 correspond to the interception of the missile, the loss of the starship, and the capture of the base, respectively.

10. −10, 235

11. Since $\frac{10}{3} = 3\frac{1}{3}$, its graph is $\frac{1}{3}$ of a unit to the right of 3.

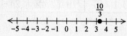

12.

13. The graph of −4.3 is $\frac{3}{10}$ of a unit to the left of −4.

14.

15.

16.

17. $\frac{5}{8}$ means $5 \div 8$, so we divide.

$$
\begin{array}{r}
0.6\,2\,5 \\
8\,\overline{)\,5.0\,0\,0} \\
4\,8 \\
\hline
2\,0 \\
1\,6 \\
\hline
4\,0 \\
4\,0 \\
\hline
0
\end{array}
$$

We have $\frac{5}{8} = 0.625$.

18. −0.125

19. We first find decimal notation for $\frac{3}{4}$. Since $\frac{3}{4}$ means $3 \div 4$, we divide.

$$
\begin{array}{r}
0.7\,5 \\
4\,\overline{)\,3.0\,0} \\
2\,8 \\
\hline
2\,0 \\
2\,0 \\
\hline
0
\end{array}
$$

Thus, $\frac{3}{4} = 0.75$, so $-\frac{3}{4} = -0.75$.

20. $0.8\overline{3}$

21. $\frac{7}{6}$ means $7 \div 6$, so we divide.

$$
\begin{array}{r}
1.1\,6\,6 \\
6\,\overline{)\,7.0\,0\,0} \\
6 \\
\hline
1\,0 \\
6 \\
\hline
4\,0 \\
3\,6 \\
\hline
4\,0 \\
3\,6 \\
\hline
4
\end{array}
$$

We have $\frac{7}{6} = 1.1\overline{6}$.

22. $0.41\overline{6}$

23. $\frac{2}{3}$ means $2 \div 3$, so we divide.

$$
\begin{array}{r}
0.6\,6\,6\,... \\
3\,\overline{)\,2.0\,0\,0} \\
1\,8 \\
\hline
2\,0 \\
1\,8 \\
\hline
2\,0 \\
1\,8 \\
\hline
2
\end{array}
$$

We have $\frac{2}{3} = 0.\overline{6}$.

24. 0.25

25. We first find decimal notation for $\frac{1}{2}$. Since $\frac{1}{2}$ means $1 \div 2$, we divide.

$$
\begin{array}{r}
0.5 \\
2\,\overline{)\,1.0} \\
1\,0 \\
\hline
0
\end{array}
$$

Thus, $\frac{1}{2} = 0.5$, so $-\frac{1}{2} = -0.5$.

26. −0.375

27. $\frac{1}{10}$ means $1 \div 10$, so we divide.

$$
\begin{array}{r}
0.1 \\
10\,\overline{)\,1.0} \\
1\,0 \\
\hline
0
\end{array}
$$

We have $\frac{1}{10} = 0.1$.

28. −0.35

29. Since 5 is to the right of 0, we have $5 > 0$.

30. >

31. Since −9 is to the left of 5, we have $-9 < 5$.

32. >

33. Since -6 is to the left of 6, we have $-6 < 6$.

34. $>$

35. Since -8 is to the left of -5, we have $-8 < -5$.

36. $<$

37. Since -5 is to the right of -11, we have $-5 > -11$.

38. $>$

39. Since -12.5 is to the left of -9.4, we have $-12.5 < -9.4$.

40. $>$

41. We convert to decimal notation.
$\frac{5}{12} = 0.41\overline{6}$ and $\frac{11}{25} = 0.44$. Thus, $\frac{5}{12} < \frac{11}{25}$.

42. $<$

43. $-2 > a$ has the same meaning as $a < -2$.

44. $9 < a$

45. $y \geq -10$ has the same meaning as $-10 \leq y$.

46. $t \leq 12$

47. $-3 \geq -11$ is true, since $-3 > -11$ is true.

48. False

49. $0 \geq 8$ is false, since neither $0 > 8$ nor $0 = 8$ is true.

50. True

51. $-8 \leq -8$ is true because $-8 = -8$ is true.

52. True

53. $|-4| = 4$ since -4 is 4 units from 0.

54. 9

55. $|17| = 17$ since 17 is 17 units from 0.

56. 3.1

57. $|5.6| = 5.6$ since 5.6 is 5.6 units from 0.

58. $\frac{2}{5}$

59. $|329| = 329$ since 329 is 329 units from 0.

60. 456

61. $\left|-\frac{9}{7}\right| = \frac{9}{7}$ since $-\frac{9}{7}$ is $\frac{9}{7}$ units from 0.

62. 8.02

63. $|0| = 0$ since 0 is 0 units from itself.

64. 1.07

65. $|x| = |-8| = 8$

66. 5

67. $-23, -4.7, 0, \frac{5}{9}, 8.31, 62$

68. 62

69. $-23, 0, 62$

70. $\pi, \sqrt{17}$

71. All are real numbers.

72. $0, 62$

73. $\begin{aligned} \frac{21}{5} \cdot \frac{1}{7} &= \frac{21 \cdot 1}{5 \cdot 7} && \text{Multiplying numerators and denominators} \\ &= \frac{3 \cdot 7 \cdot 1}{5 \cdot 7} && \text{Factoring the numerator} \\ &= \frac{3}{5} && \text{Removing a factor of 1} \end{aligned}$

74. 42

75. $5 + ab$ is equivalent to $ab + 5$ by the commutative law of addition.

$ba + 5$ is equivalent to $ab + 5$ by the commutative law of multiplication.

$5 + ba$ is equivalent to $ab + 5$ by both commutative laws.

76. $3(x + 3 + 4y)$

77. ◈

78. ◈

79. ◈

80. ◈

81. List the numbers as they occur on the number line, from left to right: $-17, -12, 5, 13$

82. $-23, -17, 0, 4$

83. Converting to decimal notation, we can write
$\frac{4}{5}, \frac{4}{3}, \frac{4}{8}, \frac{4}{6}, \frac{4}{9}, \frac{4}{2}, -\frac{4}{3}$ as
$0.8, 1.3\overline{3}, 0.5, 0.6\overline{6}, 0.4\overline{4}, 2, -1.3\overline{3}$, respectively. List the numbers (in fractional form) as they occur on the number line, from left to right:
$-\frac{4}{3}, \frac{4}{9}, \frac{4}{8}, \frac{4}{6}, \frac{4}{5}, \frac{4}{3}, \frac{4}{2}$

84. $-\frac{5}{6}, -\frac{3}{4}, -\frac{2}{3}, \frac{1}{6}, \frac{3}{8}, \frac{1}{2}$

85. $|-5| = 5$ and $|-2| = 2$, so $|-5| > |-2|$.

86. $|4| < |-7|$

87. $|-8| = 8$ and $|8| = 8$, so $|-8| = |8|$.

88. $|23| = |-23|$

89. $|-3| = 3$ and $|5| = 5$, so $|-3| < |5|$.

90. $|-19| < |-27|$

91. $|x| = 7$

x represents a number whose distance from 0 is 7. Thus, $x = 7$ or $x = -7$.

92. $-2, -1, 0, 1, 2$

93. $2 < |x| < 5$

x represents an integer whose distance from 0 is greater than 2 and also less than 5. Thus, $x = -4, -3, 3, 4$

94. $\dfrac{1}{9}$

95. $0.9\overline{9} = 3(0.3\overline{3}) = 3 \cdot \dfrac{1}{3} = \dfrac{3}{3}$

96. $\dfrac{50}{9}$

97. ◈ ▱

Exercise Set 1.5

1. Start at -7. Move 3 units to the right.

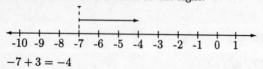

$-7 + 3 = -4$

2. -3

3. Start at -5. Move 9 units to the right.

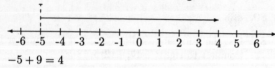

$-5 + 9 = 4$

4. 5

5. Start at -8. Move 8 units to the right.

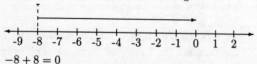

$-8 + 8 = 0$

6. 0

7. Start at -3. Move 5 units to the left.

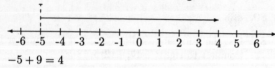

$-3 + (-5) = -8$

8. -10

9. $-15 + 0$ One number is 0. The answer is the other number. $-15 + 0 = -15$

10. -6

11. $0 + (-8)$ One number is 0. The answer is the other number. $0 + (-8) = -8$

12. -2

13. $-15 + 15$ The numbers have the same absolute value. The sum is 0. $-15 + 15 = 0$

14. 0

15. $-24 + (-17)$ Two negatives. Add the absolute values, getting 41. Make the answer negative. $-24 + (-17) = -41$

16. -42

17. $11 + (-11)$ The numbers have the same absolute value. The sum is 0. $11 + (-11) = 0$

18. 0

19. $-7 + 8$ The absolute values are 7 and 8. The difference is $8 - 7$, or 1. The positive number has the larger absolute value, so the answer is positive. $-7 + 8 = 1$

20. 3

21. $10 + (-12)$ The absolute values are 10 and 12. The difference is $12 - 10$, or 2. The negative number has the larger absolute value, so the answer is negative. $10 + (-12) = -2$

22. -9

23. $-3 + 14$ The absolute values are 3 and 14. The difference is $14 - 3$, or 11. The positive number has the larger absolute value, so the answer is positive. $-3 + 14 = 11$

24. 7

25. $-14 + (-19)$ Two negatives. Add the absolute values, getting 33. Make the answer negative. $-14 + (-19) = -33$

26. 2

27. $19 + (-19)$ The numbers has the same absolute value. The sum is 0. $19 + (-19) = 0$

28. -26

29. $23 + (-5)$ The absolute values are 23 and 5. The difference is $23 - 5$ or 18. The positive number has the larger absolute value, so the answer is positive. $23 + (-5) = 18$

30. -22

31. $-23+(-9)$ Two negatives. Add the absolute values, getting 32. Make the answer negative.
$-23+(-9) = -32$

32. 32

33. $40 + (-40)$ The numbers have the same absolute value. The sum is 0. $40 + (-40) = 0$

34. 0

35. $85 + (-65)$ The absolute values are 85 and 65. The difference is $85 - 65$, or 20. The positive number has the larger absolute value, so the answer is positive.
$85 + (-65) = 20$

36. 45

37. $-3.6 + 1.9$ The absolute values are 3.6 and 1.9. The difference is $3.6 - 1.9$, or 1.7. The negative number has the larger absolute value, so the answer is negative. $-3.6 + 1.9 = -1.7$

38. -1.8

39. $-5.4 + (-3.7)$ Two negatives. Add the absolute values, getting 9.1. Make the answer negative.
$-5.4 + (-3.7) = -9.1$

40. -13.2

41. $-\frac{4}{3} + \frac{2}{3}$ The absolute values are $\frac{4}{3}$ and $\frac{2}{3}$. The difference is $\frac{4}{3} - \frac{2}{3}$, or $\frac{2}{3}$. The negative number has the larger absolute value, so the answer is negative.
$-\frac{4}{3} + \frac{2}{3} = -\frac{2}{3}$

42. $-\frac{1}{5}$

43. $-\frac{4}{9} + \left(-\frac{6}{9}\right)$ Two negatives. Add the absolute values, getting $\frac{10}{9}$. Make the answer negative.
$-\frac{4}{9} + \left(-\frac{6}{9}\right) = -\frac{10}{9}$

44. $-\frac{8}{7}$

45. $-\frac{5}{6} + \frac{2}{3}$ The absolute values are $\frac{5}{6}$ and $\frac{2}{3}$. The difference is $\frac{5}{6} - \frac{4}{6}$, or $\frac{1}{6}$. The negative number has the larger absolute value, so the answer is negative.
$-\frac{5}{6} + \frac{2}{3} = -\frac{1}{6}$

46. $-\frac{3}{8}$

47. $-\frac{5}{8} + \left(-\frac{1}{3}\right)$ Two negatives. Add the absolute values, getting $\frac{15}{24} + \frac{8}{24}$, or $\frac{23}{24}$. Make the answer negative.
$-\frac{5}{8} + \left(-\frac{1}{3}\right) = -\frac{23}{24}$

48. $-\frac{29}{35}$

49. $35 + (-14) + (-19) + (-5)$
$= 35 + [(-14) + (-19) + (-5)]$ Using the associative law of addition
$= 35 + (-38)$ Adding the negatives
$= -3$ Adding a positive and a negative

50. -62

51. $-44 + \left(-\frac{3}{8}\right) + 95 + \left(-\frac{5}{8}\right)$
$= \left[-44 + \left(-\frac{3}{8}\right) + \left(-\frac{5}{8}\right)\right] + 95$
Using the associative law of addition
$= -45 + 95$ Adding the negatives
$= 50$ Adding a negative and a positive

52. 37.9

53. Rewording: July bill plus payment plus August bill is amount owed.
Translating: $-82 + 50 + (-37) = $ Amount owed
Since $-82 + 50 + (-37) = -32 + (-37) = -69$,
Maya owed $69 at the end of August.

54. 8 yd gain

55. Rewording:

$$\underbrace{\text{1992 loss}} \text{ plus } \underbrace{\text{1993 profit}} \text{ plus}$$

$$\downarrow \qquad \downarrow \qquad \downarrow \qquad \downarrow$$

Translating: $-28,375$ $+$ $37,425$ $+$

$$\underbrace{\text{1994 profit}} \text{ plus } \underbrace{\text{1995 loss}} \text{ plus } \underbrace{\text{1996 profit}} \text{ is}$$

$$\downarrow \qquad \downarrow \qquad \downarrow \qquad \downarrow \qquad \downarrow \qquad \downarrow$$

$95,485$ $+$ $(-19,365)$ $+$ $98,245$ $=$

$$\underbrace{\begin{array}{c}\text{total profit}\\\text{or loss.}\end{array}}$$

$$\downarrow$$

$$\begin{array}{c}\text{total profit}\\\text{or loss.}\end{array}$$

Since $-28,375+37,425+95,485+(-19,365)+98,245$

$\qquad = -47,740 + 231,155$

$\qquad = 183,415,$

the profit was \$183,415.

56. 13 mb drop

57. Rewording:

$$\underbrace{\begin{array}{c}\text{Elevation}\\\text{of base}\end{array}} \text{ plus } \underbrace{\begin{array}{c}\text{total}\\\text{height}\end{array}} \text{ is } \underbrace{\begin{array}{c}\text{elevation}\\\text{of peak.}\end{array}}$$

$$\downarrow \qquad \downarrow \qquad \downarrow \qquad \downarrow \qquad \downarrow$$

Translating: $-19,684$ $+$ $33,480$ $=$ $\begin{array}{c}\text{elevation}\\\text{of peak.}\end{array}$

Since $-19,684 + 33,480 = 13,796$, the elevation of the peak is 13,796 ft above sea level.

58. Kyle owes \$85.

59. Rewording:

$$\underbrace{\begin{array}{c}\text{Monday}\\\text{change}\end{array}} \text{ plus } \underbrace{\begin{array}{c}\text{Tuesday}\\\text{change}\end{array}} \text{ plus}$$

$$\downarrow \qquad \downarrow \qquad \downarrow \qquad \downarrow$$

Translating: $-\dfrac{1}{4}$ $+$ $\dfrac{5}{8}$ $+$

$$\underbrace{\begin{array}{c}\text{Wednesday}\\\text{change}\end{array}} \text{ is } \underbrace{\begin{array}{c}\text{total}\\\text{change.}\end{array}}$$

$$\downarrow \qquad \downarrow \qquad \downarrow$$

$$\left(-\dfrac{3}{8}\right) = \text{total change.}$$

Since $-\dfrac{1}{4} + \dfrac{5}{8} + \left(-\dfrac{3}{8}\right)$

$= \left[-\dfrac{2}{8} + \left(-\dfrac{3}{8}\right)\right] + \dfrac{5}{8}$

$= -\dfrac{5}{8} + \dfrac{5}{8}$

$= 0,$

the change in the value was \$0. That is, the value of the stock was the same at the end of the three day period as at the beginning.

60. \$85 overdrawn

61. $4a + 9a = (4 + 9)a$ Using the distributive law

$\qquad = 13a$

62. $11x$

63. $-3x + 12x = (-3 + 12)x$ Using the distributive law

$\qquad = 9x$

64. $-5m$

65. $4x + 7x = (4 + 7)x = 11x$

66. $14a$

67. $7m + (-9m) = [7 + (-9)]m = -2m$

68. $5x$

69. $-5a + (-2a) = [-5 + (-2)]a = -7a$

70. $-7n$

71. $-3 + 8x + 4 + (-10x)$

$= -3 + 4 + 8x + (-10x)$ Using the commutative law of addition

$= (-3 + 4) + [8 + (-10)]x$ Using the distributive law

$= 1 - 2x$ Adding

72. $7a + 2$

73.   Perimeter $= 8 + 5x + 9 + 7x$

$\qquad = 8 + 9 + 5x + 7x$

$\qquad = (8 + 9) + (5 + 7)x$

$\qquad = 17 + 12x$

74. $10a + 13$

75.   Perimeter $= 9 + 6n + 7 + 8n + 4n$

$\qquad = 9 + 7 + 6n + 8n + 4n$

$\qquad = (9 + 7) + (6 + 8 + 4)n$

$\qquad = 16 + 18n$

76. $19n + 11$

77. $7(3z + y + 2) = 7 \cdot 3z + 7 \cdot y + 7 \cdot 2 = 21z + 7y + 14$

78. $\dfrac{28}{3}$

79. ◈

80. ◈

81. ◈

82. ◈

83. Starting with the final value, we "undo" the rise and drop in value by adding their opposites. The result is the original value.

Rewording:
$$\underbrace{\text{Final value}}_{\downarrow} \text{ plus } \underbrace{\text{opposite of rise}}_{\downarrow} \text{ plus }$$

Translating: $64\frac{3}{8}$ + $\left(-2\frac{3}{8}\right)$ +

$$\underbrace{\text{opposite of drop}}_{\downarrow} \text{ is original value.}$$

$$3\frac{1}{4} \quad = \text{ original value.}$$

Since $64\frac{3}{8} + \left(-2\frac{3}{8}\right) + 3\frac{1}{4} = 62 + 3\frac{1}{4}$
$$= 65\frac{1}{4},$$

the stock's original value was $\$65\frac{1}{4}$.

84. $55.50

85.
$$4x + \underline{} + (-9x) + (-2y)$$
$$= 4x + (-9x) + \underline{} + (-2y)$$
$$= [4 + (-9)]x + \underline{} + (-2y)$$
$$= -5x + \underline{} + (-2y)$$

This expression is equivalent to $-5x - 7y$, so the missing term is the term which yields $-7y$ when added to $-2y$. Since $-5y + (-2y) = -7y$, the missing term is $-5y$.

86. $-15b$

87.
$$3m + 2n + \underline{} + (-2m)$$
$$= 2n + \underline{} + (-2m) + 3m$$
$$= 2n + \underline{} + (-2 + 3)m$$
$$= 2n + \underline{} + m$$

This expression is equivalent to $2n + (-6m)$, so the missing term is the term which yields $-6m$ when added to m. Since $-7m + m = -6m$, the missing term is $-7m$.

88. $-3y$

89. $P = 2l + 2w = 7x + 10$

We know $2l = 2 \cdot 5 = 10$, so $2w$ is $7x$. Then the width is a number which yields $7x$ when added to itself. Since $3.5x + 3.5x = 7x$, the width is $3.5x$, or $\frac{7}{2}x$.

90. 1 under par

Exercise Set 1.6

1. The opposite of 39 is -39 because $39 + (-39) = 0$.

2. 17

3. The opposite of -9 is 9 because $-9 + 9 = 0$.

4. $-\frac{7}{2}$

5. The opposite of -3.14 is 3.14 because $-3.14 + 3.14 = 0$.

6. -48.2

7. If $x = 23$, then $-x = -(23) = -23$. (The opposite of 23 is -23.)

8. 26

9. If $x = -\frac{14}{3}$, then $-x = -\left(-\frac{14}{3}\right) = \frac{14}{3}$.
$\left(\text{The opposite of } -\frac{14}{3} \text{ is } \frac{14}{3}.\right)$

10. $-\frac{1}{328}$

11. If $x = 0.101$, then $-x = -(0.101) = -0.101$.
(The opposite of 0.101 is -0.101.)

12. 0

13. If $x = -72$, then $-(-x) = -(-72) = 72$
(The opposite of the opposite of 72 is 72.)

14. 29

15. If $x = -\frac{2}{5}$, then $-(-x) = -\left[-\left(-\frac{2}{5}\right)\right] = -\frac{2}{5}$.
$\left(\text{The opposite of the opposite of } -\frac{2}{5} \text{ is } -\frac{2}{5}.\right)$

16. -9.1

17. When we change the sign of -1 we obtain 1.

18. 7

19. When we change the sign of 7 we obtain -7.

20. -10

21. $4 - 13 = 4 + (-13) = -9$

22. -5

23. $0 - 7 = 0 + (-7) = -7$

24. -10

25. $-7 - (-9) = -7 + 9 = 2$

26. -6

27. $-10 - (-10) = -10 + 10 = 0$

28. 0

29. $12 - 16 = 12 + (-16) = -4$

30. -5

31. $20 - 27 = 20 + (-27) = -7$

32. 26

33. $-8 - (-3) = -8 + 3 = -5$

34. 2

35. $-40 - (-40) = -40 + 40 = 0$

36. 0

37. $7 - 7 = 7 + (-7) = 0$

38. 0

39. $6 - (-6) = 6 + 6 = 12$

40. 8

41. $8 - (-3) = 8 + 3 = 11$

42. -11

43. $-6 - 8 = -6 + (-8) = -14$

44. 16

45. $-4 - (-9) = -4 + 9 = 5$

46. -16

47. $-6 - (-5) = -6 + 5 = -1$

48. -1

49. $3 - (-12) = 3 + 12 = 15$

50. 11

51. $0 - 5 = 0 + (-5) = -5$

52. -6

53. $-5 - (-2) = -5 + 2 = -3$

54. -2

55. $-7 - 14 = -7 + (-14) = -21$

56. -25

57. $0 - (-5) = 0 + 5 = 5$

58. 1

59. $-8 - 0 = -8 + 0 = -8$

60. -9

61. $3 - (-7) = 3 + 7 = 10$

62. 17

63. $2 - 25 = 2 + (-25) = -23$

64. -45

65. $-42 - 26 = -42 + (-26) = -68$

66. -81

67. $-71 - 2 = -71 + (-2) = -73$

68. -52

69. $24 - (-92) = 24 + 92 = 116$

70. 121

71. $-50 - (-50) = -50 + 50 = 0$

72. 0

73. $\dfrac{3}{8} - \dfrac{5}{8} = \dfrac{3}{8} + \left(-\dfrac{5}{8}\right) = -\dfrac{2}{8} = -\dfrac{1}{4}$

74. $-\dfrac{2}{3}$

75. $\dfrac{4}{5} - \dfrac{2}{3} = \dfrac{12}{15} - \dfrac{10}{15} = \dfrac{12}{15} + \left(-\dfrac{10}{15}\right) = \dfrac{2}{15}$

76. $-\dfrac{1}{8}$

77. $-\dfrac{3}{4} - \dfrac{2}{3} = -\dfrac{9}{12} - \dfrac{8}{12} = -\dfrac{9}{12} + \left(-\dfrac{8}{12}\right) = -\dfrac{17}{12}$

78. $-\dfrac{11}{8}$

79. $-2.8 - 0 = -2.8 + 0 = -2.8$

80. 4.94

81. $0.09 - 1 = 0.09 + (-1) = -0.91$

82. -0.911

83. $\dfrac{1}{6} - \dfrac{2}{3} = \dfrac{1}{6} - \dfrac{4}{6} = \dfrac{1}{6} + \left(-\dfrac{4}{6}\right) = -\dfrac{3}{6} = -\dfrac{1}{2}$

84. $\dfrac{1}{8}$

85. $-\dfrac{4}{7} - \left(-\dfrac{10}{7}\right) = -\dfrac{4}{7} + \dfrac{10}{7} = \dfrac{6}{7}$

86. 0

87. We subtract the smaller number from the larger.

Translate: $3.8 - (-5.2)$

Simplify: $3.8 - (-5.2) = 3.8 + 5.2 = 9$

88. $-2.1 - (-5.9)$; 3.8

89. We subtract the smaller number from the larger.

Translate: $114 - (-79)$

Simplify: $114 - (-79) = 114 + 79 = 193$

90. $23 - (-17)$; 40

91. $-21 - 37 = -21 + (-37) = -58$

92. -26

93. $9 - (-25) = 9 + 25 = 34$

94. 26

95. $-1.8 - 2.7$ is read "negative one point eight minus two point seven."

$$-1.8 - 2.7 = -1.8 + (-2.7) = -4.5$$

96. Negative two point seven minus five point nine; -8.6

97. $-250 - (-425)$ is read "negative two hundred fifty minus negative four hundred twenty-five."

$$-250 - (-425) = -250 + 425 = 175$$

98. Negative three hundred fifty minus negative one thousand; 650

99. $25 - (-12) - 7 - (-2) + 9 = 25 + 12 + (-7) + 2 + 9 = 41$

100. -22

101. $-31 + (-28) - (-14) - 17 = (-31) + (-28) + 14 + (-17) = -62$

102. 22

103. $-34 - 28 + (-33) - 44 = (-34) + (-28) + (-33) + (-44) = -139$

104. 5

105. $-93 - (-84) - 41 - (-56) = (-93) + 84 + (-41) + 56 = 6$

106. 4

107. $-7x - 4y = -7x + (-4y)$, so the terms are $-7x$ and $-4y$.

108. $7a, -9b$

109. $-5 + 3m - 6mn = -5 + 3m + (-6mn)$, so the terms are $-5, 3m$, and $-6mn$.

110. $-9, -4t, 10rt$

111.
$$4x - 7x$$
$$= 4x + (-7x) \quad \text{Adding the opposite}$$
$$= (4 + (-7))x \quad \text{Using the distributive law}$$
$$= -3x$$

112. $-11a$

113.
$$7a - 12a + 4$$
$$= 7a + (-12a) + 4 \quad \text{Adding the opposite}$$
$$= (7 + (-12))a + 4 \quad \text{Using the distributive law}$$
$$= -5a + 4$$

114. $-22x + 7$

115.
$$-8n - 9 + n$$
$$= -8n + (-9) + n \quad \text{Adding the opposite}$$
$$= -8n + n + (-9) \quad \text{Using the commutative law of addition}$$
$$= -7n - 9 \quad \text{Adding like terms}$$

116. $9n - 15$

117.
$$3x + 5 - 9x$$
$$= 3x + 5 + (-9x)$$
$$= 3x + (-9x) + 5$$
$$= -6x + 5$$

118. $3a - 5$

119.
$$2 - 6t - 9 - 2t$$
$$= 2 + (-6t) + (-9) + (-2t)$$
$$= 2 + (-9) + (-6t) + (-2t)$$
$$= -7 - 8t$$

120. $-2b - 12$

121.
$$7 + (-3x) - 9x + 1$$
$$= 7 + (-3x) + (-9x) + 1$$
$$= 7 + 1 + (-3x) + (-9x)$$
$$= 8 - 12x$$

122. $7x + 46$

123. $13x - (-2x) + 45 - (-21) = 13x + 2x + 45 + 21 = 15x + 66$

124. $15x + 39$

125. We subtract the lower temperature from the higher temperature:

$$44 - (-56) = 44 + 56 = 100$$

The temperature dropped $100°$F.

126. \$165

127. We add the elevations:

$$14,776 + (-282) = 14,494 \text{ ft}$$

The elevation of Mt. Whitney is 14,494 ft above sea level.

128. 30,340 ft

129. We subtract the smaller number from the larger:

$$-8648 - (-10,415) = -8648 + 10,415 = 1767$$

The difference in elevation is 1767 m.

130. 116 m

131. Area $= lw = (36 \text{ ft})(12 \text{ ft}) = 432 \text{ ft}^2$

132. $2 \cdot 2 \cdot 2 \cdot 2 \cdot 2 \cdot 3 \cdot 3 \cdot 3$

133. ◈

134. ◈

135. ◈

136. ◈

137. False. For example, let $m = -3$ and $n = -5$. Then $-3 > -5$, but $-3 + (-5) = -8 \not> 0$.

138. True. For example, for $m = 5$ and $n = 3$, $5 > 3$ and $5 - 3 > 0$, or $2 > 0$. For $m = -4$ and $n = -9$, $-4 > -9$ and $-4 - (-9) > 0$, or $5 > 0$.

139. False. For example, let $m = 2$ and $n = -2$. Then 2 and -2 are opposites, but $2 - (-2) = 4 \neq 0$.

140. True. For example, for $m = 4$ and $n = -4$, $4 = -(-4)$ and $4 + (-4) = 0$; for $m = -3$ and $n = 3$, $-3 = -3$ and $-3 + 3 = 0$.

141. ◈

142. ◈

Exercise Set 1.7

1. $-9 \cdot 3 = -27$ Think: $9 \cdot 3 = 27$, make the answer negative.

2. -21

3. $-8 \cdot 7 = -56$ Think: $8 \cdot 7 = 56$, make the answer negative.

4. -18

5. $8 \cdot (-3) = -24$

6. -45

7. $-9 \cdot 8 = -72$

8. -30

9. $-8 \cdot (-2) = 16$ Multiplying absolute values

10. 10

11. $-5 \cdot (-9) = 45$ Multiplying absolute values

12. 18

13. $15 \cdot (-8) = -120$

14. 120

15. $-14 \cdot 17 = -238$

16. 195

17. $-25 \cdot (-48) = 1200$

18. -1677

19. $-3.5 \cdot (-28) = 98$

20. -203.7

21. $6 \cdot (-13) = -78$

22. -63

23. $-7 \cdot (-3.1) = 21.7$

24. 12.8

25. $\frac{2}{3} \cdot \left(-\frac{3}{5}\right) = -\left(\frac{2 \cdot 3}{3 \cdot 5}\right) = -\left(\frac{2}{5} \cdot \frac{3}{3}\right) = -\frac{2}{5}$

26. $-\frac{10}{21}$

27. $-\frac{3}{8} \cdot \left(-\frac{2}{9}\right) = \frac{\cancel{3} \cdot \cancel{2} \cdot 1}{4 \cdot \cancel{2} \cdot \cancel{3} \cdot 3} = \frac{1}{12}$

28. $\frac{1}{4}$

29. $-6.3 \times 2.7 = -17.01$

30. -38.95

31. $-\frac{5}{9} \cdot \frac{3}{4} = -\frac{5 \cdot \cancel{3}}{\cancel{3} \cdot 3 \cdot 4} = -\frac{5}{12}$

32. -6

33. $\quad 3 \cdot (-7) \cdot (-2) \cdot 6$
$= -21 \cdot (-12)$ Multiplying the first two numbers and the last two numbers
$= 252$

34. 756

35. $\quad -2 \cdot (-5) \cdot (-9)$
$= 10 \cdot (-9)$ Multiplying the first two numbers
$= -90$

36. -60

37. $-\frac{1}{3} \cdot \frac{1}{4} \cdot \left(-\frac{3}{7}\right) = -\frac{1}{12} \cdot \left(-\frac{3}{7}\right) = \frac{3}{12 \cdot 7} = \frac{\cancel{3} \cdot 1}{\cancel{3} \cdot 4 \cdot 7} = \frac{1}{28}$

38. $\frac{3}{35}$

39. $-2 \cdot (-5) \cdot (-3) \cdot (-5) = 10 \cdot 15 = 150$

40. 30

41. 0, The product of 0 and any real number is 0.

42. 0

43. $(-8)(-9)(-10) = 72(-10) = -720$

44. 5040

45. $(-6)(-7)(-8)(-9)(-10) = 42 \cdot 72 \cdot (-10) =$
$3024 \cdot (-10) = -30,240$

46. $151,200$

47. $24 \div (-4) = -6$ Check: $-6 \cdot (-4) = 24$

48. -4

49. $\dfrac{36}{-9} = -4$ $-4 \cdot (-9) = 36$

50. -2

51. $\dfrac{-16}{8} = -2$ Check: $-2 \cdot 8 = -16$

52. 8

53. $\dfrac{-48}{-12} = 4$ Check: $4(-12) = -48$

54. 7

55. $\dfrac{-72}{9} = -8$ Check: $-8 \cdot 9 = -72$

56. -2

57. $-100 \div (-50) = 2$ Check: $2(-50) = -100$

58. -25

59. $-108 \div 9 = -12$ Check: $-12 \cdot 9 = -108$

60. $\dfrac{64}{7}$

61. $\dfrac{400}{-50} = -8$ Check: $-8 \cdot (-50) = 400$

62. $\dfrac{300}{13}$

63. Undefined

64. 0

65. $\dfrac{88}{-9} = -\dfrac{88}{9}$ Check: $-\dfrac{88}{9} \cdot (-9) = 88$

66. Indeterminate

67. $\dfrac{0}{-9} = 0$

68. Undefined

69. $0 \div 0$ is indeterminate.

70. 0

71. $\dfrac{-8}{3} = \dfrac{8}{-3}$ and $\dfrac{-8}{3} = -\dfrac{8}{3}$

72. $\dfrac{12}{-7}, \; -\dfrac{12}{7}$

73. $\dfrac{29}{-35} = \dfrac{-29}{35}$ and $\dfrac{29}{-35} = -\dfrac{29}{35}$

74. $\dfrac{-9}{14}, \; -\dfrac{9}{14}$

75. $-\dfrac{7}{3} = \dfrac{-7}{3}$ and $-\dfrac{7}{3} = \dfrac{7}{-3}$

76. $\dfrac{-4}{15}, \; \dfrac{4}{-15}$

77. $\dfrac{-x}{2} = \dfrac{x}{-2}$ and $\dfrac{-x}{2} = -\dfrac{x}{2}$

78. $\dfrac{-9}{a}, \; -\dfrac{9}{a}$

79. The reciprocal of $\dfrac{4}{-5}$ is $\dfrac{-5}{4}$ $\left(\text{or equivalently, } -\dfrac{5}{4}\right)$ because $\dfrac{4}{-5} \cdot \dfrac{-5}{4} = 1$.

80. $\dfrac{-9}{2}$, or $-\dfrac{9}{2}$

81. The reciprocal of $-\dfrac{47}{13}$ is $-\dfrac{13}{47}$ because $-\dfrac{47}{13} \cdot \left(-\dfrac{13}{47}\right) = 1$.

82. $-\dfrac{12}{31}$

83. The reciprocal of -10 is $\dfrac{1}{-10}$ $\left(\text{or equivalently, } -\dfrac{1}{10}\right)$ because $-10\left(\dfrac{1}{-10}\right) = 1$.

84. $\dfrac{1}{13}$

85. The reciprocal of 4.3 is $\dfrac{1}{4.3}$ because $4.3\left(\dfrac{1}{4.3}\right) = 1$.

86. $\dfrac{1}{-8.5}$, or $-\dfrac{1}{8.5}$

87. The reciprocal of $\dfrac{-9}{4}$ is $\dfrac{4}{-9}$ $\left(\text{or equivalently, } -\dfrac{4}{9}\right)$ because $\dfrac{-9}{4} \cdot \dfrac{4}{-9} = 1$.

88. $\dfrac{11}{-6}$, or $-\dfrac{11}{6}$

89. The reciprocal of -1 is $\dfrac{1}{-1}$, or -1 because $(-1)(-1) = 1$.

90. $\dfrac{1}{2}$

91. $\left(\dfrac{-7}{4}\right)\left(-\dfrac{3}{5}\right)$

$= \left(-\dfrac{7}{4}\right)\left(-\dfrac{3}{5}\right)$ Rewriting $\dfrac{-7}{4}$ as $-\dfrac{7}{4}$

$= \dfrac{21}{20}$

92. $\dfrac{5}{18}$

93.
$$\left(\dfrac{-6}{5}\right)\left(\dfrac{2}{-11}\right)$$
$$=\left(\dfrac{-6}{5}\right)\left(\dfrac{-2}{11}\right) \qquad \text{Rewriting } \dfrac{2}{-11} \text{ as } \dfrac{-2}{11}$$
$$=\dfrac{12}{55}$$

94. $\dfrac{35}{12}$

95. $\dfrac{3}{-8}+\dfrac{-5}{8}=\dfrac{-3}{8}+\dfrac{-5}{8}=\dfrac{-8}{8}=-1$

96. $-\dfrac{11}{5}$

97. $\left(\dfrac{-9}{5}\right)\left(-\dfrac{10}{7}\right)=\left(-\dfrac{9}{5}\right)\left(-\dfrac{10}{7}\right)=\dfrac{90}{35}=$
$\dfrac{\cancel{5}\cdot 18}{\cancel{5}\cdot 7}=\dfrac{18}{7}$

98. $\dfrac{5}{28}$

99. $\left(\dfrac{-3}{11}\right)+\dfrac{5}{-11}=\dfrac{-3}{11}+\dfrac{-5}{11}=\dfrac{-8}{11}$, or $-\dfrac{8}{11}$

100. $-\dfrac{13}{7}$

101. $\dfrac{7}{8}\div\left(-\dfrac{1}{2}\right)=\dfrac{7}{8}\cdot\left(-\dfrac{2}{1}\right)=-\dfrac{14}{8}=-\dfrac{7\cdot\cancel{2}}{\cancel{2}\cdot 4\cdot 1}=-\dfrac{7}{4}$

102. $-\dfrac{9}{8}$

103. $\dfrac{9}{5}\cdot\dfrac{-20}{3}=\dfrac{9}{5}\left(-\dfrac{20}{3}\right)=-\dfrac{180}{15}=-\dfrac{3\cdot 3\cdot 4\cdot \cancel{5}}{\cancel{5}\cdot \cancel{3}\cdot 1}=-12$

104. $-\dfrac{7}{36}$

105. $\left(-\dfrac{18}{7}\right)+\left(-\dfrac{3}{7}\right)=-\dfrac{21}{7}=-3$

106. -3

107. $-\dfrac{5}{9}\div\left(-\dfrac{5}{6}\right)=-\dfrac{5}{9}\cdot\left(-\dfrac{6}{5}\right)=\dfrac{30}{45}=\dfrac{\cancel{5}\cdot 2\cdot \cancel{3}}{\cancel{3}\cdot 3\cdot \cancel{5}}=\dfrac{2}{3}$

108. $\dfrac{5}{3}$

109. $-44.1\div(-6.3)=7$ Do the long division. The answer is positive.

110. -2

111. $\dfrac{-5}{9}-\dfrac{2}{9}=-\dfrac{5}{9}-\dfrac{2}{9}=-\dfrac{5}{9}+\left(-\dfrac{2}{9}\right)=-\dfrac{7}{9}$

112. $-\dfrac{5}{7}$

113.
$$\dfrac{-3}{10}+\dfrac{2}{-5}$$
$$=\dfrac{-3}{10}+\dfrac{-2}{5}$$
$$=\dfrac{-3}{10}+\dfrac{-2}{5}\cdot\dfrac{2}{2} \qquad \text{Using a common denominator of 10}$$
$$=\dfrac{-3}{10}+\dfrac{-4}{10}$$
$$=\dfrac{-7}{10}, \text{ or } -\dfrac{7}{10}$$

114. $-\dfrac{11}{9}$

115. $\dfrac{7}{10}\div\left(\dfrac{-3}{5}\right)=\dfrac{7}{10}\div\left(-\dfrac{3}{5}\right)=\dfrac{7}{10}\cdot\left(-\dfrac{5}{3}\right)=-\dfrac{35}{30}=$
$-\dfrac{7\cdot\cancel{5}}{2\cdot\cancel{5}\cdot 3}=-\dfrac{7}{6}$

116. $-\dfrac{3}{2}$

117. $\dfrac{5}{7}-\dfrac{1}{-7}=\dfrac{5}{7}-\left(-\dfrac{1}{7}\right)=\dfrac{5}{7}+\dfrac{1}{7}=\dfrac{6}{7}$

118. $\dfrac{5}{9}$

119. $\dfrac{-4}{15}+\dfrac{2}{-3}=\dfrac{-4}{15}+\dfrac{-2}{3}=\dfrac{-4}{15}+\dfrac{-2}{3}\cdot\dfrac{5}{5}=\dfrac{-4}{15}+\dfrac{-10}{15}=$
$\dfrac{-14}{15}$, or $-\dfrac{14}{15}$

120. $-\dfrac{1}{2}$

121. $\dfrac{264}{468}=\dfrac{\cancel{2}\cdot\cancel{2}\cdot 2\cdot\cancel{3}\cdot 11}{\cancel{2}\cdot\cancel{2}\cdot\cancel{3}\cdot 3\cdot 13}=\dfrac{22}{39}$

122. $12x-2y-9$

123. ◈

124. ◈

125. ◈

126. ◈

127. Consider the sum $2+3$. Its reciprocal is $\dfrac{1}{2+3}$, or $\dfrac{1}{5}$, but $\dfrac{1}{2}+\dfrac{1}{3}=\dfrac{5}{6}$.

128. $-1, 1$

129. When n is negative, $-n$ is positive, so $\dfrac{m}{-n}$ is the quotient of a negative and a positive number and, thus, is negative.

130. Positive

131. When n is negative, $-n$ is positive, so $\dfrac{-n}{m}$ is the quotient of a positive and a negative number and, thus, is negative. When m is negative, $-m$ is positive, so $-m \cdot \left(\dfrac{-n}{m}\right)$ is the product of a positive and a negative number and, thus, is negative.

132. Positive

133. $m + n$ is the sum of two negative numbers, so it is negative; $\dfrac{m}{n}$ is the quotient of two negative numbers, so it is positive. Then $(m + n) \cdot \dfrac{m}{n}$ is the product of a negative and a positive number and, thus, is negative.

134. Positive

135. a) m and n have different signs;
   b) either m or n is zero;
   c) m and n have the same sign

136. Distributive law; law of opposites; multiplicative property of 0; law of opposites

137. ◈

Exercise Set 1.8

1. $\underbrace{17 \times 17 \times 17}_{3 \text{ factors}} = 17^3$

2. 5^4

3. $\underbrace{x \cdot x \cdot x \cdot x \cdot x \cdot x \cdot x}_{7 \text{ factors}} = x^7$

4. y^6

5. $6y \cdot 6y \cdot 6y \cdot 6y = (6y)^4$

6. $(5m)^5$

7. $3^4 = 3 \cdot 3 \cdot 3 \cdot 3 = 9 \cdot 9 = 81$

8. 125

9. $(-3)^2 = (-3)(-3) = 9$

10. 49

11. $(-1)^5 = (-1)(-1)(-1)(-1)(-1) = 1 \cdot 1 \cdot (-1) = 1 \cdot (-1) = -1$

12. -1

13. $4^3 = 4 \cdot 4 \cdot 4 = 16 \cdot 4 = 64$

14. 9

15. $(-5)^4 = (-5)(-5)(-5)(-5) = 25 \cdot 25 = 625$

16. 625

17. $7^1 = 7$ (1 factor)

18. -1

19. $(2x)^4 = (2x)(2x)(2x)(2x) = 2 \cdot 2 \cdot 2 \cdot 2 \cdot x \cdot x \cdot x \cdot x = 16x^4$

20. $9x^2$

21. $(-7x)^3 = (-7x)(-7x)(-7x) = (-7)(-7)(-7)(x)(x)(x) = -343x^3$

22. $625x^4$

23. $\begin{aligned} 5 + 3 \times 7 &= 5 + 21 \quad \text{Multiplying} \\ &= 26 \quad\quad\ \text{Adding} \end{aligned}$

24. 1

25. $\begin{aligned} 8 \times 7 + 6 \times 5 &= 56 + 30 \quad \text{Multiplying} \\ &= 86 \quad\quad\ \text{Adding} \end{aligned}$

26. 51

27. $\begin{aligned} 19 - 5 \times 3 + 3 &= 19 - 15 + 3 \quad \text{Multiplying} \\ &= 4 + 3 \quad\quad\ \text{Subtracting and add-} \\ &= 7 \quad\quad\quad\ \text{ing from left to right} \end{aligned}$

28. 9

29. $\begin{aligned} 9 \div 3 + 16 \div 8 &= 3 + 2 \quad \text{Dividing} \\ &= 5 \quad\quad\ \text{Adding} \end{aligned}$

30. 28

31. $\begin{aligned} 7 + 10 - 10 \div 2 &= 7 + 10 - 5 \quad \text{Dividing} \\ &= 17 - 5 \quad \text{Adding and subtract-} \\ &= 12 \quad\quad\ \text{ing from left to right} \end{aligned}$

32. 9

33. $\begin{aligned} &\ \ 2 \cdot 5^3 \\ &= 2 \cdot 125 \quad \text{Simplifying the exponential expression} \\ &= 250 \quad\quad \text{Multiplying} \end{aligned}$

34. 24

35. $\begin{aligned} 8 - 2 \cdot 3 - 9 &= 8 - 6 - 9 \quad \text{Multiplying} \\ &= 2 - 9 \quad\quad \text{Adding and subtracting} \\ &= -7 \quad\quad\ \text{from left to right} \end{aligned}$

36. 11

37. $\begin{aligned} (8 - 2 \cdot 3) - 9 &= (8 - 6) - 9 \quad \text{Multiplying inside the parentheses} \\ &= 2 - 9 \quad\quad\ \text{Subtracting inside the parentheses} \\ &= -7 \end{aligned}$

38. -36

39. $(-24) \div (-3) \cdot \left(-\dfrac{1}{2}\right) = 8 \cdot \left(-\dfrac{1}{2}\right) = -\dfrac{8}{2} = -4$

40. 32

41. $\quad 13(-10) + 45$
$= -130 + 45 \qquad$ Multiplying
$= -85 \qquad\qquad$ Adding

42. 2

43. $2^4 + 2^3 - 10 = 16 + 8 - 10 = 24 - 10 = 14$

44. 23

45. $5^3 + 26 \cdot 71 - (16 + 25 \cdot 3) = 5^3 + 26 \cdot 71 - (16 + 75) =$
$5^3 + 26 \cdot 71 - 91 = 125 + 26 \cdot 71 - 91 = 125 + 1846 - 91 =$
$1971 - 91 = 1880$

46. 305

47. $[2 \cdot (5 - 3)]^2 = [2 \cdot 2]^2 = 4^2 = 16$

48. 76

49. $\dfrac{7 + 2}{5^2 - 4^2} = \dfrac{9}{25 - 16} = \dfrac{9}{9} = 1$

50. 2

51. $8(-7) + |6(-5)| = -56 + |-30| = -56 + 30 = -26$

52. 49

53. $19 - 5(-3) + 3 = 19 + 15 + 3 = 34 + 3 = 37$

54. 33

55. $9 \div (-3) \cdot 16 \div 8 = -3 \cdot 16 \div 8 = -48 \div 8 = -6$

56. -28

57. $20 + 4^3 \div (-8) \cdot 2 = 20 + 64 \div (-8) \cdot 2 = 20 + (-8) \cdot 2 = 20 + (-16) = 4$

58. -9000

59. $3|7 - (9 - 14)| = 3|7 - (-5)| = 3|7 + 5| = 3|12| = 3 \cdot 12 = 36$

60. 65

61. $\quad 9 - 5x = 9 - 5 \cdot 3 \qquad$ Substituting 3 for x
$= 9 - 15 \qquad\qquad$ Multiplying
$= -6 \qquad\qquad\;\;$ Subtracting

62. -1

63. $\quad 24 \div t^3$
$= 24 \div (-2)^3 \qquad$ Substituting -2 for t
$= 24 \div (-8) \qquad$ Simplifying the exponential expression
$= -3 \qquad\qquad\;\;$ Dividing

64. 16

65. $\quad 45 \div 3a = 45 \div 3 \cdot 3 \qquad$ Substituting 3 for a
$= 15 \cdot 3 \qquad\qquad$ Dividing
$= 45 \qquad\qquad\;\;$ Multiplying

66. 125

67. $\quad 5x \div 15x^2$
$= 5 \cdot 3 \div 15(3)^2 \qquad$ Substituting 3 for x
$= 5 \cdot 3 \div 15 \cdot 9 \qquad$ Simplifying the exponential expression
$= 15 \div 15 \cdot 9 \qquad\;\;$ Multiplying and dividing
$= 1 \cdot 9 \qquad\qquad\quad$ in order from
$= 9 \qquad\qquad\qquad$ left to right

68. 8

69. $-20 \div t^2 - 3(t - 1) = -20 \div (-4)^2 - 3((-4) - 1) =$
$-20 \div (-4)^2 - 3(-5) = -20 \div 16 - 3(-5) =$
$\dfrac{-20}{16} + 15 = \dfrac{-5}{4} + 15 = \dfrac{-5}{4} + \dfrac{60}{4} = \dfrac{55}{4}$

70. 20

71. $-x^2 - 5x = -(-3)^2 - 5(-3) = -9 - 5(-3) =$
$-9 + 15 = 6$

72. 24

73. $\dfrac{3a - 4a^2}{a^2 - 20} = \dfrac{3 \cdot 5 - 4(5)^2}{(5)^2 - 20} = \dfrac{3 \cdot 5 - 4 \cdot 25}{25 - 20} =$
$\dfrac{15 - 100}{5} = \dfrac{-85}{5} = -17$

74. 0

75. $-(9x + 1) = -9x - 1 \qquad$ Removing parentheses and changing the sign of each term

76. $-3x - 5$

77. $-(7 - 2x) = -7 + 2x \qquad$ Removing parentheses and changing the sign of each term

78. $-6x + 7$

79. $-(4a - 3b + 7c) = -4a + 3b - 7c$

80. $-5x + 2y + 3z$

81. $-(3x^2 + 5x - 1) = -3x^2 - 5x + 1$

82. $-8x^3 + 6x - 5$

83. $\quad 5x - (2x + 7)$
$= 5x - 2x - 7 \qquad$ Removing parentheses and changing the sign of each term
$= 3x - 7 \qquad\qquad$ Collecting like terms

84. $5y - 9$

85. $2a - (5a - 9) = 2a - 5a + 9 = -3a + 9$

86. $8n + 7$

87. $2x + 7x - (4x + 6) = 2x + 7x - 4x - 6 = 5x - 6$

88. $a - 7$

89. $9t - 5r - 2(3r + 6t) = 9t - 5r - 6r - 12t = -3t - 11r$

90. $-2m - 6n$

91. $\quad 15x - y - 5(3x - 2y + 5z)$

$= 15x - y - 15x + 10y - 25z \quad$ Multiplying each
$\qquad\qquad\qquad\qquad\qquad\qquad$ term in parentheses by -5
$= 9y - 25z$

92. $-16a + 27b - 32c$

93. $\quad 3x^2 + 7 - (2x^2 + 5) = 3x^2 + 7 - 2x^2 - 5$
$\qquad\qquad\qquad\qquad\qquad = x^2 + 2$

94. $2x^4 + 6x$

95. $\quad 5t^3 + t - 3(t + 2t^3) = 5t^3 + t - 3t - 6t^3$
$\qquad\qquad\qquad\qquad\qquad = -t^3 - 2t$

96. $2n^2 - n$

97. $\quad 12a^2 - 3ab + 5b^2 - 5(-5a^2 + 4ab - 6b^2)$
$= 12a^2 - 3ab + 5b^2 + 25a^2 - 20ab + 30b^2$
$= 37a^2 - 23ab + 35b^2$

98. $-20a^2 + 29ab + 48b^2$

99. $\quad -7t^3 - t^2 - 3(5t^3 - 3t)$
$= -7t^3 - t^2 - 15t^3 + 9t$
$= -22t^3 - t^2 + 9t$

100. $9t^4 - 45t^3 + 17t$

101. $\quad 7(x + 2) - 5(3x - 4)$
$= 7x + 14 - 15x + 20$
$= -8x + 34$

102. $-3x - 11$

103. $\quad 6(a^3 + a) - 12 - 2(a^3 + 2a - 1)$
$= 6a^3 + 6a - 12 - 2a^3 - 4a + 2$
$= 4a^3 + 2a - 10$

104. $8t$

105. $\quad 6(3x - 7) - [4(2x - 5) + 2]$
$= 6(3x - 7) - [8x - 20 + 2]$
$= 6(3x - 7) - [8x - 18]$
$= 18x - 42 - 8x + 18$
$= 10x - 24$

106. $52x - 29$

107. Let x represent "a number." Then we have $2x + 9$.

108. Let x and y represent the numbers; $\frac{1}{2}(x + y)$.

109. ◈

110. ◈

111. ◈

112. ◈

113. $\quad 5t - \{7t - [4r - 3(t - 7)] + 6r\} - 4r$
$= 5t - \{7t - [4r - 3t + 21] + 6r\} - 4r$
$= 5t - \{7t - 4r + 3t - 21 + 6r\} - 4r$
$= 5t - \{10t + 2r - 21\} - 4r$
$= 5t - 10t - 2r + 21 - 4r$
$= -5t - 6r + 21$

114. $-4z$

115. $\quad \{x - [f - (f - x)] + [x - f]\} - 3x$
$= \{x - [f - f + x] + [x - f]\} - 3x$
$= \{x - [x] + [x - f]\} - 3x$
$= \{x - x + x - f\} - 3x$
$= x - f - 3x$
$= -2x - f$

116. ◈

117. ◈

118. False

119. True; $-n + m = m + (-n) = m - n$

120. True

121. True; $n^2 - mn = n(n - m) = (n - m)n = -(-n + m)n = -(m - n)n$

122. False

123. False; let $m = 2$ and $n = 1$. Then $-2(1 - 2) = -2(-1) = 2$, but $-(2 \cdot 1 + 2^2) = -(2 + 4) = -6$.

124. True

125. True; $-n(-n - m) = n^2 + nm = n(n + m)$

Chapter 2

Equations, Inequalities, and Problem Solving

Exercise Set 2.1

1.
$$x + 7 = 20$$
$$x + 7 - 7 = 20 - 7 \quad \text{Subtracting 7 on both sides}$$
$$x = 13 \quad \text{Simplifying}$$

Check: $\dfrac{x + 7 = 20}{}$

$$13 + 7 \text{ ? } 20$$
$$20 \mid 20 \quad \text{TRUE}$$

The solution is 13.

2. 3

3.
$$x + 15 = -5$$
$$x + 15 - 15 = -5 - 15 \quad \text{Subtracting 15 on both sides}$$
$$x = -20$$

Check: $\dfrac{x + 15 = -5}{}$

$$-20 + 15 \text{ ? } -5$$
$$-5 \mid -5 \quad \text{TRUE}$$

The solution is -20.

4. 34

5.
$$x + 6 = -8$$
$$x + 6 - 6 = -8 - 6$$
$$x = -14$$

Check: $\dfrac{x + 6 = -8}{}$

$$-14 + 6 \text{ ? } -8$$
$$-8 \mid -8 \quad \text{TRUE}$$

The solution is -14.

6. -21

7.
$$-5 = x + 8$$
$$-5 - 8 = x + 8 - 8$$
$$-13 = x$$

Check: $\dfrac{-5 = x + 8}{}$

$$-5 \text{ ? } -13 + 8$$
$$-5 \mid -5 \quad \text{TRUE}$$

The solution is -13.

8. -31

9.
$$x - 9 = 6$$
$$x - 9 + 9 = 6 + 9$$
$$x = 15$$

Check: $\dfrac{x - 9 = 6}{}$

$$15 - 9 \text{ ? } 6$$
$$6 \mid 6 \quad \text{TRUE}$$

The solution is 15.

10. 13

11.
$$x - 7 = -21$$
$$x - 7 + 7 = -21 + 7$$
$$x = -14$$

Check: $\dfrac{x - 7 = -21}{}$

$$-14 - 7 \text{ ? } -21$$
$$-21 \mid -21 \quad \text{TRUE}$$

The solution is -14.

12. -11

13.
$$9 + t = 3$$
$$-9 + 9 + t = -9 + 3$$
$$t = -6$$

Check: $\dfrac{9 + t = 3}{}$

$$9 - 6 \text{ ? } 3$$
$$3 \mid 3 \quad \text{TRUE}$$

The solution is -6.

14. 18

15.
$$13 = -7 + y$$
$$7 + 13 = 7 + (-7) + y$$
$$20 = y$$

Check: $\dfrac{13 = -7 + y}{}$

$$13 \text{ ? } -7 + 20$$
$$13 \mid 13 \quad \text{TRUE}$$

The solution is 20.

16. 24

17.
$$-3 + t = -9$$
$$3 + (-3) + t = 3 + (-9)$$
$$t = -6$$

Check: $\dfrac{-3 + t = -9}{}$

$$-3 + (-6) \text{ ? } -9$$
$$-9 \mid -9 \quad \text{TRUE}$$

The solution is -6.

18. -15

19.
$$r + \frac{1}{3} = \frac{8}{3}$$
$$r + \frac{1}{3} - \frac{1}{3} = \frac{8}{3} - \frac{1}{3}$$
$$r = \frac{7}{3}$$

Check: $$r + \frac{1}{3} = \frac{8}{3}$$

$$\frac{7}{3} + \frac{1}{3} \;?\; \frac{8}{3}$$

$$\frac{8}{3} \;\Big|\; \frac{8}{3} \qquad \text{TRUE}$$

The solution is $\frac{7}{3}$.

20. $\frac{1}{4}$

21.
$$x + \frac{3}{5} = -\frac{7}{10}$$

$$x + \frac{3}{5} - \frac{3}{5} = -\frac{7}{10} - \frac{3}{5}$$

$$x = -\frac{7}{10} - \frac{3}{5} \cdot \frac{2}{2}$$

$$x = -\frac{7}{10} - \frac{6}{10}$$

$$x = -\frac{13}{10}$$

Check: $$x + \frac{3}{5} = -\frac{7}{10}$$

$$-\frac{13}{10} + \frac{3}{5} \;?\; -\frac{7}{10}$$

$$-\frac{13}{10} + \frac{6}{10} \;\Big|\;$$

$$-\frac{7}{10} \;\Big|\; -\frac{7}{10} \qquad \text{TRUE}$$

The solution is $-\frac{13}{10}$.

22. $-\frac{3}{2}$

23.
$$x - \frac{5}{6} = \frac{7}{8}$$

$$x - \frac{5}{6} + \frac{5}{6} = \frac{7}{8} + \frac{5}{6}$$

$$x = \frac{7}{8} \cdot \frac{3}{3} + \frac{5}{6} \cdot \frac{4}{4}$$

$$x = \frac{21}{24} + \frac{20}{24}$$

$$x = \frac{41}{24}$$

Check: $$x - \frac{5}{6} = \frac{7}{8}$$

$$\frac{41}{24} - \frac{5}{6} \;?\; \frac{7}{8}$$

$$\frac{41}{24} - \frac{20}{24} \;\Big|\; \frac{21}{24}$$

$$\frac{21}{24} \;\Big|\; \frac{21}{24} \qquad \text{TRUE}$$

The solution is $\frac{41}{24}$.

24. $\frac{19}{12}$

25.
$$-\frac{1}{5} + z = -\frac{1}{4}$$

$$\frac{1}{5} - \frac{1}{5} + z = \frac{1}{5} - \frac{1}{4}$$

$$z = \frac{1}{5} \cdot \frac{4}{4} - \frac{1}{4} \cdot \frac{5}{5}$$

$$z = \frac{4}{20} - \frac{5}{20}$$

$$z = -\frac{1}{20}$$

Check: $$-\frac{1}{5} + z = -\frac{1}{4}$$

$$-\frac{1}{5} + \left(-\frac{1}{20}\right) \;?\; -\frac{1}{4}$$

$$-\frac{4}{20} + \left(-\frac{1}{20}\right) \;\Big|\; -\frac{5}{20}$$

$$-\frac{5}{20} \;\Big|\; -\frac{5}{20} \qquad \text{TRUE}$$

The solution is $-\frac{1}{20}$.

26. $-\frac{5}{8}$

27.
$$m + 3.9 = 5.4$$
$$m + 3.9 - 3.9 = 5.4 - 3.9$$
$$m = 1.5$$

Check: $$m + 3.9 = 5.4$$

$$1.5 + 3.9 \;?\; 5.4$$
$$5.4 \;\Big|\; 5.4 \qquad \text{TRUE}$$

The solution is 1.5.

28. 4.7

29.
$$-9.7 = -4.7 + y$$
$$4.7 + (-9.7) = 4.7 + (-4.7) + y$$
$$-5 = y$$

Check: $$-9.7 = -4.7 + y$$

$$-9.7 \;?\; -4.7 + (-5)$$
$$-9.7 \;\Big|\; -9.7 \qquad \text{TRUE}$$

The solution is -5.

30. -10.6

31.
$$5x = 80$$
$$\frac{5x}{5} = \frac{80}{5} \qquad \text{Dividing by 5 on both sides}$$
$$1 \cdot x = 16 \qquad \text{Simplifying}$$
$$x = 16 \qquad \text{Identity property of 1}$$

Check: $$5x = 80$$

$$5 \cdot 16 \;?\; 80$$
$$80 \;\Big|\; 80 \qquad \text{TRUE}$$

The solution is 16.

32. 13

33. $5x = 45$

$\dfrac{5x}{5} = \dfrac{45}{5}$ Dividing by 5 on both sides

$1 \cdot x = 9$ Simplifying

$x = 9$ Identity property of 1

Check: $\dfrac{5x = 45}{5 \cdot 9 \;?\; 45}$

$45 \;\big|\; 45$ TRUE

The solution is 9.

34. 12

35. $84 = 7x$

$\dfrac{84}{7} = \dfrac{7x}{7}$ Dividing by 7 on both sides

$12 = 1 \cdot x$

$12 = x$

Check: $\dfrac{84 = 7x}{84 \;?\; 7 \cdot 12}$

$84 \;\big|\; 84$ TRUE

The solution is 12.

36. 7

37. $-x = 23$

$-1 \cdot x = 23$

$-1 \cdot (-1 \cdot x) = -1 \cdot 23$

$1 \cdot x = -23$

$x = -23$

Check: $\dfrac{-x = 23}{-(-23) \;?\; 23}$

$23 \;\big|\; 23$ TRUE

The solution is -23.

38. -100

39. $-x = -8$

$-1 \cdot x = -8$

$-1 \cdot (-1 \cdot x) = -1 \cdot (-8)$

$1 \cdot x = 8$

$x = 8$

Check: $\dfrac{-x = -8}{-(8) \;?\; -8}$

$-8 \;\big|\; -8$ TRUE

The solution is 8.

40. 68

41. $7x = -49$

$\dfrac{7x}{7} = \dfrac{-49}{7}$

$1 \cdot x = -7$

$x = -7$

Check: $\dfrac{7x = -49}{7(-7) \;?\; -49}$

$-49 \;\big|\; -49$ TRUE

The solution is -7.

42. -4

43. $-12x = 72$

$\dfrac{-12x}{-12} = \dfrac{72}{-12}$

$1 \cdot x = -6$

$x = -6$

Check: $\dfrac{-12x = 72}{-12(-6) \;?\; 72}$

$72 \;\big|\; 72$ TRUE

The solution is -6.

44. -7

45. $-3.4t = -20.4$

$\dfrac{-3.4t}{-3.4} = \dfrac{-20.4}{-3.4}$

$1 \cdot t = 6$

$t = 6$

Check: $\dfrac{-3.4t = -20.4}{-3.4(6) \;?\; -20.4}$

$-20.4 \;\big|\; -20.4$ TRUE

The solution is 6.

46. 8

47. $\dfrac{a}{4} = 12$

$\dfrac{1}{4} \cdot a = 12$

$4 \cdot \dfrac{1}{4} \cdot a = 4 \cdot 12$

$a = 48$

Check: $\dfrac{\dfrac{a}{4} = 12}{\dfrac{48}{4} \;?\; 12}$

$12 \;\big|\; 12$ TRUE

The solution is 48.

48. -88

49. $\dfrac{3}{4}x = 27$

$\dfrac{4}{3} \cdot \dfrac{3}{4}x = \dfrac{4}{3} \cdot 27$

$1 \cdot x = \dfrac{4 \cdot 3 \cdot 3 \cdot 3}{3 \cdot 1}$

$x = 36$

Check: $\dfrac{\dfrac{3}{4}x = 27}{\dfrac{3}{4} \cdot 36 \;?\; 27}$

$27 \;\big|\; 27$ TRUE

The solution is 36.

50. 20

51.
$$\frac{-t}{3} = 7$$
$$3 \cdot \frac{1}{3} \cdot (-t) = 3 \cdot 7$$
$$-t = 21$$
$$-1 \cdot (-1 \cdot t) = -1 \cdot 21$$
$$1 \cdot t = -21$$
$$t = -21$$

Check: $\dfrac{-t}{3} = 7$

$$\frac{-(-21)}{3} \ ? \ 7$$
$$\frac{21}{3} \ \Big|$$
$$7 \ \Big| \ 7 \quad \text{TRUE}$$

The solution is -21.

52. -54

53.
$$\frac{2}{9} = -\frac{t}{4}$$
$$\frac{2}{9} = -\frac{1}{4} \cdot t$$
$$-4\left(\frac{2}{9}\right) = -4\left(-\frac{1}{4} \cdot t\right)$$
$$-\frac{8}{9} = t$$

Check: $\dfrac{2}{9} = -\dfrac{t}{4}$

$$\frac{2}{9} \ ? \ -\frac{-8/9}{4}$$
$$\Big| \ -\left(-\frac{8}{9}\right)\left(\frac{1}{4}\right)$$
$$\Big| \ \frac{8}{36}$$
$$\frac{2}{9} \ \Big| \ \frac{2}{9} \quad \text{TRUE}$$

The solution is $-\dfrac{8}{9}$.

54. $-\dfrac{7}{9}$

55.
$$-\frac{3}{5}r = -\frac{9}{10}$$
$$-\frac{5}{3} \cdot \left(-\frac{3}{5}r\right) = -\frac{5}{3} \cdot \left(-\frac{9}{10}\right)$$
$$r = \frac{\cancel{5} \cdot \cancel{3} \cdot 3}{3 \cdot \cancel{5} \cdot 2}$$
$$r = \frac{3}{2}$$

Check: $-\dfrac{3}{5}r = -\dfrac{9}{10}$

$$-\frac{3}{5} \cdot \frac{3}{2} \ ? \ -\frac{9}{10}$$
$$-\frac{9}{10} \ \Big| \ -\frac{9}{10} \quad \text{TRUE}$$

The solution is $\dfrac{3}{2}$.

56. $\dfrac{2}{3}$

57.
$$\frac{-3r}{2} = -\frac{27}{4}$$
$$-\frac{3}{2}r = -\frac{27}{4}$$
$$-\frac{2}{3} \cdot \left(-\frac{3}{2}r\right) = -\frac{2}{3} \cdot \left(-\frac{27}{4}\right)$$
$$r = \frac{\cancel{2} \cdot \cancel{3} \cdot 3 \cdot 3}{\cancel{3} \cdot \cancel{2} \cdot 2}$$
$$r = \frac{9}{2}$$

Check: $\dfrac{-3r}{2} = -\dfrac{27}{4}$

$$-\frac{3}{2} \cdot \frac{9}{2} \ ? \ -\frac{27}{4}$$
$$-\frac{27}{4} \ \Big| \ -\frac{27}{4} \quad \text{TRUE}$$

The solution is $\dfrac{9}{2}$.

58. -1

59.
$$2.8 + t = -3.1$$
$$2.8 + t - 2.8 = -3.1 - 2.8$$
$$t = -5.9$$

The solution is -5.9.

60. 24

61.
$$-8.2x = 20.5$$
$$\frac{-8.2x}{-8.2} = \frac{20.5}{-8.2}$$
$$x = -2.5$$

The solution is -2.5.

62. -5.5

63.
$$17 = y + 29$$
$$17 - 29 = y + 29 - 29$$
$$-12 = y$$

The solution is -12.

64. -128

65.
$$a - \frac{1}{6} = -\frac{2}{3}$$
$$a - \frac{1}{6} + \frac{1}{6} = -\frac{2}{3} + \frac{1}{6}$$
$$a = -\frac{4}{6} + \frac{1}{6}$$
$$a = -\frac{3}{6}$$
$$a = -\frac{1}{2}$$

The solution is $-\dfrac{1}{2}$.

66. $-\dfrac{14}{9}$

67.
$$-24 = \frac{8x}{5}$$
$$-24 = \frac{8}{5}x$$
$$\frac{5}{8}(-24) = \frac{5}{8} \cdot \frac{8}{5}x$$
$$-\frac{5 \cdot \cancel{8} \cdot 3}{\cancel{8} \cdot 1} = x$$
$$-15 = x$$

The solution is -15.

68. $-\dfrac{1}{2}$

69.
$$-16 = -\frac{2}{3}x$$
$$-\frac{3}{2}(-16) = -\frac{3}{2}\left(-\frac{2}{3}x\right)$$
$$\frac{48}{2} = x$$
$$24 = x$$

The solution is 24.

70. $-\dfrac{19}{23}$

71. $3x + 4x = (3 + 4)x = 7x$

72. $-x + 5$

73. $3x - (4 + 2x) = 3x - 4 - 2x = x - 4$

74. $-5x - 23$

75. ◈

76. ◈

77. ◈

78. ◈

79.
$$-356.788 = -699.034 + t$$
$$699.034 + (-356.788) = 699.034 + (-699.034) + t$$
$$342.246 = t$$

The solution is 342.246.

80. -8655

81.
$$5 + x = 5 + x$$
$$5 + x - 5 = 5 + x - 5$$
$$x = x$$

$x = x$ is true for all real numbers. Thus, all real numbers are solutions.

82. No solution

83.
$$4|x| = 48$$
$$|x| = 12$$

x represents a number whose distance from 0 is 12. Thus, $x = -12$ or $x = 12$.

The solution is -12 or 12.

84. No solution

85. For all x, $0 \cdot x = 0$. Thus, all real numbers are solutions.

86. 0

87.
$$x + 4 = 5 + x$$
$$x + 4 - x = 5 + x - x$$
$$4 = 5$$

Since $4 = 5$ is false, the equation has no solution.

88. $-2, 2$

89.
$$mx = 9.4m$$
$$\frac{mx}{m} = \frac{9.4m}{m}$$
$$x = 9.4$$

The solution is 9.4.

90. 4

91.
$$\frac{7cx}{2a} = \frac{21}{a} \cdot c$$
$$\frac{7c}{2a} \cdot x = \frac{21}{a} \cdot c$$
$$\frac{2a}{7c} \cdot \frac{7c}{2a} \cdot x = \frac{2a}{7c} \cdot \frac{21}{a} \cdot \frac{c}{1}$$
$$x = \frac{2 \cdot \cancel{a} \cdot 3 \cdot \cancel{7} \cdot \cancel{c}}{\cancel{7} \cdot \cancel{c} \cdot \cancel{a} \cdot 1}$$
$$x = 6$$

The solution is 6.

92. 2

93.
$$5a = ax - 3a$$
$$5a + 3a = ax - 3a + 3a$$
$$8a = ax$$
$$\frac{8a}{a} = \frac{ax}{a}$$
$$8 = x$$

The solution is 8.

94. $-13, 13$

95.
$$x - 4720 = 1634$$
$$x - 4720 + 4720 = 1634 + 4720$$
$$x = 6354$$
$$x + 4720 = 6354 + 4720$$
$$x + 4720 = 11,074$$

96. 250

97. ◈

Exercise Set 2.2

1.
$$4x + 5 = 41$$

$4x + 5 - 5 = 41 - 5$　　Subtracting 5 on both sides

$4x = 36$　　Simplifying

$\dfrac{4x}{4} = \dfrac{36}{4}$　　Dividing by 4 on both sides

$x = 9$　　Simplifying

Check:　$\dfrac{4x + 5 = 41}{}$

$$4 \cdot 9 + 5 \ ? \ 41$$
$$36 + 5$$
$$41 \ \bigg| \ 41 \qquad \text{TRUE}$$

The solution is 9.

2. 8

3.
$$8x + 4 = 68$$

$8x + 4 - 4 = 68 - 4$　　Subtracting 4 on both sides

$8x = 64$　　Simplifying

$\dfrac{8x}{8} = \dfrac{64}{8}$　　Dividing by 8 on both sides

$x = 8$　　Simplifying

Check:　$\dfrac{8x + 4 = 68}{}$

$$8 \cdot 8 + 4 \ ? \ 68$$
$$64 + 4$$
$$68 \ \bigg| \ 68 \qquad \text{TRUE}$$

The solution is 8.

4. 9

5.
$$5x - 8 = 27$$

$5x - 8 + 8 = 27 + 8$　　Adding 8 on both sides

$5x = 35$

$\dfrac{5x}{5} = \dfrac{35}{5}$　　Dividing by 5 on both sides

$x = 7$

Check:　$\dfrac{5x - 8 = 27}{}$

$$5 \cdot 7 - 8 \ ? \ 27$$
$$35 - 8$$
$$27 \ \bigg| \ 27 \qquad \text{TRUE}$$

The solution is 7.

6. 3

7.
$$3x - 9 = 33$$
$$3x - 9 + 9 = 33 + 9$$
$$3x = 42$$
$$\dfrac{3x}{3} = \dfrac{42}{3}$$
$$x = 14$$

Check:　$\dfrac{3x - 9 = 33}{}$

$$3 \cdot 14 - 9 \ ? \ 33$$
$$42 - 9$$
$$33 \ \bigg| \ 33 \qquad \text{TRUE}$$

The solution is 14.

8. 11

9.
$$7x + 2 = -54$$
$$7x + 2 - 2 = -54 - 2$$
$$7x = -56$$
$$\dfrac{7x}{7} = \dfrac{-56}{7}$$
$$x = -8$$

Check:　$\dfrac{7x + 2 = -54}{}$

$$7(-8) + 2 \ ? \ -54$$
$$-56 + 2$$
$$-54 \ \bigg| \ -54 \qquad \text{TRUE}$$

The solution is -8.

10. -6

11.
$$-39 = 1 + 8x$$
$$-39 - 1 = 1 + 8x - 1$$
$$-40 = 8x$$
$$\dfrac{-40}{8} = \dfrac{8x}{8}$$
$$-5 = x$$

Check:　$\dfrac{-39 = 1 + 8x}{}$

$$-39 \ ? \ 1 + 8(-5)$$
$$1 - 40$$
$$-39 \ \bigg| \ -39 \qquad \text{TRUE}$$

The solution is -5.

12. -11

13.
$$9 - 4x = 37$$
$$9 - 4x - 9 = 37 - 9$$
$$-4x = 28$$
$$\dfrac{-4x}{-4} = \dfrac{28}{-4}$$
$$x = -7$$

Check:　$\dfrac{9 - 4x = 37}{}$

$$9 - 4(-7) \ ? \ 37$$
$$9 + 28$$
$$37 \ \bigg| \ 37 \qquad \text{TRUE}$$

The solution is -7.

14. -24

15.
$$-7x - 24 = -129$$
$$-7x - 24 + 24 = -129 + 24$$
$$-7x = -105$$
$$\frac{-7x}{-7} = \frac{-105}{-7}$$
$$x = 15$$

Check:
$$\begin{array}{c|c} \multicolumn{2}{c}{-7x - 24 = -129} \\ \hline -7 \cdot 15 - 24 \ ? \ -129 & \\ -105 - 24 & \\ -129 & -129 \quad \text{TRUE} \end{array}$$

The solution is 15.

16. 19

17.
$$36 = 5x + 7x$$
$$36 = 12x \qquad \text{Combining like terms}$$
$$\frac{36}{12} = \frac{12x}{12} \qquad \text{Dividing by 12 on both sides}$$
$$3 = x$$

Check:
$$\begin{array}{c|c} \multicolumn{2}{c}{36 = 5x + 7x} \\ \hline 36 \ ? \ 5 \cdot 3 + 7 \cdot 3 & \\ & 15 + 21 \\ 36 & 36 \quad \text{TRUE} \end{array}$$

The solution is 3.

18. 5

19.
$$27 - 6x = 99$$
$$27 - 6x - 27 = 99 - 27$$
$$-6x = 72$$
$$\frac{-6x}{-6} = \frac{72}{-6}$$
$$x = -12$$

Check:
$$\begin{array}{c|c} \multicolumn{2}{c}{27 - 6x = 99} \\ \hline 27 - 6(-12) \ ? \ 99 & \\ 27 + 72 & \\ 99 & 99 \quad \text{TRUE} \end{array}$$

The solution is -12.

20. 3

21.
$$4x + 3x = 42 \qquad \text{Combining like terms}$$
$$7x = 42$$
$$\frac{7x}{7} = \frac{42}{7}$$
$$x = 6$$

Check:
$$\begin{array}{c|c} \multicolumn{2}{c}{4x + 3x = 42} \\ \hline 4 \cdot 6 + 3 \cdot 6 \ ? \ 42 & \\ 24 + 18 & \\ & 42 \quad 42 \quad \text{TRUE} \end{array}$$

The solution is 6.

22. 4

23.
$$-2a + 5a = 24$$
$$3a = 24$$
$$\frac{3a}{3} = \frac{24}{3}$$
$$a = 8$$

Check:
$$\begin{array}{c|c} \multicolumn{2}{c}{-2a + 5a = 24} \\ \hline -2 \cdot 8 + 5 \cdot 8 \ ? \ 24 & \\ -16 + 40 & \\ & 24 \quad 24 \quad \text{TRUE} \end{array}$$

The solution is 8.

24. -4

25.
$$-7y - 8y = -15$$
$$-15y = -15$$
$$\frac{-15y}{-15} = \frac{-15}{-15}$$
$$y = 1$$

Check:
$$\begin{array}{c|c} \multicolumn{2}{c}{-7y - 8y = -15} \\ \hline -7 \cdot 1 - 8 \cdot 1 \ ? \ -15 & \\ -7 - 8 & \\ -15 & -15 \quad \text{TRUE} \end{array}$$

The solution is 1.

26. 3

27.
$$10.2y - 7.3y = -58$$
$$2.9y = -58$$
$$\frac{2.9y}{2.9} = \frac{-58}{2.9}$$
$$y = -\frac{58}{2.9}$$
$$y = -20$$

Check:
$$\begin{array}{c|c} \multicolumn{2}{c}{10.2y - 7.3y = -58} \\ \hline 10.2(-20) - 7.3(-20) \ ? \ -58 & \\ -204 + 146 & \\ -58 & -58 \quad \text{TRUE} \end{array}$$

The solution is -20.

28. -20

29.
$$x + \frac{1}{3}x = 8$$
$$\left(1 + \frac{1}{3}\right)x = 8$$
$$\frac{4}{3}x = 8$$
$$\frac{3}{4} \cdot \frac{4}{3}x = \frac{3}{4} \cdot 8$$
$$x = 6$$

Check:
$$x + \frac{1}{3}x = 8$$

$$6 + \frac{1}{3}\cdot 6 \ ? \ 8$$

$$6 + 2$$

$$8 \ \bigg| \ 8 \quad \text{TRUE}$$

The solution is 6.

30. 8

31.
$$8y - 35 = 3y$$

$$8y = 3y + 35 \qquad \text{Adding 35 and simplifying}$$

$$8y - 3y = 35 \qquad \text{Subtracting } 3y \text{ and simplifying}$$

$$5y = 35 \qquad \text{Collecting like terms}$$

$$\frac{5y}{5} = \frac{35}{5} \qquad \text{Dividing by 5}$$

$$y = 7$$

Check:
$$8y - 35 = 3y$$

$$8\cdot 7 - 35 \ ? \ 3\cdot 7$$

$$56 - 35 \ \bigg| \ 21$$

$$21 \ \bigg| \ 21 \quad \text{TRUE}$$

The solution is 7.

32. -3

33.
$$6x - 5 = 7 + 2x$$

$$6x - 5 - 2x = 7 + 2x - 2x \qquad \text{Subtracting } 2x \text{ on both sides}$$

$$4x - 5 = 7 \qquad \text{Simplifying}$$

$$4x - 5 + 5 = 7 + 5 \qquad \text{Adding 5 on both sides}$$

$$4x = 12 \qquad \text{Simplifying}$$

$$\frac{4x}{4} = \frac{12}{4} \qquad \text{Dividing by 4 on both sides}$$

$$x = 3$$

Check:
$$6x - 5 = 7 + 2x$$

$$6\cdot 3 - 5 \ ? \ 7 + 2\cdot 3$$

$$18 - 5 \ \bigg| \ 7 + 6$$

$$13 \ \bigg| \ 13 \quad \text{TRUE}$$

The solution is 3.

34. 5

35.
$$6x + 3 = 2x + 11$$

$$6x - 2x = 11 - 3$$

$$4x = 8$$

$$\frac{4x}{4} = \frac{8}{4}$$

$$x = 2$$

Check:
$$6x + 3 = 2x + 11$$

$$6\cdot 2 + 3 \ ? \ 2\cdot 2 + 11$$

$$12 + 3 \ \bigg| \ 4 + 11$$

$$15 \ \bigg| \ 15 \quad \text{TRUE}$$

The solution is 2.

36. 4

37.
$$5 - 2x = 3x - 7x + 25$$

$$5 - 2x = -4x + 25$$

$$4x - 2x = 25 - 5$$

$$2x = 20$$

$$\frac{2x}{2} = \frac{20}{2}$$

$$x = 10$$

Check:
$$5 - 2x = 3x - 7x + 25$$

$$5 - 2\cdot 10 \ ? \ 3\cdot 10 - 7\cdot 10 + 25$$

$$5 - 20 \ \bigg| \ 30 - 70 + 25$$

$$-15 \ \bigg| \ -40 + 25$$

$$-15 \ \bigg| \ -15 \quad \text{TRUE}$$

The solution is 10.

38. 10

39.
$$4 + 3x - 6 = 3x + 2 - x$$

$$3x - 2 = 2x + 2 \qquad \text{Combining like terms on each side}$$

$$3x - 2x = 2 + 2$$

$$x = 4$$

Check:
$$4 + 3x - 6 = 3x + 2 - x$$

$$4 + 3\cdot 4 - 6 \ ? \ 3\cdot 4 + 2 - 4$$

$$4 + 12 - 6 \ \bigg| \ 12 + 2 - 4$$

$$16 - 6 \ \bigg| \ 14 - 4$$

$$10 \ \bigg| \ 10 \quad \text{TRUE}$$

The solution is 4.

40. 0

41.
$$4y - 4 + y + 24 = 6y + 20 - 4y$$

$$5y + 20 = 2y + 20$$

$$5y - 2y = 20 - 20$$

$$3y = 0$$

$$y = 0$$

Check:
$$4y - 4 + y + 24 = 6y + 20 - 4y$$

$$4\cdot 0 - 4 + 0 + 24 \ ? \ 6\cdot 0 + 20 - 4\cdot 0$$

$$0 - 4 + 0 + 24 \ \bigg| \ 0 + 20 - 0$$

$$20 \ \bigg| \ 20 \quad \text{TRUE}$$

The solution is 0.

42. 7

43. $\frac{5}{4}x + \frac{1}{4}x = 2x + \frac{1}{2} + \frac{3}{4}x$

The number 4 is the least common denominator, so we multiply by 4 on both sides.

$$4\left(\frac{5}{4}x + \frac{1}{4}x\right) = 4\left(2x + \frac{1}{2} + \frac{3}{4}x\right)$$

$$4 \cdot \frac{5}{4}x + 4 \cdot \frac{1}{4}x = 4 \cdot 2x + 4 \cdot \frac{1}{2} + 4 \cdot \frac{3}{4}x$$

$$5x + x = 8x + 2 + 3x$$

$$6x = 11x + 2$$

$$6x - 11x = 2$$

$$-5x = 2$$

$$\frac{-5x}{-5} = \frac{2}{-5}$$

$$x = -\frac{2}{5}$$

Check:

$$\frac{5}{4}x + \frac{1}{4}x = 2x + \frac{1}{2} + \frac{3}{4}x$$

$\frac{5}{4}\left(-\frac{2}{5}\right) + \frac{1}{4}\left(-\frac{2}{5}\right)$?	$2\left(-\frac{2}{5}\right) + \frac{1}{2} + \frac{3}{4}\left(-\frac{2}{5}\right)$
$-\frac{1}{2} - \frac{1}{10}$	$-\frac{4}{5} + \frac{1}{2} - \frac{3}{10}$
$-\frac{5}{10} - \frac{1}{10}$	$-\frac{8}{10} + \frac{5}{10} - \frac{3}{10}$
$-\frac{6}{10}$	$-\frac{6}{10}$ TRUE

The solution is $-\frac{2}{5}$.

44. $\frac{1}{2}$

45. $\frac{2}{3} + \frac{1}{4}t = 6$

The number 12 is the least common denominator, so we multiply by 12 on both sides.

$$12\left(\frac{2}{3} + \frac{1}{4}t\right) = 12 \cdot 6$$

$$12 \cdot \frac{2}{3} + 12 \cdot \frac{1}{4}t = 72$$

$$8 + 3t = 72$$

$$3t = 72 - 8$$

$$3t = 64$$

$$t = \frac{64}{3}$$

Check:

$$\frac{2}{3} + \frac{1}{4}t = 6$$

$\frac{2}{3} + \frac{1}{4}\left(\frac{64}{3}\right)$? 6
$\frac{2}{3} + \frac{16}{3}$
$\frac{18}{3}$
6	6 TRUE

The solution is $\frac{64}{3}$.

46. $-\frac{2}{3}$

47. $\frac{2}{3} + 3y = 5y - \frac{2}{15}$

The number 15 is the least common denominator, so we multiply by 15 on both sides.

$$15\left(\frac{2}{3} + 3y\right) = 15\left(5y - \frac{2}{15}\right)$$

$$15 \cdot \frac{2}{3} + 15 \cdot 3y = 15 \cdot 5y - 15 \cdot \frac{2}{15}$$

$$10 + 45y = 75y - 2$$

$$10 + 2 = 75y - 45y$$

$$12 = 30y$$

$$\frac{12}{30} = y$$

$$\frac{2}{5} = y$$

Check: $\frac{2}{3} + 3y = 5y - \frac{2}{15}$

$\frac{2}{3} + 3 \cdot \frac{2}{5}$?	$5 \cdot \frac{2}{5} - \frac{2}{15}$
$\frac{2}{3} + \frac{6}{5}$	$2 - \frac{2}{15}$
$\frac{10}{15} + \frac{18}{15}$	$\frac{30}{15} - \frac{2}{15}$
$\frac{28}{15}$	$\frac{28}{15}$ TRUE

The solution is $\frac{2}{5}$.

48. -3

49. $\frac{1}{3}x + \frac{2}{5} = \frac{4}{15} + \frac{3}{5}x - \frac{2}{3}$

The number 15 is the least common denominator, so we multiply by 15 on both sides.

$$15\left(\frac{1}{3}x + \frac{2}{5}\right) = 15\left(\frac{4}{15} + \frac{3}{5}x - \frac{2}{3}\right)$$

$$15 \cdot \frac{1}{3}x + 15 \cdot \frac{2}{5} = 15 \cdot \frac{4}{15} + 15 \cdot \frac{3}{5}x - 15 \cdot \frac{2}{3}$$

$$5x + 6 = 4 + 9x - 10$$

$$5x + 6 = -6 + 9x$$

$$5x - 9x = -6 - 6$$

$$-4x = -12$$

$$\frac{-4x}{-4} = \frac{-12}{-4}$$

$$x = 3$$

Check:

$$\frac{1}{3}x + \frac{2}{5} = \frac{4}{15} + \frac{3}{5}x - \frac{2}{3}$$

$\frac{1}{3} \cdot 3 + \frac{2}{5}$	$?$	$\frac{4}{15} + \frac{3}{5} \cdot 3 - \frac{2}{3}$
$1 + \frac{2}{5}$		$\frac{4}{15} + \frac{9}{5} - \frac{2}{3}$
$\frac{5}{5} + \frac{2}{5}$		$\frac{4}{15} + \frac{27}{15} - \frac{10}{15}$
$\frac{7}{5}$		$\frac{21}{15}$
$\frac{7}{5}$		$\frac{7}{5}$ TRUE

The solution is 3.

50. -3

51.
$$2.1x + 45.2 = 3.2 - 8.4x$$
Greatest number of decimal places is 1
$$10(2.1x + 45.2) = 10(3.2 - 8.4x)$$
Multiplying by 10 to clear decimals
$$10(2.1x) + 10(45.2) = 10(3.2) - 10(8.4x)$$
$$21x + 452 = 32 - 84x$$
$$21x + 84x = 32 - 452$$
$$105x = -420$$
$$x = \frac{-420}{105}$$
$$x = -4$$

Check:

$$2.1x + 45.2 = 3.2 - 8.4x$$

$2.1(-4) + 45.2$	$?$	$3.2 - 8.4(-4)$
$-8.4 + 45.2$		$3.2 + 33.6$
36.8		36.8 TRUE

The solution is -4.

52. $\frac{5}{3}$

53.
$$1.03 - 0.6x = 0.71 - 0.2x$$
Greatest number of decimal places is 2
$$100(1.03 - 0.6x) = 100(0.71 - 0.2x)$$
Multiplying by 100 to clear decimals
$$100(1.03) - 100(0.6x) = 100(0.71) - 100(0.2x)$$
$$103 - 60x = 71 - 20x$$
$$32 = 40x$$
$$\frac{32}{40} = x$$
$$\frac{4}{5} = x, \text{ or}$$
$$0.8 = x$$

Check:

$$1.03 - 0.6x = 0.71 - 0.2x$$

$1.03 - 0.6(0.8)$	$?$	$0.71 - 0.2(0.8)$
$1.03 - 0.48$		$0.71 - 0.16$
0.55		0.55 TRUE

The solution is $\frac{4}{5}$, or 0.8.

54. 1

55.
$$\frac{2}{5}x - \frac{3}{2}x = \frac{3}{4}x + 2$$
The least common denominator is 20.
$$20\left(\frac{2}{5}x - \frac{3}{2}x\right) = 20\left(\frac{3}{4}x + 2\right)$$
$$20 \cdot \frac{2}{5}x - 20 \cdot \frac{3}{2}x = 20 \cdot \frac{3}{4}x + 20 \cdot 2$$
$$8x - 30x = 15x + 40$$
$$-22x = 15x + 40$$
$$-22x - 15x = 40$$
$$-37x = 40$$
$$\frac{-37x}{-37} = \frac{40}{-37}$$
$$x = -\frac{40}{37}$$

Check:

$$\frac{2}{5}x - \frac{3}{2}x = \frac{3}{4}x + 2$$

$\frac{2}{5}\left(-\frac{40}{37}\right) - \frac{3}{2}\left(-\frac{40}{37}\right)$	$?$	$\frac{3}{4}\left(-\frac{40}{37}\right) + 2$
$-\frac{16}{37} + \frac{60}{37}$		$-\frac{30}{37} + \frac{74}{37}$
$\frac{44}{37}$		$\frac{44}{37}$ TRUE

The solution is $-\frac{40}{37}$.

56. $\frac{32}{7}$

57. $7(2a - 1) = 21$
$$14a - 7 = 21 \quad \text{Using the distributive law}$$
$$14a = 21 + 7 \quad \text{Adding 7}$$
$$14a = 28$$
$$a = 2 \quad \text{Dividing by 14}$$

Check:

$$7(2a - 1) = 21$$

$7(2 \cdot 2 - 1)$	$?$	21
$7(4 - 1)$		
$7 \cdot 3$		
21		21 TRUE

The solution is 2.

58. 4.5 or $\frac{9}{2}$

59.
$$40 = 5(3x + 2)$$
$$40 = 15x + 10 \quad \text{Using the distributive law}$$
$$40 - 10 = 15x$$
$$30 = 15x$$
$$2 = x$$

Check:

$$40 = 5(3x + 2)$$

40	$?$	$5(3 \cdot 2 + 2)$
		$5(6 + 2)$
		$5 \cdot 8$
40		40 TRUE

The solution is 2.

60. 1

61. $2(3 + 4m) - 9 = 45$
$6 + 8m - 9 = 45$
$8m - 3 = 45$ Combining like terms
$8m = 45 + 3$
$8m = 48$
$m = 6$

Check: $\dfrac{2(3 + 4m) - 9 = 45}{}$
$2(3 + 4 \cdot 6) - 9 \ ? \ 45$
$2(3 + 24) - 9$ |
$2 \cdot 27 - 9$ |
$54 - 9$ |
45 | 45 TRUE

The solution is 6.

62. 9

63. $5r - (2r + 8) = 16$
$5r - 2r - 8 = 16$
$3r - 8 = 16$ Combining like terms
$3r = 16 + 8$
$3r = 24$
$r = 8$

Check: $\dfrac{5r - (2r + 8) = 16}{}$
$5 \cdot 8 - (2 \cdot 8 + 8) \ ? \ 16$
$40 - (16 + 8)$ |
$40 - 24$ |
16 | 16 TRUE

The solution is 8.

64. 8

65. $10 - 3(2x - 1) = 1$
$10 - 6x + 3 = 1$
$13 - 6x = 1$
$-6x = 1 - 13$
$-6x = -12$
$x = 2$

Check: $\dfrac{10 - 3(2x - 1) = 1}{}$
$10 - 3(2 \cdot 2 - 1) \ ? \ 1$
$10 - 3(4 - 1)$ |
$10 - 3 \cdot 3$ |
$10 - 9$ |
1 | 1 TRUE

The solution is 2.

66. 17

67. $3(t - 2) = 9(t + 2)$
$3t - 6 = 9t + 18$
$-6 - 18 = 9t - 3t$
$-24 = 6t$
$-4 = t$

Check: $\dfrac{3(t - 2) = 9(t + 2)}{}$
$3(-4 - 2) \ ? \ 9(-4 + 2)$
$3(-6)$ | $9(-2)$
-18 | -18 TRUE

The solution is -4.

68. $-\dfrac{5}{3}$

69. $7(5x - 2) = 6(6x - 1)$
$35x - 14 = 36x - 6$
$-14 + 6 = 36x - 35x$
$-8 = x$

Check:
$\dfrac{7(5x - 2) = 6(6x - 1)}{}$
$7(5(-8) - 2) \ ? \ 6(6(-8) - 1)$
$7(-40 - 2)$ | $6(-48 - 1)$
$7(-42)$ | $6(-49)$
-294 | -294 TRUE

The solution is -8.

70. -12

71. $19 - (2x + 3) = 2(x + 3) + x$
$19 - 2x - 3 = 2x + 6 + x$
$16 - 2x = 3x + 6$
$16 - 6 = 3x + 2x$
$10 = 5x$
$2 = x$

Check: $\dfrac{19 - (2x + 3) = 2(x + 3) + x}{}$
$19 - (2 \cdot 2 + 3) \ ? \ 2(2 + 3) + 2$
$19 - (4 + 3)$ | $2 \cdot 5 + 2$
$19 - 7$ | $10 + 2$
12 | 12 TRUE

The solution is 2.

72. 1

73. $\dfrac{1}{3}(6x + 24) - 20 = -\dfrac{1}{4}(12x - 72)$
$\dfrac{1}{3} \cdot 6x + \dfrac{1}{3} \cdot 24 - 20 = -\dfrac{1}{4} \cdot 12x - \dfrac{1}{4}(-72)$
$2x + 8 - 20 = -3x + 18$
$2x - 12 = -3x + 18$
$5x = 30$
$x = 6$

The check is left to the student. The solution is 6.

74. 5

75. $\dfrac{4}{5}(3x + 4) = 10$

The least common denominator is 5.

$5 \cdot \dfrac{4}{5}(3x + 4) = 5 \cdot 10$

$4(3x + 4) = 50$

$12x + 16 = 50$

$12x = 34$ Subtracting 16 on both sides

$x = \dfrac{34}{12}$ Dividing by 12 on both sides

$x = \dfrac{17}{6}$ Simplifying

Check:

$$\frac{4}{5}(3x+4)=10$$

$$\frac{4}{5}\left(3\cdot\frac{17}{6}+4\right) \ ? \ 10$$

$$\frac{4}{5}\left(\frac{17}{2}+4\right)$$

$$\frac{4}{5}\cdot\frac{25}{2}$$

$$10 \ \Big| \ 10 \quad \text{TRUE}$$

The solution is $\frac{17}{6}$.

76. $\frac{31}{2}$

77. $\frac{3}{2}(2x+5)+\frac{1}{4}=-\frac{7}{2}$

The least common denominator is 4.

$$4\left[\frac{3}{2}(2x+5)+\frac{1}{4}\right]=4\left(-\frac{7}{2}\right)$$

$$4\cdot\frac{3}{2}(2x+5)+4\cdot\frac{1}{4}=-14$$

$$6(2x+5)+1=-14$$

$$12x+30+1=-14$$

$$12x+31=-14$$

$$12x=-45$$

$$x=\frac{-45}{12}$$

$$x=-\frac{15}{4}$$

Check:

$$\frac{3}{2}(2x+5)+\frac{1}{4}=-\frac{7}{2}$$

$$\frac{3}{2}\left(2\left(-\frac{15}{4}\right)+5\right)+\frac{1}{4} \ ? \ -\frac{7}{2}$$

$$\frac{3}{2}\left(-\frac{15}{2}+5\right)+\frac{1}{4}$$

$$\frac{3}{2}\left(-\frac{5}{2}\right)+\frac{1}{4}$$

$$-\frac{15}{4}+\frac{1}{4}$$

$$-\frac{14}{4}$$

$$-\frac{7}{2} \ \Big| \ -\frac{7}{2} \quad \text{TRUE}$$

The solution is $-\frac{15}{4}$.

78. $\frac{16}{15}$

79. $\frac{3}{4}\left(3x-\frac{1}{2}\right)-\frac{2}{3}=\frac{1}{3}$

$$\frac{9}{4}x-\frac{3}{8}-\frac{2}{3}=\frac{1}{3}$$

Multiplying by the number 24 will clear all the fractions, so we multiply by 24 on both sides.

$$24\left(\frac{9}{4}x-\frac{3}{8}-\frac{2}{3}\right)=24\cdot\frac{1}{3}$$

$$24\cdot\frac{9}{4}x-24\cdot\frac{3}{8}-24\cdot\frac{2}{3}=8$$

$$54x-9-16=8$$

$$54x-25=8$$

$$54x=8+25$$

$$54x=33$$

$$x=\frac{33}{54}$$

$$x=\frac{11}{18}$$

The check is left to the student. The solution is $\frac{11}{18}$.

80. $-\frac{5}{32}$

81. $0.7(3x+6)=1.1-(x+2)$

$$2.1x+4.2=1.1-x-2$$

$$10(2.1x+4.2)=10(1.1-x-2) \quad \text{Clearing}$$
$$\text{decimals}$$

$$21x+42=11-10x-20$$

$$21x+42=-10x-9$$

$$21x+10x=-9-42$$

$$31x=-51$$

$$x=-\frac{51}{31}$$

The check is left to the student. The solution is $-\frac{51}{31}$.

82. $\frac{39}{14}$

83. $a+(a-3)=(a+2)-(a+1)$

$$a+a-3=a+2-a-1$$

$$2a-3=1$$

$$2a=1+3$$

$$2a=4$$

$$a=2$$

Check:

$$a+(a-3)=(a+2)-(a+1)$$

$$2+(2-3) \ ? \ (2+2)-(2+1)$$

$$2-1 \ \Big| \ 4-3$$

$$1 \ \Big| \ 1 \quad \text{TRUE}$$

The solution is 2.

84. -7.4

85. Do the long division. The answer is negative.

$$\begin{array}{r} 6.5 \\ 3.4_\wedge\overline{\smash{)}2\,2.1_\wedge 0} \\ \underline{2\,0\,4} \\ 1\,7\,0 \\ \underline{1\,7\,0} \\ 0 \end{array}$$

$$-22.1 \div 3.4 = -6.5$$

86. $7(x - 3 - 2y)$

87. Since -15 is to the left of -13 on the number line, -15 is less than -13, so $-15 < -13$.

88. -14

89. ◈

90. ◈

91. ◈

92. ◈

93. $8.43x - 2.5(3.2 - 0.7x) = -3.455x + 9.04$
$8.43x - 8 + 1.75x = -3.455x + 9.04$
$10.18x - 8 = -3.455x + 9.04$
$10.18x + 3.455x = 9.04 + 8$
$13.635x = 17.04$
$x = 1.\overline{2497}$

The solution is $1.\overline{2497}$.

94. 4.4233464

95. $-2[3(x - 2) + 4] = 4(1 - x) + 8$
$-2[3x - 6 + 4] = 4 - 4x + 8$
$-2[3x - 2] = 12 - 4x$
$-6x + 4 = 12 - 4x$
$4 - 12 = -4x + 6x$
$-8 = 2x$
$-4 = x$

The solution is -4.

96. $-\dfrac{7}{2}$

97. $3(x + 4) = 3(4 + x)$
$3x + 12 = 12 + 3x$
$3x + 12 - 12 = 12 + 3x - 12$
$3x = 3x$

Since $3x = 3x$ is true for all real numbers, all real numbers are solutions.

98. $\dfrac{837,353}{1929}$

99. $2x(x + 5) - 3(x^2 + 2x - 1) = 9 - 5x - x^2$
$2x^2 + 10x - 3x^2 - 6x + 3 = 9 - 5x - x^2$
$-x^2 + 4x + 3 = 9 - 5x - x^2$
$4x + 3 = 9 - 5x \quad \text{Adding } x^2$
$4x + 5x = 9 - 3$
$9x = 6$
$x = \dfrac{2}{3}$

The solution is $\dfrac{2}{3}$.

100. -2

101. $9 - 3x = 2(5 - 2x) - (1 - 5x)$
$9 - 3x = 10 - 4x - 1 + 5x$
$9 - 3x = 9 + x$
$9 - 9 = x + 3x$
$0 = 4x$
$0 = x$

The solution is 0.

102. 0

103. $\dfrac{x}{14} - \dfrac{5x + 2}{49} = \dfrac{3x - 4}{7}$
$98\left(\dfrac{x}{14} - \dfrac{5x + 2}{49}\right) = 98\left(\dfrac{3x - 4}{7}\right)$
$98 \cdot \dfrac{x}{14} - 98\left(\dfrac{5x + 2}{49}\right) = 42x - 56$
$7x - 10x - 4 = 42x - 56$
$-3x - 4 = 42x - 56$
$-4 + 56 = 42x + 3x$
$52 = 45x$
$\dfrac{52}{45} = x$

The solution is $\dfrac{52}{45}$.

104. -2

Exercise Set 2.3

1. We substitute 1900 for a and calculate b.
$b = 30a = 30 \cdot 1900 = 57,000$

The minimum furnace output is 57,000 Btu's.

2. 125,000 Btu's

3. We substitute 3 for s and calculate A.
$A = 6s^2 = 6 \cdot 3^2 = 6 \cdot 9 = 54$

The surface area is 54 in^2.

4. 1423

5. Substitute 30 for I and 115 for V and calculate P.
$P = I \cdot V = 30 \cdot 115 = 3450$

The wattage is 3450 watts.

6. $\dfrac{43}{3}$ m

7. Substitute 1 for t and calculate n.
$n = 0.5t^4 + 3.45t^3 - 96.65t^2 + 347.7t$
$= 0.5(1)^4 + 3.45(1)^3 - 96.65(1)^2 + 347.7(1)$
$= 0.5 + 3.45 - 96.65 + 347.7$
$= 255$

255 mg of ibuprofen remains in the bloodstream.

8. 42

9. $A = bh$

$\dfrac{A}{h} = \dfrac{bh}{h}$ Dividing by h

$\dfrac{A}{h} = b$

10. $h = \dfrac{A}{b}$

11. $d = rt$

$\dfrac{d}{t} = \dfrac{rt}{t}$ Dividing by t

$\dfrac{d}{t} = r$

12. $t = \dfrac{d}{r}$

13. $I = Prt$

$\dfrac{I}{rt} = \dfrac{Prt}{rt}$ Dividing by rt

$\dfrac{I}{rt} = P$

14. $t = \dfrac{I}{Pr}$

15. $H = 65 - m$

$H + m = 65$ Adding m

$m = 65 - H$ Subtracting H

16. $h = d + 64$

17. $P = 2l + 2w$

$P - 2w = 2l + 2w - 2w$ Subtracting $2w$

$P - 2w = 2l$

$\dfrac{P - 2w}{2} = \dfrac{2l}{2}$ Dividing by 2

$\dfrac{P - 2w}{2} = l$

18. $w = \dfrac{P - 2l}{2}$

19. $A = \pi r^2$

$\dfrac{A}{r^2} = \dfrac{\pi r^2}{r^2}$

$\dfrac{A}{r^2} = \pi$

20. $r^2 = \dfrac{A}{\pi}$

21. $A = \dfrac{1}{2} bh$

$2A = 2 \cdot \dfrac{1}{2} bh$ Multiplying by 2

$2A = bh$

$\dfrac{2A}{b} = \dfrac{bh}{b}$ Dividing by h

$\dfrac{2A}{b} = h$

22. $b = \dfrac{2A}{h}$

23. $E = mc^2$

$\dfrac{E}{c^2} = \dfrac{mc^2}{c^2}$ Dividing by c^2

$\dfrac{E}{c^2} = m$

24. $c^2 = \dfrac{E}{m}$

25. $Q = \dfrac{c + d}{2}$

$2Q = 2 \cdot \dfrac{c + d}{2}$ Multiplying by 2

$2Q = c + d$

$2Q - c = c + d - c$ Subtracting c

$2Q - c = d$

26. $p = 2Q + q$

27. $A = \dfrac{a + b + c}{3}$

$3A = 3 \cdot \dfrac{a + b + c}{3}$ Multiplying by 3

$3A = a + b + c$

$3A - a - c = a + b + c - a - c$ Subtracting a and c

$3A - a - c = b$

28. $c = 3A - a - b$

29. $M = \dfrac{A}{s}$

$s \cdot M = s \cdot \dfrac{A}{s}$ Multiplying by s

$sM = A$

30. $b = \dfrac{Pc}{a}$

31. $Ax + By = C$

$Ax + By - Ax = C - Ax$ Subtracting Ax

$By = C - Ax$

$\dfrac{By}{B} = \dfrac{C - Ax}{B}$ Dividing by B

$y = \dfrac{C - Ax}{B}$

32. $x = \dfrac{C - By}{A}$

33. $A = \dfrac{\pi r^2 S}{360}$

$\dfrac{360}{\pi r^2} \cdot A = \dfrac{360}{\pi r^2} \cdot \dfrac{\pi r^2 S}{360}$

$\dfrac{360}{\pi r^2} = S$

34. $P = \dfrac{A}{1 + rt}$

35.
$$A = \frac{1}{2}ah + \frac{1}{2}bh$$
$$2A = 2\left(\frac{1}{2}ah + \frac{1}{2}bh\right)$$
$$2A = ah + bh$$
$$2A = h(a + b)$$
$$\frac{2A}{a + b} = h$$

36. $L = \dfrac{Nr + 400W - NR}{400}$

37. We multiply from left to right:
$$7(-3)2 = (-21)2 = -42$$

38. $-\dfrac{3}{5}$

39.
$$\begin{aligned} 10 \div (-2) \cdot 5 - 4 &= -5 \cdot 5 - 4 \quad \text{Dividing} \\ &= -25 - 4 \quad \text{Multiplying} \\ &= -29 \quad \text{Subtracting} \end{aligned}$$

40. 30

41.

42.

43.

44.

45. First convert w lb to kilograms:
$$w \text{ lb} \cdot \frac{1 \text{ kg}}{2.2046 \text{ lb}} = \frac{w}{2.2046} \text{ kg}$$
Then convert h in. to centimeters:
$$h \text{ in.} \cdot \frac{1 \text{ cm}}{0.3937 \text{ in.}} = \frac{h}{0.3937} \text{ cm}$$
Now rewrite the formula using these converted values:
$$K = 917 + 6\left(\frac{w}{2.2046} + \frac{h}{0.3937} - a\right)$$

46. 35

47.
$$c = \frac{w}{a} \cdot d$$
$$ac = a \cdot \frac{w}{a} \cdot d$$
$$ac = wd$$
$$a = \frac{wd}{c}$$

48. 76.4 in.

49. To find the number of 100 meter rises in h meters we divide: $\dfrac{h}{100}$. Then
$$T = t - \frac{h}{100}.$$
Note that $12 \text{ km} = 12 \text{ km} \cdot \dfrac{1000 \text{ m}}{1 \text{ km}} = 12{,}000 \text{ m}.$
Thus, we have
$$T = t - \frac{h}{100}, \quad 0 \le h \le 12{,}000.$$

50. $y = \dfrac{z^2}{t}$

51.
$$ac = bc + d$$
$$ac - bc = d$$
$$c(a - b) = d$$
$$c = \frac{d}{a - b}$$

52. $t = \dfrac{rs}{q - r}$

53.
$$3a = c - a(b + d)$$
$$3a = c - ab - ad$$
$$3a + ab + ad = c$$
$$a(3 + b + d) = c$$
$$a = \frac{c}{3 + b + d}$$

54.

Exercise Set 2.4

1. $32\% = 32 \times 0.01$ Replacing % by $\times 0.01$
     $= 0.32$

2. 0.49

3. $7\% = 7 \times 0.01$ Replacing % by $\times 0.01$
     $= 0.07$

4. 0.913

5. $24.1\% = 24.1 \times 0.01 = 0.241$

6. 0.02

7. $0.46\% = 0.46 \times 0.01 = 0.0046$

8. 0.048

9. 4.54

First move the decimal point   4.54.
two places to the right;
then write a % symbol: 454%

10. 100%

11. 0.998

First move the decimal point   0.99.8
two places to the right;
then write a % symbol: 99.8%

12. 73%

13. 2   (Note: 2 = 2.00)

First move the decimal point   2.00.
two places to the right;
then write a % symbol: 200%

14. 0.57%

15. 1.34

First move the decimal point   1.34.

two places to the right;   └─↑

then write a % symbol: 134%

16. 920%

17. 0.0068

First move the decimal point   0.00.68

two places to the right;   └─↑

then write a % symbol: 0.68%

18. 67.5%

19. $\frac{3}{8}$ $\left(\text{Note: } \frac{3}{8} = 0.375\right)$

First move the decimal point   0.37.5

two places to the right;   └─↑

then write a % symbol: 37.5%

20. 75%

21. $\frac{7}{25}$ $\left(\text{Note: } \frac{7}{25} = 0.28\right)$

First move the decimal point   0.28.

two places to the right;   └─↑

then write a % symbol: 28%

22. 80%

23. $\frac{2}{3}$ $\left(\text{Note: } \frac{2}{3} = 0.66\overline{6}\right)$

First move the decimal point   0.66.$\overline{6}$

two places to the right;   └─↑

then write a % symbol: 66.$\overline{6}$%

Since $0.\overline{6} = \frac{2}{3}$, this can also be expressed as $66\frac{2}{3}$%.

24. $83.\overline{3}$%, or $83\frac{1}{3}$%

25. *Translate.*

$$\underbrace{\text{What percent}}_{y} \text{ of } \underset{\downarrow}{75} \underset{\cdot}{} \underset{=}{75} \underset{39}{39}$$

What percent of 75 is 39?

$y \quad \cdot \quad 75 \quad = \quad 39$

We solve the equation and then convert to percent notation.

$$y \cdot 75 = 39$$

$$y = \frac{39}{75}$$

$$y = 0.52 = 52\%$$

The answer is 52%.

26. 12.5%

27. *Translate.*

What percent of 125 is 30?

$y \quad \cdot \quad 125 \quad = \quad 30$

We solve the equation and then convert to percent notation.

$$y \cdot 125 = 30$$

$$y = \frac{30}{125}$$

$$y = 0.24 = 24\%$$

The answer is 24%.

28. 19%

29. *Translate.*

18 is 30% of what number?

$18 \quad = \quad 30\% \quad \cdot \quad y$

We solve the equation.

$$18 = 0.3y \qquad (30\% = 0.3)$$

$$\frac{18}{0.3} = y$$

$$60 = y$$

The answer is 60.

30. 85

31. *Translate.*

0.3 is 12% of what number?

$0.3 \quad = \quad 12\% \quad \cdot \quad y$

We solve the equation.

$$0.3 = 0.12y \qquad (12\% = 0.12)$$

$$\frac{0.3}{0.12} = y$$

$$2.5 = y$$

The answer is 2.5.

32. 4

33. *Translate.*

What number is 65% of 420?

$y \quad = \quad 65\% \quad \cdot \quad 420$

We solve the equation.

$$y = 0.65 \cdot 420 \qquad (65\% = 0.65)$$

$$y = 273 \qquad \text{Multiplying}$$

The answer is 273.

34. 10,000

35. *Translate.*

What percent of 60 is 75?

$y \quad \cdot \quad 60 \quad = \quad 75$

We solve the equation and then convert to percent notation.

$$y \cdot 60 = 75$$
$$y = \frac{75}{60}$$
$$y = 1.25 = 125\%$$

The answer is 125%.

36. 225%

37. *Translate*.

What is 2% of 40?
$$\downarrow \quad \downarrow \quad \downarrow \quad \downarrow \quad \downarrow$$
$$x \quad = \quad 2\% \quad \cdot \quad 40$$

We solve the equation.

$$x = 0.02 \cdot 40 \qquad (2\% = 0.02)$$
$$x = 0.8 \qquad \text{Multiplying}$$

The answer is 0.8.

38. 0.8

39. *Translate*.

2 is what percent of 40?
$$\downarrow \downarrow \qquad \downarrow \qquad \downarrow \quad \downarrow$$
$$2 = \qquad y \qquad \cdot \quad 40$$

We solve the equation and convert to percent notation.

$$2 = y \cdot 40$$
$$\frac{2}{40} = y$$
$$0.05 = y, \text{ or } 5\% = y$$

The answer is 5%.

40. 2000

41. If $n =$ the number of women who had babies in good or excellent health, we have:

n is 95% of 300.
$$\downarrow \downarrow \quad \downarrow \quad \downarrow \quad \downarrow$$
$$n = 0.95 \cdot 300$$
$$n = 285$$

285 women had babies in good or excellent health.

42. 16

43. Let $p =$ the number of people who voted in the 1996 presidential election, in millions. Then we have:

45.3 is 49% of p.
$$\downarrow \quad \downarrow \quad \downarrow \quad \downarrow \downarrow$$
$$45.3 = 0.49 \cdot p$$
$$\frac{45.3}{0.49} = p$$
$$92.4 \approx p$$

About 92.4 million people voted in the 1996 presidential election.

44. \$88.9 billion

45. Let $p =$ the percentage by which the average annual healthcare bill grew from 1985 to 1993. Then we have:

668 is what percent of 1108?
$$\downarrow \quad \downarrow \qquad \downarrow \qquad \downarrow \quad \downarrow$$
$$668 = \qquad p \qquad \cdot \quad 1108$$
$$\frac{668}{1108} = p$$
$$0.60 \approx p, \text{ or }$$
$$60\% \approx p$$

The average annual healthcare bill grew about 60% from 1985 to 1993.

46. 3.75%

47. Let $b =$ the number of brochures the business can expect to be opened and read. Then we have:

b is 78% of 9500.
$$\downarrow \downarrow \quad \downarrow \quad \downarrow \quad \downarrow$$
$$b = 0.78 \cdot 9500$$
$$b = 7410$$

The business can expect 7410 brochures to be opened and read.

48. 27

49. Let $p =$ the percent of people who will catch the cold. Then we have:

56 is what percent of 800?
$$\downarrow \downarrow \qquad \downarrow \qquad \downarrow \quad \downarrow$$
$$56 = \qquad p \qquad \cdot \quad 800$$
$$\frac{56}{800} = p$$
$$0.07 = p, \text{ or }$$
$$7\% = p$$

7% will catch the cold.

50. 86.4%

51. Let $h =$ the percent that were hits. Then we have:

13 is what percent of 25?
$$\downarrow \downarrow \qquad \downarrow \qquad \downarrow \quad \downarrow$$
$$13 = \qquad h \qquad \cdot \quad 25$$
$$\frac{13}{25} = h$$
$$0.52 = h, \text{ or }$$
$$52\% = h$$

52% were hits.

52. \$36

53. When the sales tax is 5%, the total amount paid is 105% of the cost of the merchandise. Let $c =$ the cost of the building materials. Then we have:

\$987 is 105% of c.
$$\downarrow \quad \downarrow \quad \downarrow \quad \downarrow \downarrow$$
$$987 = 1.05 \cdot c$$

$$\frac{987}{1.05} = c$$
$$940 = c$$

The price of the building materials was \$940.

54. 16%

55. When the sales tax is 5%, the total amount paid is 105% of the cost of the merchandise. Let $c =$ the amount the charity owes, or the cost of the sump pump without tax. Then we have:

\$145.90 is 105% of c.

$$145.90 = 1.05 \cdot c$$
$$\frac{145.90}{1.05} = c$$
$$138.95 \approx c$$

The charity owes \$138.95.

56. \$148.50

57. A self-employed person must earn 120% as much as a non-self-employed person. Let $a =$ the amount Trey would need to earn, in dollars per hour, on his own for a comparable income. Then we have:

a is 120% of \$12.

$$a = 1.2 \cdot 12$$
$$a = 14.4$$

Trey would need to earn \$14.40 per hour on his own.

58. \$18 per hour

59. The number of calories in a serving of Light Style Bread is 85% of the number of calories in a serving of regular bread. Let $c =$ the number of calories in a serving of regular bread. Then we have:

140 calories is 85% of c.

$$140 = 0.85 \cdot c$$
$$\frac{140}{0.85} = c$$
$$165 \approx c$$

There are about 165 calories in a serving of regular bread.

60. About 58 calories

61. We divide:

$$\begin{array}{r} 0.6\,8 \\ 25\overline{)1\,7.0\,0} \\ \underline{1\,5\,0} \\ 2\,0\,0 \\ \underline{2\,0\,0} \\ 0 \end{array}$$

Decimal notation for $\frac{17}{25}$ is 0.68.

62. -90

63. $-45.8 - (-32.6) = -45.8 + 32.6 = -13.2$

64. $-21a + 12b$

65. ◈

66. ◈

67. ◈

68. ◈

69. Let $x =$ Claude's pre-tax earnings. Then his taxes are 26% of x, or $0.26x$, and his post-tax earnings are $x - 0.26x$, or $0.74x$. Then the percentage of his post-tax earnings represented by his taxes is given by

$$\frac{0.26x}{0.74x} = \frac{0.26}{0.74} \approx 0.35, \text{ or } 35\%.$$

70. 18,500

71.
$$a = 130\% \cdot b$$
$$a = 1.3b$$
$$\frac{a}{1.3} = b$$
$$\frac{1}{1.3}a = b$$
$$0.77a \approx b$$

Thus, b is about 77% of a.

72. Rollie's Music: \$12.83; Sound Warp: \$12.97

73. The new price is 125% of the old price. Let $p =$ the new price. Then we have:

p is 125% of \$20,800.

$$p = 1.25 \cdot 20,800$$
$$p = 26,000$$

Now let $x =$ the percent of the new price represented by the old price. We have:

\$20,800 is what percent of \$26,000.

$$20,800 = x \cdot 26,000$$
$$\frac{20,800}{26,000} = x$$
$$0.8 = x, \text{ or}$$
$$80\% = x$$

The old price is $100\% - 80\%$, or 20% lower than the new price.

Exercise Set 2.5

1. *Familiarize*. Let x = the number. Then "three less than twice a number" translates to $2x - 3$.

***Translate*.**

$$\underbrace{\text{Three less than twice a number}}_{2x - 3} \underbrace{\text{is}}_{=} \underbrace{19.}_{19}$$

***Carry out*.** We solve the equation.

$$
\begin{aligned}
2x - 3 &= 19 \\
2x &= 22 \quad \text{Adding 3} \\
x &= 11 \quad \text{Dividing by 2}
\end{aligned}
$$

***Check*.** Twice, or two times, 11 is 22. Three less than 22 is 19. The answer checks.

***State*.** The number is 11.

2. 8

3. *Familiarize*. Let a = the number. Then "five times the sum of 3 and some number" translates to $5(a+3)$.

***Translate*.**

$$\underbrace{\text{Five times the sum of 3 and some number}}_{5(a + 3)} \underbrace{\text{is}}_{=} \underbrace{70.}_{70}$$

***Carry out*.** We solve the equation.

$$
\begin{aligned}
5(a + 3) &= 70 \\
5a + 15 &= 70 \quad \text{Using the distributive law} \\
5a &= 55 \quad \text{Subtracting 15} \\
a &= 11 \quad \text{Dividing by 5}
\end{aligned}
$$

***Check*.** The sum of 3 and 11 is 14, and $5 \cdot 14 = 70$. The answer checks.

***State*.** The number is 11.

4. 13

5. *Familiarize*. Let p = the regular price of the CD player. At 20% off, Doug paid 80% of the regular price.

***Translate*.**

$$\underbrace{72}_{72} = \underbrace{0.8}_{0.8\%} \cdot \underbrace{\text{the regular price.}}_{p}$$

***Carry out*.** We solve the equation.

$$
\begin{aligned}
72 &= 0.8p \\
\frac{72}{0.8} &= p \quad \text{Dividing by 0.8} \\
90 &= p
\end{aligned}
$$

***Check*.** 80% of $90, or 0.8($90), is $72. The answer checks.

***State*.** The regular price was $90.

6. $48

7. *Familiarize*. Let x = the first integer. Then $x+1$ = the second integer, and $x + 2$ = the third integer.

***Translate*.**

$$\underbrace{\text{The sum of three consecutive integers}}_{x + (x + 1) + (x + 2)} \underbrace{\text{is}}_{=} \underbrace{48.}_{48}$$

***Carry out*.** We solve the equation.

$$
\begin{aligned}
x + (x + 1) + (x + 2) &= 48 \\
3x + 3 &= 48 \quad \text{Combining like terms} \\
3x &= 45 \quad \text{Subtracting 3} \\
x &= 15 \quad \text{Dividing by 3}
\end{aligned}
$$

If x is 15, then $x + 1$ is 16 and $x + 2 = 17$.

***Check*.** 15, 16, and 17 are consecutive integers, and $15 + 16 + 17 = 48$. The result checks.

***State*.** The numbers are 15, 16, and 17.

8. 19, 21

9. *Familiarize*. Let x = the first even number. Then $x + 2$ = the next even number.

***Translate*.**

$$\underbrace{\text{The sum of two consecutive even numbers}}_{x + (x + 2)} \underbrace{\text{is}}_{=} \underbrace{50.}_{50}$$

***Carry out*.** We solve the equation.

$$
\begin{aligned}
x + (x + 2) &= 50 \\
2x + 2 &= 50 \quad \text{Combining like terms} \\
2x &= 48 \quad \text{Subtracting 2} \\
x &= 24 \quad \text{Dividing by 2}
\end{aligned}
$$

If x is 24, then $x + 2$ is 26.

***Check*.** 24 and 26 are consecutive even numbers, and $24 + 26 = 50$. The result checks.

***State*.** The numbers are 24 and 26.

10. 52, 54

11. *Familiarize*. Let x = the smaller odd integer. Then $x + 2$ = the next odd integer.

***Translate*.** We reword the problem.

$$\underbrace{\text{Smaller odd integer}}_{x} + \underbrace{\text{next odd integer}}_{(x + 2)} \underbrace{\text{is}}_{=} \underbrace{128.}_{128}$$

***Carry out*.** We solve the equation.

$$
\begin{aligned}
x + (x + 2) &= 128 \\
2x + 2 &= 128 \quad \text{Combining like terms} \\
2x &= 126 \quad \text{Subtracting 2} \\
x &= 63 \quad \text{Dividing by 2}
\end{aligned}
$$

If x is 63, then $x + 2$ is 65.

Check. 63 and 65 are consecutive odd integers, and their sum is 128. The answer checks.

State. The integers are 63 and 65.

12. $\dfrac{10}{3}$ km

13. *Familiarize.* Let d = the musher's distance from Nome, in miles. Then $2d$ = the distance from Anchorage, in miles. This is the number of miles the musher has completed. The sum of the two distances is the length of the race, 1049 miles.

Translate.

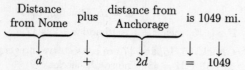

Distance from Nome	plus	distance from Anchorage	is 1049 mi.
$\downarrow$	$\downarrow$	$\downarrow$	$\downarrow\quad\downarrow$
d	$+$	$2d$	$=\quad 1049$

Carry out. We solve the equation.

$$d + 2d = 1049$$
$$3d = 1049 \qquad \text{Combining like terms}$$
$$d = \frac{1049}{3}$$

If $d = \dfrac{1049}{3}$, then $2d = 2 \cdot \dfrac{1049}{3} = \dfrac{2098}{3} = 699\dfrac{1}{3}$.

Check. $\dfrac{2098}{3}$ is twice $\dfrac{1049}{3}$, and $\dfrac{1049}{3} + \dfrac{2098}{3} = \dfrac{3147}{3} = 1049$. The result checks.

State. The musher has traveled $699\dfrac{1}{3}$ miles.

14. 30 m, 90 m, 360 m

15. *Familiarize.* Let a = the amount spent to remodel bathrooms, in billions of dollars. Then $2a$ = the amount spent to remodel kitchens. The sum of these two amounts is $35 billion.

Translate.

Amount spent on bathrooms	plus	amount spent on kitchens	is $35 billion.
$\downarrow$	$\downarrow$	$\downarrow$	$\downarrow\quad\downarrow$
a	$+$	$2a$	$=\quad 35$

Carry out. We solve the equation.

$$a + 2a = 35$$
$$3a = 35 \qquad \text{Combining like terms}$$
$$a = \frac{35}{3}, \text{ or } 11\frac{2}{3}$$

If $a = \dfrac{35}{3}$, then $2a = 2 \cdot \dfrac{35}{3} = \dfrac{70}{3} = 23\dfrac{1}{3}$.

Check. $\dfrac{70}{3}$ is twice $\dfrac{35}{3}$, and $\dfrac{35}{3} + \dfrac{70}{3} = \dfrac{105}{3} = 35$. The answer checks.

State. \$$11\dfrac{2}{3}$ billion was spent to remodel bathrooms, and \$$23\dfrac{1}{3}$ billion was spent to remodel kitchens.

16. 30°, 90°, 60°

17. *Familiarize.* We draw a picture. We let x = the measure of the first angle. Then $4x$ = the measure of the second angle, and $(x + 4x) - 45$, or $5x - 45$ = the measure of the third angle.

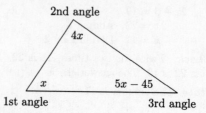

2nd angle

$4x$

x $5x - 45$

1st angle   3rd angle

Recall that the measures of the angles of any triangle add up to 180°.

Translate.

Measure of first angle	+	measure of second angle	+
$\downarrow$	$\downarrow$	$\downarrow$	$\downarrow$
x	$+$	$4x$	$+$

measure of third angle	is 180°.
$\downarrow$	$\downarrow\quad\downarrow$
$(5x - 45)$	$=\quad 180$

Carry out. We solve the equation.

$$x + 4x + (5x - 45) = 180$$
$$10x - 45 = 180$$
$$10x = 225$$
$$x = 22.5$$

Possible answers for the angle measures are as follows:

First angle: $x = 22.5°$
Second angle: $4x = 4(22.5) = 90°$
Third angle: $5x - 45 = 5(22.5) - 45$
$$= 112.5 - 45 = 67.5°$$

Check. Consider 22.5°, 90°, and 67.5°. The second is four times the first, and the third is 45° less than five times the first. The sum is 180°. These numbers check.

State. The measure of the first angle is 22.5°, the measure of the second angle is 90°, and the measure of the third angle is 67.5°.

18. 95°

19. *Familiarize.* Let x = the measure of the first angle. Then $4x$ = the measure of the second angle, and $x + 4x + 5 = 5x + 5$ = the measure of the third angle. Recall that the sum of the measures of the angles of a triangle is 180°.

Translate.

Measure of first angle $+$ measure of second angle $+$

$$\downarrow \qquad \downarrow \qquad \downarrow \qquad \downarrow$$
$$x \qquad + \qquad 4x \qquad +$$

measure of third angle is 180°.

$$\downarrow \qquad \downarrow \quad \downarrow$$
$$(5x + 5) \quad = \quad 180$$

Carry out. We solve the equation.

$$x + 4x + (5x + 5) = 180$$
$$10x + 5 = 180$$
$$10x = 175$$
$$x = 17.5$$

If x is 17.5, then $4x$ is 70 and $5x + 5$ is 92.5.

Check. Consider 17.5°, 70°, and 92.5°. The second is four times the first, and the third is 5° more than the sum of the other two. The sum is 180°. These numbers check.

State. The measure of the second angle is 70°.

20. $360,000

21. *Familiarize.* The page numbers are consecutive integers. If we let p = the smaller number, then $p + 1$ = the larger number.

Translate. We reword the problem.

First integer $+$ Second integer $= 285$

$$\downarrow \qquad \downarrow \qquad \downarrow \qquad \downarrow \downarrow$$
$$x \qquad + \qquad (x + 1) \qquad = 285$$

Carry out. We solve the equation.

$$x + (x + 1) = 285$$
$$2x + 1 = 285 \qquad \text{Combining like terms}$$
$$2x = 284 \qquad \text{Adding } -1$$
$$x = 142 \qquad \text{Dividing by 2}$$

Check. If $x = 142$, then $x + 1 = 143$. These are consecutive integers, and $142 + 143 = 285$. The answer checks.

State. The page numbers are 142 and 143.

22. 140 and 141

23. *Familiarize.* Let s = the length of the shortest side, in mm. Then $s + 2$ and $s + 4$ represent the lengths of the other two sides. The perimeter is the sum of the lengths of the sides.

Translate.

Length of first side plus length of second side plus

$$\downarrow \qquad \downarrow \qquad \downarrow \qquad \downarrow$$
$$s \qquad + \qquad (s + 2) \qquad +$$

length of third side $= 195$ mm

$$\downarrow \qquad \downarrow \quad \downarrow$$
$$(s + 4) \quad = \quad 195$$

Carry out. We solve the equation.

$$s + (s + 2) + (s + 4) = 195$$
$$3s + 6 = 195$$
$$3s = 189$$
$$s = 63$$

If s is 63, then $s + 2$ is 65 and $s + 4$ is 67.

Check. The numbers 63, 65, and 67 are consecutive odd integers. Their sum is 195. These numbers check.

State. The lengths of the sides of the triangle are 63 mm, 65 mm, and 67 mm.

24. 130 mm, 132 mm, 134 mm

25. *Familiarize.* We draw a picture. Let w = the width of the rectangle in feet. Then $w + 100$ = the length.

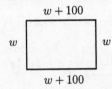

The perimeter of a rectangle is the sum of the lengths of the sides. The area is the product of the length and the width.

Translate. We use the definition of perimeter to write an equation that will allow us to find the width and length.

Width$+$Width$+$ Length $+$ Length $=$Perimeter.

$$\downarrow \ \downarrow \ \downarrow \ \downarrow \qquad \downarrow \qquad \downarrow \qquad \downarrow \qquad \downarrow \qquad \downarrow$$
$$w \ + \ w \ +(w + 100)+(w + 100)= \quad 860$$

Carry out. We solve the equation.

$$w + w + (w + 100) + (w + 100) = 860$$
$$4w + 200 = 860$$
$$4w = 660$$
$$w = 165$$

If $w = 165$, then $w + 100 = 165 + 100 = 265$, and the area is $265(165) = 43,725$.

Check. The length is 100 ft more than the width. The perimeter is $165 + 165 + 265 + 265 = 860$ ft. This checks. To check the area we recheck the computation. This also checks.

State. The width of the rectangle is 165 ft, the length is 265 ft, and the area is 43,725 ft^2.

26. Width: 100 ft, length: 160 ft

27. *Familiarize*. We draw a picture. Let l = the length of the state, in miles. Then $l - 90$ = the width.

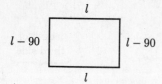

The perimeter is the sum of the lengths of the sides.

***Translate*.** We use the definition of perimeter to write an equation.

$$\underbrace{\text{Width}} + \underbrace{\text{Width}} + \underbrace{\text{Length}} + \underbrace{\text{Length}} \text{ is } 1280.$$
$$(l - 90) + (l - 90) + l + l = 1280$$

***Carry out*.** We solve the equation.

$$(l - 90) + (l - 90) + l + l = 1280$$
$$4l - 180 = 1280$$
$$4l = 1460$$
$$l = 365$$

Then $l - 90 = 275$.

***Check*.** The width, 275 mi, is 90 mi less than the length, 365 mi. The perimeter is 275 mi + 275 mi + 365 mi + 365 mi, or 1280 mi. This checks.

***State*.** The length is 365 mi, and the width is 275 mi.

28. Length: 27.9 cm, width: 21.6 cm

29. *Familiarize*. Let x = the original investment. Interest earned in 1 year is found by taking 6% of the original investment. Then 6% of x, or $0.06x$ = the interest. The amount in the account at the end of the year is the sum of the original investment and the interest earned.

***Translate*.**

$$\underbrace{\text{Original investment}} \text{ plus } \underbrace{\text{interest earned}} \text{ is } \$6996.$$
$$x + 0.06x = 6996$$

***Carry out*.** We solve the equation.

$$x + 0.06x = 6996$$
$$1 \cdot x + 0.06x = 6996$$
$$1.06x = 6996$$
$$x = \frac{6996}{1.06}$$
$$x = 6600$$

***Check*.** 6% of $6600 is $396. Adding this to $6600 we get $6996. This checks.

***State*.** The original investment was $6600.

30. $6540

31. *Familiarize*. The total cost is the daily charge plus the mileage charge. The mileage charge is the cost per mile times the number of miles driven. Let m = the number of miles that can be driven for $80.

***Translate*.** We reword the problem.

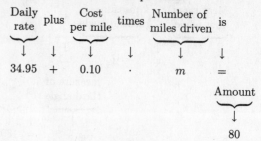

***Carry out*.** We solve the equation.

$$34.95 + 0.10m = 80$$
$$100(34.95 + 0.10m) = 100(80) \quad \text{Clearing decimals}$$
$$3495 + 10m = 8000$$
$$10m = 4505$$
$$m = 450.5$$

***Check*.** The mileage cost is found by multiplying 450.5 by $0.10 obtaining $45.05. Then we add $45.05 to $34.95, the daily rate, and get $80.

***State*.** The businessperson can drive 450.5 mi on the car-rental allotment.

32. 460.5 mi

33. *Familiarize*. Let x = the measure of one angle. Then $90 - x$ = the measure of its complement.

***Translate*.**

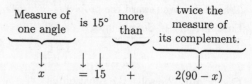

***Carry out*.** We solve the equation.

$$x = 15 + 2(90 - x)$$
$$x = 15 + 180 - 2x$$
$$x = 195 - 2x$$
$$3x = 195$$
$$x = 65$$

If x is 65, then $90 - x$ is 25.

***Check*.** The sum of the angle measures is 90°. Also, 65° is 15° more than twice its complement, 25°. The answer checks.

***State*.** The angle measures are 65° and 25°.

34. 105°, 75°

35. *Familiarize*. We will use the equation $R = -0.028t + 20.8$ where R is in seconds and t is the number of years since 1920. We want to find t when $R = 18.0$ sec.

***Translate*.**

$$\underbrace{\text{Record}} \quad \text{is} \quad \underbrace{18.0 \text{ sec.}}$$
$$-0.028t + 20.8 = 18.0$$

Carry out.

$$-0.028t + 20.8 = 18.0$$
$$1000(-0.028t + 20.8) = 1000(18.0) \quad \text{Clearing the}$$
$$\text{decimals}$$
$$-28t + 20{,}800 = 18{,}000$$
$$-28t = -2800$$
$$t = 100$$

Check. Substitute 100 for t in the given equation:

$$R = -0.028(100) + 20.8 = -2.8 + 20.8 = 18.0$$

This checks.

State. The record will be 18.0 sec 100 years after 1920, or in 2020.

36. 160

37. $5a + 10b - 45 = 5 \cdot a + 5 \cdot 2b - 5 \cdot 9 = 5(a + 2b - 9)$

38. $3(x - 4y + 20)$

39. $7x - 3(8 - 2x) + 12 = 7x - 24 + 6x + 12 = 13x - 12$

40. $-10x + 30$

41.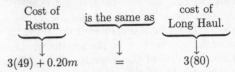

42. ◈

43. ◈

44. ◈

45. Familiarize. Let $s =$ the number of points Stacy would have scored if her three-pointer had been good. Had it been good, the Blazers would have scored $77 + 3$, or 80, points. Since it wasn't good, Stacy actually scored $s - 3$ points.

Translate. We write an equation corresponding to the situation that would have resulted if the shot had been good.

$$\underbrace{\text{Stacy's}}_{\downarrow} \quad \underbrace{\text{is half of}}_{\downarrow \ \downarrow \ \downarrow} \quad \underbrace{\text{80 points.}}_{\downarrow}$$
$$s \qquad\quad = \frac{1}{2} \cdot \qquad 80$$

Carry out. We carry out the calculation.

$$s = \frac{1}{2} \cdot 80$$
$$s = 40$$

If s is 40, then $s - 3$ is 37.

Check. If the Blazers had scored 80 points and Stacy had scored 40 points, Stacy would have scored half of the Blazer's points. Since the three-pointer wasn't good, Stacy actually scored $40 - 3$, or 37 points. The result checks.

State. Stacy scored 37 points in the loss.

46. $0.19

47. Familiarize. Let $m =$ the number of miles Ed drives. We will find the mileage for which the cost of the two rentals is the same. Then we will examine a mileage less than this and also a mileage greater than this to determine guidelines for the least expensive rental.

Translate.

$$\underbrace{\text{Cost of}}_{\text{Reston}} \quad \underbrace{\text{is the same as}} \quad \underbrace{\text{cost of}}_{\text{Long Haul.}}$$
$$\downarrow \qquad\qquad \downarrow \qquad\qquad \downarrow$$
$$3(49) + 0.20m \quad = \quad\quad 3(80)$$

Carry out. We solve the equation.

$$3(49) + 0.20m = 3(80)$$
$$147 + 0.20m = 240$$
$$0.20m = 93$$
$$m = 465$$

Thus, if Ed drives 465 miles there is no difference in the cost of the rentals.

Note that a three-day rental from Long Haul costs $3(\$80)$, or $240, regardless of the mileage. We find the cost of a Reston rental for a mileage less than 465 mi and also for a mileage greater than 465 mi.

For 460 mi: $3(\$49) + \$0.20(460) = \$239$

For 470 mi: $3(\$49) + \$0.20(470) = \$241$

Thus, the Reston rental is less for mileages less than 465 mi and more expensive for mileages greater than 465 mi.

Check. If Ed drives 465 mi, the Reston rental costs $3(\$49) + \$0.20(465)$, or $240, and the Long Haul rental costs $3(\$80)$, or $240, also, so the cost is the same for 465 mi. Go over the computations for mileages less than and greater than 465 mi. The result checks.

State. If Ed drives less than 465 mi the Reston rental is less expensive. If he drives more than 465 mi the Long Haul rental is less expensive.

48. 19

49. Familiarize. Let $s =$ one score. Then four score $= 4s$ and four score and seven $= 4s + 7$.

Translate. We reword .

$$\underbrace{1776} \ \text{plus} \ \underbrace{\text{four score and seven}} \ \text{is} \ \underbrace{1863}$$
$$\downarrow \qquad \downarrow \qquad\qquad \downarrow \qquad\qquad \downarrow \quad \downarrow$$
$$1776 \ + \qquad (4s + 7) \qquad = \ 1863$$

Carry out. We solve the equation.

$$1776 + (4s + 7) = 1863$$
$$4s + 1783 = 1863$$
$$4s = 80$$
$$s = 20$$

Check. If a score is 20 years, then four score and seven represents 87 years. Adding 87 to 1776 we get 1863. This checks.

State. A score is 20.

50. 16, 4

51. Familiarize. Let $x =$ the first odd number. Then the next three odd numbers are $x + 2$, $x + 4$, and $x + 6$. The sum of measures of the angles of an n-sided polygon is given by the formula $(n-2) \cdot 180°$. Thus, the sum of the measures of the angles of a quadrilateral is $(4-2) \cdot 180°$, or $2 \cdot 180°$, or $360°$.

Translate.

$$\underbrace{\text{The sum of the measures of the angles}} \quad \text{is} \quad 360°.$$

$$x + (x+2) + (x+4) + (x+6) = 360$$

Carry out. We solve the equation.

$$x + (x+2) + (x+4) + (x+6) = 360$$
$$4x + 12 = 360$$
$$4x = 348$$
$$x = 87$$

If x is 87, then $x + 2$ is 89, $x + 4$ is 91, and $x + 6$ is 93.

Check. The numbers 87, 89, 91, and 93 are consecutive odd numbers. The sum of the numbers is 360. The answer checks.

State. The measures of the angles are 87°, 89°, 91°, and 93°.

52. 104°, 106°, 108°, 110°, 112°

53. Familiarize. Let $a =$ the original number of apples in the basket.

Translate.

$$\underbrace{\text{One third of the apples}} + \underbrace{\text{one fourth of the apples}} +$$
$$\frac{1}{3}a \quad + \quad \frac{1}{4}a \quad +$$

$$\underbrace{\text{one eighth of the apples}} + \underbrace{\text{one fifth of the apples}} + \underbrace{10 \text{ apples}} +$$
$$\frac{1}{8}a \quad + \quad \frac{1}{5}a \quad + \quad 10 \quad +$$

$$\underbrace{1 \text{ apple}} \text{ is } \underbrace{\text{the original number of apples.}}$$
$$1 \quad = \quad a$$

Carry out. We solve the equation. Note that the LCD is 120.

$$\frac{1}{3}a + \frac{1}{4}a + \frac{1}{8}a + \frac{1}{5}a + 10 + 1 = a$$
$$\frac{1}{3}a + \frac{1}{4}a + \frac{1}{8}a + \frac{1}{5}a + 11 = a$$
$$120\left(\frac{1}{3}a + \frac{1}{4}a + \frac{1}{8}a + \frac{1}{5}a + 11\right) = 120 \cdot a$$
$$40a + 30a + 15a + 24a + 1320 = 120a$$
$$109a + 1320 = 120a$$
$$1320 = 11a$$
$$120 = a$$

Check. $\frac{1}{3} \cdot 120 = 40$, $\frac{1}{4} \cdot 120 = 30$, $\frac{1}{8} \cdot 120 = 15$, and $\frac{1}{5} \cdot 120 = 24$. Then $40 + 30 + 15 + 24 + 10 + 1 = 120$. The result checks.

State. There were originally 120 apples in the basket.

54. 30

55. Familiarize. Let $s =$ Luke's weekly salary. Then Luke spends 12% of s, or $0.12s$ on dining out, and 55% of this amount is given by $0.55(0.12s)$. This is the amount spent on fast-food.

Translate.

$$\underbrace{\text{Amount spent on fast-food}} \text{ is } \$39.60.$$
$$0.55(0.12s) \quad = \quad 39.60$$

Carry out.

$$0.55(0.12s) = 39.60$$
$$0.066s = 39.60$$
$$s = 600$$

Check. If Luke's weekly salary is $600, then he spends $0.12(\$600)$, or \$72, dining out. Then he spends $0.55(\$72)$, or \$39.60 on fast-food. The answer checks.

State. Luke's weekly salary is \$600.

56. Length: 12 cm, width: 9 cm

57. Familiarize. Let $h =$ the height of the triangle. We know that the base is 8 in. Recall that the area of a triangle is given by the formula $A = \frac{1}{2}bh$.

Translate.

$$\underbrace{\text{Area}} \text{ is } \underbrace{2.9047 \text{ in}^2.}$$
$$\frac{1}{2} \cdot 8 \cdot h = 2.9047$$

Carry out. We solve the equation.

$$\frac{1}{2} \cdot 8 \cdot h = 2.9047$$
$$4h = 2.9047$$
$$h = 0.726175$$

Check. The area of a triangle whose base is 8 in. and whose height is 0.726175 in. is $\frac{1}{2}(8)(0.726175)$, or 2.9047. The answer checks.

State. The height of the triangle is 0.726175 in.

58. 76

59. *Familiarize*. Let $n =$ the number of CD's purchased. Assume that two or more CD's were purchased. Then the first CD costs \$8.49 and the total cost of the remaining $n-1$ CD's is \$3.99$(n-1)$. The shipping and handling costs are \$2.47 for the first CD, \$2.28 for the second, and a total of \$1.99$(n-2)$ for the remaining $n-2$ CD's. Then the total cost of the shipment is \$8.49 $+$ \$3.99$(n-1)$ $+$ \$2.47 $+$ \$2.28 $+$ \$1.99$(n-2)$.

Translate.

$$\underbrace{\text{Total cost of shipment}} \qquad \text{was \$65.07.}$$
$$8.49+3.99(n-1)+2.47+2.28+1.99(n-2) \;=\; 65.07$$

Carry out. We solve the equation.

$$8.49 + 3.99(n-1) + 2.47 + 2.28 + 1.99(n-2) = 65.07$$
$$8.49+3.99n-3.99+2.47+2.28+1.99n-3.98 = 65.07$$
$$5.27 + 5.98n = 65.07$$
$$5.98n = 59.80$$
$$n = 10$$

Check. If 10 CD's are purchased, the total cost of the CD's is \$8.49 + \$3.99(9) = \$44.40. The total shipping and handling costs are \$2.47 + \$2.28 + \$1.99(8) = \$20.67. Then the total cost of the order is \$44.40 + \$20.67 = \$65.07.

State. There were 10 CD's in the shipment.

60.

61.

Exercise Set 2.6

1. $x > -3$

  a) Since $4 > -3$ is true, 4 is a solution.

  b) Since $0 > -3$ is true, 0 is a solution.

  c) Since $-4.1 > -3$ is false, -4.1 is not a solution.

  d) Since $-3.9 > -3$ is false, -3.9 is not a solution.

  e) Since $5.6 > -3$ is true, 5.6 is a solution.

2. a) Yes, b) No, c) Yes, d) Yes, e) No

3. $x \geq 6$

  a) Since $-6 \geq 6$ is false, -6 is not a solution.

  b) Since $0 \geq 6$ is false, 0 is not a solution.

  c) Since $6 \geq 6$ is true, 6 is a solution.

  d) Since $6.01 \geq 6$ is true, 6.01 is a solution.

  e) Since $-3\frac{1}{2} \geq 6$ is false, $-3\frac{1}{2}$ is not a solution.

4. a) Yes, b) Yes, c) Yes, d) No, e) Yes

5. The solutions of $x \geq 6$ are shown by shading the point 6 and all points to the right of 6. The closed circle at 6 indicates that 6 is part of the graph.

$$x \geq 6$$
<-1 0 1 2 3 4 5 6 7 8 9->

6.
$$y < 0$$
<-5 -4 -3 -2 -1 0 1 2 3 4 5->

7. The solutions of $t < -3$ are those numbers less than -3. They are shown on the graph by shading all points to the left of -3. The open circle at -3 indicates that -3 is not part of the graph.

$$t < -3$$
<-5 -4 -3 -2 -1 0 1 2 3 4 5->

8.
$$y > 5$$
<-1 0 1 2 3 4 5 6 7 8 9->

9. The solutions of $m > -4$ are those numbers greater than -4. They are shown on the graph by shading all points to the right of -4. The open circle at 4 indicates that -4 is not part of the graph.

$$m > -4$$
<-5 -4 -3 -2 -1 0 1 2 3 4 5->

10.
$$p \leq 3$$
<-5 -4 -3 -2 -1 0 1 2 3 4 5->

11. In order to be a solution of the inequality $-3 < x \leq 5$, a number must be a solution of both $-3 < x$ and $x \leq 5$. The solution set is graphed as follows:

$$-3 < x \leq 5$$
<-4 -3 -2 -1 0 1 2 3 4 5 6->

The open circle at -3 means that -3 is not part of the graph. The closed circle at 5 means that 5 is part of the graph.

12.
$$-5 \leq x < 2$$
<-6 -5 -4 -3 -2 -1 0 1 2 3 4->

13. In order to be a solution of the inequality $0 < x < 3$, a number must be a solution of both $0 < x$ and $x < 3$. The solution set is graphed as follows:

$$0 < x < 3$$
<-5 -4 -3 -2 -1 0 1 2 3 4 5->

The open circles at 0 and at 3 mean that 0 and 3 are not part of the graph.

14.
$$-5 \leq x \leq 0$$
<-7 -6 -5 -4 -3 -2 -1 0 1 2 3->

15. All points to the right of -1 are shaded. The open circle at -1 indicates that -1 is not part of the graph. Using set-builder notation we have $\{x|x > -1\}$.

16. $\{x|x < 3\}$

17. The point 2 and all points to the left of 2 are shaded. Using set-builder notation we have $\{x|x \leq 2\}$.

18. $\{x|x \geq -2\}$

19. All points to the left of -2 are shaded. The open circle at -2 indicates that -2 is not part of the graph. Using set-builder notation we have $\{x|x < -2\}$.

20. $\{x|x > 1\}$

21. The point 0 and all points to the right of 0 are shaded. Using set-builder notation we have $\{x|x \geq 0\}$.

22. $\{x|x \leq 0\}$

23.
$$y + 1 > 7$$
$$y + 1 - 1 > 7 - 1 \quad \text{Adding } -1$$
$$y > 6 \quad \text{Simplifying}$$

The solution set is $\{y|y > 6\}$. The graph is as follows:

24. $\{y|y > 3\}$

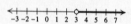

25.
$$x + 8 \leq -10$$
$$x + 8 - 8 \leq -10 - 8 \quad \text{Subtracting } 8$$
$$x \leq -18 \quad \text{Simplifying}$$

The solution set is $\{x|x \leq -18\}$. The graph is as follows:

26. $\{x|x \leq -21\}$

27.
$$x - 4 < 6$$
$$x - 4 + 4 < 6 + 4$$
$$x < 10$$

The solution set is $\{x|x < 10\}$. The graph is as follows:

28. $\{x|x < 17\}$

29.
$$x - 6 \geq 2$$
$$x - 6 + 6 \geq 2 + 6$$
$$x \geq 8$$

The solution set is $\{x|x \geq 8\}$. The graph is as follows:

30. $\{x|x \geq 13\}$

31.
$$y - 7 > -12$$
$$y - 7 + 7 > -12 + 7$$
$$y > -5$$

The solution set is $\{y|y > -5\}$. The graph is as follows:

32. $\{y|y > -6\}$

33.
$$2x + 4 \leq x + 9$$
$$2x + 4 - 4 \leq x + 9 - 4 \quad \text{Adding } -4$$
$$2x \leq x + 5 \quad \text{Simplifying}$$
$$2x - x \leq x + 5 - x \quad \text{Adding } -x$$
$$x \leq 5 \quad \text{Simplifying}$$

The solution set is $\{x|x \leq 5\}$. The graph is as follows:

34. $\{x|x \leq -3\}$

35.
$$4x - 6 \geq 3x - 1$$
$$4x - 6 + 6 \geq 3x - 1 + 6 \quad \text{Adding } 6$$
$$4x \geq 3x + 5$$
$$4x - 3x \geq 3x + 5 - 3x \quad \text{Adding } -3x$$
$$x \geq 5$$

The solution set is $\{x|x \geq 5\}$.

36. $\{x|x \geq 20\}$

37.
$$y + \frac{1}{3} \le \frac{5}{6}$$
$$y + \frac{1}{3} - \frac{1}{3} \le \frac{5}{6} - \frac{1}{3}$$
$$y \le \frac{5}{6} - \frac{2}{6}$$
$$y \le \frac{3}{6}$$
$$y \le \frac{1}{2}$$

The solution set is $\left\{y \middle| y \le \frac{1}{2}\right\}$.

38. $\left\{x \middle| x \le \frac{1}{4}\right\}$

39.
$$x - \frac{1}{8} > \frac{1}{2}$$
$$x - \frac{1}{8} + \frac{1}{8} > \frac{1}{2} + \frac{1}{8}$$
$$x > \frac{4}{8} + \frac{1}{8}$$
$$x > \frac{5}{8}$$

The solution set is $\left\{x \middle| x > \frac{5}{8}\right\}$.

40. $\left\{y \middle| y > \frac{7}{12}\right\}$

41.
$$-9x + 17 > 17 - 8x$$
$$-9x + 17 - 17 > 17 - 8x - 17 \quad \text{Adding } -17$$
$$-9x > -8x$$
$$-9x + 9x > -8x + 9x \qquad \text{Adding } 9x$$
$$0 > x$$

The solution set is $\{x | x < 0\}$.

42. $\{x | x < 0\}$

43.
$$5x < 35$$
$$\frac{1}{5} \cdot 5x < \frac{1}{5} \cdot 35 \quad \text{Multiplying by } \frac{1}{5}$$
$$x < 7$$

The solution set is $\{x | x < 7\}$. The graph is as follows:

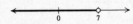

44. $\{x | x \ge 4\}$

45.
$$9y \le 81$$
$$\frac{1}{9} \cdot 9y \le \frac{1}{9} \cdot 81 \quad \text{Multiplying by } \frac{1}{9}$$
$$y \le 9$$

The solution set is $\{y | y \le 9\}$. The graph is as follows:

46. $\{x | x > 24\}$

47.
$$-7x \le 13$$
$$-\frac{1}{7} \cdot (-7x) \ge -\frac{1}{7} \cdot 13 \quad \text{Multiplying by } -\frac{1}{7}$$
$$\text{The symbol has to be reversed.}$$
$$x \ge -\frac{13}{7} \quad \text{Simplifying}$$

The solution set is $\left\{x \middle| x \ge -\frac{13}{7}\right\}$.

48. $\left\{y \middle| y < \frac{17}{8}\right\}$

49.
$$12x > -36$$
$$\frac{1}{12} \cdot 12x > \frac{1}{12} \cdot (-36)$$
$$x > -3$$

The solution set is $\{x | x > -3\}$. The graph is as follows:

50. $\{x | x > 4\}$

51.
$$7y \ge -2$$
$$\frac{1}{7} \cdot 7y \ge \frac{1}{7}(-2) \quad \text{Multiplying by } \frac{1}{7}$$
$$y \ge -\frac{2}{7}$$

The solution set is $\left\{y \middle| y \ge -\frac{2}{7}\right\}$.

52. $\left\{x \middle| x > -\frac{3}{5}\right\}$

53.
$$-5x < -17$$
$$-\frac{1}{5} \cdot (-5x) > -\frac{1}{5} \cdot (-17) \quad \text{Multiplying by } -\frac{1}{5}$$
$$\text{The symbol has to be reversed.}$$
$$x > \frac{17}{5} \qquad \text{Simplifying}$$

The solution set is $\left\{x \middle| x > \frac{17}{5}\right\}$.

54. $\left\{ y \middle| y \ge \dfrac{14}{3} \right\}$

55.
$$-4y \le \frac{1}{3}$$
$$-\frac{1}{4} \cdot (-4y) \ge -\frac{1}{4} \cdot \frac{1}{3}$$

The symbol has to be reversed.

$$y \ge -\frac{1}{12}$$

The solution set is $\left\{ y \middle| y \ge -\dfrac{1}{12} \right\}$.

56. $\left\{ x \middle| x \le -\dfrac{1}{10} \right\}$

57.
$$-\frac{8}{5} \le -2x$$
$$-\frac{1}{2} \cdot \left(-\frac{8}{5} \right) \ge -\frac{1}{2} \cdot (-2x)$$
$$\frac{8}{10} \ge x$$
$$\frac{4}{5} \ge x, \text{ or } x \le \frac{4}{5}$$

The solution set is $\left\{ x \middle| \dfrac{4}{5} \ge x \right\}$, or $\left\{ x \middle| x \le \dfrac{4}{5} \right\}$.

58. $\left\{ y \middle| y < \dfrac{1}{16} \right\}$

59.
$$5 + 3x < 32$$
$$5 + 3x - 5 < 32 - 5 \quad \text{Adding } -5$$
$$3x < 27 \qquad\qquad \text{Simplifying}$$
$$x < 9 \qquad\qquad \text{Multiplying by } \frac{1}{3}$$

The solution set is $\{x | x < 9\}$.

60. $\{y | y < 8\}$

61.
$$6 + 5y \ge 26$$
$$6 + 5y - 6 \ge 26 - 6 \quad \text{Adding } -6$$
$$5y \ge 20$$
$$y \ge 4 \qquad\qquad \text{Multiplying by } \frac{1}{5}$$

The solution set is $\{y | y \ge 4\}$.

62. $\{x | x \ge 8\}$

63.
$$3x - 5 \le 13$$
$$3x - 5 + 5 \le 13 + 5 \quad \text{Adding } 5$$
$$3x \le 18$$
$$\frac{1}{3} \cdot 3x \le \frac{1}{3} \cdot 18 \quad \text{Multiplying by } \frac{1}{3}$$
$$x \le 6$$

The solution set is $\{x | x \le 6\}$.

64. $\{y | y \le 6\}$

65.
$$13x - 7 < -46$$
$$13x - 7 + 7 < -46 + 7$$
$$13x < -39$$
$$\frac{1}{13} \cdot 13x < \frac{1}{13} \cdot (-39)$$
$$x < -3$$

The solution set is $\{x | x < -3\}$.

66. $\{y | y < -6\}$

67.
$$16 < 4 - 3y$$
$$16 - 4 < 4 - 3y - 4 \quad \text{Adding } -4$$
$$12 < -3y$$
$$-\frac{1}{3} \cdot 12 > -\frac{1}{3} \cdot (-3y) \quad \text{Multiplying by } -\frac{1}{3}$$

The symbol has to be reversed.

$$-4 > y$$

The solution set is $\{y | -4 > y\}$, or $\{y | y < -4\}$.

68. $\{x | x < -2\}$

69.
$$39 > 3 - 9x$$
$$39 - 3 > 3 - 9x - 3 \quad \text{Adding } -3$$
$$36 > -9x$$
$$-\frac{1}{9} \cdot 36 < -\frac{1}{9} \cdot (-9x) \quad \text{Multiplying by } -\frac{1}{9}$$

The symbol has to be reversed.

$$-4 < x$$

The solution set is $\{x | -4 < x\}$, or $\{x | x > -4\}$.

70. $\{y | y > -5\}$

71.
$$3 - 6y > 23$$
$$-3 + 3 - 6y > -3 + 23$$
$$-6y > 20$$
$$-\frac{1}{6} \cdot (-6y) < -\frac{1}{6} \cdot 20$$

The symbol has to be reversed.

$$y < -\frac{20}{6}$$
$$y < -\frac{10}{3}$$

The solution set is $\left\{ y \middle| y < -\dfrac{10}{3} \right\}$.

72. $\{y | y < -3\}$

73.
$$-3 < 8x + 7 - 7x$$
$$-3 < x + 7 \qquad\qquad \text{Collecting like terms}$$
$$-3 - 7 < x + 7 - 7$$
$$-10 < x$$

The solution set is $\{x | -10 < x\}$, or $\{x | x > -10\}$.

74. $\{x | x > -13\}$

75.
$$6 - 4y > 4 - 3y$$
$$6 - 4y + 4y > 4 - 3y + 4y \quad \text{Adding } 4y$$
$$6 > 4 + y$$
$$-4 + 6 > -4 + 4 + y \quad \text{Adding } -4$$
$$2 > y, \text{ or } y < 2$$

The solution set is $\{y | 2 > y\}$, or $\{y | y < 2\}$.

76. $\{y|y < 2\}$

77.
$$5 - 9y \le 2 - 8y$$
$$5 - 9y + 9y \le 2 - 8y + 9y$$
$$5 \le 2 + y$$
$$-2 + 5 \le -2 + 2 + y$$
$$3 \le y, \text{ or } y \ge 3$$
The solution set is $\{y|3 \le y\}$, or $\{y|y \ge 3\}$.

78. $\{y|y \ge 2\}$

79.
$$33 - 12x < 4x + 97$$
$$33 - 12x - 97 < 4x + 97 - 97$$
$$-64 - 12x < 4x$$
$$-64 - 12x + 12x < 4x + 12x$$
$$-64 < 16x$$
$$-4 < x$$
The solution set is $\{x|-4 < x\}$, or $\{x|x > -4\}$.

80. $\left\{x\middle|x < \dfrac{9}{5}\right\}$

81.
$$2.1x + 43.2 > 1.2 - 8.4x$$
$$10(2.1x + 43.2) > 10(1.2 - 8.4x) \quad \text{Multiplying by}$$
$$\qquad\qquad\qquad\qquad\qquad 10 \text{ to clear decimals}$$
$$21x + 432 > 12 - 84x$$
$$21x + 84x > 12 - 432 \quad \text{Adding } 84x \text{ and}$$
$$\qquad\qquad\qquad\qquad -432$$
$$105x > -420$$
$$x > -4 \quad \text{Multiplying by } \dfrac{1}{105}$$
The solution set is $\{x|x > -4\}$.

82. $\left\{y\middle|y \le \dfrac{5}{3}\right\}$

83.
$$0.7n - 15 + n \ge 2n - 8 - 0.4n$$
$$1.7n - 15 \ge 1.6n - 8 \quad \text{Collecting like terms}$$
$$10(1.7n - 15) \ge 10(1.6n - 8) \quad \text{Multiplying by 10}$$
$$17n - 150 \ge 16n - 80$$
$$17n - 16n \ge -80 + 150 \quad \text{Adding } -16n \text{ and}$$
$$\qquad\qquad\qquad\qquad 150$$
$$n \ge 70$$
The solution set is $\{n|n \ge 70\}$

84. $\{t|t > 1\}$

85.
$$\frac{x}{3} - 4 \le 1$$
$$3\left(\frac{x}{3} - 4\right) \le 3 \cdot 1 \quad \text{Multiplying by 3 to}$$
$$\qquad\qquad\qquad\qquad \text{to clear the fraction}$$
$$x - 12 \le 3 \quad \text{Simplifying}$$
$$x \le 15 \quad \text{Adding 12}$$
The solution set is $\{x|x \le 15\}$.

86. $\{x|x > 2\}$

87.
$$\frac{y}{5} + 2 \le \frac{3}{5}$$
$$5\left(\frac{y}{5} + 2\right) \le 5 \cdot \frac{3}{5} \quad \text{Clearing fractions}$$
$$y + 10 \le 3$$
$$y \le -7 \quad \text{Adding } -10$$
The solution set is $\{y|y \le -7\}$.

88. $\{x|x \ge -25\}$

89.
$$3(2y - 3) < 27$$
$$6y - 9 < 27 \quad \text{Removing parentheses}$$
$$6y < 36 \quad \text{Adding 9}$$
$$y < 6 \quad \text{Multiplying by } \frac{1}{6}$$
The solution set is $\{y|y < 6\}$.

90. $\{y|y > 5\}$

91.
$$3(t - 2) \ge 9(t + 2)$$
$$3t - 6 \ge 9t + 18$$
$$3t - 9t > 18 + 6$$
$$-6t \ge 24$$
$$t \le -4 \quad \text{Multiplying by } -\frac{1}{6} \text{ and}$$
$$\qquad\qquad\qquad \text{reversing the symbol}$$
The solution set is $\{t|t \le -4\}$.

92. $\left\{t\middle|t < -\dfrac{5}{3}\right\}$

93.
$$3(r - 6) + 2 < 4(r + 2) - 21$$
$$3r - 18 + 2 < 4r + 8 - 21$$
$$3r - 16 < 4r - 13$$
$$-16 + 13 < 4r - 3r$$
$$-3 < r, \text{ or } r > -3$$
The solution set is $\{r|r > -3\}$.

94. $\{t|t > -12\}$

95.
$$\frac{2}{3}(2x - 1) \ge 10$$
$$\frac{3}{2} \cdot \frac{2}{3}(2x - 1) \ge \frac{3}{2} \cdot 10 \quad \text{Multiplying by } \frac{3}{2}$$
$$2x - 1 \ge 15$$
$$2x \ge 16$$
$$x \ge 8$$
The solution set is $\{x|x \ge 8\}$.

96. $\{x|x \le 7\}$

97. $\frac{3}{4}\left(3x - \frac{1}{2}\right) - \frac{2}{3} < \frac{1}{3}$

$\qquad \frac{3}{4}\left(3x - \frac{1}{2}\right) < 1 \quad$ Adding $\frac{2}{3}$

$\qquad \frac{9}{4}x - \frac{3}{8} < 1 \quad$ Removing parentheses

$\qquad 8 \cdot \left(\frac{9}{4}x - \frac{3}{8}\right) < 8 \cdot 1 \quad$ Clearing fractions

$\qquad\qquad 18x - 3 < 8$

$\qquad\qquad\qquad 18x < 11$

$\qquad\qquad\qquad\quad x < \frac{11}{18}$

The solution set is $\left\{x\middle| x < \frac{11}{18}\right\}$.

98. $\left\{x\middle| x > -\frac{5}{32}\right\}$

99. $5 - 3^2 + (8 - 2)^2 \cdot 4 = 5 - 3^2 + 6^2 \cdot 4$

$\qquad\qquad = 5 - 9 + 36 \cdot 4$

$\qquad\qquad = 5 - 9 + 144$

$\qquad\qquad = -4 + 144$

$\qquad\qquad = 140$

100. 41

101. $5(2x - 4) - 3(4x + 1) = 10x - 20 - 12x - 3 =$

$\qquad -2x - 23$

102. $-1 + 37x$

103. ◈

104. ◈

105. ◈

106. ◈

107. $6[4 - 2(6 + 3t)] > 5[3(7 - t) - 4(8 + 2t)] - 20$

$\qquad 6[4 - 12 - 6t] > 5[21 - 3t - 32 - 8t] - 20$

$\qquad 6[-8 - 6t] > 5[-11 - 11t] - 20$

$\qquad -48 - 36t > -55 - 55t - 20$

$\qquad -48 - 36t > -75 - 55t$

$\qquad -36t + 55t > -75 + 48$

$\qquad\qquad 19t > -27$

$\qquad\qquad\quad t > -\frac{27}{19}$

The solution set is $\left\{t\middle| t > -\frac{27}{19}\right\}$.

108. $\left\{x\middle| x \leq \frac{5}{6}\right\}$

109. $-(x + 5) \geq 4a - 5$

$\qquad -x - 5 \geq 4a - 5$

$\qquad\quad -x \geq 4a - 5 + 5$

$\qquad\quad -x \geq 4a$

$\qquad -1(-x) \leq -1 \cdot 4a$

$\qquad\qquad x \leq -4a$

The solution set is $\{x|x \leq -4a\}$.

110. $\{x|x > 7\}$

111. $\qquad y < ax + b \quad$ Assume $a > 0$.

$\qquad y - b < ax$

$\qquad \frac{y - b}{a} < x \qquad$ Since $a > 0$, the inequality symbol stays the same.

The solution set is $\left\{x\middle| x > \frac{y - b}{a}\right\}$.

112. $\left\{x\middle| x < \frac{y - b}{a}\right\}$

113. $|x| < 3$

a) Since $|3.2| = 3.2$, and $3.2 < 3$ is false, 3.2 is not a solution.

b) Since $|-2| = 2$ and $2 < 3$ is true, -2 is a solution.

c) Since $|-3| = 3$ and $3 < 3$ is false, -3 is not a solution.

d) Since $|-2.9| = 2.9$ and $2.9 < 3$ is true, -2.9 is a solution.

e) Since $|3| = 3$ and $3 < 3$ is false, 3 is not a solution.

f) Since $|1.7| = 1.7$ and $1.7 < 3$ is true, 1.7 is a solution.

114.

Exercise Set 2.7

1. Let n represent the number. Then we have

$\qquad n < 9$.

2. Let n represent the number; $n \geq 5$

3. Let b represent the weight of the bag, in pounds. Then we have

$\qquad b \geq 2$.

4. Let p represent the number of people who attended the concert; $75 < p < 100$

5. Let s represent the average speed, in mph. Then we have

$\qquad 90 < s < 110$.

6. Let n represent the number of people who attended the Million Man March; $n \geq 400,000$

7. Let a represent the number of people who attended the Million Man March. Then we have

$\qquad a \leq 1,200,000$.

8. Let a represent the amount of acid, in liters; $a \leq 40$

9. Let c represent the cost, per gallon, of gasoline. Then we have

$\qquad c \geq \$1.20$.

10. Let t represent the temperature; $t \leq -2$

11. *Familiarize.* The average of the four scores is their sum divided by the number of tests, 4. We let s represent Nadia's score on the last test.

Translate. The average of the four scores is given by

$$\frac{82 + 76 + 78 + s}{4}.$$

Since this average must be at least 80, this means that it must be greater than or equal to 80. Thus, we can translate the problem to the inequality

$$\frac{82 + 76 + 78 + s}{4} \geq 80.$$

Carry out. We first multiply by 4 to clear the fraction.

$$4\left(\frac{82 + 76 + 78 + s}{4}\right) \geq 4 \cdot 80$$
$$82 + 76 + 78 + s \geq 320$$
$$236 + s \geq 320$$
$$s \geq 84$$

Check. As a partial check, we show that Nadia can get a score of 84 on the fourth test and have an average of at least 80:

$$\frac{82 + 76 + 78 + 84}{4} = \frac{320}{4} = 80.$$

State. Scores of 84 and higher will earn Nadia at least a B.

12. 97 and higher

13. *Familiarize.* Let m represent the number of miles driven. Then the cost of those miles is $\$0.39m$. The total cost is the daily rate plus the mileage cost. The total cost cannot exceed $250. In other words the total cost must be less or equal to the budgeted amount of $250.

Translate.

Daily rate	+	Mileage cost	≤	Budget
↓	↓	↓	↓	↓
44.95	+	$0.39m$	≤	250

Carry out.

$$44.95 + 0.39m \leq 250$$
$$4495 + 39m \leq 25{,}000 \quad \text{Clearing decimals}$$
$$39m \leq 20{,}505$$
$$m \leq \frac{20{,}505}{39}$$
$$m \leq 525.8 \quad \text{Rounding to the near-est tenth}$$

Check. We can check to see if the solution set seems reasonable.

When $m = 525$, the total cost is

$44.95 + 0.39(525)$, or $249.70.

When $m = 525.8$, the total cost is

$44.95 + 0.39(525.8)$, or $250.01.

When $m = 526$, the total cost is

$44.95 + 0.39(526)$, or $250.09.

From these calculations it would appear that the solution is correct considering that rounding occurred.

State. To stay within the budget, the mileage must not exceed 525.8 mi.

14. Mileages less than or equal to 341.4 mi

15. *Familiarize.* Let m represent the length of a telephone call, in minutes.

Translate.

$0.75 charge	plus	charge for time used	is at least	$3.00.
↓	↓	↓	↓	↓
0.75	+	$0.45m$	≥	3

Carry out.

$$0.75 + 0.45m \geq 3$$
$$0.45m \geq 2.25$$
$$m \geq 5$$

Check. As a partial check, we show that if a call lasts 5 minutes it costs at least $3.00:

$$\$0.75 + \$0.45(5) = \$0.75 + \$2.25 = \$3.00$$

State. Simon's calls last at least 5 minutes each.

16. At least 3.5 hr

17. *Familiarize.* Let $s =$ the number of servings Dale eats on Saturday.

Translate.

Average number of fruit servings	is at least	5.
↓	↓	↓
$\dfrac{4+6+7+4+6+4+s}{7}$	≥	5

Carry out. We first multiply by 7 to clear the fraction.

$$7\left(\frac{4+6+7+4+6+4+s}{7}\right) \geq 7 \cdot 5$$
$$4+6+7+4+6+4+s \geq 35$$
$$31 + s \geq 35$$
$$s \geq 4$$

Check. As a partial check, we show that Dale can eat 4 servings of fruit on Saturday and average at least 5 servings per day for the week:

$$\frac{4+6+7+4+6+4+4}{7} = \frac{35}{7} = 5$$

State. Dale should eat at least 4 servings of fruit on Saturday.

18. At least 8

19. *Familiarize.* Let $d =$ the depth of the well, in feet. Then the cost on the pay-as-you-go plan is $500 + \$8d$. The cost of the guaranteed-water plan is $4000.

We want to find the values of d for which the pay-as-you-go plan costs less than the guaranteed-water plan.

Translate.

$$\underbrace{\text{Cost of pay-as-you-go plan}} \quad \underbrace{\text{is less than}} \quad \underbrace{\begin{array}{c}\text{cost of}\\ \text{guaranteed-}\\ \text{water plan}\end{array}}$$

$$\downarrow \qquad\qquad \downarrow \qquad\qquad \downarrow$$
$$500 + 8d \qquad < \qquad 4000$$

Carry out.

$$500 + 8d < 4000$$
$$8d < 3500$$
$$d < 437.5$$

Check. We check to see that the solution is reasonable.

When $d = 437$, $\$500 + \$8 \cdot 437 = \$3996 < \$4000$

When $d = 437.5$, $\$500 + \$8(437.5) = \$4000$

When $d = 438$, $\$500 + \$8(438) = \$4004 > \$4000$

From these calculations, it appears that the solution is correct.

State. It would save a customer money to use the pay-as-you-go plan for a well of 437.5 ft or less.

20. For calls of less than 4 units of time

21. Familiarize. We first make a drawing. We let l represent the length.

The area is the length times the width, or $4l$.

Translate.

$$\underbrace{\text{Area}} \quad \underbrace{\text{is less than}} \quad \underbrace{86 \text{ cm}^2}.$$
$$\downarrow \qquad \downarrow \qquad \downarrow$$
$$4l \qquad < \qquad 86$$

Carry out.

$$4l < 86$$
$$l < 21.5$$

Check. We check to see if the solution seems reasonable.

When $l = 22$, the area is $22 \cdot 4$, or 88 cm^2.

When $l = 21.5$, the area is $21.5(4)$, or 86 cm^2.

When $l = 21$, the area is $21 \cdot 4$, or 84 cm^2.

From these calculations, it would appear that the solution is correct.

State. The area will be less than 86 cm^2 for lengths less than 21.5 cm.

22. 16.5 yd or more

23. Familiarize. Let $v =$ the blue book value of the car. Since the car was repaired, we know that \$8500 does not exceed $0.8v$ or, in other words, $0.8v$ is at least \$8500.

Translate.

$$\underbrace{\begin{array}{c}80\% \text{ of the}\\ \text{blue book value}\end{array}} \quad \underbrace{\text{is at least}} \quad \$8500.$$
$$\downarrow \qquad\qquad \downarrow \qquad\qquad \downarrow$$
$$0.8v \qquad\qquad \geq \qquad\qquad 8500$$

Carry out.

$$0.8v \geq 8500$$
$$v \geq \frac{8500}{0.8}$$
$$v \geq 10,625$$

Check. As a partial check, we show that 80% of \$10,625 is at least \$8500:

$$0.8(\$10,625) = \$8500$$

State. The blue book value of the car was at least \$10,625.

24. More than \$16,800

25. Familiarize. $R = -0.075t + 3.85$

In the formula R represents the world record and t represents the years since 1930. When $t = 0$ (1930), the record was $-0.075 \cdot 0 + 3.85$, or 3.85 minutes. When $t = 2$ (1932), the record was $-0.075(2) + 3.85$, or 3.7 minutes. For what values of t will $-0.075t + 3.85$ be less than 3.5?

Translate. The record is to be less than 3.5. We have the inequality

$$R < 3.5.$$

To find the t values which satisfy this condition we substitute $-0.075t + 3.85$ for R.

$$-0.075t + 3.85 < 3.5$$

Carry out.

$$-0.075t + 3.85 < 3.5$$
$$-0.075t < 3.5 - 3.85$$
$$-0.075t < -0.35$$
$$t > \frac{-0.35}{-0.075}$$
$$t > 4\frac{2}{3}$$

Check. We check to see if the solution set we obtained seems reasonable.

When $t = 4\frac{1}{2}$, $R = -0.075(4.5) + 3.85$, or 3.5125.

When $t = 4\frac{2}{3}$, $R = -0.075\left(\frac{14}{3}\right) + 3.85$, or 3.5.

When $t = 4\frac{3}{4}$, $R = -0.075(4.75) + 3.85$, or 3.49375.

Since $r = 3.5$ when $t = 4\frac{2}{3}$ and R decreases as t increases, R will be less than 3.5 when t is greater than $4\frac{2}{3}$.

State. The world record will be less than 3.5 minutes when t is greater than $4\frac{2}{3}$ years $\left(\text{more than } 4\frac{2}{3}\right.$ years after 1930$\Big)$. This occurs in years after 1934.

26. Years after 1984

27. *Familiarize.* Let $w =$ the number of weeks it takes for the puppy's weight to exceed $22\frac{1}{2}$ lb.

Translate.

Initial weight	plus	amount gained in w weeks	exceeds	$22\frac{1}{2}$ lb.
↓	↓	↓	↓	↓
9	+	$\frac{3}{4}w$	>	$22\frac{1}{2}$

Carry out. We solve the inequality.

$$9 + \frac{3}{4}w > 22\frac{1}{2}$$
$$\frac{3}{4}w > 13\frac{1}{2}$$
$$\frac{3}{4}w > \frac{27}{2} \quad \left(13\frac{1}{2} = \frac{27}{2}\right)$$
$$w > 18 \quad \text{Multiplying by } \frac{4}{3}$$

Check. We check to see if the solution seems reasonable.

When $w = 17$, $9 + \frac{3}{4} \cdot 17 = 21\frac{3}{4}$.

When $w = 18$, $9 + \frac{3}{4} \cdot 18 = 22\frac{1}{2}$.

When $w = 19$, $9 + \frac{3}{4} \cdot 19 = 23\frac{1}{4}$.

It would appear that the solution is correct.

State. The puppy's weight will exceed $22\frac{1}{2}$ lb when the puppy is more than 18 weeks old.

28. Dates at least 6 weeks after July 1

29. *Familiarize.* We will use the formula $F = \frac{9}{5}C + 32$.

Translate.

Fahrenheit temperature	is above	98.6°.
↓	↓	↓
F	>	98.6

Substituting $\frac{9}{5}C + 32$ for F, we have

$$\frac{9}{5}C + 32 > 98.6.$$

Carry out. We solve the inequality.

$$\frac{9}{5}C + 32 > 98.6$$
$$\frac{9}{5}C > 66.6$$
$$C > \frac{333}{9}$$
$$C > 37$$

Check. We check to see if the solution seems reasonable.

When $C = 36$, $\frac{9}{5} \cdot 36 + 32 = 96.8$.

When $C = 37$, $\frac{9}{5} \cdot 37 + 32 = 98.6$.

When $C = 38$, $\frac{9}{5} \cdot 38 + 32 = 100.4$.

It would appear that the solution is correct, considering that rounding occurred.

State. The human body is feverish for Celsius temperatures greater than 37°.

30. Temperatures less than 31.3° C

31. *Familiarize.* The average number of calls per week is the sum of the calls for the three weeks divided by the number of weeks, 3. We let c represent the number of calls made during the third week.

Translate. The average of the three weeks is given by

$$\frac{17 + 22 + c}{3}.$$

Since the average must be at least 20, this means that it must be greater than or equal to 20. Thus, we can translate the problem to the inequality

$$\frac{17 + 22 + c}{3} \geq 20.$$

Carry out. We first multiply by 3 to clear the fraction.

$$3\left(\frac{17 + 22 + c}{3}\right) \geq 3 \cdot 20$$
$$17 + 22 + c \geq 60$$
$$39 + c \geq 60$$
$$c \geq 21$$

Check. Suppose c is a number greater than or equal to 21. Then by adding 17 and 22 on both sides of the inequality we get

$$17 + 22 + c \geq 17 + 22 + 21$$
$$17 + 22 + c \geq 60$$

so

$$\frac{17 + 22 + c}{3} \geq \frac{60}{3}, \text{ or } 20.$$

State. 21 calls or more will maintain an average of at least 20 for the three-week period.

32. Numbers less than or equal to 0

33. *Familiarize.* We first make a drawing. We let w represent the width.

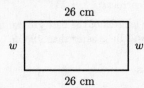

The perimeter is $P = 2l + 2w$, or $2 \cdot 26 + 2w$, or $52 + 2w$.

Translate.

$$\underbrace{\text{The perimeter}}_{\downarrow} \quad \underbrace{\text{is greater than}}_{\downarrow} \quad \underbrace{\text{80 cm.}}_{\downarrow}$$
$$52 + 2w \qquad\qquad > \qquad\qquad 80$$

Carry out.

$$52 + 2w > 80$$
$$2w > 28$$
$$w > 14$$

Check. We check to see if the solution seems reasonable.

When $w = 13$, $P = 52 + 2 \cdot 13$, or 78 cm.

When $w = 14$, $P = 52 + 2 \cdot 14$, or 80 cm.

When $w = 15$, $P = 52 + 2 \cdot 15$, or 82 cm.

From these calculations, it appears that the solution is correct.

State. Widths greater than 14 cm will make the perimeter greater than 80 cm.

34. 92 ft or more; 92 ft or less

35. *Familiarize.* We first make a drawing. We let b represent the length of the base. Then the lengths of the other sides are $b - 2$ and $b + 3$.

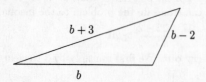

The perimeter is the sum of the lengths of the sides or $b + b - 2 + b + 3$, or $3b + 1$.

Translate.

$$\underbrace{\text{The perimeter}}_{\downarrow} \quad \underbrace{\text{is greater than}}_{\downarrow} \quad \underbrace{\text{19 cm.}}_{\downarrow}$$
$$3b + 1 \qquad\qquad > \qquad\qquad 19$$

Carry out.

$$3b + 1 > 19$$
$$3b > 18$$
$$b > 6$$

Check. We check to see if the solution seems reasonable.

When $b = 5$, the perimeter is $3 \cdot 5 + 1$, or 16 cm.

When $b = 6$, the perimeter is $3 \cdot 6 + 1$, or 19 cm.

When $b = 7$, the perimeter is $3 \cdot 7 + 1$, or 22 cm.

From these calculations, it would appear that the solution is correct.

State. For lengths of the base greater than 6 cm the perimeter will be greater than 19 cm.

36. $\dfrac{35}{3}$ ft or less

37. *Familiarize.* Let $h =$ the number of hours George worked. Then $h + 3 =$ the number of hours Joan worked.

Translate.

$$\underbrace{\substack{\text{George's} \\ \text{hours}}}_{\downarrow} + \underbrace{\substack{\text{Joan's} \\ \text{hours}}}_{\downarrow} \quad \underbrace{\text{are more than}}_{\downarrow} \quad \underbrace{\text{27 hours.}}_{\downarrow}$$
$$h \quad + \quad (h + 3) \qquad > \qquad\qquad 27$$

Carry out.

$$h + (h + 3) > 27$$
$$2h + 3 > 27$$
$$2h > 24$$
$$h > 12$$

If h is at least 12, then $h + 3$ is at least 15.

Check. We check to see if the solution seems reasonable.

When $h = 11$, together they work

$$11 + (11 + 3), \text{ or } 25 \text{ hr.}$$

When $h = 12$, together they work

$$12 + (12 + 3), \text{ or } 27 \text{ hr.}$$

When $h = 13$, together they work

$$13 + (13 + 3), \text{ or } 29 \text{ hr.}$$

From these calculations, it would appear that the solution is correct.

State. George worked at least 12 hours, and Joan worked at least 15 hours.

38. $49.02

39. *Familiarize.* Let $w =$ the weight of the small box, in pounds. Then $w + 1$ and $w + 2$ represent the weights of the medium and large boxes, respectively.

Translate.

$$\underbrace{\substack{\text{Weight of} \\ \text{small box}}}_{\downarrow} + \underbrace{\substack{\text{weight of} \\ \text{medium box}}}_{\downarrow} +$$
$$w \qquad + \qquad (w + 1) \qquad +$$

$$\underbrace{\substack{\text{weight of} \\ \text{large box}}}_{\downarrow} \quad \underbrace{\text{is at most}}_{\downarrow} \quad \underbrace{\text{30 lb.}}_{\downarrow}$$
$$(w + 2) \qquad\qquad \leq \qquad\qquad 30$$

Carry out.

$$w + (w + 1) + (w + 2) \leq 30$$
$$3w + 3 \leq 30$$
$$3w \leq 27$$
$$w \leq 9$$

Check. As a partial check, we show that if the weight of the small box is 9 lb, the total weight of the three boxes is no more than 30 lb:

$$9 + 10 + 11 = 30$$

State. The small box weighs 9 lb or less.

40. 64 km or more

41. *Familiarize*. Let r = the amount of fat in a serving of the regular cookies, in grams. If a reduced fat cookie has at least 25% less fat than a regular cookie, then it has at most 75% as much fat as the regular cookie.

Translate.

4 g of fat	is at most	75%	of	the amount of fat in a regular cookie.
↓	↓	↓	↓	↓
4	≤	0.75	·	r

Carry out.
$$4 \le 0.75r$$
$$5.\overline{3} \le r, \text{ or}$$
$$5\frac{1}{3} \le r$$

Check. As a partial check, we show that 4 g of fat does not exceed 75% of $5\frac{1}{3}$ g of fat:
$$0.75\left(5\frac{1}{3}\right) = 0.75\left(\frac{16}{3}\right) = 4$$

State. A regular cookie contains at least $5\frac{1}{3}$ g of fat.

42. At least 16 g per serving

43. *Familiarize*. Let b = the height, in cm. Recall that the area of a triangle with base b and height h is given by $\frac{1}{2}bh$. In this case, we have $\frac{1}{2} \cdot 16 \cdot h$.

Translate.

Area of triangle	is at least	72 cm².
↓	↓	↓
$\frac{1}{2} \cdot 16 \cdot h$	≥	72

Carry out.
$$\frac{1}{2} \cdot 16 \cdot h \ge 72$$
$$8h \ge 72$$
$$h \ge 9$$

Check. As a partial check we show that if the height is 9 cm, the area of the triangle is at least 72 cm²:
$$\frac{1}{2} \cdot 16 \cdot 9 = 72 \text{ cm}^2$$

State. Heights of at least 9 cm will guarantee that the triangle's area is at least 72 cm².

44. 4 cm or less

45. *Familiarize*. We will use the formula
$$P = 0.1522Y - 298.592.$$

Translate. We have the inequality $P \ge 6$. To find the years that satisfy this condition we substitute $0.1522Y - 298.592$ for P:
$$0.1522Y - 298.592 \ge 6$$

Carry out.
$$0.1522Y - 298.592 \ge 6$$
$$0.1522Y \ge 304.592$$
$$Y \ge 2001 \quad \text{Rounding}$$

Check. We check to see if the solution seems reasonable.

When $Y = 2000$, $P = 0.1522(2000) - 298.592$, or about \$5.81.

When $Y = 2001$, $P = 0.1522(2001) - 298.592$, or about \$5.96.

When $Y = 2002$, $P = 0.1522(2002) - 298.592$, or about \$6.11.

From these calculations, it would appear that the solution is correct considering that rounding occurred.

State. From about 2001 on, the average price of a movie ticket will be at least \$6.

46. $215\frac{5}{27}$ mi or less

47.
$$-3 + 2(-5)^2(-3) - 7$$
$$= -3 + 2(25)(-3) - 7 \quad \text{Evaluating the exponential expression}$$
$$= -3 - 150 - 7 \quad \text{Multiplying}$$
$$= -160 \quad \text{Subtracting}$$

48. $4a^2 - 2$

49.
$$9x - 5 + 4x^2 - 2 - 13x$$
$$= 4x^2 + 9x - 13x - 5 - 2$$
$$= 4x^2 + (9 - 13)x - 5 - 2$$
$$= 4x^2 - 4x - 7$$

50. $-17x + 18$

51.
$$5ab + 9b - 8ab - 12a$$
$$= 5ab - 8ab + 9b - 12a$$
$$= (5 - 8)ab + 9b - 12a$$
$$= -3ab + 9b - 12a$$

52. $-4a^2b - 4ab + 2ab^2$

53. ◈

54. ◈

55. ◈

56. ◈

57. *Familiarize*. Let x = the smaller odd integer. Then $x + 2$ = the larger.

Translate.

The sum of two consecutive odd integers	is less than	100.
↓	↓	↓
$x + (x + 2)$	<	100

Carry out.

$$x + (x + 2) < 100$$
$$2x + 2 < 100$$
$$2x < 98$$
$$x < 49$$

The largest odd integer less than 49 is 47. Thus, a possible solution gives the integers 47 and 49.

Check. We check to see if the solution seems reasonable.

$$47 + 49 = 98 < 100$$

$$49 + 51 = 100 \quad \text{(This sum is not less than 100.)}$$

State. The largest pair of consecutive odd integers whose sum is less than 100 is 47 and 49.

58. 8 cm or less

59. *Familiarize*. We use the formula $F = \frac{9}{5}C + 32$.

Translate. We are interested in temperatures such that $5° < F < 15°$. Substituting for F, we have:

$$5 < \frac{9}{5}C + 32 < 15$$

Carry out.

$$5 < \frac{9}{5}C + 32 < 15$$
$$5 \cdot 5 < 5\left(\frac{9}{5}C + 32\right) < 5 \cdot 15$$
$$25 < 9C + 160 < 75$$
$$-135 < 9C < -85$$
$$-15 < C < -9\frac{4}{9}$$

Check. The check is left to the student.

State. Green ski wax works best for temperatures between $-15°$ C and $-9\frac{4}{9}°$ C.

60. More than 6 hr

61. *Familiarize*. Let $s =$ the gross sales. Then $s - 10,000 =$ the gross sales over \$10,000. Plan A pays $600 + 0.04s$ per month, and Plan B pays $800 + 0.06(s - 10,000)$ per month.

Translate.

Plan A pays more than Plan B pays.

$$600 + 0.04s > 800 + 0.06(s - 10,000)$$

Carry out.

$$600 + 0.04s > 800 + 0.06(s - 10,000)$$
$$600 + 0.04s > 800 + 0.06s - 600$$
$$600 + 0.04s > 200 + 0.06s$$
$$600 - 200 > 0.06s - 0.04s$$
$$400 > 0.02s$$
$$20,000 > s$$

Check. The check is left to the student.

State. For gross sales less than \$20,000, Plan A is better than Plan B.

62. Between 5 and 9 hours

63. *Familiarize*. Let $f =$ the fat content of a serving of regular tortilla chips, in grams. A product that contains 60% less fat than another product has 40% of the fat content of that product. If Reduced Fat Tortilla Pops cannot be labeled lowfat, then they contain at least 3 g of fat.

Translate.

40% of the fat content of regular tortilla chips is at least 3 grams of fat

$$0.4 \cdot f \geq 3$$

Carry out.

$$0.4f \geq 3$$
$$f \geq 7.5$$

Check. As a partial check, we show that 40% of 7.5 g is not less than 3 g.

$$0.4(7.5) = 3$$

State. A serving of regular tortilla chips contains at least 7.5 g of fat.

64. ◈

65. ◈

Chapter 3

Introduction to Graphing

Exercise Set 3.1

1. We go to the top of the bar that is above the body weight 160 lb. Then we move horizontally from the top of the bar to the vertical scale listing numbers of drinks. It appears approximately 5 drinks will give a 160 lb person a blood-alcohol level of 0.10%.

2. Approximately 3 drinks

3. From $3\frac{1}{2}$ on the vertical scale we move horizontally until we reach a bar whose top is above the horizontal line on which we are moving. The first such bar corresponds to a body weight of 120 lb. Thus, we can conclude an individual weighs at least 120 lb if $3\frac{1}{2}$ drinks are consumed without reaching a blood-alcohol level of 0.10%.

4. The individual weighs more than 160 lb.

5. *Familiarize*. The total amount Leila pays in federal income tax is

   $0.18 \times \$31,200$, or $\$5616$.

   The pie chart indicates that 18% of tax dollars are spent on social programs. We let $y =$ the amount of Leila's taxable income that will be spent on social programs.

   Translate. We reword the problem.

   What is 18% of \$5616?
   $\downarrow$ $\downarrow$ $\downarrow$ $\downarrow$ $\downarrow$
   y $=$ 18% $\cdot$ 5616

   Carry out.

   $y = 0.18 \cdot 5616 = 1010.88$

   Check. We go over the calculations again. The result checks.

   State. \$1010.88 of Leila's taxable income will be spent on social programs.

6. \$374.40

7. *Familiarize*. The total amount the Caseys pay in federal income tax is

   $0.23 \cdot \$101,500$, or $\$23,345$.

   The pie chart indicates that 35% of tax dollars are spent on social security/medicare. We let $y =$ the amount of the Casey's taxable income that will be spent on social security/medicare.

   Translate. We reword the problem.

   What is 35% of \$23,345?
   $\downarrow$ $\downarrow$ $\downarrow$ $\downarrow$ $\downarrow$
   y $=$ 35% $\cdot$ 23,345

Carry out.

   $y = 0.35 \cdot 23,345 = 8170.75$

Check. We go over the calculations again. The result checks.

State. \$8170.75 of the Casey's taxable income will be spent on social security/medicare.

8. \$556.80

9. *Familiarize*. From the pie chart we see that 9.3% of solid waste is plastic. We let $x =$ the amount of plastic, in millions of tons, in the waste generated in 1993.

   Translate. We reword the problem.

   What is 9.3% of 206.9?
   $\downarrow$ $\downarrow$ $\downarrow$ $\downarrow$ $\downarrow$
   x $=$ 9.3% $\cdot$ 206.9

   Carry out.

   $x = 0.093 \cdot 206.9 \approx 19.2$

   Check. We can repeat the calculation. The result checks.

   State. In 1993, about 19.2 million tons of waste was plastic.

10. 1.7 lb

11. *Familiarize*. From the pie chart we see that 6.6% of solid waste is glass. From Exercise 10 we know that the average American generates 4.5 lb of solid waste per day. Then the amount of this that is glass is

    $0.066(4.5)$, or 0.297 lb

    We let $x =$ the amount of glass, in pounds, that the average American recycles each day.

    Translate. We reword the problem.

    What is 23% of 0.297?
    $\downarrow$ $\downarrow$ $\downarrow$ $\downarrow$ $\downarrow$
    x $=$ 23% $\cdot$ 0.297

    Carry out.

    $x = 0.23(0.297) \approx 0.07$

    Check. We go over the calculations again. The result checks.

    State. The average American recycles about 0.07 lb of glass each day.

12. 0.08 lb

13. We locate 1965 on the horizontal scale and then move up to the line representing public education expenditures. At that point we move left to the vertical scale and read the information we are seeking. Approximately 4% of the GNP was spent on public education in 1965.

14. Approximately 8%

15. We locate 10% on the vertical scale and then move right until the line representing health care expenditures is reached. At that point we can move down to the horizontal scale and read the information we are seeking. Health care costs represented about 10% of the GNP in 1982.

16. 1990

17. *Familiarize*. From the graph we see that about 4.5% of GNP was spent on public education in 1980. We let p = the amount spent on public education in 1980, in billions of dollars.

Translate. Reword the problem.

What is 4.5% of 2732?
p = 4.5% · 2732

Carry out.

$p = 0.045 \cdot 2732 \approx 123$

Check. We go over the calculations again. The result checks.

State. About $123 billion was spent on public education in 1980.

18. $1012 billion

19. *Familiarize*. From the graph we see that about 5% of GNP was spent on public education in 1970. We let y = the GNP in 1970, in billions of dollars.

Translate. We reword the problem.

50.8 is 5% of what number?
50.8 = 5% · y

Carry out. We solve the equation.

$$50.8 = 0.05y$$
$$\frac{50.8}{0.05} = y$$
$$1016 = y$$

Check. 5% of 1016 is 50.8. The answer checks.

State. The GNP in 1970 was about $1016 billion.

20. $5344 billion

21. Starting at the origin:

(1,2) is 1 unit right and 2 units up;

(−2,3) is 2 units left and 3 units up;

(4,−1) is 4 units right and 1 unit down;

(−5,−3) is 5 units left and 3 units down;

(4,0) is 4 units right and 0 units up or down;

(0,−2) is 0 units right or left and 2 units down.

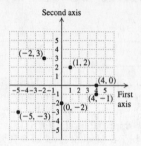

22.

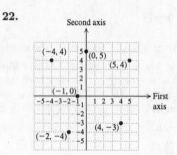

23. Starting at the origin:

(4,4) is 4 units right and 4 units up;

(−2,4) is 2 units left and 4 units up;

(5,−3) is 5 units right and 3 units down;

(−5,−5) is 5 units left and 5 units down;

(0,4) is 0 units right or left and 4 units up;

(0,−4) is 0 units right or left and 4 units down;

(3,0) is 3 units right and 0 units up or down;

(−4,0) is 4 units left and 0 units up or down.

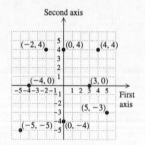

24.

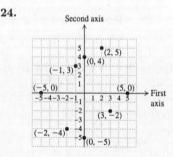

25.

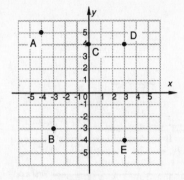

Point A is 4 units left and 5 units up. The coordinates of A are $(-4, 5)$.

Point B is 3 units left and 3 units down. The coordinates of B are $(-3, -3)$.

Point C is 0 units right or left and 4 units up. The coordinates of C are $(0,4)$.

Point D is 3 units right and 4 units up. The coordinates of D are $(3,4)$.

Point E is 3 units right and 4 units down. The coordinates of E are $(3, -4)$.

26. $A: (3,3)$, $B: (0,-4)$, $C: (-5,0)$, $D: (-1,-1)$, $E: (2,0)$

27.

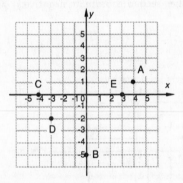

Point A is 4 units right and 1 unit up. The coordinates of A are $(4,1)$.

Point B is 0 units right or left and 5 units down. The coordinates of B are $(0, -5)$.

Point C is 4 units left and 0 units up or down. The coordinates of C are $(-4, 0)$.

Point D is 3 units left and 2 units down. The coordinates of D are $(-3, -2)$.

Point E is 3 units right and 0 units up or down. The coordinates of E are $(3,0)$.

28. $A: (-5,1)$, $B: (0,5)$, $C: (5,3)$, $D: (0,-1)$, $E: (2,-4)$

29. Since the first coordinate is negative and the second coordinate positive, the point $(-5,3)$ is located in quadrant II.

30. II

31. Since the first coordinate is positive and the second coordinate negative, the point $(100, -1)$ is in quadrant IV.

32. IV

33. Since both coordinates are negative, the point $(-6, -29)$ is in quadrant III.

34. III

35. Since both coordinates are positive, the point $(3.8, 9.2)$ is in quadrant I.

36. I

37. In quadrant III, first coordinates are always <u>negative</u> and second coordinates are always <u>negative</u>.

38. Second, first

39. Use the horizontal axis to represent years and the vertical axis to represent percents. Plot the points $(1970, 45\%)$, $(1980, 51\%)$, and $(1990, 54\%)$ and connect adjacent pairs of points with line segments.

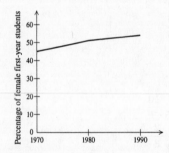

40.

Percentage of first-year college students believing "Activities of married women are best confined to home and family"

41. Use the horizontal axis to represent years and the vertical axis to represent pounds. Plot the points $(1975, 125.8)$, $(1985, 124.9)$, and $(1994, 114.8)$ and connect adjacent pairs of points with line segments.

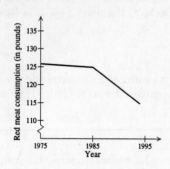

42.

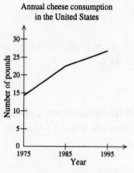

43.
$$\frac{2}{9} + \frac{2}{3} = \frac{2}{9} + \frac{2}{3} \cdot \frac{3}{3}$$
$$= \frac{2}{9} + \frac{6}{9}$$
$$= \frac{8}{9}$$

44. $\frac{1}{12}$

45.
$$\frac{3}{5} \cdot \frac{10}{9} = \frac{3 \cdot 10}{5 \cdot 9}$$
$$= \frac{3 \cdot 2 \cdot 5}{5 \cdot 3 \cdot 3}$$
$$= \frac{\cancel{3} \cdot 2 \cdot \cancel{5}}{\cancel{5} \cdot \cancel{3} \cdot 3}$$
$$= \frac{2}{3}$$

46. $\frac{13}{15}$

47.
$$\frac{3}{7} - \frac{4}{5}$$
$$= \frac{3}{7} \cdot \frac{5}{5} - \frac{4}{5} \cdot \frac{7}{7} \qquad \text{Using 35 as the common denominator}$$
$$= \frac{15}{35} - \frac{28}{35}$$
$$= -\frac{13}{35}$$

48. $-\frac{2}{15}$

49. ◈

50. ◈

51. ◈

52. ◈

53.

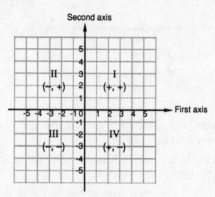

If the first coordinate is negative, then the point must be in either quadrant II or quadrant III.

54. I or II

55. If the first and second coordinates are opposites, then they have different signs. Thus, the point must be in either quadrant II (first coordinate negative, second coordinate positive) or in quadrant IV (first coordinate positive, second coordinate negative.)

56. I or III

57.

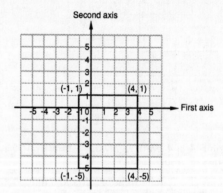

The coordinates of the fourth vertex are $(-1, -5)$.

58. $(5,2)$, $(-7, 2)$, or $(3, -8)$

59. Answers may vary.

We select eight points such that the sum of the coordinates for each point is 7.

$(0, 7)$	$0 + 7 = 7$
$(1, 6)$	$1 + 6 = 7$
$(2, 5)$	$2 + 5 = 7$
$(3, 4)$	$3 + 4 = 7$
$(4, 3)$	$4 + 3 = 7$
$(5, 2)$	$5 + 2 = 7$
$(6, 1)$	$6 + 1 = 7$
$(7, 0)$	$7 + 0 = 7$

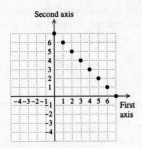

60. Answers may vary.

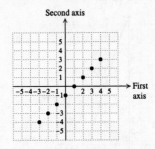

61. Plot the three given points and observe that the co-ordinates of the fourth vertex are $(5, 3)$

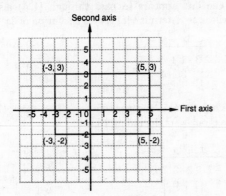

The length of the rectangle is 8 units, and the width is 5 units.

$$P = 2l + 2w$$

$$P = 2 \cdot 8 + 2 \cdot 5 = 16 + 10 = 26 \text{ units}$$

62. $\dfrac{65}{2}$ sq units

63. Latitude 32.5° North,
Longitude 64.5° West

64. Latitude 27° North,
Longitude 81° West

65. ◈

Exercise Set 3.2

1. We substitute 2 for x and 9 for y (alphabetical order of variables).

$$\begin{array}{c|c} y = 4x + 3 \\ \hline 9 \; ? \; 4 \cdot 2 + 3 \\ 8 + 3 \\ 9 \quad | \quad 11 \qquad \text{FALSE} \end{array}$$

Since $9 = 11$ is false, the pair $(2, 9)$ is not a solution.

2. Yes

3. We substitute 4 for x and 2 for y.

$$\begin{array}{c|c} 2x + 3y = 12 \\ \hline 2 \cdot 4 + 3 \cdot 2 \; ? \; 12 \\ 8 + 6 \\ 14 \quad | \quad 12 \quad \text{FALSE} \end{array}$$

Since $14 = 12$ is false, the pair $(4, 2)$ is not a solution.

4. No

5. We substitute 3 for a and -1 for b.

$$\begin{array}{c|c} 3a - 4b = 13 \\ \hline 3 \cdot 3 - 4(-1) \; ? \; 13 \\ 9 + 4 \\ 13 \quad | \quad 13 \quad \text{TRUE} \end{array}$$

Since $13 = 13$ is true, the pair $(3, -1)$ is a solution.

6. Yes

7. To show that a pair is a solution, we substitute, replacing x with the first coordinate and y with the second coordinate in each pair.

$$\begin{array}{c|c} y = x - 5 \\ \hline 2 \; ? \; 7 - 5 \\ 2 \quad | \quad 2 \qquad \text{TRUE} \end{array} \qquad \begin{array}{c|c} y = x - 5 \\ \hline -4 \; ? \; -1 - 5 \\ -4 \quad | \quad -4 \qquad \text{TRUE} \end{array}$$

In each case the substitution results in a true equation. Thus, $(7, 2)$ and $(1, -4)$ are both solutions of $y = x - 5$. We graph these points and sketch the line passing through them.

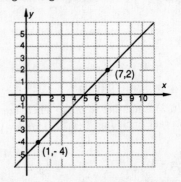

The line appears to pass through $(0, -5)$ also. We check to determine if $(0, -5)$ is a solution of $y = x - 5$.

$$y = x - 5$$

$$\frac{-5 \ ? \ 0 - 5}{-5 \ | \ -5} \quad \text{TRUE}$$

Thus, $(0, -5)$ is another solution. There are other correct answers, including $(-1, -6)$, $(2, -3)$, $(3, -2)$, $(4, -1)$, $(5, 0)$, and $(6, 1)$.

8.
$$y = x + 3 \qquad\qquad y = x + 3$$
$$\frac{2 \ ? \ -1 + 3}{2 \ | \ 2} \ \text{TRUE} \qquad \frac{7 \ ? \ 4 + 3}{7 \ | \ 7} \ \text{TRUE}$$

$(0, 3)$; answers may vary

9. To show that a pair is a solution, we substitute, replacing x with the first coordinate and y with the second coordinate in each pair.

$$y = \frac{1}{2}x + 3 \qquad\qquad y = \frac{1}{2}x + 3$$

$$\frac{5 \ ? \ \frac{1}{2} \cdot 4 + 3}{ \ | \ 2 + 3} \qquad \frac{2 \ ? \ \frac{1}{2}(-2) + 3}{ \ | \ -1 + 3}$$
$$\frac{}{5 \ | \ 5} \ \text{TRUE} \qquad \frac{}{2 \ | \ 2} \ \text{TRUE}$$

In each case the substitution results in a true equation. Thus, $(4, 5)$ and $(-2, 2)$ are both solutions of $y = \frac{1}{2}x + 3$. We graph these points and sketch the line passing through them.

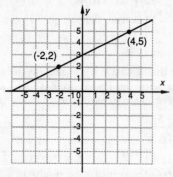

The line appears to pass through $(0, 3)$ also. We check to determine if $(0, 3)$ is a solution of $y = \frac{1}{2}x + 3$.

$$y = \frac{1}{2}x + 3$$

$$\frac{3 \ ? \ \frac{1}{2} \cdot 0 + 3}{3 \ | \ 3} \quad \text{TRUE}$$

Thus, $(0, 3)$ is another solution. There are other correct answers, including $(-6, 0)$, $(-4, 1)$, $(2, 4)$, and $(6, 6)$.

10.
$$y = \frac{1}{2}x - 1 \qquad\qquad y = \frac{1}{2}x - 1$$

$$\frac{2 \ ? \ \frac{1}{2} \cdot 6 - 1}{ \ | \ 3 - 1} \qquad \frac{-1 \ ? \ \frac{1}{2} \cdot 0 - 1}{-1 \ | \ -1} \ \text{TRUE}$$
$$\frac{}{2 \ | \ 2} \ \text{TRUE}$$

$(2, 0)$; answers may vary

11. To show that a pair is a solution, we substitute, replacing x with the first coordinate and y with the second coordinate in each pair.

$$3x + y = 7 \qquad\qquad 3x + y = 7$$
$$\frac{3 \cdot 2 + 1 \ ? \ 7}{6 + 1 \ |} \qquad \frac{3 \cdot 4 - 5 \ ? \ 7}{12 - 5 \ |}$$
$$\frac{}{7 \ | \ 7} \ \text{TRUE} \qquad \frac{}{7 \ | \ 7} \ \text{TRUE}$$

In each case the substitution results in a true equation. Thus, $(2, 1)$ and $(4, -5)$ are both solutions of $3x + y = 7$. We graph these points and sketch the line passing through them.

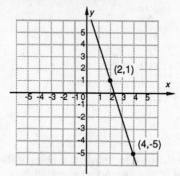

The line appears to pass through $(1, 4)$ also. We check to determine if $(1, 4)$ is a solution of $3x + y = 7$.

$$3x + y = 7$$
$$\frac{3 \cdot 1 + 4 \ ? \ 7}{3 + 4 \ |}$$
$$\frac{}{7 \ | \ 7} \ \text{TRUE}$$

Thus, $(1, 4)$ is another solution. There are other correct answers, including $(3, -2)$.

12.
$$x + 2y = 5 \qquad\qquad x + 2y = 5$$
$$\frac{-1 + 2 \cdot 3 \ ? \ 5}{-1 + 6 \ |} \qquad \frac{7 + 2(-1) \ ? \ 5}{7 - 2 \ |}$$
$$\frac{}{5 \ | \ 5} \ \text{TRUE} \qquad \frac{}{5 \ | \ 5} \ \text{TRUE}$$

$(1, 2)$; answers may vary

13. To show that a pair is a solution, we substitute, replacing x with the first coordinate and y with the second coordinate in each pair.

$$4x - 2y = 10$$
$$\frac{4 \cdot 0 - 2(-5) \ ? \ 10}{10 \ | \ 10} \ \text{TRUE}$$

$$4x - 2y = 10$$
$$\frac{4 \cdot 4 - 2 \cdot 3 \ ? \ 10}{16 - 6 \ |}$$
$$\frac{}{10 \ | \ 10} \ \text{TRUE}$$

In each case the substitution results in a true equation. Thus, $(0, -5)$ and $(4, 3)$ are both solutions of $4x - 2y = 10$. We graph these points and sketch the line passing through them.

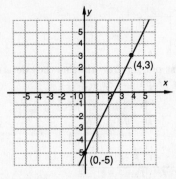

The line appears to pass through $(2, -1)$ also. We check to determine if $(2, -1)$ is a solution of $4x - 2y = 10$.

$$\begin{array}{c|c} 4x - 2y = 10 \\ \hline 4 \cdot 2 - 2(-1) \ ? \ 10 \\ 8 + 2 \\ 10 \ \big| \ 10 \quad \text{TRUE} \end{array}$$

Thus, $(2, -1)$ is another solution. There are other correct answers, including $(1, -3)$, $(2, -1)$, $(3, 1)$, and $(5, 5)$.

14.
$$\begin{array}{c|c} 6x - 3y = 3 \\ \hline 6 \cdot 1 - 3 \cdot 1 \ ? \ 3 \\ 6 - 3 \\ 3 \ \big| \ 3 \quad \text{TRUE} \end{array}$$

$$\begin{array}{c|c} 6x - 3y = 3 \\ \hline 6(-1) - 3(-3) \ ? \ 3 \\ -6 + 9 \\ 3 \ \big| \ 3 \quad \text{TRUE} \end{array}$$

$(0, -1)$; answers may vary

15. $y = x + 1$

The equation is in the form $y = mx + b$. The y-intercept is $(0, 1)$. We find two other pairs.

When $x = 3$, $\quad y = 3 + 1 = 4$.
When $x = -5$, $\quad y = -5 + 1 = -4$.

x	y
0	1
3	4
-5	-4

Plot these points, draw the line they determine, and label the graph $y = x + 1$.

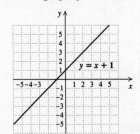

16.

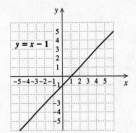

17. $y = x$

The equation is equivalent to $y = x + 0$. The y-intercept is $(0, 0)$. We find two other points.

When $x = -2$, $\quad y = -2$.
When $x = 3$, $\quad\ \ y = 3$.

x	y
0	0
-2	-2
3	3

Plot these points, draw the line they determine, and label the graph $y = x$.

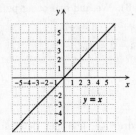

18.

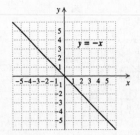

19. $y = \dfrac{1}{2}x$

The equation is equivalent to $y = \dfrac{1}{2}x + 0$. The y-intercept is $(0, 0)$. We find two other points.

When $x = -4$, $y = \dfrac{1}{2}(-4) = -2$.

When $x = 4$, $y = \dfrac{1}{2} \cdot 4 = 2$.

x	y
0	0
-4	-2
4	2

Plot these points, draw the line they determine, and label the graph $y = \dfrac{1}{2}x$.

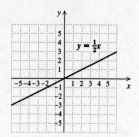

20.

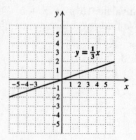

21. $y = x - 3$

The equation is equivalent to $y = x + (-3)$. The y-intercept is $(0, -3)$. We find two other points.

When $x = -2$, $y = -2 - 3 = -5$.

When $x = 4$, $y = 4 - 3 = 1$.

x	y
0	-3
-2	-5
4	1

Plot these points, draw the line they determine, and label the graph $y = x - 3$.

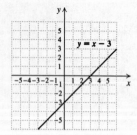

22.

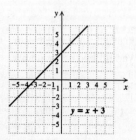

23. $y = 3x - 2 = 3x + (-2)$

The y-intercept is $(0, -2)$. We find two other points.

When $x = -2$, $y = 3(-2) + 2 = -6 + 2 = -4$.

When $x = 1$, $y = 3 \cdot 1 + 2 = 3 + 2 = 5$.

x	y
0	-2
-2	-4
1	5

Plot these points, draw the line they determine, and label the graph $y = 3x + 2$.

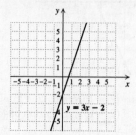

24.

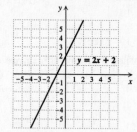

25. $y = \dfrac{1}{2}x + 1$

The y-intercept is $(0, 1)$. We find two other points using multiples of 2 for x to avoid fractions.

When $x = -4$, $y = \dfrac{1}{2}(-4) + 1 = -2 + 1 = -1$.

When $x = 4$, $y = \dfrac{1}{2} \cdot 4 + 1 = 2 + 1 = 3$.

x	y
0	1
-4	-1
4	3

Plot these points, draw the line they determine, and label the graph $y = \dfrac{1}{2}x + 1$.

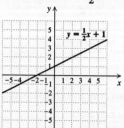

26.

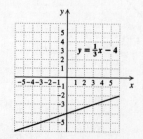

27. $x + y = -5$

$\qquad y = -x - 5$

$\qquad y = -x + (-5)$

The y-intercept is $(0, -5)$. We find two other points.

When $x = -4$, $y = -(-4) - 5 = 4 - 5 = -1$.

When $x = -1$, $y = -(-1) - 5 = 1 - 5 = -4$.

x	y
0	-5
-4	-1
-1	-4

Plot these points, draw the line they determine, and label the graph $x + y = -5$.

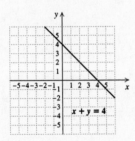

28.

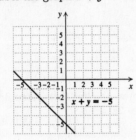

29. $y = \dfrac{5}{3}x - 2 = \dfrac{5}{3}x + (-2)$

The y-intercept is $(0, -2)$. We find two other points using multiples of 3 for x to avoid fractions.

When $x = -3$, $y = \dfrac{5}{3}(-3) - 2 = -5 - 2 = -7$.

When $x = 3$, $y = \dfrac{5}{3} \cdot 3 - 2 = 5 - 2 = 3$.

x	y
0	-2
-3	-7
3	3

Plot these points, draw the line they determine, and label the graph $y = \dfrac{5}{3}x - 2$.

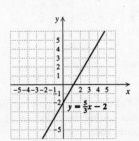

30.

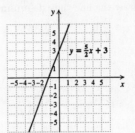

31. $x + 2y = 8$

$\qquad 2y = -x + 8$

$\qquad y = -\dfrac{1}{2}x + 4$

The y-intercept is $(0, 4)$. We find two other points using multiples of 2 for x to avoid fractions.

When $x = -2$, $y = -\dfrac{1}{2}(-2) + 4 = 1 + 4 = 5$.

When $x = 4$, $y = -\dfrac{1}{2} \cdot 4 + 4 = -2 + 4 = 2$.

x	y
0	4
-2	5
4	2

Plot these points, draw the line they determine, and label the graph $x + 2y = 8$.

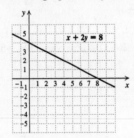

32.

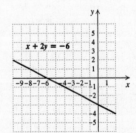

33. $y = \dfrac{3}{2}x + 1$

The y-intercept is $(0, 1)$. We find two other points using multiples of 2 for x to avoid fractions.

When $x = -4$, $y = \dfrac{3}{2}(-4) + 1 = -6 + 1 = -5$.

When $x = 2$, $y = \dfrac{3}{2} \cdot 2 + 1 = 3 + 1 = 4$.

x	y
0	1
-4	-5
2	4

Plot these points, draw the line they determine, and label the graph $y = \dfrac{3}{2}x + 1$.

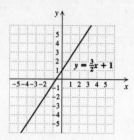

34.

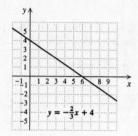

35. $8x - 4y = 12$

$-4y = -8x + 12$

$y = 2x - 3$

$y = 2x + (-3)$

The y-intercept is $(0, -3)$. We find two other points.

When $x = -1$, $y = 2(-1) - 3 = -2 - 3 = -5$.

When $x = 3$, $y = 2 \cdot 3 - 3 = 6 - 3 = 3$.

x	y
0	-3
-1	-5
3	3

Plot these points, draw the line they determine, and label the graph $8x - 4y = 12$.

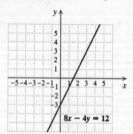

36.

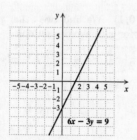

37. $8y + 2x = -4$

$8y = -2x - 4$

$y = -\dfrac{1}{4}x - \dfrac{1}{2}$

$y = -\dfrac{1}{4}x + \left(-\dfrac{1}{2}\right)$

The y-intercept is $\left(0, -\dfrac{1}{2}\right)$. We find two other points.

When $x = -2$, $y = -\dfrac{1}{4}(-2) - \dfrac{1}{2} = \dfrac{1}{2} - \dfrac{1}{2} = 0$.

When $x = 2$, $y = -\dfrac{1}{4} \cdot 2 - \dfrac{1}{2} = -\dfrac{1}{2} - \dfrac{1}{2} = -1$.

x	y
0	$-\dfrac{1}{2}$
-2	0
2	-1

Plot these points, draw the line they determine, and label the graph $8y + 2x = -4$.

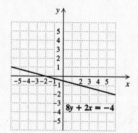

38.

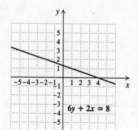

39. We graph $v = -\dfrac{3}{4}t + 6$. Since time cannot be negative in this application, we select only nonnegative values for t.

If $t = 0$, $v = -\dfrac{3}{4} \cdot 0 + 6 = 6$.

If $t = 4$, $v = -\dfrac{3}{4} \cdot 4 + 6 = -3 + 6 = 3$.

If $t = 8$, $v = -\dfrac{3}{4} \cdot 8 + 6 = -6 + 6 = 0$.

t	v
0	6
4	3
8	0

We plot the points and draw the graph. Since the value of the program cannot be negative, the graph stops at the horizontal axis.

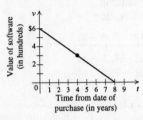

To estimate the program's value after 4 years, we find the second coordinate associated with 4. Actually, we did this when we found ordered pairs to graph. However, if we hadn't, we would locate the point on the line that is above 4 and then find the value on the vertical axis that corresponds to that point. That value is 3, so the program is worth $300 after 4 years.

40.

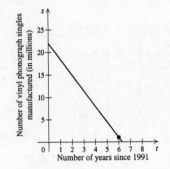

1 million

41. We graph $t + w = 15$, or $w = -t + 15$. Since time cannot be negative in this application, we select only nonnegative values for t.

If $t = 0$, $w = -0 + 15 = 15$.

If $t = 2$, $w = -2 + 15 = 13$.

If $t = 5$, $w = -5 + 15 = 10$.

t	w
0	15
2	13
5	10

We plot the points and draw the graph. Since the likelihood of death cannot be negative, the graph stops at the horizontal axis.

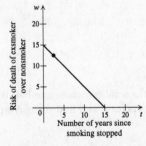

To estimate how much more likely it is for Sandy to die from lung cancer than Polly, we find the second coordinate associated with $2\frac{1}{2}$. Locate the point on

the line that is above $2\frac{1}{2}$ and then find the value on the vertical axis that corresponds to that point. That value is about $12\frac{1}{2}$, so it is $12\frac{1}{2}$ times more likely for Sandy to die from lung cancer than Polly.

42.

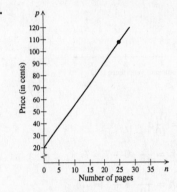

108¢, or $1.08

43. We graph $T = \frac{6}{5}c + 1$. Since the number of credits cannot be negative, we select only nonnegative values for c.

If $c = 5$, $T = \frac{6}{5} \cdot 5 + 1 = 6 + 1 = 7$.

If $c = 10$, $T = \frac{6}{5} \cdot 10 + 1 = 12 + 1 = 13$.

If $c = 15$, $T = \frac{6}{5} \cdot 15 + 1 = 18 + 1 = 19$.

c	T
5	7
10	13
15	19

We plot the points and draw the graph.

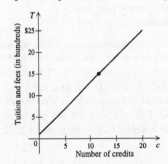

Four three-credit courses total $4 \cdot 3$, or 12, credits. To estimate the cost of tuition and fees for a student who is registered for 12 credits, we find the second coordinate associated with 12. Locate the point on the line that is above 12 and then find the value on the vertical axis that corresponds to that point. That value is about 15, so tuition and fees will cost about $1500.

44.

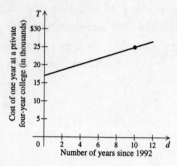

$25,000

45. We graph $n = \frac{1}{10}d + 7$.

When $d = 0$, $n = \frac{1}{10} \cdot 0 + 7 = 7$.

When $d = 5$, $n = \frac{1}{10} \cdot 5 + 7 = 7.5$.

When $d = 10$, $n = \frac{1}{10} \cdot 10 + 7 = 8$.

d	n
0	7
5	7.5
10	8

We plot the points and draw the graph.

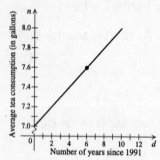

To estimate tea consumption in 1997, we first note that 1997 is 6 years after 1991. Then we find the second coordinate associated with 6. Locate the point on the line that is above 6 and find the value on the vertical axis that corresponds to that point. That value is about 7.6, so about 7.6 gal of tea were consumed by the average U.S. consumer in 1997.

46.

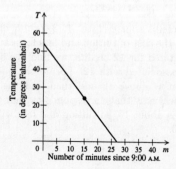

24°F

47. $3x - 7 = -34$
$\quad\ \ \ 3x = -27$ Adding 7
$\quad\ \ \ \ \ x = -9$ Dividing by 3

Check: $\dfrac{3x - 7 = -34}{\begin{array}{c|c} 3(-9) - 7\ ?\ -34 \\ -27 - 7 \\ \quad -34 \end{array}}$

$\quad\ \ \ \ \ -34\ \bigg|\ -34$ TRUE

The solution is -9.

48. $-\dfrac{2}{7}$

49. $2(x - 9) + 4 = 2 - 3x$
$\quad 2x - 18 + 4 = 2 - 3x$
$\quad\ \ 2x - 14 = 2 - 3x$
$\quad\ \ 5x - 14 = 2$ Adding $3x$
$\quad\quad\quad\ 5x = 16$ Adding 14
$\quad\quad\quad\ \ x = \dfrac{16}{5}$

Check: $2(x - 9) + 4 = 2 - 3x$

$2\left(\dfrac{16}{5} - 9\right) + 4\ ?\ 2 - 3 \cdot \dfrac{16}{5}$

$2\left(-\dfrac{29}{5}\right) + 4\ \bigg|\ 2 - \dfrac{48}{5}$

$-\dfrac{58}{5} + 4\ \bigg|\ -\dfrac{38}{5}$

$\phantom{-\dfrac{58}{5} + 4\ }\ -\dfrac{38}{5}\ \bigg|\ -\dfrac{38}{5}$ TRUE

The solution is $\dfrac{16}{5}$.

50. $p = \dfrac{w}{q + 1}$

51. $Ax + By = C$
$\quad\quad\ By = C - Ax$ Subtracting Ax
$\quad\quad\ \ y = \dfrac{C - Ax}{B}$ Dividing by B

52. $Q = 2A - T$

53. ◈

54. ◈

55. ◈

56. ◈

57. $x + y = 7$
$\quad\quad\ y = -x + 7$

Beginning with the smallest whole number, 0, substitute whole numbers for x and find the corresponding y-value.

When $x = 0$, $y = -0 + 7 = 7$.
When $x = 1$, $y = -1 + 7 = 6$.
When $x = 2$, $y = -2 + 7 = 5$.
When $x = 3$, $y = -3 + 7 = 4$.
When $x = 4$, $y = -4 + 7 = 3$.
When $x = 5$, $y = -5 + 7 = 2$.
When $x = 6$, $y = -6 + 7 = 1$.
When $x = 7$, $y = -7 + 7 = 0$.

For x-values greater than 7, the corresponding y-values are not whole numbers. Thus, we have all the whole-number solutions. They are $(0,7)$, $(1,6)$, $(2,5)$, $(3,4)$, $(4,3)$, $(5,2)$, $(6,1)$, and $(7,0)$. We use the points to graph the equation.

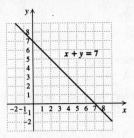

58. $(0,9)$, $(1,8)$, $(2,7)$, $(3,6)$, $(4,5)$, $(5,4)$, $(6,3)$, $(7,2)$, $(8,1)$, $(9,0)$.

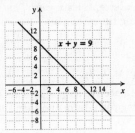

59.

60. $y = -x + 5$

61. Note that the sum of the coordinates of each point on the graph is 2. Thus, we have $x + y = 2$, or $y = -x + 2$.

62. $y = x + 2$

63. Note that each y-coordinate is 3 times the corresponding x-coordinate. Thus, we have $y = 3x$.

64.

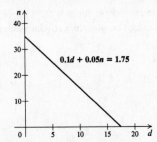

5 dimes, 25 nickels; 10 dimes, 15 nickels; 12 dimes, 11 nickels

65. The equation is $25d + 5l = 225$.

Since the number of dinners cannot be negative, we choose only nonnegative values of d when graphing the equation. The graph stops at the horizontal axis since the number of lunches cannot be negative.

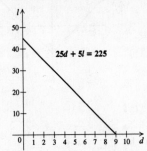

We see that three points on the graph are $(1,40)$, $(5,20)$, and $(8,5)$. Thus, three combinations of dinners and lunches that total $225 are

1 dinner, 40 lunches,

5 dinners, 20 lunches,

8 dinners, 5 lunches.

66.

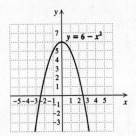

67. $y = \frac{1}{2}x^3$

x	y
-3	$-\frac{27}{2}$
-2	-4
-1	$-\frac{1}{2}$
0	0
1	$\frac{1}{2}$
2	4
3	$\frac{27}{2}$

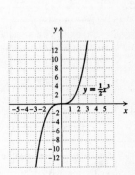

68.

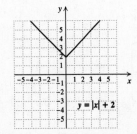

69. $y = |x| - 3$

x	y
-3	0
-2	-1
-1	-2
0	-3
1	-2
2	-1
3	0

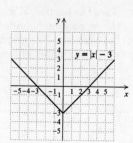

70.

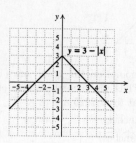

71.　$y = -2.8x + 3.5$

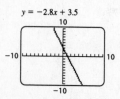

72.　$y = 4.5x + 2.1$

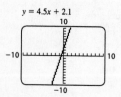

73.　$y = 2.8x - 3.5$

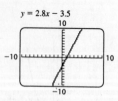

74.　$y = -4.5x - 2.1$

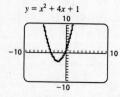

75.　$y = x^2 + 4x + 1$

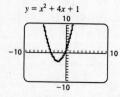

76.　$y = -x^2 + 4x - 7$

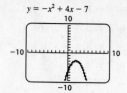

77. ◎

Exercise Set 3.3

1. (a) The graph crosses the y-axis at $(0, 5)$, so the y-intercept is $(0, 5)$.

　(b) The graph crosses the x-axis at $(2, 0)$, so the x-intercept is $(2, 0)$.

2. (a) $(0, 3)$;　(b) $(4, 0)$

3. (a) The graph crosses the y-axis at $(0, -4)$, so the y-intercept is $(0, -4)$.

　(b) The graph crosses the x-axis at $(3, 0)$, so the x-intercept is $(3, 0)$.

4. (a) $(0, 5)$;　(b) $(-3, 0)$

5. $3x + 5y = 15$

　(a) To find the y-intercept, let $x = 0$. This is the same as ignoring the x-term and then solving.

$$5y = 15$$
$$y = 3$$

　The y-intercept is $(0, 3)$.

　(b) To find the x-intercept, let $y = 0$. This is the same as ignoring the y-term and then solving.

$$3x = 15$$
$$x = 5$$

　The x-intercept is $(5, 0)$.

6. (a) $(0,10)$; (b) $(4, 0)$

7. $7x - 2y = 28$

　(a) To find the y-intercept, let $x = 0$. This is the same as ignoring the x-term and then solving.

$$-2y = 28$$
$$y = -14$$

　The y-intercept is $(0, -14)$.

　(b) To find the x-intercept, let $y = 0$. This is the same as ignoring the y-term and then solving.

$$7x = 28$$
$$x = 4$$

　The x-intercept is $(4, 0)$.

8. (a) $(0, -6)$; (b) $(8, 0)$

9. $-4x + 3y = 10$

   (a) To find the y-intercept, let $x = 0$. This is the same as ignoring the x-term and then solving.

$$3y = 10$$
$$y = \frac{10}{3}$$

   The y-intercept is $\left(0, \frac{10}{3}\right)$.

   (b) To find the x-intercept, let $y = 0$. This is the same as ignoring the y-term and then solving.

$$-4x = 10$$
$$x = -\frac{5}{2}$$

   The x-intercept is $\left(-\frac{5}{2}, 0\right)$.

10. (a) $\left(0, \frac{7}{3}\right)$; (b) $\left(-\frac{7}{2}, 0\right)$

11. $6x - 3 = 9y$

   $6x - 9y = 3$ Writing the equation in the form $Ax + By = C$

   (a) To find the y-intercept, let $x = 0$. This is the same as ignoring the x-term and then solving.

$$-9y = 3$$
$$y = -\frac{1}{3}$$

   The y-intercept is $\left(0, -\frac{1}{3}\right)$.

   (b) To find the x-intercept, let $y = 0$. This is the same as ignoring the y-term and then solving.

$$6x = 3$$
$$x = \frac{1}{2}$$

   The x-intercept is $\left(\frac{1}{2}, 0\right)$.

12. (a) $\left(0, \frac{1}{2}\right)$; (b) $\left(-\frac{1}{3}, 0\right)$

13. $2x + y = 6$

   Find the y-intercept:

$$y = 6 \qquad \text{Ignoring the } x\text{-term}$$

   The y-intercept is $(0, 6)$.

   Find the x-intercept:

$$2x = 6 \quad \text{Ignoring the } y\text{-term}$$
$$x = 3$$

   The x-intercept is $(3, 0)$.

   To find a third point we replace x with 2 and solve for y.

$$2 \cdot 2 + y = 6$$
$$4 + y = 6$$
$$y = 2$$

The point $(2, 2)$ appears to line up with the intercepts, so we draw the graph.

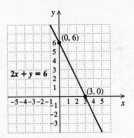

14.

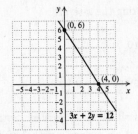

15. $6x + 9y = 18$

   Find the y-intercept:

$$9y = 18 \quad \text{Ignoring the } x\text{-term}$$
$$y = 2$$

   The y-intercept is $(0, 2)$.

   Find the x-intercept:

$$6x = 18 \quad \text{Ignoring the } y\text{-term}$$
$$x = 3$$

   The x-intercept is $(3, 0)$.

   To find a third point we replace x with -3 and solve for y.

$$6(-3) + 9y = 18$$
$$-18 + 9y = 18$$
$$9y = 36$$
$$y = 4$$

The point $(-3, 4)$ appears to line up with the intercepts, so we draw the graph.

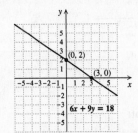

16.

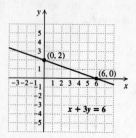

17. $-x + 3y = 9$

Find the y-intercept:

$$3y = 9 \quad \text{Ignoring the } x\text{-term}$$
$$y = 3$$

The y-intercept is $(0, 3)$.

Find the x-intercept:

$$-x = 9 \quad \text{Ignoring the } y\text{-term}$$
$$x = -9$$

The x-intercept is $(-9, 0)$.

To find a third point we replace x with 3 and solve for y.

$$-3 + 3y = 9$$
$$3y = 12$$
$$y = 4$$

The point $(3, 4)$ appears to line up with the intercepts, so we draw the graph.

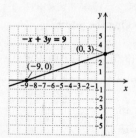

18.

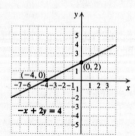

19. $2x - y = 8$

Find the y-intercept:

$$-y = 8 \quad \text{Ignoring the } x\text{-term}$$
$$y = -8$$

The y-intercept is $(0, -8)$.

Find the x-intercept:

$$2x = 8 \quad \text{Ignoring the } y\text{-term}$$
$$x = 4$$

The x-intercept is $(4, 0)$.

To find a third point we replace x with 2 and solve for y.

$$2 \cdot 2 - y = 8$$
$$4 - y = 8$$
$$-y = 4$$
$$y = -4$$

The point $(2, -4)$ appears to line up with the intercepts, so we draw the graph.

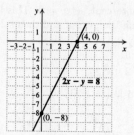

20.

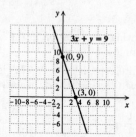

21. $6 - 3y = 9x$

We can leave the equation in the given form or rewrite it in the form $Ax + By = C$. We will use the given form.

To find the y-intercept, let $x = 0$.

$$6 - 3y = 9 \cdot 0$$
$$6 - 3y = 0$$
$$-3y = -6$$
$$y = 2$$

The y-intercept is $(0, 2)$.

Find the x-intercept:

$$6 = 9x \quad \text{Ignoring the } y\text{-term}$$
$$\frac{2}{3} = x$$

The x-intercept is $\left(\frac{2}{3}, 0\right)$.

To find a third point we replace x with 2 and solve for y.

$$6 - 3y = 9 \cdot 2$$
$$6 - 3y = 18$$
$$-3y = 12$$
$$y = -4$$

The point $(2, -4)$ appears to line up with the intercepts, so we draw the graph.

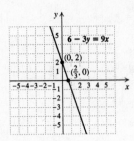

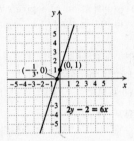

22.

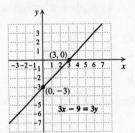

24.

23. $5x - 10 = 5y$

We can leave the equation in the given form or rewrite it in the form $Ax + By = C$. We will use the given form.

Find the y-intercept:

$$-10 = 5y \quad \text{Ignoring the } x\text{-term}$$
$$-2 = y$$

The y-intercept is $(0, -2)$.

To find the x-intercept, let $y = 0$.

$$5x - 10 = 5 \cdot 0$$
$$5x - 10 = 0$$
$$5x = 10$$
$$x = 2$$

The x-intercept is $(2, 0)$.

To find a third point we replace x with 5 and solve for y.

$$5 \cdot 5 - 10 = 5y$$
$$25 - 10 = 5y$$
$$15 = 5y$$
$$3 = y$$

The point $(5, 3)$ appears to line up with the intercepts, so we draw the graph.

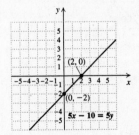

25. $2x - 5y = 10$

Find the y-intercept:

$$-5y = 10 \quad \text{Ignoring the } x\text{-term}$$
$$y = -2$$

The y-intercept is $(0, -2)$.

Find the x-intercept:

$$2x = 10 \quad \text{Ignoring the } y\text{-term}$$
$$x = 5$$

The x-intercept is $(5, 0)$.

To find a third point we replace x with -5 and solve for y.

$$2(-5) - 5y = 10$$
$$-10 - 5y = 10$$
$$-5y = 20$$
$$y = -4$$

The point $(-5, -4)$ appears to line up with the intercepts, so we draw the graph.

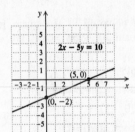

26.

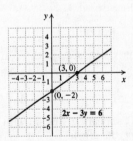

27. $2x + 6y = 12$

Find the y-intercept:

$$6y = 12 \quad \text{Ignoring the } x\text{-term}$$
$$y = 2$$

The y-intercept is $(0, 2)$.

Find the x-intercept:

$$2x = 12 \quad \text{Ignoring the } y\text{-term}$$
$$x = 6$$

The x-intercept is $(6, 0)$.

To find a third point we replace x with 3 and solve for y.

$$2 \cdot 3 + 6y = 12$$
$$6 + 6y = 12$$
$$6y = 6$$
$$y = 1$$

The point $(3, 1)$ appears to line up with the intercepts, so we draw the graph.

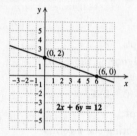

28.

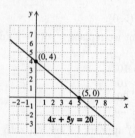

29. $x - 1 = y$

We can leave the equation in the given form or rewrite it in the form $Ax + By = C$. We will use the given form.

Find the y-intercept:

$$-1 = y \qquad \text{Ignoring the } x\text{-term}$$

The y-intercept is $(0, -1)$.

To find the x-intercept, let $y = 0$.

$$x - 1 = 0$$
$$x = 1$$

The x-intercept is $(1, 0)$.

To find a third point we replace x with -3 and solve for y.

$$-3 - 1 = y$$
$$-4 = y$$

The point $(-3, -4)$ appears to line up with the intercepts, so we draw the graph.

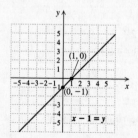

30.

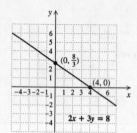

31. $2x - 1 = y$

We can leave the equation in the given form or rewrite it in the form $Ax + By = C$. We will use the given form.

Find the y-intercept:

$$-1 = y \qquad \text{Ignoring the } x\text{-term}$$

The y-intercept is $(0, -1)$.

To find the x-intercept, let $y = 0$.

$$2x - 1 = 0$$
$$2x = 1$$
$$x = \frac{1}{2}$$

The x-intercept is $\left(\frac{1}{2}, 0 \right)$.

To find a third point we replace x with 3 and solve for y.

$$2 \cdot 3 - 1 = y$$
$$6 - 1 = y$$
$$5 = y$$

The point $(3, 5)$ appears to line up with the intercepts, so we draw the graph.

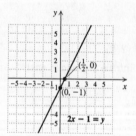

32.

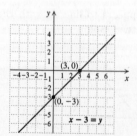

33. $4x - 3y = 12$

Find the y-intercept:

$$-3y = 12 \qquad \text{Ignoring the } x\text{-term}$$
$$y = -4$$

The y-intercept is $(0, -4)$.

Find the x-intercept:

$$4x = 12 \quad \text{Ignoring the } y\text{-term}$$
$$x = 3$$

The x-intercept is $(3, 0)$.

To find a third point we replace x with 6 and solve for y.

$$4 \cdot 6 - 3y = 12$$
$$24 - 3y = 12$$
$$-3y = -12$$
$$y = 4$$

The point $(6, 4)$ appears to line up with the intercepts, so we draw the graph.

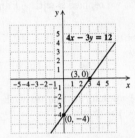

34.

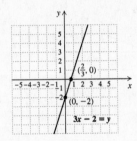

35. $7x + 2y = 6$

Find the y-intercept:

$$2y = 6 \quad \text{Ignoring the } x\text{-term}$$
$$y = 3$$

The y-intercept is $(0, 3)$.

Find the x-intercept:

$$7x = 6 \quad \text{Ignoring the } y\text{-term}$$
$$x = \frac{6}{7}$$

The x-intercept is $\left(\frac{6}{7}, 0\right)$.

To find a third point we replace x with 2 and solve for y.

$$7 \cdot 2 + 2y = 6$$
$$14 + 2y = 6$$
$$2y = -8$$
$$y = -4$$

The point $(2, -4)$ appears to line up with the intercepts, so we draw the graph.

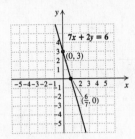

36.

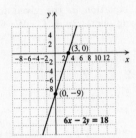

37. $y = -4 - 4x$

We can leave the equation in the given form or rewrite it in the form $Ax + By = C$. We will use the given form.

Find the y-intercept:

$$y = -4 \qquad \text{Ignoring the } x\text{-term}$$

The y-intercept is $(0, -4)$.

To find the x-intercept, let $y = 0$.

$$0 = -4 - 4x$$
$$4x = -4$$
$$x = -1$$

The x-intercept is $(-1, 0)$.

To find a third point we replace x with -2 and solve for y.

$$y = -4 - 4(-2)$$
$$y = -4 + 8$$
$$y = 4$$

The point $(-2, 4)$ appears to line up with the intercepts, so we draw the graph.

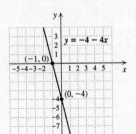

38.

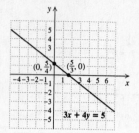

39. $-3x = 6y - 2$

We can leave the equation in the given form or rewrite it in the form $Ax + By = C$. We will use the given form.

To find the y-intercept, let $x = 0$.

$$-3 \cdot 0 = 6y - 2$$
$$0 = 6y - 2$$
$$2 = 6y$$
$$\frac{1}{3} = y$$

The y-intercept is $\left(0, \frac{1}{3}\right)$.

Find the x-intercept:

$$-3x = -2 \quad \text{Ignoring the } y\text{-term}$$
$$x = \frac{2}{3}$$

The x-intercept is $\left(\frac{2}{3}, 0\right)$.

To find a third point we replace x with -4 and solve for y.

$$-3(-4) = 6y - 2$$
$$12 = 6y - 2$$
$$14 = 6y$$
$$\frac{7}{3} = y$$

The point $\left(-4, \frac{7}{3}\right)$ appears to line up with the intercepts, so we draw the graph.

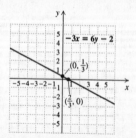

40.

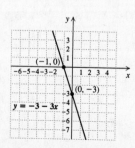

41. $3 = 2x - 5y$

Find the y-intercept:

$$3 = -5y \quad \text{Ignoring the } x\text{-term}$$
$$-\frac{3}{5} = y$$

The y-intercept is $\left(0, -\frac{3}{5}\right)$.

Find the x-intercept:

$$3 = 2x \quad \text{Ignoring the } y\text{-term}$$
$$\frac{3}{2} = x$$

The x-intercept is $\left(\frac{3}{2}, 0\right)$.

To find a third point we replace x with -1 and solve for y.

$$3 = 2(-1) - 5y$$
$$3 = -2 - 5y$$
$$5 = -5y$$
$$-1 = y$$

The point $(-1, -1)$ appears to line up with the intercepts, so we draw the graph.

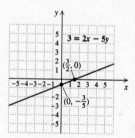

42.

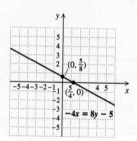

43. $x + 2y = 0$

Find the y-intercept:

$$2y = 0 \quad \text{Ignoring the } x\text{-term}$$
$$y = 0$$

The y-intercept is $(0, 0)$. Note that this is also the x-intercept.

In order to graph the line, we will find a second point.

When $x = 4$, $4 + 2y = 0$
$$2y = -4$$
$$y = -2.$$

Thus, a second point is $(4, -2)$.

To find a third point we replace x with -2 and solve for y.

$$-2 + 2y = 0$$
$$2y = 2$$
$$y = 1$$

The point $(-2, 1)$ appears to line up with the other two points, so we draw the graph.

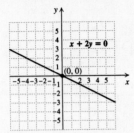

44.

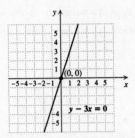

45. Note that every point on the horizontal line passing through $(0, -1)$ has -1 as the y-coordinate. Thus, the equation of the line is $y = -1$.

46. $x = -1$

47. Note that every point on the vertical line passing through $(4, 0)$ has 4 as the x-coordinate. Thus, the equation of the line is $x = 4$.

48. $y = -5$

49. Note that every point on the horizontal line passing through $(0, 0)$ has 0 as the y-coordinate. Thus, the equation of the line is $y = 0$.

50. $x = 0$

51. $x = -3$

Any ordered pair $(-3, y)$ is a solution. The variable x must be -3, but the y variable can be any number we choose. A few solutions are listed below. Plot these points and draw the line.

x	y
-3	-2
-3	0
-3	4

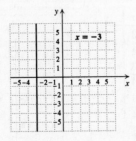

52.

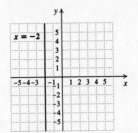

53. $y = 2$

Any ordered pair $(x, 2)$ is a solution. The variable y must be 2, but the x variable can be any number we choose. A few solutions are listed below. Plot these points and draw the line.

x	y
-3	2
0	2
2	2

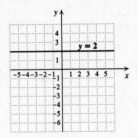

54.

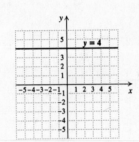

55. $x = 7$

Any ordered pair $(7, y)$ is a solution. The variable x must be 7, but the y variable can be any number we choose. A few solutions are listed below. Plot these points and draw the line.

x	y
7	-1
7	4
7	5

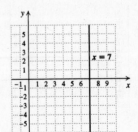

56.

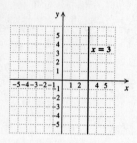

57. $x = 0$

Any ordered pair $(0, y)$ is a solution. The variable x must be 0, but the y variable can be any number we choose. A few solutions are listed below. Plot these points and draw the line.

x	y
0	-5
0	-1
0	3

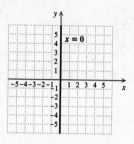

58.

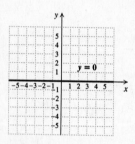

59. $y = \dfrac{5}{2}$

Any ordered pair $\left(x, \dfrac{5}{2}\right)$ is a solution. A few solutions are listed below. Plot these points and draw the line.

x	y
-3	$\dfrac{5}{2}$
0	$\dfrac{5}{2}$
4	$\dfrac{5}{2}$

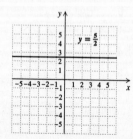

60.

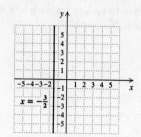

61. $-3y = -15$

$\qquad y = 5 \qquad$ Solving for y

Any ordered pair $(x, 5)$ is a solution. A few solutions are listed below. Plot these points and draw the line.

x	y
-3	5
0	5
2	5

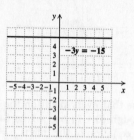

62.

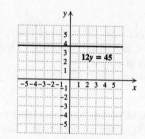

63. $4x + 3 = 0$

$\qquad 4x = -3$

$\qquad x = -\dfrac{3}{4} \qquad$ Solving for x

Any ordered pair $\left(-\dfrac{3}{4}, y\right)$ is a solution. A few solutions are listed below. Plot these points and draw the line.

x	y
$-\dfrac{3}{4}$	-2
$-\dfrac{3}{4}$	0
$-\dfrac{3}{4}$	3

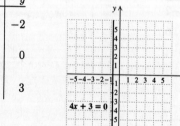

64.

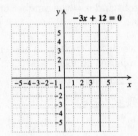

65. $18 - 3y = 0$

$$-3y = -18$$
$$y = 6 \qquad \text{Solving for } y$$

Any ordered pair $(x, 6)$ is a solution. A few solutions are listed below. Plot these points and draw the line.

x	y
-4	6
0	6
2	6

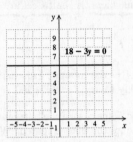

66.

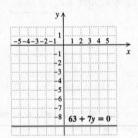

67. We begin by factoring 98 in any way we can:

$$98 = 2 \cdot 49$$

The factor 49 is not prime, so we factor it again:

$$98 = 2 \cdot 49 = 2 \cdot 7 \cdot 7$$

Both 2 and 7 are prime, so the prime factorization of 98 is $2 \cdot 7 \cdot 7$.

68. $2 \cdot 2 \cdot 2 \cdot 2 \cdot 3 \cdot 5$

69. We begin by factoring 275 in any way we can:

$$275 = 25 \cdot 11$$

The factor 25 is not prime, so we factor it again:

$$275 = 25 \cdot 11 = 5 \cdot 5 \cdot 11$$

Both 5 and 11 are prime, so the prime factorization of 275 is $5 \cdot 5 \cdot 11$.

70. $\dfrac{2}{5}$

71. $\dfrac{12}{84} = \dfrac{12 \cdot 1}{12 \cdot 7} = \dfrac{\cancel{12} \cdot 1}{\cancel{12} \cdot 7} = \dfrac{1}{7}$

72. $\dfrac{5}{3}$

73. ◈

74. ◈

75. ◈

76. ◈

77. The x-axis is a horizontal line, so it is of the form $y = b$. All points on the x-axis are of the form $(x, 0)$, so b must be 0 and the equation is $y = 0$.

78. $x = 0$

79. Since the y-coordinate of the point of intersection must be 6 and x must equal y, the point of intersection is $(6, 6)$.

80. $(-3, -3)$

81. Consider an equation of the form $y = mx + b$. The y-intercept is $(0, 5)$, so we substitute 5 for b:

$$y = mx + 5$$

Another point on the line is $(2, 0)$, so we substitute 2 for x and 0 for y and solve for m:

$$0 = m \cdot 2 + 5$$
$$-5 = 2m$$
$$-\frac{5}{2} = m$$

The equation is $y = -\dfrac{5}{2}x + 5$, or $5x + 2y = 10$.

82. $y = \dfrac{5}{3}x + 5$, or $5x - 3y = -15$

83. A line parallel to the x-axis has an equation of the form $y = b$. Since the y-coordinate of one point on the line is -4, then $b = -4$ and the equation is $y = -4$.

84. -3

85. Substitute -4 for x and 0 for y.

$$3(-4) + C = 5 \cdot 0$$
$$-12 + C = 0$$
$$C = 12$$

86. -24

87. ◈

88. $(0, 25)$; $\left(\dfrac{50}{3}, 0\right)$, or $(16.\overline{6}, 0)$

89. Find the y-intercept:

$$-7y = 80 \qquad \text{Covering the } x\text{-term}$$
$$y = -\frac{80}{7} = -11.\overline{428571}$$

The y-intercept is $\left(0, -\dfrac{80}{7}\right)$, or $(0, -11.\overline{428571})$.

Find the x-intercept:

$2x = 80$ Covering the y-term

$x = 40$

The x-intercept is $(40, 0)$.

90. $(0, -9); (45, 0)$

91. From the equation we see that the y-intercept is $(0, -15)$.

To find the x-intercept, let $y = 0$.

$0 = 1.3x - 15$

$15 = 1.3x$

$\frac{15}{1.3} = x$

$\frac{150}{13} = x$, or

$11.\overline{538461} = x$

The x-intercept is $\left(\frac{150}{13}, 0\right)$, or $(11.\overline{538461}, 0)$.

92. $\left(0, -\frac{1}{20}\right)$, or $(0, -0.05); \left(\frac{1}{25}, 0\right)$, or $(0.04, 0)$

93. Find the y-intercept.

$25y = 1$ Covering the x-term

$y = \frac{1}{25}$, or 0.04

The y-intercept is $\left(0, \frac{1}{25}\right)$, or $(0, 0.04)$.

Find the x-intercept:

$50x = 1$ Covering the y-term

$x = \frac{1}{50}$, or 0.02

The x-intercept is $\left(\frac{1}{50}, 0\right)$, or $(0.02, 0)$.

Exercise Set 3.4

1. Let x and y represent the two numbers; then $x + y = 19$.

2. Let x and y represent the two numbers; $x + y = 39$.

3. Let x and y represent the two numbers. We translate.

A number plus twice another number is 65.

$x + 2y = 65$

4. Let x and y represent the two numbers; $x + 2y = 93$.

5. Let a = the amount Justine paid and n = the amount the necklaces were worth. We reword and translate.

The amount Justine paid is twice the amount the necklaces were worth.

$a = 2n$

6. Let p = the total profit and n = the number of flower pots sold; $p = 5.25n$.

7. Let d = the distance Jason travels, in meters, and t = the number of minutes he jogs.

The distance Jason travels is 220 times the number of minutes he jogs.

$d = 220t$

8. Let m = the number of careless mistakes Eva makes and b = the number her brother makes; $m = \frac{1}{2}b$.

9. Let f = Frank's age and c = Cecilia's age. We reword and translate.

Frank's age is half Cecilia's age less 3.

$f = \frac{1}{2}c - 3$

10. Let l = the amount Lois earns and r = Roberta's weekly salary; $l = 3r - 200$.

11. Let c = the cost of the bike rental and t = the number of hours the bike is rented. We translate.

The cost of the bike rental is \$8 plus \$2 per hour.

$c = 8 + 2t$

12. Let c = the total cost of the photocopies, in dollars, and p = the number of pages copied; $c = 3 + 0.05p$.

13. Let d = the depth of the lake, in feet, and w = the number of weeks. We reword and translate.

The depth of the lake is 94 ft minus 3 times the number of weeks.

$d = 94 - 3w$

14. Let s = the politician's savings, in dollars, and w = the number of weeks; $s = 75,000 - 8000w$.

15. *Familiarize.* We let the horizontal axis represent mileage and the vertical axis represent cost with m = the number of miles driven and c = the cost, in dollars.

Translate. Since the cost of a rental is \$59.95 plus 45¢ for each mile, and since m miles are to be driven, we have

$c = 59.95 + 0.45m$

Carry out. We make a table of values using some convenient choices for m, and then we draw the graph.

When $m = 50$, $c = 59.95 + 0.45(50) = 82.45$.

When $m = 150$, $c = 59.95 + 0.45(150) = 127.45$.

When $m = 300$, $c = 59.95 + 0.45(300) = 194.95$.

Mileage	Cost
50	$82.45
150	$127.45
300	$194.95

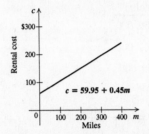

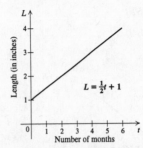

To estimate how far the van can be driven on a budget of $150, locate $150 on the vertical axis, move horizontally to the graphed line, and then move down to the horizontal axis. It appears that about 200 miles can be driven on a budget of $150.

Check. If 200 miles are driven, the cost will be

$$c = 59.95 + 0.45(200)$$
$$= 59.95 + 90$$
$$= 149.95$$

Since this is close to $150, our estimate is fairly accurate.

State. On a budget of $150, the van can be driven about 200 miles.

16.

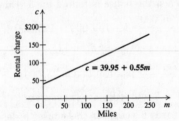

About 150 miles

17. Familiarize and Translate. We use the given formula, $L = \frac{1}{2}t + 1$.

Carry out. We make a table of values using some convenient choices for t, and then we draw the graph.

When $t = 0$, $L = \frac{1}{2} \cdot 0 + 1 = 1$.

When $t = 4$, $L = \frac{1}{2} \cdot 4 + 1 = 2 + 1 = 3$.

When $t = 6$, $L = \frac{1}{2} \cdot 6 + 1 = 3 + 1 = 4$.

Months	Length
0	1
4	3
6	4

To estimate how long it will take for Tina's hair to be $2\frac{3}{4}$ inches long, locate $2\frac{3}{4}$ on the vertical axis, move horizontally to the graphed line, and then move down to the horizontal axis. It appears that it will take about $3\frac{1}{2}$ months.

Check. After $3\frac{1}{2}$ months, the length of Tina's hair will be

$$L = \frac{1}{2}\left(3\frac{1}{2}\right) + 1$$
$$= \frac{1}{2} \cdot \frac{7}{2} + 1$$
$$= \frac{7}{4} + 1 = \frac{11}{4} = 2\frac{3}{4}$$

Our estimate was accurate.

State. Tina's hair will be $2\frac{3}{4}$ inches long $3\frac{1}{2}$ months after the haircut.

18.

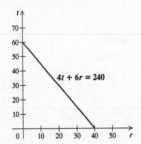

About 25 pounds

19. Familiarize. Let s = weekly sales, in dollars, and p = weekly wages, in dollars.

Translate. The wages are $200 plus 8% of the sales, s, so we have

$$p = 200 + 0.08s$$

Carry out. We make a table of values, using some convenient choices for s, and then we draw the graph.

When $s = 2000$, $p = 200 + 0.08(2000) = 360$.

When $s = 4000$, $p = 200 + 0.08(4000) = 520$.

When $s = 8000$, $p = 200 + 0.08(8000) = 840$.

Sales	Wages
$2000	$360
$4000	$520
$8000	$840

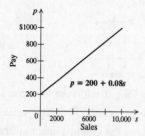

To estimate the sales when the wages are $790, locate $790 on the vertical axis, move horizontally to the graphed line, and then move downward to the horizontal axis. It appears that sales are about $7400 when wages are $790.

Check. If sales are $7400, the wages will be

$$p = 200 + 0.08(7400)$$
$$= 200 + 592$$
$$= 792$$

Since this is close to $790, our estimate of $7400 is fairly accurate.

State. When a week's pay is $790, the saleperson's sales are about $7400.

20.

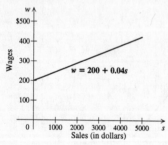

About $4400

21. **Familiarize**. Let V = the depreciated value of the house, in dollars, and t = the number of years the house is rented.

Translate. The depreciated value is the original value, $150,000, less $\frac{1}{18}$ of $150,000 for each year the house is rented. We have

$$V = 150,000 - \frac{1}{18} \cdot 150,000 \cdot t, \text{ or}$$

$$V = -\frac{25,000}{3}t + 150,000.$$

Carry out. We make a table of values, using some convenient choices for t, and then we draw the graph.

When $t = 0$, $V = -\dfrac{25,000}{3} \cdot 0 + 150,000 = 150,000$.

When $t = 6$, $V = -\dfrac{25,000}{3} \cdot 6 + 150,000 = 100,000$.

When $t = 12$, $V = -\dfrac{25,000}{3} \cdot 12 + 150,000 = 50,000$.

Years	Value
0	$150,000
6	$100,000
12	$50,000

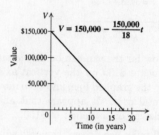

To estimate how long it takes the house to depreciate in value from $125,000 to $75,000 we first find how long the house has been rented when its value is $125,000. From the graph we see that this is about 3 years. Then we use the graph to find how long the house has been rented when its value is $75,000. We see that this is about 9 years. Finally, we find the difference in these times:

$$9 - 3 = 6 \text{ years}$$

Check. When the house is rented for 3 years,

$$V = -\frac{25,000}{3} \cdot 3 + 150,000 = 125,000.$$

When the house is rented for 9 years,

$$v = -\frac{25,000}{3} \cdot 9 + 150,000 = 75,000.$$

Since $9 - 3 = 6$, our estimate is accurate.

State. It takes 6 years for the house to depreciate in value from $125,000 to $75,000.

22.

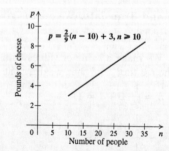

About 33

23. **Familiarize**. Let t = the number of 15-min units of time and c = the cost of parking, in dollars.

Translate. The parking fee is $3 plus 50¢ for each 15-min unit of time, so we have

$$c = 3 + 0.5t.$$

Carry out. We make a table of values, using some convenient choices for t, and then we draw the graph.

When $t = 2$, $c = 3 + 0.5(2) = 4$.

When $t = 6$, $c = 3 + 0.5(6) = 6$.

When $t = 10$, $c = 3 + 0.5(10) = 8$.

Units of time	Fee
2	$4
6	$6
10	$8

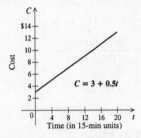

To estimate how long someone was parked when the fee was $7.50, locate $7.50 on the vertical axis, move horizontally to the graphed line, and then move down to the horizontal axis. It appears that a car can be parked for 9 15-min units of time for $7.50. This is $9 \cdot 15$, or 135 min, or 2 hr 15 min.

Check. If a vehicle is parked for 9 15-min units of time, the fee will be

$$c = 3 + 0.5(9)$$
$$= 3 + 4.5$$
$$= 7.5$$

Since this is $7.50, our estimate was accurate.

State. A vehicle was parked for 135 min, or 2 hr, 15 min, when charged $7.50.

24.

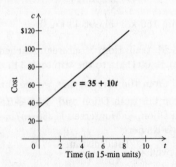

105 min, or 1 hr, 45 min

25. *Familiarize*. Let $p =$ the number of pages and $c =$ the cost of a copy of the report, in dollars.

Translate. The cost of a copy of the report is $2.25 plus 5¢ per page so we have

$$c = 2.25 + 0.05p.$$

Carry out. We make a table of values, using convenient choices for p, and then draw the graph.

When $p = 25$, $c = 2.25 + 0.05(25) = 3.50$.

When $p = 50$, $c = 2.25 + 0.05(50) = 4.75$.

When $p = 100$, $c = 2.25 + 0.05(100) = 7.25$.

Number of pages	Cost
25	$3.50
50	$4.75
100	$7.25

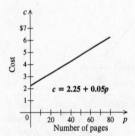

To estimate the length of a report that cost $5.20 per copy, locate $5.20 on the vertical axis, move horizontally to the graphed line, and then move down to the horizontal axis. It appears that a report of about 60 pages costs $5.20 per copy.

Check. If the report has 60 pages, the cost will be

$$c = 2.25 + 0.05(60)$$
$$= 2.25 + 3$$
$$= 5.25$$

Since this is close to $5.20, our estimate of 60 pages is fairly accurate.

State. A report that costs $5.20 per copy is about 60 pages long.

26.

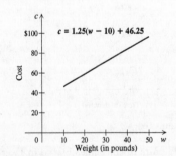

4 pounds

27. *Familiarize*. Let $t =$ the time of the descent, in minutes, and $a =$ the altitude, in feet.

Translate. The altitude is 32,200 ft less 3000 ft for each minute of the descent. We have

$$a = 32,200 - 3000t.$$

Carry out. We make a table of values using some convenient choices for t and then we draw the graph.

When $t = 2$, $a = 32,200 - 3000 \cdot 2 = 26,200$.

When $t = 6$, $a = 32,200 - 3000 \cdot 6 = 14,200$.

When $t = 10$, $a = 32,200 - 3000 \cdot 10 = 2200$.

Time of Descent	Altitude
2	26, 200
6	14, 200
10	2200

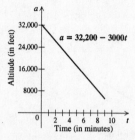

The descent lasts until the plane lands in Denver where the altitude is 5200 ft. Thus, to estimate how long the descent will last, we locate 5200 on the vertical axis, move horizontally to the graphed line, and then move down to the horizontal axis. It appears that the plane is at 5200 ft after about 9 minutes.

Check. If the descent lasts 9 minutes, the altitude is

$$a = 32,200 - 3000 \cdot 9$$
$$= 32,200 - 27,000$$
$$= 5200$$

We see that the estimate of 9 minutes is accurate.

State. The descent will last 9 minutes.

28.

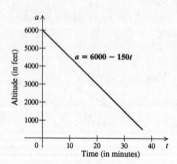

About 37 minutes

29. The ordered pairs in the table created in Exercise 17 can be used to determine the rate at which Tina's hair is growing.

Months	Length
0	1
4	3
6	4

Note that as the time changes from 0 months to 4 months, the length of Tina's hair changes from 1 inch to 3 inches. That is, a change in length of $3 - 1$, or 2 inches corresponds to a time change of $4 - 0$, or 4 months. The length is changing at the rate of

$$\frac{2 \text{ inches}}{4 \text{ months}}, \text{ or } \frac{1}{2} \text{ inch per month.}$$

30. $0.55 per mile

31. From the train schedule we see that the train that left Tonnerre at 1:10 P.M. reached Laroche at 2:15 P.M. Thus, the ride took 1 hr, 5 min.

We look at the vertical axis to determine the distance between Laroche and Tonnerre. Laroche is 159 km from Dijon and Tonnerre is 118 km from Dijon, so Laroche is $159 - 118 = 41$ km from Tonnerre.

When finding the average speed, we express 1 hr, 5 min as $1 \text{ hr} + \frac{5}{60} \text{ hr} = 1\frac{1}{12} \text{ hr} = \frac{13}{12} \text{ hr}$.

$$\text{Average speed} = \frac{\text{Total distance}}{\text{Total time}}$$
$$= \frac{41 \text{ km}}{13/12 \text{ hr}}$$
$$= 41 \text{ km} \cdot \frac{12}{13} \text{ hr}$$
$$\approx 38 \text{ km/h}$$

32. 2 hr, 15 min; 80 km; about 36 km/h

33. From the train schedule we see that the train that left Dijon at 11:35 A.M. reached Montereau at 4:20 P.M. Thus, the ride took 4 hr, 45 min.

Looking at the vertical axis, we see that Montereau is 235 km from Dijon.

When finding the average speed, we express 4 hr, 45 min as 4.75 hr.

$$\text{Average speed} = \frac{\text{Total distance}}{\text{Total time}}$$
$$= \frac{235 \text{ km}}{4.75 \text{ hr}}$$
$$\approx 49 \text{ km/h}$$

34. 3 hr, 40 min; 156 km; about 43 km/h

35. The 8:25 A.M. train from Montereau reached Nuits-s-Ravière at 12:00 P.M., so the trip took 3 hr, 35 min, or $3\frac{7}{12}$ hr. From the vertical axis we see that Montereau is 235 km from Dijon and Nuits-s-Ravière is 87 km from Dijon, so Montereau is $235 - 87 = 148$ km from Nuits-s-Ravière.

$$\text{Average speed} = \frac{\text{Total distance}}{\text{Total time}}$$
$$= \frac{148 \text{ km}}{3\frac{7}{12} \text{ hr}}$$
$$\approx 41 \text{ km/h}$$

The longest horizontal segment on the line representing the 8:25 train from Montereau to Nuits-s-Ravière spans a 25 minute period of time, so this is the longest stop.

36. About 31 km/h; 5

37. Of all the segments representing trains from Paris to Tonnerre, the one with the steepest slant represents the 7:15 P.M. train. The steeper the slant, the greater the rate of travel, so the 7:15 P.M. train is the quickest way to travel from Paris to Tonnerre by train.

38. The 2:05 P.M; the segment representing that train is the steepest.

39. The fastest train is represented by the segment with the steepest slant. We see that the fastest train for Dijon left Tonnerre at 3:05 P.M.

40. 57 km/h

41.
$$s = vt + d$$
$$s - d = vt \qquad \text{Subtracting } d$$
$$\frac{s - d}{v} = t \qquad \text{Dividing by } v$$

42. 25

43. $2(4x + 5y - 3z) = 2 \cdot 4x + 2 \cdot 5y - 2 \cdot 3z = 8x + 10y - 6z$

44. $3(x + 6y - 2z)$

45.

46.

47.

48.

49. Answers may vary. Janet's rate of travel for the first 2.5 km is greater than her rate the last 2.5 km because her running speed is faster than her walking speed. Thus, the line will slant up more steeply for the first half of the trip than for the second half.

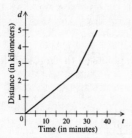

50.

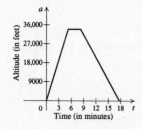

51. Let s = weekly sales and w = weekly wages. For $0 \le s \le 2000$, $w = 140 + 0.13s$. For $s > 2000$, $w = 140 + 0.13(2000) + 0.2(s - 2000)$, or $w = 0.2s$. We draw the graph.

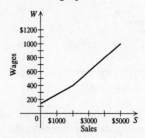

Locate 2700 on the horizontal axis, move up to the graphed line, and then move horizontally to the vertical axis. We estimate the wages to be about $550 when a salesperson sells $2700 in merchandise in one week.

52.

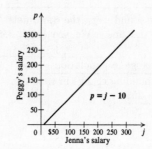

53.

54.

55. We graph $c = 3 + 0.5t$, where t represents the number of 15-min units of time, as a series of steps. The cost is constant within each 15-min unit of time. Thus,

for $0 < t \le 1$, $c = 3 + 0.5(1) = \$3.50$;

for $1 < t \le 2$, $c = 3 + 0.5(2) = \$4.00$;

for $2 < t \le 3$, $c = 3 + 0.5(3) = \$4.50$;

and so on. We draw the graph. An open circle at a point indicates that the point is not on the graph.

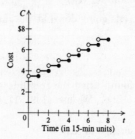

56.

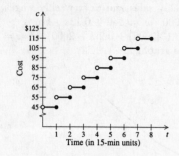

57. Graph $y = 219 + 0.035x$. Use Zoom and Trace to find the x-coordinate that corresponds to the y-coordinate 370.03. We find that the sales total was about \$4315.14.

58. 249

Exercise Set 3.5

1. The rate can be found using the coordinates of any two points on the line. We use $(1992, 255)$ and $(1994, 261)$.

$$\text{Rate} = \frac{\text{change in population}}{\text{corresponding change in time}}$$
$$= \frac{261 - 255}{1994 - 1992}$$
$$= \frac{6}{2}$$
$$= 3 \text{ million people per year}$$

2. 15 calories per minute

3. The rate can be found using the coordinates of any two points on the line. We use $(1993, 292)$ and $(1994, 282)$.

$$\text{Rate} = \frac{\text{change in outlay}}{\text{corresponding change in time}}$$
$$= \frac{282 - 292}{1994 - 1993}$$
$$= \frac{-10}{1}$$
$$= -\$10 \text{ billion per year}$$

That is, the outlay is going down at the rate of \$10 billion per year.

4. \$100 per year

5. The rate can be found using the coordinates of any two points on the line. We use $(1991, 11,400)$ and $(1994, 13,800)$.

$$\text{Rate} = \frac{\text{change in cost}}{\text{corresponding change in time}}$$
$$= \frac{13,800 - 11,400}{1994 - 1991}$$
$$= \frac{2400}{3}$$
$$= \$800 \text{ per year}$$

6. Going down at the rate of 31 thousand farms per year

7. We can use any two points on the line, such as $(0, 1)$ and $(4, 4)$.

$$m = \frac{\text{change in } y}{\text{change in } x}$$
$$= \frac{4 - 1}{4 - 0} = \frac{3}{4}$$

8. $\frac{2}{3}$

9. We can use any two points on the line, such as $(1, 0)$ and $(3, 3)$.

$$m = \frac{\text{change in } y}{\text{change in } x}$$
$$= \frac{3 - 0}{3 - 1} = \frac{3}{2}$$

10. $\frac{1}{3}$

11. We can use any two points on the line, such as $(-3, -4)$ and $(0, -3)$.

$$m = \frac{\text{change in } y}{\text{change in } x}$$
$$= \frac{-3 - (-4)}{0 - (-3)} = \frac{1}{3}$$

12. 3

13. We can use any two points on the line, such as $(0, 2)$ and $(2, 0)$.

$$m = \frac{\text{change in } y}{\text{change in } x}$$
$$= \frac{2 - 0}{0 - 2} = \frac{2}{-2} = -1$$

14. $-\frac{1}{2}$

15. We can use any two points on the line, such as $(-1, 3)$ and $(3, -3)$.

$$m = \frac{\text{change in } y}{\text{change in } x}$$
$$= \frac{-3 - 3}{3 - (-1)} = \frac{-6}{4} = -\frac{3}{2}$$

16. $-\frac{1}{3}$

17. We can use any two points on the line, such as $(2, 4)$ and $(4, 0)$.

$$m = \frac{\text{change in } y}{\text{change in } x}$$
$$= \frac{0 - 4}{4 - 2} = \frac{-4}{2} = -2$$

18. 0

19. This is a vertical line, so the slope is undefined. If we did not recognize this, we could use any two points on the line and attempt to compute the slope. We use $(2, 0)$ and $(2, 2)$.

$$m = \frac{\text{change in } y}{\text{change in } x}$$

$$= \frac{2 - 0}{2 - 2}$$

$$= \frac{2}{0} \qquad \text{Undefined}$$

20. $-\dfrac{1}{4}$

21. $(2, 3)$ and $(5, -1)$

$$m = \frac{-1 - 3}{5 - 2} = \frac{-4}{3} = -\frac{4}{3}$$

22. $\dfrac{2}{3}$

23. $(-2, 4)$ and $(3, 0)$

$$m = \frac{4 - 0}{-2 - 3} = \frac{4}{-5} = -\frac{4}{5}$$

24. $-\dfrac{5}{6}$

25. $(4, 0)$ and $(5, 7)$

$$m = \frac{0 - 7}{4 - 5} = \frac{-7}{-1} = 7$$

26. $\dfrac{2}{3}$

27. $(0, 8)$ and $(-3, 10)$

$$m = \frac{8 - 10}{0 - (-3)} = \frac{8 - 10}{0 + 3} = \frac{-2}{3} = -\frac{2}{3}$$

28. $-\dfrac{1}{2}$

29. $(-2, 3)$ and $(-6, 5)$

$$m = \frac{5 - 3}{-6 - (-2)} = \frac{2}{-6 + 2} = \frac{2}{-4} = -\frac{1}{2}$$

30. $-\dfrac{11}{8}$

31. $\left(-2, \dfrac{1}{2}\right)$ and $\left(-5, \dfrac{1}{2}\right)$

$$m = \frac{\frac{1}{2} - \frac{1}{2}}{-2 - (-5)} = \frac{\frac{1}{2} - \frac{1}{2}}{-2 + 5} = \frac{0}{3} = 0$$

32. 0

33. $(9, -4)$ and $(9, -7)$

$$m = \frac{-4 - (-7)}{9 - 9} = \frac{-4 + 7}{9 - 9} = \frac{3}{0} \quad \text{(undefined)}$$

The slope is undefined.

34. Undefined

35. The line $x = -3$ is a vertical line. The slope is undefined.

36. Undefined

37. The line $y = 4$ is a horizontal line. A horizontal line has slope 0.

38. 0

39. The line $x = 9$ is a vertical line. The slope is undefined.

40. Undefined

41. The line $y = -9$ is a horizontal line. A horizontal line has slope 0.

42. 0

43. The grade is expressed as a percent.

$$m = \frac{106}{1325} = 0.08 = 8\%$$

44. 0.05

45. The grade is expressed as a percent.

$$m = \frac{1}{12} \approx 0.083 \approx 8.3\%$$

46. 7%

47. $m = \dfrac{2.4}{8.2} = \dfrac{12}{41}$, or about 29%

48. 8%

49. Longs Peak rises $14{,}255 - 9600 = 4655$ ft.

$$m = \frac{4655}{15{,}840} \approx 0.29 \approx 29\%$$

50. 64%

51. $3^2 - 5^3 = 9 - 125 = -116$

52. 324

53. $(-1)^{17} = -1$

54. $-\dfrac{7}{5}$

55. $3x = 1 - 5x$

$\qquad 8x = 1 \qquad$ Adding $5x$

$\qquad x = \dfrac{1}{8} \qquad$ Dividing by 8

The solution is $\dfrac{1}{8}$.

56. $\{x | x \geq 3\}$

57. ◈

58. ◈

59. ◈

60. ◈

61. If the line never enters the first quadrant, then two points on the line are $(5, -6)$ and $(0, b)$, $-6 \le b \le 0$. The slope is of the form $m = \dfrac{-6 - b}{5 - 0}$, or $\dfrac{-6 - b}{5}$, $-6 \le b \le 0$. Then $-\dfrac{6}{5} \le m \le 0$, so the numbers the line could have for its slope are $\left\{ m \,\middle|\, -\dfrac{6}{5} \le m \le 0 \right\}$.

62. $\left\{ m \,\middle|\, m \ge \dfrac{4}{3} \right\}$

63. $x + y = 18$

$y = 18 - x$

The slope is $\dfrac{y}{x}$, or $\dfrac{18 - x}{x}$.

64. 27 candles per hour

65. Let $t =$ the number of units each tick mark on the vertical axis represents. Note that the graph drops 4 units for every 3 units of horizontal change. Then we have:

$$\dfrac{-4t}{3} = -\dfrac{2}{3}$$

$$-4t = -2 \quad \text{Multiplying by 3}$$

$$t = \dfrac{1}{2} \quad \text{Dividing by } -4$$

Each tick mark on the vertical axis represents $\dfrac{1}{2}$ unit.

66. $\dfrac{1}{4}$

67. First we find Ryan's rate. Then we double it to find Marcy's rate. Note that 50 minutes $= \dfrac{50}{60}$ hr $= \dfrac{5}{6}$ hr.

$$\text{Ryan's rate} = \dfrac{\text{change in number of bushels picked}}{\text{corresponding change in time}}$$

$$= \dfrac{5\frac{1}{2} - 4 \text{ bushels}}{\frac{5}{6} \text{ hr}}$$

$$= \dfrac{1\frac{1}{2} \text{ bushels}}{\frac{5}{6} \text{ hr}}$$

$$= \dfrac{3}{2} \cdot \dfrac{6}{5} \dfrac{\text{bushels}}{\text{hr}}$$

$$= \dfrac{9}{5} \text{ bushels per hour, or}$$

$$1.8 \text{ bushels per hour}$$

Then Marcy's rate is $2(1.8) = 3.6$ bushels per hour.

68. ◈

69. ◈

Chapter 4
Polynomials

Exercise Set 4.1

1. $m^4 \cdot m^6 = m^{4+6} = m^{10}$

2. 3^7

3. $8^5 \cdot 8^9 = 8^{5+9} = 8^{14}$

4. n^{23}

5. $x^4 \cdot x^3 = x^{4+3} = x^7$

6. y^{16}

7. $5^7 \cdot 5^0 = 5^{7+0} = 5^7$

8. t^{16}

9. $(3y)^4(3y)^8 = (3y)^{4+8} = (3y)^{12}$

10. $(2t)^{25}$

11. $(5t)(5t)^6 = (5t)^1(5t)^6 = (5t)^{1+6} = (5t)^7$

12. $8x$

13. $(a^2b^7)(a^3b^2) = a^2b^7a^3b^2$ Using an associative law
$= a^2a^3b^7b^2$ Using a commutative law
$= a^5b^9$ Adding exponents

14. $(m-3)^9$

15. $(x+1)^5(x+1)^7 = (x+1)^{5+7} = (x+1)^{12}$

16. $a^{12}b^4$

17. $r^3 \cdot r^7 \cdot r^2 = r^{3+7+2} = r^{12}$

18. s^{11}

19. $(xy^4)(xy)^3 = (xy^4)(x^3y^3)$
$= x \cdot x^3 \cdot y^4 \cdot y^3$
$= x^{1+3}y^{4+3}$
$= x^4y^7$

20. a^7b^5

21. $\dfrac{7^5}{7^2} = 7^{5-2} = 7^3$ Subtracting exponents

22. 4^4

23. $\dfrac{x^{15}}{x^3} = x^{15-3} = x^{12}$ Subtracting exponents

24. a^8

25. $\dfrac{y^9}{y^5} = y^{9-5} = y^4$

26. x

27. $\dfrac{(5a)^7}{(5a)^6} = (5a)^{7-6} = (5a)^1 = 5a$

28. $3m$

29. $\dfrac{(x+y)^8}{(x+y)^3} = (x+y)^{8-3} = (x+y)^5$

30. $(a-b)^9$

31. $\dfrac{18m^5}{6m^2} = \dfrac{18}{6}m^{5-2} = 3m^3$

32. $5n^4$

33. $\dfrac{a^9b^7}{a^2b} = \dfrac{a^9}{a^2} \cdot \dfrac{b^7}{b^1} = a^{9-2}b^{7-1} = a^7b^6$

34. r^7s^7

35. $\dfrac{m^9n^8}{m^0n^4} = \dfrac{m^9}{m^0} \cdot \dfrac{n^8}{n^4} = m^{9-0}n^{8-4} = m^9n^4$

36. a^8b^9

37. When $x = 13$, $x^0 = 13^0 = 1$. (Any nonzero number raised to the 0 power is 1.)

38. 1

39. When $x = -4$, $5x^0 = 5(-4)^0 = 5 \cdot 1 = 5$.

40. 7

41. For any $n \neq 0$, $n^0 = 1$. (Any nonzero number raised to the 0 power is 1.)

42. 1

43. $9^1 - 9^0 = 9 - 1 = 8$

44. -6

45. $(x^3)^4 = x^{3 \cdot 4} = x^{12}$ Multiplying exponents

46. a^{24}

47. $(5^8)^2 = 5^{8 \cdot 2} = 5^{16}$ Multiplying exponents

48. 2^{15}

49. $(m^7)^5 = m^{7 \cdot 5} = m^{35}$

50. n^{18}

51. $(a^{25})^3 = a^{25 \cdot 3} = a^{75}$

52. a^{75}

53. $(7x)^2 = 7^2 \cdot x^2 = 49x^2$

54. $25a^2$

55. $(-2a)^3 = (-2)^3 a^3 = -8a^3$

56. $-27x^3$

57. $(4m^3)^2 = 4^2(m^3)^2 = 16m^6$

58. $25n^8$

59. $(a^2 b)^7 = (a^2)^7(b^7) = a^{14} b^7$

60. $x^9 y^{36}$

61. $(a^3 b^2)^5 = (a^3)^5(b^2)^5 = a^{15} b^{10}$

62. $m^{24} n^{30}$

63. $(-5x^4 y^5)^2 = (-5)^2(x^4)^2(y^5)^2 = 25x^8 y^{10}$

64. $81a^{20} b^{28}$

65. $\left(\dfrac{a}{4}\right)^3 = \dfrac{a^3}{4^3} = \dfrac{a^3}{64}$ Raising the numerator and the denominator to the third power

66. $\dfrac{81}{x^4}$

67. $\left(\dfrac{7}{5a}\right)^2 = \dfrac{7^2}{(5a)^2} = \dfrac{49}{5^2 a^2} = \dfrac{49}{25a^2}$

68. $\dfrac{64x^3}{27}$

69. $\left(\dfrac{a^4}{b^3}\right)^5 = \dfrac{(a^4)^5}{(b^3)^5} = \dfrac{a^{20}}{b^{15}}$

70. $\dfrac{x^{35}}{y^{14}}$

71. $\left(\dfrac{y^3}{2}\right)^2 = \dfrac{(y^3)^2}{2^2} = \dfrac{y^6}{4}$

72. $\dfrac{a^{15}}{27}$

73. $\left(\dfrac{x^2 y}{z^3}\right)^4 = \dfrac{(x^2 y)^4}{(z^3)^4} = \dfrac{(x^2)^4(y^4)}{z^{12}} = \dfrac{x^8 y^4}{z^{12}}$

74. $\dfrac{x^{15}}{y^{10} z^5}$

75. $\left(\dfrac{a^3}{-2b^5}\right)^4 = \dfrac{(a^3)^4}{(-2b^5)^4} = \dfrac{a^{12}}{(-2)^4(b^5)^4} = \dfrac{a^{12}}{16b^{20}}$

76. $\dfrac{x^{20}}{81y^{12}}$

77. $\left(\dfrac{2a^2}{3b^4}\right)^3 = \dfrac{(2a^2)^3}{(3b^4)^3} = \dfrac{2^3(a^2)^3}{3^3(b^4)^3} = \dfrac{8a^6}{27b^{12}}$

78. $\dfrac{9x^{10}}{16y^6}$

79. $\left(\dfrac{4x^3 y^5}{3z^7}\right)^2 = \dfrac{(4x^3 y^5)^2}{(3z^7)^2} = \dfrac{4^2(x^3)^2(y^5)^2}{3^2(z^7)^2} = \dfrac{16x^6 y^{10}}{9z^{14}}$

80. $\dfrac{125a^{21}}{8b^{15} c^3}$

81. $3s + 3t + 24 = 3s + 3t + 3 \cdot 8 = 3(s + t + 8)$

82. $-7(x + 2)$

83. $9x + 2y - 4x - 2y = (9 - 4)x + (2 - 2)y = 5x$

84. 37.5%

85. Graph: $y = x - 5$

We make a table of solutions of the equation. Since the equation is in the form $y = mx + b$, we see that the y-intercept is $(0, -5)$. We find two other pairs.

When $x = 2$, $y = 2 - 5 = -3$.

When $x = 5$, $y = 5 - 5 = 0$.

x	y
0	-5
2	-3
5	0

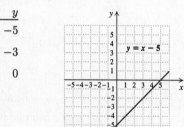

86.

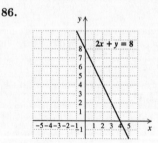

87.

88. ◈

89. ◈

90. ◈

91. $(y^{2x})(y^{3x}) = y^{2x+3x} = y^{5x}$

92. a^{2k}

93. $\dfrac{a^{6t}(a^{7t})}{a^{9t}} = \dfrac{a^{6t+7t}}{a^{9t}} = \dfrac{a^{13t}}{a^{9t}} = a^{13t-9t} = a^{4t}$

94. 2

95. Since the bases are the same, the expression with the larger exponent is larger. Thus, $3^5 > 3^4$.

96. $<$

97. Since the exponents are the same, the expression with the larger base is larger. Thus, $4^3 < 5^3$.

98. $<$

99. $9^7 = (3^2)^7 = 3^{14}$

When bases are the same, the expression with the larger exponent is larger. Thus, $3^{14} > 3^{13}$, or $9^7 > 3^{13}$.

100. $>$

101. Choose any number except 0.

For example, let $a = 1$. Then $(a + 5)^2 = (1 + 5)^2 = 6^2 = 36$, but $a^2 + 5^2 = 1^2 + 5^2 = 1 + 25 = 26$.

102. Let $x = 1$; then $3x^2 = 3 \cdot 1^2 = 3 \cdot 1 = 3$, but $(3x)^2 = (3 \cdot 1)^2 = 3^2 = 9$.

103. Choose any number except $\dfrac{7}{6}$. For example let $a = 0$. Then $\dfrac{0 + 7}{7} = \dfrac{7}{7} = 1$, but $a = 0$.

104. Let $t = -1$; then $\dfrac{t^6}{t^2} = \dfrac{(-1)^6}{(-1)^2} = \dfrac{1}{1} = 1$, but $t^3 = (-1)^3 = -1$.

105.
$$\frac{t^{38}}{t^x} = t^x$$

$t^{38-x} = t^x$ Using the quotient rule

$38 - x = x$ The exponents are the same.

$$38 = 2x$$
$$19 = x$$

The solution is 19.

106. $15,638.03

107. We substitute 20,800 for P, 4.5% for r, and 6 for t.
$$A = P(1 + r)^t$$
$$A = 20,800(1 + 4.5\%)^6$$
$$A = 20,800(1 + 0.045)^6$$
$$A = 20,800(1.045)^6$$
$$A \approx 27,087.01$$

There is $27,087.01 in the account at the end of 6 years.

108. ◈

Exercise Set 4.2

1. $3x^4 - 7x^3 + x - 5 = 3x^4 + (-7x^3) + x + (-5)$

The terms are $3x^4$, $-7x^3$, x, and -5.

2. $5a^3$, $4a^2$, $-9a$, -7

3. $-t^4 + 2t^3 - 5t^2 + 3 = (-t^4) + 2t^3 + (-5t^2) + 3$

The terms are $-t^4$, $2t^3$, $-5t^2$, and 3.

4. n^5, $-4n^3$, $2n$, -8

5. $7x^3 - 5x$

Term	Coefficient	Degree
$7x^3$	7	3
$-5x$	-5	1

6.

Term	Coefficient	Degree
$9a^3$	9	3
$-4a^2$	-4	2

7. $9t^2 - 3t + 4$

Term	Coefficient	Degree
$9t^2$	9	2
$-3t$	-3	1
4	4	0

8.

Term	Coefficient	Degree
$7x^4$	7	4
$5x$	5	1
-3	-3	0

9. $5a^4 + 9a + a^3$

Term	Coefficient	Degree
$5a^4$	5	4
$9a$	9	1
a^3	1	3

10.

Term	Coefficient	Degree
$6t^5$	6	5
$-3t^2$	-3	2
$-t$	-1	1

11. $x^4 - x^3 + 4x - 3$

Term	Coefficient	Degree
x^4	1	4
$-x^3$	-1	3
$4x$	4	1
-3	-3	0

12.

Term	Coefficient	Degree
$3a^4$	3	4
$-a^3$	-1	3
a	1	1
-9	-9	0

13. $4a^3 + 7a^5 + a^2$

a)

Term	$4a^3$	$7a^5$	a^2
Degree	3	5	2

b) The term of highest degree is $7a^5$. This is the leading term. Then the leading coefficient is 7.

c) Since the term of highest degree is $7a^5$, the degree of the polynomial is 5.

14. (a) 1, 2, 6; (b) $3x^6$, 3; (c) 6

15. $2t + 3 + 4t^2$

a)

Term	$2t$	3	$4t^2$
Degree	1	0	2

b) The term of highest degree is $4t^2$. This is the leading term. Then the leading coefficient is 4.

c) Since the term of highest degree is $4t^2$, the degree of the polynomial is 2.

16. (a) 2, 0, 5; (b) $2a^5$, 2; (c) 5

17. $-5x^4 + x^2 - x + 3$

a)

Term	$-5x^4$	x^2	$-x$	3
Degree	4	2	1	0

b) The term of highest degree is $-5x^4$. This is the leading term. Then the leading coefficient is -5.

c) Since the term of highest degree is $-5x^4$, the degree of the polynomial is 4.

18. (a) 3, 2, 1, 0; (b) $-7x^3$, -7; (c) 3

19. $9a - a^4 + 3 + 2a^3$

a)

Term	$9a$	$-a^4$	3	$2a^3$
Degree	1	4	0	3

b) The term of highest degree is $-a^4$. This is the leading term. Then the leading coefficient is -1.

c) Since the term of highest degree is $-a^4$, the degree of the polynomial is 4.

20. (a) 1, 5, 3, 0; (b) $2x^5$, 2; (c) 5

21. $3x^2 + 8x^5 - 4x^3 + 6 - \frac{1}{2}x^4$

Term	Coefficient	Degree of Term	Degree of Polynomial
$8x^5$	8	5	
$-\frac{1}{2}x^4$	$-\frac{1}{2}$	4	
$-4x^3$	-4	3	5
$3x^2$	3	2	
6	6	0	

22.

Term	Coefficient	Degree of Term	Degree of Polynomial
$-7x^4$	-7	4	
$6x^3$	6	3	
$-3x^2$	-3	2	4
$8x$	8	1	
-2	-2	0	

23. Three monomials are added, so $x^2 - 10x + 25$ is a trinomial.

24. Monomial

25. The polynomial $x^3 - 7x^2 + 2x - 4$ is none of these because it is composed of four monomials.

26. Binomial

27. Two monomials are added, so $4x^2 - 15$ is a binomial.

28. Trinomial

29. The polynomial $40x$ is a monomial because it is the product of a constant and a variable raised to a whole number power.

30. None of these

31. $3x^2 + 5x + 4x^2 = (3+4)x^2 + 5x = 7x^2 + 5x$

32. $7a^2 + 8a$

33. $3a^4 - 2a + 2a + a^4 = (3+1)a^4 + (-2+2)a = 4a^4 + 0a = 4a^4$

34. $4b^5$

35. $2x^2 - 6x + 3x + 4x^2 = (2+4)x^2 + (-6+3)x = 6x^2 - 3x$

36. $\frac{3}{4}x^5 - 2x - 42$

37. $\frac{1}{3}x^3 + 2x - \frac{1}{6}x^3 + 4 - 16 =$

$\left(\frac{1}{3} - \frac{1}{6}\right)x^3 + 2x + (4 - 16) =$

$\frac{1}{6}x^3 + 2x - 12$

38. x^4

39. $8x^2 + 2x^3 - 3x^3 - 4x^2 - 4x^2 =$

$(8 - 4 - 4)x^2 + (2 - 3)x^3 =$

$0x^2 - x^3 = -x^3$

40. $\frac{15}{16}x^3 - \frac{7}{6}x^2$

41. $\frac{1}{5}x^4 + \frac{1}{5} - 2x^2 + \frac{1}{10} - \frac{3}{15}x^4 + 2x^2 - \frac{3}{10} =$

$\left(\frac{1}{5} - \frac{3}{15}\right)x^4 + (-2 + 2)x^2 + \left(\frac{1}{5} + \frac{1}{10} - \frac{3}{10}\right) =$

$\left(\frac{3}{15} - \frac{3}{15}\right)x^4 + 0x^2 + \left(\frac{2}{10} + \frac{1}{10} - \frac{3}{10}\right) =$

$0x^4 + 0x^2 + 0 = 0$

42. $x^6 + x^4$

43. $-1 + 5x^3 - 3 - 7x^3 + x^4 + 5 = 1 - 2x^3 + x^4 =$

$x^4 - 2x^3 + 1$

44. $13x^3 - 9x + 8$

45. $2x - \frac{5}{6} + 4x^3 + x + \frac{1}{3} - 2x =$

$x - \frac{1}{2} + 4x^3 = 4x^3 + x - \frac{1}{2}$

46. $-\frac{1}{3}a^3 - 4a^2 + \frac{1}{3}$

47. $-7x + 5 = -7 \cdot 3 + 5$

$= -21 + 5$

$= -16$

48. -6

49. $2x^2 - 3x + 7 = 2 \cdot 3^2 - 3 \cdot 3 + 7$

$= 2 \cdot 9 - 3 \cdot 3 + 7$

$= 18 - 9 + 7$

$= 16$

50. 27

51. $5x + 7 = 5(-2) + 7$

$= -10 + 7$

$= -3$

52. 13

53. $x^2 - 3x + 1 = (-2)^2 - 3(-2) + 1$

$= 4 - 3(-2) + 1$

$= 4 + 6 + 1$

$= 11$

54. -15

55. $\quad -3x^3 + 7x^2 - 4x - 5$

$= -3(-2)^3 + 7(-2)^2 - 4(-2) - 5$

$= -3(-8) + 7 \cdot 4 - 4(-2) - 5$

$= 24 + 28 + 8 - 5$

$= 55$

56. -5

57. Locate 10 on the horizontal axis. From there move vertically to the graph and then horizontally to the M-axis. This locates an M-value of about 9. Thus, about 9 words were memorized in 10 minutes.

58. About 17

59. Locate 8 on the horizontal axis. From there move vertically to the graph and then horizontally to the M-axis. This locates an M-value of about 6. Thus, the value of $-0.001t^3 + 0.1t^2$ for $t = 8$ is approximately 6.

60. About 13

61. Locate 13 on the horizontal axis. It is halfway between 12 and 14. From there move vertically to the graph and then horizontally to the M-axis. This locates an M-value of about 15. Thus, the value of $-0.001t^3 + 0.1t^2$ when t is 13 is approximately 15.

62. About 5

63. $11.12t^2 = 11.12(10)^2 = 11.12(100) = 1112$

A skydiver has fallen approximately 1112 ft 10 seconds after jumping from a plane.

64. 3091 ft

65. $0.4r^2 - 40r + 1039 = 0.4(18)^2 - 40(18) + 1039 =$

$0.4(324) - 720 + 1039 = 129.6 - 720 + 1039 =$

448.6

There are approximately 449 accidents daily involving an 18-year-old driver.

66. 399

67. Evaluate the polynomial for $x = 75$:

$280x - 0.4x^2 = 280 \cdot 75 - 0.4(75)^2 =$

$21,000 - 0.4(5625) = 21,000 - 2250 = 18,750$

The total revenue is $18,750.

68. $24,000

69. Evaluate the polynomial for $x = 500$:
$5000 + 0.6(500)^2 = 5000 + 0.6(250,000) =$
$5000 + 150,000 = 155,000$
The total cost is \$155,000.

70. \$258,500

71. $2\pi r = 2(3.14)(10)$ Substituting 3.14 for π
and 10 for r
$= 62.8$
The circumference is 62.8 cm.

72. 31.4 ft

73. $\pi r^2 = 3.14(5)^2$ Substituting 3.14 for π
and 5 for r
$= 3.14(25)$
$= 78.5$
The area is 78.5 m^2.

74. 314 in^2

75. $3(x+2) = 5x - 9$
$3x + 6 = 5x - 9$ Removing parentheses
$6 = 2x - 9$ Subtracting $3x$ on both sides
$15 = 2x$ Adding 9 on both sides
$\dfrac{15}{2} = x$ Dividing by 2 on both sides

The solution is $\dfrac{15}{2}$.

76. 274 and 275

77. **Familiarize**. Let $x =$ the cost per mile of gasoline in dollars. Then the total cost of the gasoline for the year was $14,800x$.

Translate.

$$\underbrace{\text{Cost of insurance}} + \underbrace{\text{cost of registration and oil}} + \underbrace{\text{cost of gasoline}} = \$2011.$$

$$972 + 114 + 14,800x = 2011$$

Carry out. We solve the equation.
$972 + 114 + 14,800x = 2011$
$1086 + 14,800x = 2011$
$14,800x = 925$
$x = 0.0625$

Check. If gasoline cost \$0.0625 per mile, then the total cost of the gasoline was $14,800(\$0.0625)$, or \$925. Then the total auto expense was \$972+\$114+\$925, or \$2011. The answer checks.

State. Gasoline cost \$0.0625, or 6.25¢ per mile.

78. $b = \dfrac{cx + r}{a}$

79. ◈

80. ◈

81. ◈

82. ◈

83. Answers may vary. Use an ax^5-term, where a is an integer, and 3 other terms with different degrees, each less than degree 5, and integer coefficients. Three answers are $-6x^5 + 14x^4 - x^2 + 11$, $x^5 - 8x^3 + 3x + 1$, and $23x^5 + 2x^4 - x^2 + 5x$.

84. $0.2y^4 - y + \dfrac{5}{2}$; answers may vary

85. $(5m^5)^2 = 5^2 m^{5\cdot 2} = 25m^{10}$
The degree is 10.

86. $9y^4$, $-\dfrac{3}{2}y^4$, $4.2y^4$; answers may vary

87.
$\dfrac{9}{2}x^8 + \dfrac{1}{9}x^2 + \dfrac{1}{2}x^9 + \dfrac{9}{2}x + \dfrac{9}{2}x^9 + \dfrac{8}{9}x^2 + \dfrac{1}{2}x - \dfrac{1}{2}x^8$
$= \left(\dfrac{1}{2} + \dfrac{9}{2}\right)x^9 + \left(\dfrac{9}{2} - \dfrac{1}{2}\right)x^8 + \left(\dfrac{1}{9} + \dfrac{8}{9}\right)x^2 + \left(\dfrac{9}{2} + \dfrac{1}{2}\right)x$
$= \dfrac{10}{2}x^9 + \dfrac{8}{2}x^8 + \dfrac{9}{9}x^2 + \dfrac{10}{2}x$
$= 5x^9 + 4x^8 + x^2 + 5x$

88. $3x^6$

89. For $s = 18$:
$s^2 - 50s + 675 = 18^2 - 50(18) + 675 = 324 - 900 + 675 = 99$
$-s^2 + 50s - 675 = -(18)^2 + 50(18) - 675 = -324 + 900 - 675 = -99$

For $s = 25$:
$s^2 - 50s + 675 = 25^2 - 50(25) + 675 = 625 - 1250 + 675 = 50$
$-s^2 + 50s - 675 = -(25)^2 + 50(25) - 675 = -625 + 1250 - 675 = -50$

For $s = 32$:
$s^2 - 50s + 675 = 32^2 - 50(32) + 675 = 1024 - 1600 + 675 = 99$
$-s^2 + 50s - 675 = -(32)^2 + 50(32) - 675 = -1024 + 1600 - 675 = -99$

90. 50

91. When $t = 3$, $-t^2 + 10t - 18 = -3^2 + 10 \cdot 3 - 18 =$
$-9 + 30 - 18 = 3$.
When $t = 4$, $-t^2 + 10t - 18 = -4^2 + 10 \cdot 4 - 18 =$
$-16 + 40 - 18 = 6$.
When $t = 5$, $-t^2 + 10t - 18 = -5^2 + 10 \cdot 5 - 18 =$
$-25 + 50 - 18 = 7$.
When $t = 6$, $-t^2 + 10t - 18 = -6^2 + 10 \cdot 6 - 18 =$
$-36 + 60 - 18 = 6$.
When $t = 7$, $-t^2 + 10t - 18 = -7^2 + 10 \cdot 7 - 18 =$
$-49 + 70 - 18 = 3$.

We complete the table. Then we plot the points and connect them with a smooth curve.

t	$-t^2 + 10t - 18$
3	3
4	6
5	7
6	6
7	3

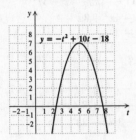

92.

t	$-t^2 + 6t - 4$
1	1
2	4
3	5
4	4
5	1

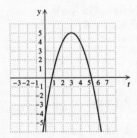

93. When $d = 0$, $-0.0064d^2 + 0.8d + 2 =$
$-0.0064(0)^2 + 0.8(0) + 2 = 0 + 0 + 2 = 2$.
When $d = 30$, $-0.0064(30)^2 + 0.8(30) + 2 =$
$-5.76 + 24 + 2 = 20.24$.
When $d = 60$, $-0.0064(60)^2 + 0.8(60) + 2 =$
$-23.04 + 48 + 2 = 26.96$.
When $d = 90$, $-0.0064(90)^2 + 0.8(90) + 2 =$
$-51.84 + 72 + 2 = 22.16$.
When $d = 120$, $-0.0064(120)^2 + 0.8(120) + 2 =$
$-92.16 + 96 + 2 = 5.84$.

We complete the table. Then we plot the points and connect them with a smooth curve.

d	$-0.0064d^2 + 0.8d + 2$
0	2
30	20.24
60	26.96
90	22.16
120	5.84

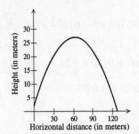

94. $x^3 - 2x^2 - 6x + 3$

Exercise Set 4.3

1. $(5x + 3) + (-7x + 1) = (5 - 7)x + (3 + 1) = -2x + 4$

2. $-5x + 6$

3. $(-6x + 2) + (x^2 + x - 3) =$
$x^2 + (-6 + 1)x + (2 - 3) = x^2 - 5x - 1$

4. $x^2 + 3x - 5$

5. $(3x^2 - 5x + 10) + (2x^2 + 8x - 40) =$
$(3 + 2)x^2 + (-5 + 8)x + (10 - 40) = 5x^2 + 3x - 30$

6. $6x^4 + 3x^3 + 4x^2 - 3x + 2$

7. $(1.2x^3 + 4.5x^2 - 3.8x) + (-3.4x^3 - 4.7x^2 + 23) =$
$(1.2 - 3.4)x^3 + (4.5 - 4.7)x^2 - 3.8x + 23 =$
$-2.2x^3 - 0.2x^2 - 3.8x + 23$

8. $2.8x^4 - 0.6x^2 + 1.8x - 3.2$

9. $(3 + 5x + 7x^2 + 8x^3) + (8 - 3x + 9x^2 - 8x^3) =$
$(3 + 8) + (5 - 3)x + (7 + 9)x^2 + (8 - 8)x^3 =$
$11 + 2x + 16x^2 + 0x^3 = 11 + 2x + 16x^2$

10. $2x^4 + 3x + 3x^2 + 4 - 3x^3$, or
$2x^4 - 3x^3 + 3x^2 + 3x + 4$

11. $(9x^8 - 7x^4 + 2x^2 + 5) + (8x^7 + 4x^4 - 2x) =$
$9x^8 + 8x^7 + (-7 + 4)x^4 + 2x^2 - 2x + 5 =$
$9x^8 + 8x^7 - 3x^4 + 2x^2 - 2x + 5$

12. $4x^5 + 9x^2 + 1$

13. $\left(\dfrac{1}{4}x^4 + \dfrac{2}{3}x^3 + \dfrac{5}{8}x^2 + 7\right) + \left(-\dfrac{3}{4}x^4 + \dfrac{3}{8}x^2 - 7\right) =$

$\left(\dfrac{1}{4} - \dfrac{3}{4}\right)x^4 + \dfrac{2}{3}x^3 + \left(\dfrac{5}{8} + \dfrac{3}{8}\right)x^2 + (7 - 7) =$

$-\dfrac{2}{4}x^4 + \dfrac{2}{3}x^3 + \dfrac{8}{8}x^2 + 0 =$

$-\dfrac{1}{2}x^4 + \dfrac{2}{3}x^3 + x^2$

14. $\dfrac{2}{15}x^9 - \dfrac{2}{5}x^5 + \dfrac{1}{4}x^4 - \dfrac{1}{2}x^2 + 7$

15. $(0.02x^5 - 0.2x^3 + x + 0.08) + (-0.01x^5 + x^4 - 0.8x - 0.02) =$

$(0.02 - 0.01)x^5 + x^4 - 0.2x^3 + (1 - 0.8)x + (0.08 - 0.02) =$

$0.01x^5 + x^4 - 0.2x^3 + 0.2x + 0.06$

16. $0.1x^6 + 0.02x^3 + 0.22x + 0.55$

17. $-3x^4 + 6x^2 + 2x - 1$
   $\underline{\ -3x^2 + 2x + 1}$
   $-3x^4 + 3x^2 + 4x + 0$
   $-3x^4 + 3x^2 + 4x$

18. $-4x^3 + 4x^2 + 6x$

19. Rewrite the problem so the coefficients of like terms have the same number of decimal places.

   $0.15x^4 + 0.10x^3 - 0.90x^2$
   $ - 0.01x^3 + 0.01x^2 + x$
   $1.25x^4 + 0.11x^2 + 0.01$
   $ 0.27x^3 + 0.99$
   $\underline{-0.35x^4 + 15.00x^2 - 0.03}$
   $1.05x^4 + 0.36x^3 + 14.22x^2 + x + 0.97$

20. $1.3x^4 + 0.35x^3 + 9.53x^2 + 2x + 0.96$

21. Two equivalent expressions for the opposite of $-x^2 + 9x - 4$ are

a) $-(-x^2 + 9x - 4)$ and

b) $x^2 - 9x + 4$. (Changing the sign of every term)

22. $-(-4x^3 - 5x^2 + 2x),\ 4x^3 + 5x^2 - 2x$

23. Two equivalent expressions for the opposite of $12x^4 - 3x^3 + 3$ are

a) $-(12x^4 - 3x^3 + 3)$ and

b) $-12x^4 + 3x^3 - 3$. (Changing the sign of every term)

24. $-(4x^3 - 6x^2 - 8x + 1),\ -4x^3 + 6x^2 + 8x - 1$

25. We change the sign of every term inside parentheses.

   $-(8x - 9) = -8x + 9$

26. $6x - 5$

27. We change the sign of every term inside parentheses.

   $-(4x^2 - 3x + 2) = -4x^2 + 3x - 2$

28. $6a^3 - 2a^2 + 9a - 1$

29. We change the sign of every term inside parentheses.

   $-\left(-4x^4 + 6x^2 + \dfrac{3}{4}x - 8\right) = 4x^4 - 6x^2 - \dfrac{3}{4}x + 8$

30. $5x^4 - 4x^3 + x^2 - 0.9$

31. $(7x + 4) - (-2x + 1)$
   $= 7x + 4 + 2x - 1$ Changing the sign of every term inside parentheses
   $= 9x + 3$

32. $7x + 2$

33. $(-6x + 2) - (x^2 + x - 3) = -6x + 2 - x^2 - x + 3$
   $= -x^2 - 7x + 5$

34. $x^2 - 13x + 13$

35. $(6x^4 + 3x^3 - 1) - (4x^2 - 3x + 3)$
   $= 6x^4 + 3x^3 - 1 - 4x^2 + 3x - 3$
   $= 6x^4 + 3x^3 - 4x^2 + 3x - 4$

36. $-3x^3 + x^2 + 2x - 3$

37. $(1.2x^3 + 4.5x^2 - 3.8x) - (-3.4x^3 - 4.7x^2 + 23)$
   $= 1.2x^3 + 4.5x^2 - 3.8x + 3.4x^3 + 4.7x^2 - 23$
   $= 4.6x^3 + 9.2x^2 - 3.8x - 23$

38. $-1.8x^4 - 0.6x^2 - 1.8x + 4.6$

39. $(7x^3 - 2x^2 + 6) - (7x^2 + 2x - 4)$
   $= 7x^3 - 2x^2 + 6 - 7x^2 - 2x + 4$
   $= 7x^3 - 9x^2 - 2x + 10$

40. $-2x^5 - 6x^4 + x + 2$

41. $(6x^2 + 2x) - (-3x^2 - 7x + 8)$
   $= 6x^2 + 2x + 3x^2 + 7x - 8$
   $= 9x^2 + 9x - 8$

42. $7x^3 + 3x^2 + 2x - 1$

43. $\dfrac{5}{8}x^3 - \dfrac{1}{4}x - \dfrac{1}{3} - \left(-\dfrac{1}{8}x^3 + \dfrac{1}{4}x - \dfrac{1}{3}\right)$
   $= \dfrac{5}{8}x^3 - \dfrac{1}{4}x - \dfrac{1}{3} + \dfrac{1}{8}x^3 - \dfrac{1}{4}x + \dfrac{1}{3}$
   $= \dfrac{6}{8}x^3 - \dfrac{2}{4}x$
   $= \dfrac{3}{4}x^3 - \dfrac{1}{2}x$

44. $\dfrac{3}{5}x^3 - 0.11$

45. $(0.08x^3 - 0.02x^2 + 0.01x) - (0.02x^3 + 0.03x^2 - 1)$
   $= 0.08x^3 - 0.02x^2 + 0.01x - 0.02x^3 - 0.03x^2 + 1$
   $= 0.06x^3 - 0.05x^2 + 0.01x + 1$

46. $0.1x^4 - 0.9$

47.
$$x^2 + 5x + 6$$
$$-(x^2 + 2x \quad)$$

$$\begin{array}{ll} x^2 + 5x + 6 & \text{Changing signs and} \\ \underline{-x^2 - 2x} & \text{removing parentheses} \\ 3x + 6 & \text{Adding} \end{array}$$

48. $-x^2 + 1$

49.
$$5x^4 + 6x^3 - 9x^2$$
$$-(-6x^4 - 6x^3 \qquad + 8x + 9)$$

$$\begin{array}{ll} 5x^4 + 6x^3 - 9x^2 & \text{Changing signs and} \\ \underline{6x^4 + 6x^3 \qquad -8x - 9} & \text{removing parentheses} \\ 11x^4 + 12x^3 - 9x^2 - 8x - 9 & \text{Adding} \end{array}$$

50. $5x^4 - 6x^3 - x^2 + 5x + 15$

51.
$$3x^4 + 6x^2 + 8x - 1$$
$$-(4x^5 - 6x^4 \qquad - 8x - 7)$$

$$\begin{array}{ll} 3x^4 + 6x^2 + 8x - 1 & \text{Changing signs and} \\ \underline{-4x^5 + 6x^4 \qquad + 8x + 7} & \text{removing parentheses} \\ -4x^5 + 9x^4 + 6x^2 + 16x + 6 & \text{Adding} \end{array}$$

52. $-4x^5 - 6x^3 + 8x^2 - 5x - 2$

53. a)

Familiarize. The area of a rectangle is the product of the length and the width.

Translate. The sum of the areas is found as follows:

$$\begin{array}{c} \text{Area} \\ \text{of } A \end{array} + \begin{array}{c} \text{Area} \\ \text{of } B \end{array} + \begin{array}{c} \text{Area} \\ \text{of } C \end{array} + \begin{array}{c} \text{Area} \\ \text{of } D \end{array}$$
$$= 3x \cdot x + x \cdot x + 4 \cdot x + x \cdot x$$

Carry out. We collect like terms.

$$3x^2 + x^2 + 4x + x^2 = 5x^2 + 4x$$

Check. We can go over our calculations. We can also assign some value to x, say 2, and carry out the computation of the area in two ways.

Sum of areas: $3 \cdot 2 \cdot 2 + 2 \cdot 2 + 4 \cdot 2 + 2 \cdot 2 =$
$$12 + 4 + 8 + 4 = 28$$

Substituting in the polynomial:
$$5(2)^2 + 4 \cdot 2 = 20 + 8 = 28$$

Since the results are the same, our solution is probably correct.

State. A polynomial for the sum of the areas is $5x^2 + 4x$.

b) For $x = 5$: $5x^2 + 4x = 5 \cdot 5^2 + 4 \cdot 5 =$
$$5 \cdot 25 + 4 \cdot 5 = 125 + 20 = 145$$

When $x = 5$, the sum of the areas is 145 square units.
For $x = 7$: $5x^2 + 4x = 5 \cdot 7^2 + 4 \cdot 7 =$
$$5 \cdot 49 + 4 \cdot 7 = 245 + 28 = 273$$

When $x = 7$, the sum of the areas is 273 square units.

54. (a) $r^2\pi + 13\pi$; (b) 38π, 140.69π

55.

Familiarize. The perimeter is the sum of the lengths of the sides.

Translate. The sum of the lengths is found as follows:

$$3y + 7y + (2y + 3) + 5 + 7 + 2y + 7 + 3$$

Carry out. We collect like terms.

$$(3 + 7 + 2 + 2)y + (3 + 5 + 7 + 7 + 3) = 14y + 25$$

Check. We can go over our calculations. We can also assign some value to y, say 3, and carry out the computation of the perimeter in two ways.

Sum of lengths: $3 \cdot 3 + 7 \cdot 3 + (2 \cdot 3 + 3) + 5 + 7 + 2 \cdot 3 + 7 + 3 =$
$$9 + 21 + 9 + 5 + 7 + 6 + 7 + 3 = 67$$

Substituting in the polynomial:
$$14 \cdot 3 + 25 = 42 + 25 = 67$$

Since the results are the same, our solution is probably correct.

State. A polynomial for the perimeter of the figure is $14y + 25$.

56. $11\frac{1}{2}a + 12$

57.

The area of the figure can be found by adding the areas of the four rectangles A, B, C, and D. The area of a rectangle is the product of the length and the width.

$$
\begin{array}{ccccccc}
\text{Area} & + & \text{Area} & + & \text{Area} & + & \text{Area} \\
\text{of } A & & \text{of } B & & \text{of } C & & \text{of } D \\
= 9 \cdot r & + & 11 \cdot 9 & + & r \cdot r & + & 11 \cdot r \\
= 9r & + & 99 & + & r^2 & + & 11r
\end{array}
$$

An algebraic expression for the area of the figure is $9r + 99 + r^2 + 11r$.

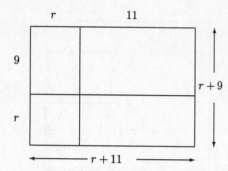

The length and width of the figure can be expressed as $r + 11$ and $r + 9$, respectively. The area of this figure (a rectangle) is the product of the length and width. An algebraic expression for the area is $(r + 11) \cdot (r + 9)$.

The algebraic expressions $9r + 99 + r^2 + 11r$ and $(r + 11) \cdot (r + 9)$ represent the same area.

58. $(x + 3)^2$, $x^2 + 3x + 3x + 9$

59.

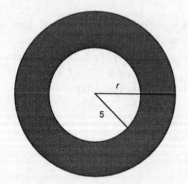

Familiarize. Recall that the area of a circle is the product of π and the square of the radius, r^2.

$$A = \pi r^2$$

Translate.

$$
\begin{array}{ccc}
\text{Area of circle} & - & \text{Area of circle} \\
\text{with radius } r & & \text{with radius } 5 \\
\pi \cdot r^2 & - & \pi \cdot 5^2
\end{array}
\begin{array}{c}
= \begin{array}{c}\text{Shaded}\\ \text{area}\end{array} \\
= \text{Shaded area}
\end{array}
$$

Carry out. We simplify the expression.

$$\pi \cdot r^2 - \pi \cdot 5^2 = \pi r^2 - 25\pi$$

Check. We can go over our calculations. We can also assign some value to r, say 7, and carry out the computation in two ways.

Difference of areas: $\pi \cdot 7^2 - \pi \cdot 5^2 = 49\pi - 25\pi = 24\pi$

Substituting in the polynomial: $\pi \cdot 7^2 - 25\pi = 49\pi - 25\pi = 24\pi$

Since the results are the same, our solution is probably correct.

State. A polynomial for the shaded area is $\pi r^2 - 25\pi$.

60. $m^2 - 40$

61. *Familiarize*. We label the figure with additional information.

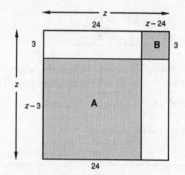

Translate.

Area of shaded sections = Area of A + Area of B

Area of shaded sections = $24(z - 3) + 3(z - 24)$

Carry out. We simplify the expression.

$24(z - 3) + 3(z - 24) = 24z - 72 + 3z - 72 = 27z - 144$

Check. We can go over the calculations. We can also assign some value to z, say 30, and carry out the computation in two ways.

Sum of areas:

$$24 \cdot 27 + 3 \cdot 6 = 648 + 18 = 666$$

Substituting in the polynomial:

$$27 \cdot 30 - 144 = 810 - 144 = 666$$

Since the results are the same, our solution is probably correct.

State. A polynomial for the shaded area is $27z - 144$.

62. $\pi x^2 - 2x^2$

63.

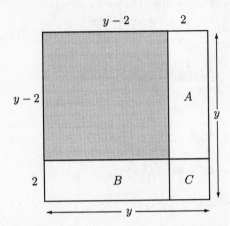

The shaded area is $(y-2)^2$. We find it as follows:

$$\begin{array}{c} \text{Shaded} \\ \text{area} \end{array} = \begin{array}{c} \text{Area of} \\ \text{square} \end{array} - \begin{array}{c} \text{Area} \\ \text{of } A \end{array} - \begin{array}{c} \text{Area} \\ \text{of } B \end{array} - \begin{array}{c} \text{Area} \\ \text{of } C \end{array}$$

$$(y-2)^2 = y^2 - 2(y-2) - 2(y-2) - 2\cdot 2$$

$$(y-2)^2 = y^2 - 2y + 4 - 2y + 4 - 4$$

$$(y-2)^2 = y^2 - 4y + 4$$

64. $100 - 40x + 4x^2$

65.

$$1.5x - 2.7x = 23 - 5.6x$$

$$10(1.5x - 2.7x) = 10(23 - 5.6x) \quad \text{Clearing decimals}$$

$$15x - 27x = 230 - 56x$$

$$-12x = 230 - 56x \quad \text{Collecting like terms}$$

$$44x = 230 \quad \text{Adding } 56x$$

$$x = \frac{230}{44} \quad \text{Dividing by 44}$$

$$x = \frac{115}{22} \quad \text{Simplifying}$$

The solution is $\frac{115}{22}$.

66. 1

67.

$$8(x-2) = 16$$

$$8x - 16 = 16 \quad \text{Multiplying to remove parentheses}$$

$$8x = 32 \quad \text{Adding 16}$$

$$x = 4 \quad \text{Dividing by 8}$$

The solution is 4.

68. $-\dfrac{76}{3}$

69.

$$3x - 7 \leq 5x + 13$$

$$-2x - 7 \leq 13 \quad \text{Subtracting } 5x$$

$$-2x \leq 20 \quad \text{Adding 7}$$

$$x \geq -10 \quad \begin{array}{l}\text{Dividing by } -2 \text{ and} \\ \text{reversing the inequality} \\ \text{symbol}\end{array}$$

The solution set is $\{x | x \geq -10\}$.

70. $\{x | x < 0\}$

71. ◈

72. ◈

73. ◈

74. ◈

75.

$$(5a^2 - 8a) + (7a^2 - 9a - 13) - (7a - 5)$$
$$= 5a^2 - 8a + 7a^2 - 9a - 13 - 7a + 5$$
$$= 12a^2 - 24a - 8$$

76. $5x^2 - 9x - 1$

77.

$$(-8y^2 - 4) - (3y + 6) - (2y^2 - y)$$
$$= -8y^2 - 4 - 3y - 6 - 2y^2 + y$$
$$= -10y^2 - 2y - 10$$

78. $4x^3 - 5x^2 + 6$

79.

$$(-y^4 - 7y^3 + y^2) + (-2y^4 + 5y - 2) - (-6y^3 + y^2)$$
$$= -y^4 - 7y^3 + y^2 - 2y^4 + 5y - 2 + 6y^3 - y^2$$
$$= -3y^4 - y^3 + 5y - 2$$

80. $2 + x + 2x^2 + 4x^3$

81.

$$(345.099x^3 - 6.178x) - (94.508x^3 - 8.99x)$$
$$= 345.099x^3 - 6.178x - 94.508x^3 + 8.99x$$
$$= 250.591x^3 + 2.812x$$

82. $2x^2 + 36x$

83. *Familiarize*. The surface area is $2lw + 2lh + 2wh$, where l = length, w = width, and h = height of the rectangular solid. Here we have $l = 7$, $w = a$, and $h = 4$.

Translate. We substitute in the formula above.

$$2\cdot 7\cdot a + 2\cdot 7\cdot 4 + 2\cdot a\cdot 4$$

Carry out. We simplify the expression.

$$2\cdot 7\cdot a + 2\cdot 7\cdot 4 + 2\cdot a\cdot 4$$
$$= 14a + 56 + 8a$$
$$= 22a + 56$$

Check. We can go over the calculations. We can also assign some value to a, say 6, and carry out the computation in two ways.

Using the formula: $2\cdot 7\cdot 6 + 2\cdot 7\cdot 4 + 2\cdot 6\cdot 4 = 84 + 56 + 48 = 188$

Substituting in the polynomial: $22\cdot 6 + 56 = 132 + 56 = 188$

Since the results are the same, our solution is probably correct.

State. A polynomial for the surface area is $22a + 56$.

84. (a) $-x^2 + 280x - 5000$; (b) \$10,375; (c) \$13,000

85. ◈

Exercise Set 4.4

1. $(3x^4)8 = (3 \cdot 8)x^4 = 24x^4$

2. $28x^3$

3. $(-x^2)(-x) = (-1 \cdot x^2)(-1 \cdot x) = (-1)(-1)(x^2 \cdot x) = x^3$

4. $-x^7$

5. $(-x^5)(x^3) = (-1 \cdot x^5)(1x^3) = (-1)(1)(x^5 \cdot x^3) = -x^8$

6. x^8

7. $(7t^5)(4t^3) = (7 \cdot 4)(t^5 \cdot t^3) = 28t^8$

8. $30a^4$

9. $(-0.1x^6)(0.2x^4) = (-0.1)(0.2)(x^6 \cdot x^4) = -0.02x^{10}$

10. $-0.12x^9$

11. $\left(-\frac{1}{5}x^3\right)\left(-\frac{1}{3}x\right) = \left(-\frac{1}{5}\right)\left(-\frac{1}{3}\right)(x^3 \cdot x) = \frac{1}{15}x^4$

12. $-\frac{1}{20}x^{12}$

13. $19t^2 \cdot 0 = 0$ Any number multiplied by 0 is 0.

14. $5n^3$

15. $(3x^2)(-4x^3)(2x^6) = (3)(-4)(2)(x^2 \cdot x^3 \cdot x^6) = -24x^{11}$

16. $60y^{12}$

17. $3x(-x+5) = 3x(-x) + 3x(5)$
$= -3x^2 + 15x$

18. $8x^2 - 12x$

19. $4x(x+1) = 4x(x) + 4x(1)$
$= 4x^2 + 4x$

20. $3x^2 + 6x$

21. $(x+7)5x = x \cdot 5x + 7 \cdot 5x$
$= 5x^2 + 35x$

22. $3x^2 - 18x$

23. $x^2(x^3+1) = x^2(x^3) + x^2(1)$
$= x^5 + x^2$

24. $-2x^5 + 2x^3$

25. $3x(2x^2 - 6x + 1) = 3x(2x^2) + 3x(-6x) + 3x(1)$
$= 6x^3 - 18x^2 + 3x$

26. $-8x^4 + 24x^3 + 20x^2 - 4x$

27. $4x^2(3x+6) = 4x^2(3x) + 4x^2(6)$
$= 12x^3 + 24x^2$

28. $-10x^3 + 5x^2$

29. $-6x^2(x^2+x) = -6x^2(x^2) - 6x^2(x)$
$= -6x^4 - 6x^3$

30. $-4x^4 + 4x^3$

31. $\frac{2}{3}a^4\left(6a^5 - 12a^3 - \frac{5}{8}\right)$
$= \frac{2}{3}a^4(6a^5) - \frac{2}{3}a^4(12a^3) - \frac{2}{3}a^4\left(\frac{5}{8}\right)$
$= \frac{12}{3}a^9 - \frac{24}{3}a^7 - \frac{10}{24}a^4$
$= 4a^9 - 8a^7 - \frac{5}{12}a^4$

32. $6t^{11} - 9t^9 + \frac{9}{7}t^5$

33. $(x+6)(x+3) = (x+6)x + (x+6)3$
$= x \cdot x + 6 \cdot x + x \cdot 3 + 6 \cdot 3$
$= x^2 + 6x + 3x + 18$
$= x^2 + 9x + 18$

34. $x^2 + 7x + 10$

35. $(x+5)(x-2) = (x+5)x + (x+5)(-2)$
$= x \cdot x + 5 \cdot x + x(-2) + 5(-2)$
$= x^2 + 5x - 2x - 10$
$= x^2 + 3x - 10$

36. $x^2 + 4x - 12$

37. $(x-4)(x-3) = (x-4)x + (x-4)(-3)$
$= x \cdot x - 4 \cdot x + x(-3) - 4(-3)$
$= x^2 - 4x - 3x + 12$
$= x^2 - 7x + 12$

38. $x^2 - 10x + 21$

39. $(x+3)(x-3) = (x+3)x + (x+3)(-3)$
$= x \cdot x + 3 \cdot x + x(-3) + 3(-3)$
$= x^2 + 3x - 3x - 9$
$= x^2 - 9$

40. $x^2 - 36$

41. $(5-x)(5-2x) = (5-x)5 + (5-x)(-2x)$
$= 5 \cdot 5 - x \cdot 5 + 5(-2x) - x(-2x)$
$= 25 - 5x - 10x + 2x^2$
$= 25 - 15x + 2x^2$

42. $18 + 12x + 2x^2$

43. $\left(t+\frac{3}{2}\right)\left(t+\frac{4}{3}\right) = \left(t+\frac{3}{2}\right)t + \left(t+\frac{3}{2}\right)\left(\frac{4}{3}\right)$
$= t \cdot t + \frac{3}{2} \cdot t + t \cdot \frac{4}{3} + \frac{3}{2} \cdot \frac{4}{3}$
$= t^2 + \frac{3}{2}t + \frac{4}{3}t + 2$
$= t^2 + \frac{9}{6}t + \frac{8}{6}t + 2$
$= t^2 + \frac{17}{6}t + 2$

44. $a^2 + \frac{21}{10}a - 1$

45.
$$\left(\frac{1}{4}a + 2\right)\left(\frac{3}{4}a - 1\right)$$
$$= \left(\frac{1}{4}a + 2\right)\left(\frac{3}{4}a\right) + \left(\frac{1}{4}a + 2\right)(-1)$$
$$= \frac{1}{4}a\left(\frac{3}{4}a\right) + 2 \cdot \frac{3}{4}a + \frac{1}{4}a(-1) + 2(-1)$$
$$= \frac{3}{16}a^2 + \frac{3}{2}a - \frac{1}{4}a - 2$$
$$= \frac{3}{16}a^2 + \frac{6}{4}a - \frac{1}{4}a - 2$$
$$= \frac{3}{16}a^2 + \frac{5}{4}a - 2$$

46. $\frac{6}{25}t^2 - \frac{1}{5}t - 1$

47. Illustrate $x(x + 5)$ as the area of a rectangle with width x and length $x + 5$.

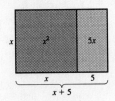

48.

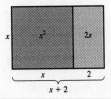

49. Illustrate $(x + 1)(x + 2)$ as the area of a rectangle with width $x + 1$ and length $x + 2$.

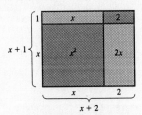

50.

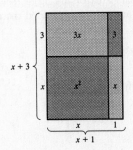

51. Illustrate $(x + 5)(x + 3)$ as the area of a rectangle with length $x + 5$ and width $x + 3$.

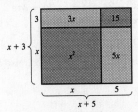

52.

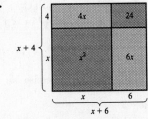

53. Illustrate $(3x + 2)(3x + 2)$ as the area of a square with sides of length $3x + 2$.

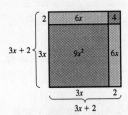

54.

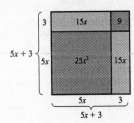

55.
$$(x^2 - x + 3)(x + 1)$$
$$= (x^2 - x + 3)x + (x^2 - x + 3)1$$
$$= x^2 \cdot x - x \cdot x + 3 \cdot x + x^2 \cdot 1 - x \cdot 1 + 3 \cdot 1$$
$$= x^3 - x^2 + 3x + x^2 - x + 3$$
$$= x^3 + 2x + 3$$

A partial check can be made by selecting a convenient replacement for x, say 1, and comparing the values of the original expression and the result.

$$(1^2 - 1 + 3)(1 + 1) \qquad 1^3 + 2 \cdot 1 + 3$$
$$= (1 - 1 + 3)(1 + 1) \qquad = 1 + 2 + 3$$
$$= 3 \cdot 2 \qquad\qquad = 6$$
$$= 6$$

Since the value of both expressions is 6, the multiplication is very likely correct.

56. $x^3 - 3x + 2$

57.
$$(2a + 5)(a^2 - 3a + 2)$$
$$= (2a + 5)a^2 - (2a + 5)(3a) + (2a + 5)2$$
$$= 2a \cdot a^2 + 5 \cdot a^2 - 2a \cdot 3a - 5 \cdot 3a + 2a \cdot 2 + 5 \cdot 2$$
$$= 2a^3 + 5a^2 - 6a^2 - 15a + 4a + 10$$
$$= 2a^3 - a^2 - 11a + 10$$

A partial check can be made as in Exercise 55.

58. $3t^3 - 11t^2 - 17t + 4$

59.
$$(y^2 - 3)(2y^3 + y + 1)$$
$$= (y^2 - 3)(2y^3) + (y^2 - 3)y + (y^2 - 3)(1)$$
$$= y^2 \cdot 2y^3 - 3 \cdot 2y^3 + y^2 \cdot y - 3 \cdot y + y^2 \cdot 1 - 3 \cdot 1$$
$$= 2y^5 - 6y^3 + y^3 - 3y + y^2 - 3$$
$$= 2y^5 - 5y^3 + y^2 - 3y - 3$$

A partial check can be made as in Exercise 55.

60. $5a^5 + 7a^3 - a^2 - 6a - 2$

61.
$$(5x^3 - 7x^2 + 1)(x - 3x^2)$$
$$= (5x^3 - 7x^2 + 1)x - (5x^3 - 7x^2 + 1)(3x^2)$$
$$= 5x^3 \cdot x - 7x^2 \cdot x + 1 \cdot x - 5x^3 \cdot 3x^2 + 7x^2 \cdot 3x^2 - 1 \cdot 3x^2$$
$$= 5x^4 - 7x^3 + x - 15x^5 + 21x^4 - 3x^2$$
$$= -15x^5 + 26x^4 - 7x^3 - 3x^2 + x$$

A partial check can be made in Exercise 55.

62. $8x^5 - 6x^3 - 6x^2 - 5x - 3$

63.

$x^2 - 3x + 2$	Line up like terms
$x^2 + x + 1$	in columns
$x^2 - 3x + 2$	Multiplying by 1
$x^3 - 3x^2 + 2x$	Multiplying by x
$x^4 - 3x^3 + 2x^2$	Multiplying by x^2
$x^4 - 2x^3 \qquad - x + 2$	

A partial check can be made as in Exercise 55.

64. $x^4 + 4x^3 - 3x^2 + 16x - 3$

65.

$2x^2 + 3x - 4$	
$2x^2 + x - 2$	
$-4x^2 - 6x + 8$	Multiplying by -2
$2x^3 + 3x^2 - 4x$	Multiplying by x
$4x^4 + 6x^3 - 8x^2$	Multiplying by $2x^2$
$4x^4 + 8x^3 - 9x^2 - 10x + 8$	Collecting like terms

A partial check can be made as in Exercise 55.

66. $4x^4 - 12x^3 - 5x^2 + 17x + 6$

67. We will multiply horizontally while still aligning like terms.
$$(x + 1)(x^3 + 7x^2 + 5x + 4)$$

$= x^4 + 7x^3 + 5x^2 + 4x$	Multiplying by x
$+ \ x^3 + 7x^2 + 5x + 4$	Multiplying by 1
$= x^4 + 8x^3 + 12x^2 + 9x + 4$	

A partial check can be made as in Exercise 55.

68. $x^4 + 7x^3 + 19x^2 + 21x + 6$

69. We will multiply horizontally while still aligning like terms.
$$\left(x - \frac{1}{2}\right)\left(2x^3 - 4x^2 + 3x - \frac{2}{5}\right)$$

$$= 2x^4 - 4x^3 + 3x^2 - \frac{2}{5}x$$
$$\qquad - x^3 + 2x^2 - \frac{3}{2}x + \frac{1}{5}$$
$$\overline{2x^4 - 5x^3 + 5x^2 - \frac{19}{10}x + \frac{1}{5}}$$

A partial check can be made as in Exercise 55.

70. $6x^4 - 10x^3 - 9x^2 - \frac{7}{6}x + \frac{1}{6}$

71.
$$10 - 2 + (-6)^2 \div 3 \cdot 2$$
$$= 10 - 2 + 36 \div 3 \cdot 2$$
$$= 10 - 2 + 12 \cdot 2$$
$$= 10 - 2 + 24$$
$$= 32$$

72.

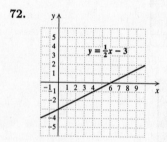

73.
$$15x - 18y + 12 = 3 \cdot 5x - 3 \cdot 6y + 3 \cdot 4$$
$$= 3(5x - 6y + 4)$$

74. $\dfrac{23}{19}$

75. ◈

76. ◈

77. ◈

78. ◈

79. The shaded area is the area of the large rectangle, $6y(14y - 5)$ less the area of the unshaded rectangle, $3y(3y + 5)$. We have:
$$6y(14y - 5) - 3y(3y + 5)$$
$$= 84y^2 - 30y - 9y^2 - 15y$$
$$= 75y^2 - 45y$$

80. $78t^2 + 40t$

81. Let $n =$ the missing number. Label the figure with the known areas.

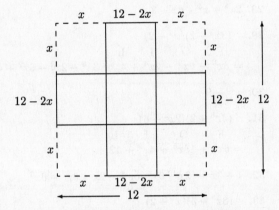

Then the area of the figure is $x^2 + 2x + nx + 2n$. This is equivalent to $x^2 + 7x + 10$, so we have $2x + nx = 7x$ and $2n = 10$. Solving either equation for n, we find that the missing number is 5.

82. 5

83.

The dimensions of the box are $12 - 2x$ by $12 - 2x$ by x. The volume is the product of the dimensions (volume = length $\times$ width $\times$ height):

$$\text{Volume} = (12 - 2x)(12 - 2x)x$$
$$= (144 - 48x + 4x^2)x$$
$$= 144x - 48x^2 + 4x^3, \text{ or}$$
$$4x^3 - 48x^2 + 144x$$

The outside surface area is the sum of the area of the bottom and the areas of the four sides. The dimensions of the bottom are $12 - 2x$ by $12 - 2x$, and the dimensions of each side are x by $12 - 2x$.

$$\begin{aligned}\text{Surface area} &= \text{Area of bottom} + \\ &\quad 4 \cdot \text{Area of each side} \\ &= (12 - 2x)(12 - 2x) + 4 \cdot x(12 - 2x) \\ &= 144 - 24x - 24x + 4x^2 + 48x - 8x^2 \\ &= 144 - 48x + 4x^2 + 48x - 8x^2 \\ &= 144 - 4x^2, \text{ or } -4x^2 + 144\end{aligned}$$

84. $x^3 - 5x^2 + 8x - 4$

85.
$$\begin{aligned}V &= (x + 2)^3 \\ &= (x + 2)(x + 2)(x + 2) \\ &= (x^2 + 2x + 2x + 4)(x + 2) \\ &= (x^2 + 4x + 4)(x + 2) \\ &= x^3 + 4x^2 + 4x + 2x^2 + 8x + 8 \\ &= x^3 + 6x^2 + 12x + 8\end{aligned}$$

The volume of the cube is $x^3 + 6x^2 + 12x + 8 \text{ cm}^3$.

86. 8 ft by 16 ft

87.
$$\begin{aligned}&(x + 3)(x + 6) + (x + 3)(x + 6) \\ &= (x + 3)x + (x + 3)6 + (x + 3)x + (x + 3)6 \\ &= x^2 + 3x + 6x + 18 + x^2 + 3x + 6x + 18 \\ &= 2x^2 + 18x + 36\end{aligned}$$

88. $2x^2 - 18x + 28$

89.
$$\begin{aligned}&(x + 5)^2 - (x - 3)^2 \\ &= (x + 5)(x + 5) - (x - 3)(x - 3) \\ &= (x + 5)x + (x + 5)5 - [(x - 3)x - (x - 3)3] \\ &= x^2 + 5x + 5x + 25 - (x^2 - 3x - 3x + 9) \\ &= x^2 + 10x + 25 - (x^2 - 6x + 9) \\ &= x^2 + 10x + 25 - x^2 + 6x - 9 \\ &= 16x + 16\end{aligned}$$

90. $2x^2 - 20x + 52$

91. ▮

Exercise Set 4.5

1. $(x + 5)(x^2 + 1)$
$$\quad\; \text{F} \qquad \text{O} \qquad \text{I} \qquad \text{L}$$
$$= x \cdot x^2 + x \cdot 1 + 5 \cdot x^2 + 5 \cdot 1$$
$$= x^3 + x + 5x^2 + 5, \text{ or } x^3 + 5x^2 + x + 5$$

2. $x^3 - x^2 - 3x + 3$

3. $(x^3 + 6)(x + 2)$
$$\quad\; \text{F} \qquad \text{O} \qquad \text{I} \qquad \text{L}$$
$$= x^3 \cdot x + x^3 \cdot 2 + 6 \cdot x + 6 \cdot 2$$
$$= x^4 + 2x^3 + 6x + 12$$

4. $x^5 + 12x^4 + 2x + 24$

5. $(y + 2)(y - 3)$
$$\quad\; \text{F} \qquad \text{O} \qquad \text{I} \qquad \text{L}$$
$$= y \cdot y + y \cdot (-3) + 2 \cdot y + 2 \cdot (-3)$$
$$= y^2 - 3y + 2y - 6$$
$$= y^2 - y - 6$$

6. $a^2 + 4a + 4$

7. $(3x + 2)(3x + 3)$
$$\quad\; \text{F} \qquad \text{O} \qquad \text{I} \qquad \text{L}$$
$$= 3x \cdot 3x + 3x \cdot 3 + 2 \cdot 3x + 2 \cdot 3$$
$$= 9x^2 + 9x + 6x + 6$$
$$= 9x^2 + 15x + 6$$

8. $8x^2 + 10x + 2$

9. $(5x - 6)(x + 2)$
$\quad\quad$ F $\quad\quad$ O $\quad\quad$ I $\quad\quad$ L
$= 5x \cdot x + 5x \cdot 2 + (-6) \cdot x + (-6) \cdot 2$
$= 5x^2 + 10x - 6x - 12$
$= 5x^2 + 4x - 12$

10. $t^2 - 81$

11. $(3t - 1)(3t + 1)$
$\quad\quad$ F $\quad\quad$ O $\quad\quad$ I $\quad\quad$ L
$= 3t \cdot 3t + 3t \cdot 1 + (-1) \cdot 3t + (-1) \cdot 1$
$= 9t^2 + 3t - 3t - 1$
$= 9t^2 - 1$

12. $4m^2 + 12m + 9$

13. $(2x - 7)(x - 1)$
$\quad\quad$ F $\quad\quad$ O $\quad\quad$ I $\quad\quad$ L
$= 2x \cdot x + 2x \cdot (-1) + (-7) \cdot x + (-7) \cdot (-1)$
$= 2x^2 - 2x - 7x + 7$
$= 2x^2 - 9x + 7$

14. $6x^2 - x - 1$

15. $\left(p - \dfrac{1}{4}\right)\left(p + \dfrac{1}{4}\right)$
$\quad\quad$ F $\quad\quad$ O $\quad\quad$ I $\quad\quad$ L
$= p \cdot p + p \cdot \dfrac{1}{4} + \left(-\dfrac{1}{4}\right) \cdot p + \left(-\dfrac{1}{4}\right) \cdot \dfrac{1}{4}$
$= p^2 + \dfrac{1}{4}p - \dfrac{1}{4}p - \dfrac{1}{16}$
$= p^2 - \dfrac{1}{16}$

16. $q^2 + \dfrac{3}{2}q + \dfrac{9}{16}$

17. $(x - 0.1)(x + 0.1)$
$\quad\quad$ F $\quad\quad$ O $\quad\quad$ I $\quad\quad$ L
$= x \cdot x + x \cdot (0.1) + (-0.1) \cdot x + (-0.1)(0.1)$
$= x^2 + 0.1x - 0.1x - 0.01$
$= x^2 - 0.01$

18. $x^2 - 0.1x - 0.12$

19. $(2x^2 + 6)(x + 1)$
$\quad\quad$ F $\quad\quad$ O $\quad\quad$ I $\quad\quad$ L
$= 2x^3 + 2x^2 + 6x + 6$

20. $4x^3 - 2x^2 + 6x - 3$

21. $(-2x + 1)(x + 6)$
$\quad\quad$ F $\quad\quad$ O $\quad\quad$ I $\quad\quad$ L
$= -2x^2 - 12x + x + 6$
$= -2x^2 - 11x + 6$

22. $6x^2 - 4x - 16$

23. $(a + 7)(a + 7)$
$\quad\quad$ F $\quad\quad$ O $\quad\quad$ I $\quad\quad$ L
$= a^2 + 7a + 7a + 49$
$= a^2 + 14a + 49$

24. $4y^2 + 28y + 49$

25. $(1 + 3t)(1 - 5t)$
$\quad\quad$ F $\quad\quad$ O $\quad\quad$ I $\quad\quad$ L
$= 1 - 5t + 3t - 15t^2$
$= 1 - 2t - 15t^2$

26. $-3x^2 - 5x - 2$

27. $(x^2 + 3)(x^3 - 1)$
$\quad\quad$ F $\quad\quad$ O $\quad\quad$ I $\quad\quad$ L
$= x^5 - x^2 + 3x^3 - 3$, or $x^5 + 3x^3 - x^2 - 3$

28. $2x^5 + x^4 - 6x - 3$

29. $(3x^2 - 2)(x^4 - 2)$
$\quad\quad$ F $\quad\quad$ O $\quad\quad$ I $\quad\quad$ L
$= 3x^6 - 6x^2 - 2x^4 + 4$, or $3x^6 - 2x^4 - 6x^2 + 4$

30. $x^{20} - 9$

31. $(3x^5 + 2)(2x^2 + 6)$
$\quad\quad$ F $\quad\quad$ O $\quad\quad$ I $\quad\quad$ L
$= 6x^7 + 18x^5 + 4x^2 + 12$

32. $1 + 3x^2 - 2x - 6x^3$, or $1 - 2x + 3x^2 - 6x^3$

33. $(8x^3 + 5)(x^2 + 2)$
$\quad\quad$ F $\quad\quad$ O $\quad\quad$ I $\quad\quad$ L
$= 8x^5 + 16x^3 + 5x^2 + 10$

34. $20 - 8x^2 - 10x + 4x^3$, or $20 - 10x - 8x^2 + 4x^3$

35. $(4x^2 + 3)(x - 3)$
$\quad\quad$ F $\quad\quad$ O $\quad\quad$ I $\quad\quad$ L
$= 4x^3 - 12x^2 + 3x - 9$

36. $14x^2 - 53x + 14$

37. $(x + 8)(x - 8)$ $\quad$ Product of sum and difference of the same two terms
$= x^2 - 8^2$
$= x^2 - 64$

38. $x^2 - 1$

39. $(2x + 1)(2x - 1)$ $\quad$ Product of sum and difference of the same two terms
$= (2x)^2 - 1^2$
$= 4x^2 - 1$

40. $x^4 - 1$

41. $(5m - 2)(5m + 2)$ $\quad$ Product of sum and difference of the same two terms
$= (5m)^2 - 2^2$
$= 25m^2 - 4$

42. $9x^8 - 4$

43. $\quad (2x^2 + 3)(2x^2 - 3)$ Product of sum and difference of the same two terms

$= (2x^2)^2 - 3^2$

$= 4x^4 - 9$

44. $36x^{10} - 25$

45. $\quad (3x^4 - 1)(3x^4 + 1)$

$= (3x^4)^2 - 1^2$

$= 9x^8 - 1$

46. $t^4 - 0.04$

47. $\quad (x^6 - x^2)(x^6 + x^2)$

$= (x^6)^2 - (x^2)^2$

$= x^{12} - x^4$

48. $4x^6 - 0.09$

49. $\quad (x^4 + 3x)(x^4 - 3x)$

$= (x^4)^2 - (3x)^2$

$= x^8 - 9x^2$

50. $\dfrac{9}{16} - 4x^6$

51. $\quad (2y^8 + 3)(2y^8 - 3)$

$= (2y^8)^2 - 3^2$

$= 4y^{16} - 9$

52. $m^2 - \dfrac{4}{9}$

53. $\quad (x + 2)^2$

$= x^2 + 2 \cdot x \cdot 2 + 2^2$ Square of a binomial

$= x^2 + 4x + 4$

54. $4x^2 - 4x + 1$

55. $\quad (3x^2 + 1)$ Square of a binomial

$= (3x^2)^2 + 2 \cdot 3x^2 \cdot 1 + 1^2$

$= 9x^4 + 6x^2 + 1$

56. $9x^2 + \dfrac{9}{2}x + \dfrac{9}{16}$

57. $\quad \left(a - \dfrac{2}{5}\right)^2$ Square of a binomial

$= a^2 - 2 \cdot a \cdot \dfrac{2}{5} + \left(\dfrac{2}{5}\right)^2$

$= a^2 - \dfrac{4}{5}a + \dfrac{4}{25}$

58. $4a^2 - \dfrac{4}{5}a + \dfrac{1}{25}$

59. $(x^2 + 3)^2 = (x^2)^2 + 2 \cdot x^2 \cdot 3 + 3^2$

$= x^4 + 6x^2 + 9$

60. $64x^2 - 16x^3 + x^4$

61. $(2 - 3x^4)^2 = 2^2 - 2 \cdot 2 \cdot 3x^4 + (3x^4)^2$

$= 4 - 12x^4 + 9x^8$

62. $36x^6 - 24x^3 + 4$

63. $(5 + 6t^2)^2 = 5^2 + 2 \cdot 5 \cdot 6t^2 + (6t^2)^2$

$= 25 + 60t^2 + 36t^4$

64. $9p^4 - 6p^3 + p^2$

65. $(7x - 0.3)^2 = (7x)^2 - 2(7x)(0.3) + (0.3)^2$

$= 49x^2 - 4.2x + 0.09$

66. $16a^2 - 4.8a + 0.36$

67. $\quad 5a^3(2a^2 - 1)$

$= 5a^3 \cdot 2a^2 - 5a^3 \cdot 1$ Multiplying each term of the binomial by the monomial

$= 10a^5 - 5a^3$

68. $a^3 - a^2 - 10a + 12$

69. $\quad (x^2 - 5)(x^2 + x - 1)$

$= x^4 + x^3 - x^2$ Multiplying horizontally

$\quad\quad -5x^2 - 5x + 5$ and aligning like terms

$= x^4 + x^3 - 6x^2 - 5x + 5$

70. $27x^6 - 9x^5$

71. $\quad (3 - 2x^3)^2$

$= 3^2 - 2 \cdot 3 \cdot 2x^3 + (2x^3)^2$ Squaring a binomial

$= 9 - 12x^3 + 4x^6$

72. $x^2 - 8x^4 + 16x^6$

73. $\quad 4x(x^2 + 6x - 3)$

$= 4x \cdot x^2 + 4x \cdot 6x + 4x(-3)$ Multiplying each term of the trinomial by the monomial

$= 4x^3 + 24x^2 - 12x$

74. $-8x^6 + 48x^3 + 72x$

75. $\quad \left(2x^2 - \dfrac{1}{2}\right)\left(2x^2 - \dfrac{1}{2}\right)$ Squaring a binomial

$= (2x^2)^2 - 2 \cdot 2x^2 \cdot \dfrac{1}{2} + \left(\dfrac{1}{2}\right)^2$

$= 4x^4 - 2x^2 + \dfrac{1}{4}$

76. $x^4 - 2x^2 + 1$

77. $\quad (-1 + 3p)(1 + 3p)$

$= (3p - 1)(3p + 1)$ Product of the sum and difference of the same two terms

$= (3p)^2 - 1^2$

$= 9p^2 - 1$, or $-1 + 9p^2$

78. $-9q^2 + 4$, or $4 - 9q^2$

79.
$$3t^2(5t^3 - t^2 + t)$$
$$= 3t^2 \cdot 5t^3 + 3t^2(-t^2) + 3t^2 \cdot t \quad \text{Multiplying each term of the trinomial by the monomial}$$
$$= 15t^5 - 3t^4 + 3t^3$$

80. $-5x^5 - 40x^4 + 45x^3$

81.
$$(6x^4 - 3)^2 \quad \text{Squaring a binomial}$$
$$= (6x^4)^2 - 2 \cdot 6x^4 \cdot 3 + 3^2$$
$$= 36x^8 + 36x^4 + 9$$

82. $64a^2 + 80a + 25$

83.
$$(3x+2)(4x^2+5) \quad \text{Product of two binomials; use FOIL}$$
$$= 3x \cdot 4x^2 + 3x \cdot 5 + 2 \cdot 4x^2 + 2 \cdot 5$$
$$= 12x^3 + 15x + 8x^2 + 10, \text{ or}$$
$$12x^3 + 8x^2 + 15x + 10$$

84. $6x^4 - 3x^2 - 63$

85.
$$(8 - 6x^4)^2 \quad \text{Squaring a binomial}$$
$$= 8^2 - 2 \cdot 8 \cdot 6x^4 + (6x^4)^2$$
$$= 64 - 96x^4 + 36x^8$$

86. $\frac{2}{9}t^4 + 3t^2 - 5$

87.
$$(a+1)(a^2-a+1)$$
$$\begin{array}{r} = a^3 - a^2 + a \quad \text{Multiplying horizontally} \\ \underline{a^2 - a + 1} \quad \text{and aligning like terms} \\ a^3 \qquad\qquad + 1 \end{array}$$

88. $x^3 - 125$

89.

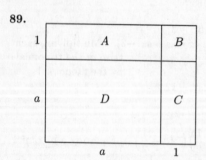

We can find the shaded area in two ways.

Method 1: The figure is a square with side $a + 1$, so the area is $(a + 1)^2 = a^2 + 2a + 1$.

Method 2: We add the areas of A, B, C, and D.
$1 \cdot a + 1 \cdot 1 + 1 \cdot a + a \cdot a = a + 1 + a + a^2 = a^2 + 2a + 1$.

Either way we find that the total shaded area is $a^2 + 2a + 1$.

90. $x^2 + 6x + 9$

91.

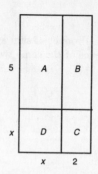

We can find the shaded area in two ways.

Method 1: The figure is a rectangle with dimensions $x + 5$ by $x + 2$, so the area is
$$(x + 5)(x + 2) = x^2 + 2x + 5x + 10 = x^2 + 7x + 10.$$
Method 2: We add the areas of A, B, C, and D.
$$5 \cdot x + 2 \cdot 5 + 2 \cdot x + x \cdot x = 5x + 10 + 2x + x^2 = x^2 + 7x + 10.$$
Either way, we find that the area is $x^2 + 7x + 10$.

92. $t^2 + 7t + 12$

93.

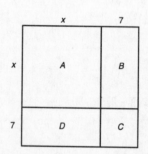

We can find the shaded area in two ways.

Method 1: The figure is a square with side $x + 7$, so the area is $(x + 7)^2 = x^2 + 14x + 49$.

Method 2: We add the areas of A, B, C, and D.
$$x \cdot x + x \cdot 7 + 7 \cdot 7 + 7 \cdot x = x^2 + 7x + 49 + 7x = x^2 + 14x + 49.$$
Either way, we find that the total shaded area is $x^2 + 14x + 49$.

94. $a^2 + 10a + 25$

95.

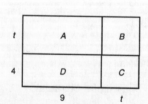

We can find the shaded area in two ways.

Method 1: The figure is a rectangle with dimensions $t + 9$ by $t + 4$, so the area is
$$(t + 9)(t + 4) = t^2 + 4t + 9t + 36 = t^2 + 13t + 36$$
Method 2: We add the areas of A, B, C, and D.
$$9 \cdot t + t \cdot t + 4 \cdot t + 4 \cdot 9 = 9t + t^2 + 4t + 36 = t^2 + 13t + 36.$$
Either way, we find that the total shaded area is $t^2 + 13t + 36$.

96. $a^2 + 8a + 7$

97.

We can find the shaded area in two ways.

Method 1: The figure is a square with side $3x + 4$, so the area is $(3x + 4)^2 = 9x^2 + 24x + 16$.

Method 2: We add the areas of A, B, C, and D.

$3x \cdot 3x + 3x \cdot 4 + 4 \cdot 4 + 3x \cdot 4 = 9x^2 + 12x + 16 + 12x = 9x^2 + 24x + 16$.

Either way, we find that the total shaded area is $9x^2 + 24x + 16$.

98. $25t^2 + 20t + 4$

99. We draw a square with side $x + 6$.

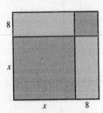

100.

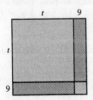

101. We draw a square with side $t + 9$.

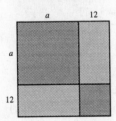

102.

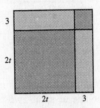

103. We draw a square with side $4a + 1$.

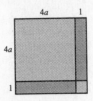

104.

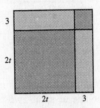

105. *Familiarize*. Let $t =$ the number of watts used by the television set. Then $10t =$ the number of watts used by the lamps, and $40t =$ the number of watts used by the air conditioner.

Translate.

$$\underbrace{\text{Lamp watts}}_{10t} + \underbrace{\begin{array}{c}\text{Air}\\\text{conditioner}\\\text{watts}\end{array}}_{40t} + \underbrace{\begin{array}{c}\text{Television}\\\text{watts}\end{array}}_{t} = \underbrace{\begin{array}{c}\text{Total}\\\text{watts}\end{array}}_{2550}$$

Solve. We solve the equation.

$$10t + 40t + t = 2550$$
$$51t = 2550$$
$$t = 50$$

The possible solution is:

Television, t: 50 watts

Lamps, $10t$: $10 \cdot 50$, or 500 watts

Air conditioner, $40t$: $40 \cdot 50$, or 2000 watts

Check. The number of watts used by the lamps, 500, is 10 times 50, the number used by the television. The number of watts used by the air conditioner, 2000, is 40 times 50, the number used by the television. Also, $50 + 500 + 2000 = 2550$, the total wattage used.

State. The television uses 50 watts, the lamps use 500 watts, and the air conditioner uses 2000 watts.

106. $\dfrac{28}{27}$

107.
$$ab - c = ad$$
$$ab = ad + c$$
$$ab - ad = c$$
$$a(b - d) = c$$
$$a = \frac{c}{b - d}$$

108. IV

109.

110. ◈

111. ◈

112. ◈

113.
$$5x(3x-1)(2x+3)$$
$$= 5x(6x^2 + 7x - 3) \quad \text{Using FOIL}$$
$$= 30x^3 + 35x^2 - 15x$$

114. $16x^4 - 81$

115.
$$[(a-5)(a+5)]^2$$
$$= (a^2 - 25)^2 \quad \text{Finding the product of a sum}$$
$$\qquad\qquad\qquad \text{and difference of same two terms}$$
$$= a^4 - 50a^2 + 625 \quad \text{Squaring a binomial}$$

116. $a^4 - 18a^2 + 81$

117.
$$(3t^4 - 2)^2 1(3t^4 + 2)^2$$
$$= [(3t^4 - 2)(3t^4 + 2)]^2$$
$$= (9t^8 - 4)^2$$
$$= 81t^{16} - 72t^8 + 16$$

118. $1050.4081x^2 + 348.0834x + 28.8369$

119. $18 \times 22 = (20-2)(20+2) = 20^2 - 2^2 =$
$400 - 4 = 396$

120. $(100-7)(100+7) = 10,000 - 49 = 9951$

121.
$$(x+2)(x-5) = (x+1)(x-3)$$
$$x^2 - 5x + 2x - 10 = x^2 - 3x + x - 3$$
$$x^2 - 3x - 10 = x^2 - 2x - 3$$
$$-3x - 10 = -2x - 3 \quad \text{Adding } -x^2$$
$$-3x + 2x = 10 - 3 \quad \text{Adding } 2x \text{ and } 10$$
$$-x = 7$$
$$x = -7$$
The solution is -7.

122. 0

123. If $l = $ the length, then $l+1 = $ the height, and $l - 1 = $ the width. Recall that the volume of a rectangular solid is given by length × width × height.
$$\text{Volume} = l(l-1)(l+1) = l(l^2 - 1) = l^3 - l$$

124. $w^3 + 3w^2 + 2w$

125.

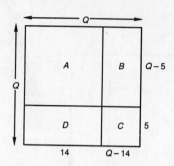

The dimensions of the shaded area, B, are $Q - 14$ by $Q - 5$, so one expression is $(Q-14)(Q-5)$.

To find another expression we find the area of regions B and C together and subtract the area of region C. The region consisting of B and C together has dimensions Q by $Q - 14$, so its area is $Q(Q-14)$. Region C has dimensions 5 by $Q - 14$, so its area is $5(Q - 14)$. Then another expression for the shaded area, B, is $Q(Q - 14) - 5(Q - 14)$.

It is possible to find other equivalent expressions also.

126. $F^2 - (F-7)(F-17)$, $24F - 119$

127.

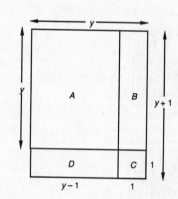

The dimensions of the shaded area, regions A and D together, are $y+1$ by $y-1$ so the area is $(y+1)(y-1)$.

To find another expression we add the areas of regions A and D. The dimensions of region A are y by $y - 1$, and the dimensions of region D are $y - 1$ by 1, so the sum of the areas is $y(y-1) + (y-1)(1)$, or $y(y-1) + y - 1$.

It is possible to find other equivalent expressions also.

128. 10, 11, 12

129. a)

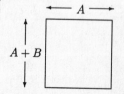

The area of the entire rectangle is $A(A+B)$, or $A^2 + AB$.

b)

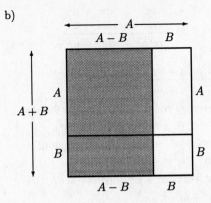

The sum of the areas of the two unshaded rectangles is $A \cdot B + B \cdot B$, or $AB + B^2$.

c) Area in part (a) - area in part (b)
$$= A^2 + AB - (AB + B^2)$$
$$= A^2 + AB - AB - B^2$$
$$= A^2 - B^2$$

d) The area of the shaded region is $(A+B)(A-B) = A^2 - B^2$. This is the same as the polynomial found in part (c).

130.

Exercise Set 4.6

1. We replace x by 3 and y by -2.
$$x^2 - 3y^2 + 2xy = 3^2 - 3(-2)^2 + 2 \cdot 3(-2) =$$
$$9 - 12 - 12 = -15.$$

2. 53

3. We replace x by 2, y by -3, and z by -1.
$$xyz^2 - z = 2(-3)(-1)^2 - (-1) = -6 + 1 = -5$$

4. -1

5. Evaluate the polynomial for $h = 160$ and $A = 50$.
$$0.041h - 0.018A - 2.69$$
$$= 0.041(160) - 0.018(50) - 2.69$$
$$= 6.56 - 0.9 - 2.69$$
$$= 2.97$$

The woman's lung capacity is 2.97 liters.

6. 3.715 liters

7. Evaluate the polynomial for $h = 50$, $v = 40$, and $t = 2$.
$$h + vt - 4.9t^2$$
$$= 50 + 40 \cdot 2 - 4.9(2)^2$$
$$= 50 + 80 - 19.6$$
$$= 110.4$$

The rocket will be 110.4 m above the ground 2 seconds after blast off.

8. 205.9 m

9. Evaluate the polynomial for $h = 4$, $r = \frac{3}{4}$, and $\pi \approx 3.14$.
$$2\pi rh + \pi r^2 \approx 2(3.14)\left(\frac{3}{4}\right)(4) + (3.14)\left(\frac{3}{4}\right)^2$$
$$\approx 2(3.14)\left(\frac{3}{4}\right)(4) + (3.14)\left(\frac{9}{16}\right)$$
$$\approx 18.84 + 1.76625$$
$$\approx 20.60625$$

The surface area is about 20.60625 in^2.

10. 63.78125 in^2

11. $x^3y - 2xy + 3x^2 - 5$

Term	Coefficient	Degree	
x^3y	1	4	(Think: $x^3y = x^3y^1$)
$-2xy$	-2	2	(Think: $-2xy = -2x^1y^1$)
$3x^2$	3	2	
-5	-5	0	(Think: $-5 = -5x^0$)

The degree of the polynomial is the degree of the term of highest degree. The term of highest degree is x^3y. Its degree is 4, so the degree of the polynomial is 4.

12. Coefficients: 5, -1, 15, 1

Degrees: 3, 2, 1, 0; 3

13. $17x^2y^3 - 3x^3yz - 7$

Term	Coefficient	Degree	
$17x^2y^3$	17	5	
$-3x^3yz$	-3	5	(Think: $-3x^3yz = -3x^3y^1z^1$)
-7	-7	0	(Think: $-7 = -7x^0$)

The terms of highest degree are $17x^2y^3$ and $-3x^3yz$. Each has degree 5. The degree of the polynomial is 5.

14. Coefficients: 6, -1, 8, -1

Degrees: 0, 2, 4, 5; 5

15. $5a + b - 4a - 3b = (5 - 4)a + (1 - 3)b = a - 2b$

16. $y - 7$

17. $3x^2y - 2xy^2 + x^2$

There are no like terms, so none of the terms can be collected.

18. $m^3 + 2m^2n - 3m^2 + 3mn^2$

19. $2u^2v - 3uv^2 + 6u^2v - 2uv^2$
$$= (2 + 6)u^2v + (-3 - 2)uv^2$$
$$= 8u^2v - 5uv^2$$

20. $-2x^2 - 4xy - 2y^2$

21.
$$8uv + 3av + 14au + 7av$$
$$= 8uv + (3+7)av + 14au$$
$$= 8uv + 10av + 14au$$

22. $3x^2y + 3z^2y + 3xy^2$

23.
$$(2x^2 - xy + y^2) + (-x^2 - 3xy + 2y^2)$$
$$= (2-1)x^2 + (-1-3)xy + (1+2)y^2$$
$$= x^2 - 4xy + 3y^2$$

24. $-4r^3 + 2rs - 9s^2$

25.
$$(7a^4 - 5ab + 6ab^2) - (9a^4 + 3ab - ab^2)$$
$$= 7a^4 - 5ab + 6ab^2 - 9a^4 - 3ab + ab^2$$
$$\qquad\qquad\qquad\text{Adding the opposite}$$
$$= (7-9)a^4 + (-5-3)ab + (6+1)ab^2$$
$$= -2a^4 - 8ab + 7ab^2$$

26. $3r + s - 4$

27.
$$(b^3a^2 - 2b^2a^3 + 3ba + 4) + (b^2a^3 - 4b^3a^2 + 2ba - 1)$$
$$= (1-4)b^3a^2 + (-2+1)b^2a^3 + (3+2)ba + (4-1)$$
$$= -3b^3a^2 - b^2a^3 + 5ba + 3$$

28. $-x^2 - 8xy - y^2$

29.
$$(x^3 - y^3) - (-2x^3 + x^2y - xy^2 + 2y^3)$$
$$= x^3 - y^3 + 2x^3 - x^2y + xy^2 - 2y^3$$
$$= 3x^3 - 3y^3 - x^2y + xy^2, \text{ or}$$
$$\quad 3x^3 - x^2y + xy^2 - 3y^3$$

30. $2ab$

31.
$$(3y^4x^2 + 2y^3x - 3y) - (2y^4x^2 + 2y^3x - 4y - 2x)$$
$$= 3y^4x^2 + 2y^3x - 3y - 2y^4x^2 - 2y^3x + 4y + 2x$$
$$= (3-2)y^4x^2 + (2-2)y^3x + (-3+4)y + 2x$$
$$= y^4x^2 + 0y^3x + y + 2x$$
$$= y^4x^2 + y + 2x$$

32. $15a^2b - 4ab$

33.
$$(4x + 5y) + (-5x + 6y) - (7x + 3y)$$
$$= 4x + 5y - 5x + 6y - 7x - 3y$$
$$= (4 - 5 - 7)x + (5 + 6 - 3)y$$
$$= -8x + 8y$$

34. $-5b$

35. $\overset{\text{F}\qquad\text{O}\qquad\text{I}\qquad\text{L}}{(3z - u)(2z + 3u) = 6z^2 + 9zu - 2uz - 3u^2}$
$$= 6z^2 + 7zu - 3u^2$$

36. $a^4b^2 - 7a^2b + 10$

37. $\overset{\text{F}\qquad\text{O}\qquad\text{I}\qquad\text{L}}{(xy + 7)(xy - 4) = x^2y^2 - 4xy + 7xy - 28 - 28}$
$$= x^2y^2 - 3xy - 28$$

38. $a^6 - b^2c^2$

39.
$$
\begin{array}{r}
m^2 \ - \ mn \ + \ n^2 \\
m^2 \ + \ mn \ + \ n^2 \\
\hline
m^2n^2 - mn^3 + n^4 \\
m^3n - m^2n^2 + mn^3 \\
m^4 - m^3n + m^2n^2 \\
\hline
m^4 \qquad + m^2n^2 \qquad + n^4
\end{array}
$$
This exercise could also be done as follows:
$$(m^2 + n^2 - mn)(m^2 + mn + n^2)$$
$$= [(m^2 + n^2) - mn][(m^2 + n^2) + mn]$$
$$= (m^2 + n^2)^2 - (mn)^2$$
$$= m^4 + 2m^2n^2 + n^4 - m^2n^2$$
$$= m^4 + m^2n^2 + n^4$$

40. $y^6x + y^4 + y^4x + 2y^2 + 1$

41.
$$
\begin{array}{r}
a^2 \ + \ ab \ + \ b^2 \\
a \ - \ b \\
\hline
- \ a^2b - ab^2 - b^3 \\
a^3 + a^2b + ab^2 \\
\hline
a^3 \qquad\qquad - b^3
\end{array}
$$

42. $12x^2y^2 + 2xy - 2$

43. $(m^3n + 8)(m^3n - 6)$
$$\overset{\text{F}\qquad\text{O}\qquad\text{I}\qquad\text{L}}{= m^6n^2 - 6m^3n + 8m^3n - 48}$$
$$= m^6n^2 + 2m^3n - 48$$

44. $12 - c^2d^2 - c^4d^4$

45. $(6x - 2y)(5x - 3y)$
$$\overset{\text{F}\qquad\text{O}\qquad\text{I}\qquad\text{L}}{= 30x^2 - 18xy - 10xy + 6y^2}$$
$$= 30x^2 - 28xy + 6y^2$$

46. $m^3 + m^2n - mn^2 - n^3$

47. $(pq + 0.2)(0.4pq - 0.1)$
$$\overset{\text{F}\qquad\text{O}\qquad\text{I}\qquad\text{L}}{= 0.4p^2q^2 - 0.1pq + 0.08pq - 0.02}$$
$$= 0.4p^2q^2 - 0.02pq - 0.02$$

48. $x^5y^5 - x^2y^2 + x^9y^9 - x^6y^6, \text{ or}$
$$x^9y^9 - x^6y^6 + x^5y^5 - x^2y^2$$

49.
$$(x + h)^2$$
$$= x^2 + 2xh + h^2 \quad [(A+B)^2 = A^2 + 2AB + B^2]$$

50. $9a^2 + 12ab + 4b^2$

51.
$$(r^3t^2 - 4)^2$$
$$= (r^3t^2)^2 - 2 \cdot r^3t^2 \cdot 4 + 4^2$$
$$\qquad\qquad\qquad [(A-B)^2 = A^2 - 2AB + B^2]$$
$$= r^6t^4 - 8r^3t^2 + 16$$

52. $9a^4b^2 - 6a^2b^3 + b^4$

53. $(c^2 - d)(c^2 + d) = (c^2)^2 - d^2$
$$= c^4 - d^2$$

54. $p^6 - 25q^2$

55. $(ab + cd^2)(ab - cd^2) = (ab)^2 - (cd^2)^2$
$$= a^2b^2 - c^2d^4$$

56. $x^2y^2 - p^2q^2$

57. $(x + y - 3)(x + y + 3)$
$$= [(x + y) - 3][(x + y) + 3]$$
$$= (x + y)^2 - 3^2$$
$$= x^2 + 2xy + y^2 - 9$$

58. $x^2 - y^2 - 2yz - z^2$

59. $[a + b + c][a - (b + c)]$
$$= [a + (b + c)][a - (b + c)]$$
$$= a^2 - (b + c)^2$$
$$= a^2 - (b^2 + 2bc + c^2)$$
$$= a^2 - b^2 - 2bc - c^2$$

60. $a^2 - b^2 - 2bc - c^2$

61. The figure is a rectangle with dimensions $a + b$ by $a + c$. Its area is $(a + b)(a + c) = a^2 + ac + ab + bc$.

62. $x^2 + 2xy + y^2$

63. The figure is a parallelogram with base $x + z$ and height $x - z$. Thus the area is $(x + z)(x - z) = x^2 - z^2$.

64. $\frac{1}{2}a^2b^2 - 2$

65. The figure is a square with side $x + y + z$. Thus the area is
$$(x + y + z)^2$$
$$= [(x + y) + z]^2$$
$$= (x + y)^2 + 2(x + y)(z) + z^2$$
$$= x^2 + 2xy + y^2 + 2xz + 2yz + z^2.$$

66. $a^2 + 2ac + c^2 + ad + cd + ab + bc + bd$

67. The figure is a triangle with base $x + 2y$ and height $x - y$. Thus the area is $\frac{1}{2}(x + 2y)(x - y) = \frac{1}{2}(x^2 + xy - 2y^2) = \frac{1}{2}x^2 + \frac{1}{2}xy - y^2$.

68. $m^2 - n^2$

69. We draw a rectangle with dimensions $r + s$ by $u + v$.

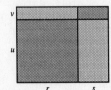

70.

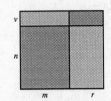

71. We draw a rectangle with dimensions $a + b + c$ by $a + d + f$.

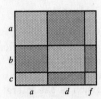

72.

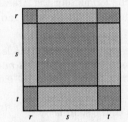

73. Locate June, 1994, on the horizontal scale. Then move up to the line representing white office paper and left to the vertical scale to read the information being sought. In June, 1994, the price being paid for white office paper was $50 per ton.

74. December, 1995

75. Locate the highest point on the line representing newsprint. Then move down to the horizontal scale to read the information being sought. The value of newsprint peaked in June, 1995.

76. December, 1991

77. Find the 6-month period for which the line slants up most steeply from left to right. We see that the price paid for newsprint increased the most during the 6-month period from June, 1994, to December, 1994.

78. June, 1995, to December, 1995

79.

80. ◈

81. ◈

82. ◈

83. The unshaded region is a circle with radius $a - b$. Then the shaded area is the area of a circle with radius a less the area of a circle with radius $a - b$. Thus, we have:

Shaded area $= \pi a^2 - \pi(a-b)^2$
$= \pi a^2 - \pi(a^2 - 2ab + b^2)$
$= \pi a^2 - \pi a^2 + 2\pi ab - \pi b^2$
$= 2\pi ab - \pi b^2$

84. $4xy - 4y^2$

85. The shaded area is the area of a square with side a less the areas of 4 squares with side b. Thus, the shaded area is $a^2 - 4 \cdot b^2$, or $a^2 - 4b^2$.

86. $\pi x^2 + 2xy$

87. The lateral surface area of the outer portion of the solid is the lateral surface area of a right circular cylinder with radius n and height h. The lateral surface area of the inner portion is the lateral surface area of a right circular cylinder with radius m and height h. Recall that the formula for the lateral surface area of a right circular cylinder with radius r and height h is $2\pi rh$.

The surface area of the top is the area of a circle with radius n less the area of a circle with radius m. The surface area of the bottom is the same as the surface area of the top.

Thus, the surface area of the solid is
$$2\pi nh + 2\pi mh + 2\pi n^2 - 2\pi m^2.$$

88. $2x^2 + 4xh - 2\pi r^2 + 2\pi rh$

89. Replace t with 2 and multiply.
$P(1+r)^2$
$= P(1 + 2r + r^2)$
$= P + 2Pr + Pr^2$

90. $P - 2Pr + Pr^2$

91. ◈

Exercise Set 4.7

1. $\dfrac{32x^5 - 16x}{8} = \dfrac{32x^5}{8} - \dfrac{16x}{8}$
$= \dfrac{32}{8}x^5 - \dfrac{16}{8}x$ Dividing coefficients
$= 4x^5 - 2x$

To check, we multiply the quotient by 8:
$(4x^5 - 2x)8 = 32x^5 - 16x$
The answer checks.

2. $2a^4 - \dfrac{1}{2}a^2$

3. $\dfrac{u - 2u^2 + u^7}{u}$
$= \dfrac{u}{u} - \dfrac{2u^2}{u} + \dfrac{u^7}{u}$
$= 1 - 2u + u^6$

Check: We multiply.
$$\begin{array}{r} 1 - 2u + u^6 \\ u \\ \hline u - 2u^2 + u^7 \end{array}$$

4. $50x^4 - 7x^3 + x$

5. $(15t^3 - 24t^2 + 6t) \div (3t)$
$= \dfrac{15t^3 - 24t^2 + 6t}{3t}$
$= \dfrac{15t^3}{3t} - \dfrac{24t^2}{3t} + \dfrac{6t}{3t}$
$= 5t^2 - 8t + 2$

Check: We multiply.
$$\begin{array}{r} 5t^2 - 8t + 2 \\ 3t \\ \hline 15t^3 - 24t^2 + 6t \end{array}$$

6. $5t^2 - 3t - 6$

7. $(35x^6 - 20x^4 - 5x^2) \div (-5x^2)$
$= \dfrac{35x^6 - 20x^4 - 5x^2}{-5x^2}$
$= \dfrac{35x^6}{-5x^2} - \dfrac{20x^4}{-5x^2} - \dfrac{5x^2}{-5x^2}$
$= -7x^4 - (-4x^2) - (-1)$
$= -7x^4 + 4x^2 + 1$

Check: We multiply.
$$\begin{array}{r} -7x^4 + 4x^2 + 1 \\ -5x^2 \\ \hline 35x^6 - 20x^4 - 5x^2 \end{array}$$

8. $-2x^4 - 4x^3 + 1$

9. $(24x^5 - 40x^4 + 6x^3) \div (4x^3)$
$= \dfrac{24x^5 - 40x^4 + 6x^3}{4x^3}$
$= \dfrac{24x^5}{4x^3} - \dfrac{40x^4}{4x^3} + \dfrac{6x^3}{4x^3}$
$= 6x^2 - 10x + \dfrac{3}{2}$

Check: We multiply.
$$\begin{array}{r} 6x^2 - 10x + \dfrac{3}{2} \\ 4x^3 \\ \hline 24x^5 - 40x^4 + 6x^3 \end{array}$$

10. $2x^3 - 3x^2 - \dfrac{1}{3}$

11. $\dfrac{8x^2 - 10x + 1}{2}$

$= \dfrac{8x^2}{2} - \dfrac{10x}{2} + \dfrac{1}{2}$

$= 4x^2 - 5x + \dfrac{1}{2}$

Check: We multiply.

$$4x^2 - 5x + \dfrac{1}{2}$$
$$\underline{\hspace{4cm} 2}$$
$$8x^2 - 10x + 1$$

12. $2x^2 + x - \dfrac{2}{3}$

13. $\dfrac{2x^3 + 6x^2 + 4x}{2x}$

$= \dfrac{2x^3}{2x} + \dfrac{6x^2}{2x} + \dfrac{4x}{2x}$

$= x^2 + 3x + 2$

Check: We multiply.

$$x^2 + 3x + 2$$
$$\underline{\hspace{4.5cm} 2x}$$
$$2x^3 + 6x^2 + 4x$$

14. $2x^2 - 3x + 5$

15. $\dfrac{9r^2s^2 + 3r^2s - 6rs^2}{-3rs}$

$= \dfrac{9r^2s^2}{-3rs} + \dfrac{3r^2s}{-3rs} - \dfrac{6rs^2}{-3rs}$

$= -3rs - r + 2s$

Check: We multiply.

$$-3rs - r + 2s$$
$$\underline{\hspace{4.5cm} -3rs}$$
$$9r^2s^2 + 3r^2s - 6rs^2$$

16. $1 - 2x^2y + 3x^4y^5$

17.
$$
\begin{array}{r}
x + 6 \\
x - 2 \overline{)\, x^2 + 4x - 12 } \\
\underline{x^2 - 2x } \\
6x - 12 \leftarrow (x^2 + 4x) - (x^2 - 2x) = 6x \\
\underline{6x - 12} \\
0 \leftarrow (6x - 12) - (6x - 12) = 0
\end{array}
$$

The answer is $x + 6$.

18. $x - 2$

19.
$$
\begin{array}{r}
x - 5 \\
x - 5 \overline{)\, x^2 - 10x - 25 } \\
\underline{x^2 - 5x } \\
-5x - 25 \leftarrow (x^2 - 10x) - (x^2 - 5x) = \\
-5x \\
\underline{-5x + 25} \\
-50 \leftarrow (-5x - 25) - (-5x + 25)
\end{array}
$$

The answer is $x - 5 + \dfrac{-50}{x - 5}$, or $x - 5 - \dfrac{50}{x - 5}$.

20. $x + 4 + \dfrac{-32}{x + 4}$

21.
$$
\begin{array}{r}
2x - 1 \\
x + 6 \overline{)\, 2x^2 + 11x - 5 } \\
\underline{2x^2 + 12x } \\
-x - 5 \leftarrow (2x^2 + 11x) - (2x^2 + 12x) = \\
-x \\
\underline{-x - 6} \\
1 \leftarrow (-x - 5) - (-x - 6) = 1
\end{array}
$$

The answer is $2x - 1 + \dfrac{1}{x + 6}$.

22. $3x + 4 + \dfrac{-5}{x - 2}$

23.
$$
\begin{array}{r}
x - 3 \\
x + 3 \overline{)\, x^2 + 0x - 9 } \leftarrow \text{Filling in the missing term} \\
\underline{x^2 + 3x } \\
-3x - 9 \leftarrow x^2 - (x^2 + 3x) = -3x \\
\underline{-3x - 9} \\
0 \leftarrow (-3x - 9) - (-3x - 9)
\end{array}
$$

The answer is $x - 3$.

24. $x - 5$

25.
$$
\begin{array}{r}
x + 4 \\
3x - 1 \overline{)\, 3x^2 + 11x - 4 } \\
\underline{3x^2 - x } \\
12x - 4 \leftarrow (3x^2 + 11x) - (3x^2 - x) = \\
12x \\
\underline{12x - 4} \\
0 \leftarrow (12x - 4) - (12x - 4) = 0
\end{array}
$$

The answer is $x + 4$.

26. $2x + 3$

27.
$$
\begin{array}{r}
2x^2 - 7x + 4 \\
4x + 3 \overline{)\, 8x^3 - 22x^2 - 5x + 12 } \\
\underline{8x^3 + 6x^2 } \\
-28x^2 - 5x \leftarrow (8x^3 - 22x^2) - \\
(8x^3 + 6x^2) = -28x^2 \\
\underline{-28x^2 - 21x} \\
16x + 12 \leftarrow (-28x^2 - 5x) - \\
(-28x^2 - 21x) = 16x \\
\underline{16x + 12} \\
0 \leftarrow (16x + 12) - (16x + 12)
\end{array}
$$

The answer is $2x^2 - 7x + 4$.

28. $x^2 - 3x + 1$

29.
$$
\begin{array}{r}
x^2 + 1 \\
x^2 - 3 \overline{)\, x^4 + 0x^3 - 2x^2 + 4x - 5 } \leftarrow \text{Writing in the} \\
\underline{x^4 - 3x^2 } \text{missing term} \\
x^2 + 4x - 5 \leftarrow (x^4 - 2x^2) - \\
(x^4 - 3x^2) = x^2 \\
\underline{x^2 - 3} \\
4x - 2 \leftarrow (x^2 + 4x - 5) - \\
(x^2 - 3) = 4x - 2
\end{array}
$$

The answer is $x^2 + 1 + \dfrac{4x - 2}{x^2 - 3}$.

30. $x^2 - 1 + \dfrac{3x - 1}{x^2 + 5}$

31.
$$
\begin{array}{r}
t^2 - 2t + 3 \\
t + 1 \overline{\smash{\big)}\ t^3 - t^2 + t - 1} \\
\underline{t^3 + t^2} \\
-2t^2 + t \quad \leftarrow (t^3 - t^2) - (t^3 + t^2) = -2t^2 \\
\underline{-2t^2 - 2t} \\
3t - 1 \quad \leftarrow (-2t^2 + t) - \\
(-2t^2 - 2t) = 3t \\
\underline{3t + 3} \\
-4
\end{array}
$$

The answer is $t^2 - 2t + 3 + \dfrac{-4}{t + 1}$, or

$t^2 - 2t + 3 - \dfrac{4}{t + 1}$.

32. $t^2 + 1$

33.
$$
\begin{array}{r}
3x^2 \qquad - 3 \\
2x^2 + 1 \overline{\smash{\big)}\ 6x^4 + 0x^3 - 3x^2 + x - 4} \leftarrow \text{Writing in the} \\
\underline{6x^4 \qquad + 3x^2} \qquad\qquad \text{missing term} \\
-6x^2 + x - 4 \leftarrow (6x^4 - 3x^2) - \\
(6x^4 + 3x^2) = -6x^2 \\
\underline{-6x^2 \qquad - 3} \\
x - 1 \leftarrow (-6x^2 + x - 4) - \\
(-6x^2 - 3) = x - 1
\end{array}
$$

The answer is $3x^2 - 3 + \dfrac{x - 1}{2x^2 + 1}$.

34. $2x^2 + 1 + \dfrac{-x}{2x^2 - 3}$

35. ***Familiarize***. Let $w =$ the width. Then $w + 15 =$ the length. We draw a picture.

$$w + 15$$

$$w \ \boxed{} \ w$$

$$w + 15$$

We will use the fact that the perimeter is 640 ft to find w (the width). Then we can find $w + 15$ (the length) and multiply the length and the width to find the area.

Translate.

$$\text{Width} + \text{Width} + \text{Length} + \text{Length} = \text{Perimeter}$$
$$w + w + (w + 15) + (w + 15) = 640$$

Carry out.
$$
\begin{aligned}
w + w + (w + 15) + (w + 15) &= 640 \\
4w + 30 &= 640 \\
4w &= 610 \\
w &= 152.5
\end{aligned}
$$

If the width is 152.5, then the length is $152.5 + 15$, or 167.5. The area is $(167.5)(152.5)$, or $25,543.75$ ft^2.

Check. The length, 167.5 ft, is 15 ft greater than the width, 152.5 ft. The perimeter is $152.5 + 152.5 + 167.5 + 167.5$, or 640 ft. We should also recheck the computation we used to find the area. The answer checks.

State. The area is $25,543.75$ ft^2.

36. 2

37.
$$
\begin{aligned}
2x &> 12 + 7x \\
-5x &> 12 \qquad \text{Subtracting } 7x \\
x &< -\frac{12}{5} \qquad \text{Dividing by } -5 \text{ and reversing the} \\
&\qquad\qquad\quad \text{inequality symbol}
\end{aligned}
$$

The solution set is $\left\{ x \middle| x < -\dfrac{12}{5} \right\}$.

38.

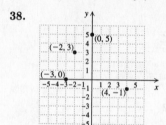

39. To plot a point for which both coordinates are negative, start at the origin and move to the left and then down. Such a point lies in quadrant III.

40.

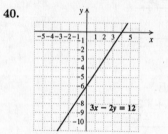

41. ◈

42. ◈

43. ◈

44. ◈

45.
$$
\begin{aligned}
&(45x^{8k} + 30x^{6k} - 60x^{4k}) \div (3x^{2k}) \\
&= \frac{45x^{8k} + 30x^{6k} - 60x^{4k}}{3x^{2k}} \\
&= \frac{45x^{8k}}{3x^{2k}} + \frac{30x^{6k}}{3x^{2k}} - \frac{60x^{4k}}{3x^{2k}} \\
&= 15x^{8k-2k} + 10x^{6k-2k} - 20x^{4k-2k} \\
&= 15x^{6k} + 10x^{4k} - 20x^{2k}
\end{aligned}
$$

46. $5a^{6k} - 16a^{3k} + 14$

47.

$$\begin{array}{r} y^3 - ay^2 + a^2y - a^3 \\ y+a \overline{\smash{\big)}\, y^4 + 0y^3 + 0y^2 + 0y + a^2} \\ \underline{y^4 + ay^3} \\ -ay^3 \\ \underline{-ay^3 - a^2y^2} \\ a^2y^2 \\ \underline{a^2y^2 + a^3y} \\ -a^3y + a^2 \\ \underline{-a^3y \qquad - a^4} \\ a^2 + a^4 \end{array}$$

The answer is $y^3 - ay^2 + a^2y - a^3 + \dfrac{a^2 + a^4}{y+a}$.

48. $a + 3 + \dfrac{5}{5a^2 - 7a - 2}$

49.

$$\begin{array}{r} 5y + 2 \\ 3y^2 - 5y - 2 \overline{\smash{\big)}\, 15y^3 - 19y^2 - 30y + 7} \\ \underline{15y^3 - 25y^2 - 10y} \\ 6y^2 - 20y + 7 \\ \underline{6y^2 - 10y - 4} \\ -10y + 11 \end{array}$$

The answer is $5y + 2 + \dfrac{-10y + 11}{3y^2 - 5y - 2}$.

50. $2x^2 + x - 3$

51. $(x-3)^2 + (5x-8) = x^2 - 6x + 9 + 5x - 8 = x^2 - x + 1$

$$\begin{array}{r} 5x^5 + 5x^4 - 8x^2 - 8x + 2 \\ x^2 - x + 1 \overline{\smash{\big)}\, 5x^7 + 0x^6 + 0x^5 - 3x^4 + 0x^3 + 2x^2 - 10x + 2} \\ \underline{5x^7 - 5x^6 + 5x^5} \\ 5x^6 - 5x^5 - 3x^4 \\ \underline{5x^6 - 5x^5 + 5x^4} \\ -8x^4 + 2x^2 \\ \underline{-8x^4 + 8x^3 - 8x^2} \\ -8x^3 + 10x^2 - 10x \\ \underline{-8x^3 + 8x^2 - 8x} \\ 2x^2 - 2x + 2 \\ \underline{2x^2 - 2x + 2} \\ 0 \end{array}$$

The answer is $5x^5 + 5x^4 - 8x^2 - 8x + 2$.

52. $3a^{2h} + 2a^h - 5$

53.

$$\begin{array}{r} x - 3 \\ x - 1 \overline{\smash{\big)}\, x^2 - 4x + c} \\ \underline{x^2 - x} \\ -3x + c \\ \underline{-3x + 3} \\ c - 3 \end{array}$$

We set the remainder equal to 0.

$$c - 3 = 0$$
$$c = 3$$

Thus, c must be 3.

54. -2

55.

$$\begin{array}{r} c^2x + (2c + c^2) \\ x - 1 \overline{\smash{\big)}\, c^2x^2 + 2cx + 1} \\ \underline{c^2x^2 - c^2x} \\ (2c + c^2)x + 1 \\ \underline{(2c + c^2)x - (2c + c^2)} \\ 1 + (2c + c^2) \end{array}$$

We set the remainder equal to 0.

$$c^2 + 2c + 1 = 0$$
$$(c+1)^2 = 0$$
$$c + 1 = 0 \quad or \quad c + 1 = 0$$
$$c = -1 \quad or \qquad c = -1$$

Thus, c must be -1.

Exercise Set 4.8

1. $6^{-2} = \dfrac{1}{6^2} = \dfrac{1}{36}$

2. $\dfrac{1}{2^4} = \dfrac{1}{16}$

3. $10^{-4} = \dfrac{1}{10^4} = \dfrac{1}{10,000}$

4. $\dfrac{1}{5^3} = \dfrac{1}{125}$

5. $(-2)^{-6} = \dfrac{1}{(-2)^6} = \dfrac{1}{64}$

6. $\dfrac{1}{(-3)^4} = \dfrac{1}{81}$

7. $a^{-5} = \dfrac{1}{a^5}$

8. $\dfrac{1}{x^2}$

9. $\dfrac{1}{y^{-4}} = y^4$

10. t^7

11. $\dfrac{1}{z^{-9}} = z^9$

12. h^8

13. $7^{-1} = \dfrac{1}{7^1} = \dfrac{1}{7}$

14. $\dfrac{3}{2}$

15. $\left(\dfrac{1}{4}\right)^{-2} = \dfrac{1}{\left(\dfrac{1}{4}\right)^2} = \dfrac{1}{\dfrac{1}{16}} = 1 \cdot \dfrac{16}{1} = 16$

16. $\dfrac{1}{\left(\dfrac{4}{5}\right)^2} = \dfrac{25}{16}$

17. $\frac{1}{4^3} = 4^{-3}$

18. 5^{-2}

19. $\frac{1}{t^6} = t^{-6}$

20. y^{-2}

21. $\frac{1}{a^4} = a^{-4}$

22. t^{-5}

23. $\frac{1}{p^8} = p^{-8}$

24. m^{-12}

25. $\frac{1}{5} = \frac{1}{5^1} = 5^{-1}$

26. 8^{-1}

27. $\frac{1}{t} = \frac{1}{t^1} = t^{-1}$

28. m^{-1}

29. $2^{-5} \cdot 2^8 = 2^{-5+8} = 2^3$, or 8

30. 5

31. $x^{-2} \cdot x = x^{-2+1} = x^{-1}$, or $\frac{1}{x}$

32. 1

33. $x^{-7} \cdot x^{-6} = x^{-13}$, or $\frac{1}{x^{13}}$

34. y^{-13}, or $\frac{1}{y^{13}}$

35. $\frac{m^6}{m^{12}} = m^{6-12} = m^{-6}$, or $\frac{1}{m^6}$

36. p^{-1}, or $\frac{1}{p}$

37. $\frac{(8x)^6}{(8x)^{10}} = (8x)^{6-10} = (8x)^{-4}$, or $\frac{1}{(8x)^4}$

38. $(9t)^{-7}$, or $\frac{1}{(9t)^7}$

39. $\frac{18^9}{18^9} = 18^{9-9} = 18^0 = 1$

40. 1

41. $(a^{-5}b^{-7})(a^{-3}b^{-6}) = a^{-5+(-3)}b^{-7+(-6)} = a^{-8}b^{-13}$, or $\frac{1}{a^8 b^{13}}$

42. $x^{-5}y^{-9}$, or $\frac{1}{x^5 y^9}$

43. $\frac{x^7}{x^{-2}} = x^{7-(-2)} = x^9$

44. t^{11}

45. $\frac{z^{-6}}{z^{-2}} = z^{-6-(-2)} = z^{-4}$, or $\frac{1}{z^4}$

46. y^{-4}, or $\frac{1}{y^4}$

47. $\frac{a^{-6}}{a^{-10}} = a^{-6-(-10)} = a^4$

48. y^5

49. $\frac{x}{x^{-1}} = x^{1-(-1)} = x^2$

50. x^5

51. $(a^{-3})^5 = a^{-3\cdot5} = a^{-15}$, or $\frac{1}{a^{15}}$

52. x^{-30}, or $\frac{1}{x^{30}}$

53. $(a^{-5})^{-6} = a^{-5(-6)} = a^{30}$

54. x^{12}

55. $(n^{-2})^8 = n^{-2\cdot8} = n^{-16}$, or $\frac{1}{n^{16}}$

56. m^{-21}, or $\frac{1}{m^{21}}$

57. $(mn)^{-5} = m^{-5}n^{-5}$, or $\frac{1}{m^5 n^5}$

58. $a^{-3}b^{-3}$, or $\frac{1}{a^3 b^3}$

59. $(4xy)^{-2} = 4^{-2}x^{-2}y^{-2}$, or $\frac{1}{16x^2 y^2}$

60. $5^{-2}a^{-2}b^{-2}$, or $\frac{1}{25a^2 b^2}$

61. $(3a^{-4})^4 = 3^4(a^{-4})^4 = 81a^{-16}$, or $\frac{81}{a^{16}}$

62. $36x^{-10}$, or $\frac{36}{x^{10}}$

63. $(t^5 x^3)^{-4} = (t^5)^{-4}(x^3)^{-4} = t^{-20}x^{-12}$, or $\frac{1}{t^{20}x^{12}}$

64. $x^{-12}y^{-15}$, or $\frac{1}{x^{12}y^{15}}$

65. $(x^{-2}y^{-7})^{-5} = (x^{-2})^{-5}(y^{-7})^{-5} = x^{10}y^{35}$

66. $x^{24}y^8$

67. $(x^3 y^{-4} z^{-5})(x^{-4}y^{-2}z^9) = x^{3+(-4)}y^{-4+(-2)}z^{-5+9} = x^{-1}y^{-6}z^4$, or $\frac{z^4}{xy^6}$

68. $a^{-8}b^5c^4$, or $\dfrac{b^5c^4}{a^8}$

69. $(m^{-4}n^7p^3)(m^9n^{-2}p^{-10}) =$

$m^{-4+9}n^{7+(-2)}p^{3+(-10)} = m^5n^5p^{-7}$, or $\dfrac{m^5n^5}{p^7}$

70. $t^{-14}p^3m^6$, or $\dfrac{p^3m^6}{t^{14}}$

71. $\left(\dfrac{y^2}{2}\right)^{-2} = \dfrac{(y^2)^{-2}}{2^{-2}} = \dfrac{y^{-4}}{2^{-2}} = \dfrac{2^2}{y^4} = \dfrac{4}{y^4}$

72. $\dfrac{a^{-8}}{3^{-2}}$, or $\dfrac{9}{a^8}$

73. $\left(\dfrac{3}{a^2}\right)^4 = \dfrac{3^4}{(a^2)^4} = \dfrac{81}{a^8}$

74. $\dfrac{49}{x^{14}}$

75. $\left(\dfrac{x^2y}{z^4}\right)^3 = \dfrac{(x^2)^3y^3}{(z^4)^3} = \dfrac{x^6y^3}{z^{12}}$

76. $\dfrac{m^3}{n^{12}p^3}$

77. $\left(\dfrac{a^2b}{cd^3}\right)^{-5} = \dfrac{(a^2)^{-5}b^{-5}}{c^{-5}(d^3)^{-5}} = \dfrac{a^{-10}b^{-5}}{c^{-5}d^{-15}}$, or $\dfrac{c^5d^{15}}{a^{10}b^5}$

78. $\dfrac{2^{-3}a^{-6}}{3^{-3}b^{-12}}$, or $\dfrac{27b^{12}}{8a^6}$

79. 9.12×10^4

Since the exponent is positive, the decimal point will move to the right.

9.1200.   The decimal point moves right 4 places.

$9.12 \times 10^4 = 91,200$

80. 892

81. 6.92×10^{-3}

Since the exponent is negative, the decimal point will move to the left.

.006.92   The decimal point moves left 3 places.

$6.92 \times 10^{-3} = 0.00692$

82. 0.000726

83. 2.04×10^8

Since the exponent is positive, the decimal point will move to the right.

2.04000000.

8 places

$2.04 \times 10^8 = 204,000,000$

84. $13,500,000$

85. 8.764×10^{-10}

Since the exponent is negative, the decimal point will move to the left.

0.0000000008.764

10 places

$8.764 \times 10^{-10} = 0.0000000008764$

86. 0.009043

87. $10^7 = 1 \times 10^7$

Since the exponent is positive, the decimal point will move to the right.

1.0000000.

7 places

$10^7 = 10,000,000$

88. $10,000$

89. $10^{-4} = 1 \times 10^{-4}$

Since the exponent is negative, the decimal point will move to the left.

.0001.

4 places

$10^{-4} = 0.0001$

90. 0.0000001

91. $370,000 = 3.7 \times 10^n$

To write 3.7 as 370,000 we move the decimal point 5 places to the right. Thus, n is 5 and

$370,000 = 3.7 \times 10^5$.

92. 7.15×10^4

93. $0.00583 = 5.83 \times 10^n$

To write 5.83 as 0.00583 we move the decimal point 3 places to the left. Thus, n is -3 and

$0.00583 = 5.83 \times 10^{-3}$.

94. 8.14×10^{-2}

95. $78,000,000,000 = 7.8 \times 10^n$

To write 7.8 as 78,000,000,000 we move the decimal point 10 places to the right. Thus, n is 10 and

$78,000,000,000 = 7.8 \times 10^{10}$.

96. 3.7×10^{12}

97. $907,000,000,000,000,000 = 9.07 \times 10^n$

To write 9.07 as 907,000,000,000,000,000 we move the decimal point 17 places to the right. Thus, n is 17 and

$907,000,000,000,000,000 = 9.07 \times 10^{17}$.

98. 1.68×10^{14}

99. $0.00000486 = 4.86 \times 10^n$

To write 4.86 as 0.00000486 we move the decimal point 6 places to the left. Thus, n is -6 and
$$0.00000486 = 4.86 \times 10^{-6}.$$

100. 2.75×10^{-10}

101. $0.000000018 = 1.8 \times 10^n$

To write 1.8 as 0.000000018 we move the decimal point 8 places to the left. Thus, n is -8 and
$$0.000000018 = 1.8 \times 10^{-8}.$$

102. 2×10^{-11}

103. $10,000,000 = 1 \times 10^n$, or 10^n

To write 1 as 10,000,000 we move the decimal point 7 places to the right. Thus, n is 7 and
$$10,000,000 = 10^7.$$

104. 10^{11}

105. $(4 \times 10^7)(2 \times 10^5) = (4 \cdot 2) \times (10^7 \cdot 10^5)$
$$= 8 \times 10^{7+5} \quad \text{Adding exponents}$$
$$= 8 \times 10^{12}$$

106. 6.46×10^5

107. $(3.8 \times 10^9)(6.5 \times 10^{-2}) = (3.8 \cdot 6.5) \times (10^9 \cdot 10^{-2})$
$$= 24.7 \times 10^7$$

The answer is not yet in scientific notation since 24.7 is not a number between 1 and 10. We convert to scientific notation.
$$24.7 \times 10^7 = (2.47 \times 10) \times 10^7 = 2.47 \times 10^8$$

108. 6.106×10^{-11}

109. $(8.7 \times 10^{-12})(4.5 \times 10^{-5})$
$$= (8.7 \cdot 4.5) \times (10^{-12} \cdot 10^{-5})$$
$$= 39.15 \times 10^{-17}$$

The answer is not yet in scientific notation since 39.15 is not a number between 1 and 10. We convert to scientific notation.
$$39.15 \times 10^{-17} = (3.915 \times 10) \times 10^{-17} = 3.915 \times 10^{-16}$$

110. 2.914×10^{-6}

111. $\dfrac{8.5 \times 10^8}{3.4 \times 10^{-5}} = \dfrac{8.5}{3.4} \times \dfrac{10^8}{10^{-5}}$
$$= 2.5 \times 10^{8-(-5)}$$
$$= 2.5 \times 10^{13}$$

112. 2.24×10^{-7}

113. $(3.0 \times 10^6) \div (6.0 \times 10^9) = \dfrac{3.0 \times 10^6}{6.0 \times 10^9}$
$$= \dfrac{3.0}{6.0} \times \dfrac{10^6}{10^9}$$
$$= 0.5 \times 10^{6-9}$$
$$= 0.5 \times 10^{-3}$$

The answer is not yet in scientific notation because 0.5 is not between 1 and 10. We convert to scientific notation.
$$0.5 \times 10^{-3} = (5.0 \times 10^{-1}) \times 10^{-3} =$$
$$5.0 \times 10^{-4}$$

114. 9.375×10^2

115. $\dfrac{7.5 \times 10^{-9}}{2.5 \times 10^{12}} = \dfrac{7.5}{2.5} \times \dfrac{10^{-9}}{10^{12}}$
$$= 3.0 \times 10^{-9-12}$$
$$= 3.0 \times 10^{-21}$$

116. 5×10^{-24}

117. *Familiarize.* There are 365 days in one year. Express 3 million and 365 in scientific notation.

$$3 \text{ million} = 3,000,000 = 3 \times 10$$
$$365 = 3.65 \times 10^2$$

Let $a =$ the amount of orange juice Americans consume in one year, in gallons.

Translate. We reword the problem.

What is daily consumption times number of days in a year?

$$a = (3 \times 10^6) \times (3.65 \times 10^2)$$

Carry out. We do the computation.
$$a = (3 \times 10^6) \times (3.65 \times 10^2)$$
$$a = (3 \times 3.65) \times (10^6 \times 10^2)$$
$$a = 10.95 \times 10^8$$
$$a = (1.095 \times 10) \times 10^8$$
$$a = 1.095 \times 10^9$$

Check. We review the computation. Also, the answer seems reasonable since it is larger than 3 million.

State. Americans consume 1.095×10^9 gallons of orange juice in one year.

118. 1.512×10^{10} cubic feet, 1.324512×10^{14} cubic feet

119. *Familiarize.* Express 200 million and 1 million in scientific notation:

$$200 \text{ million} = 200,000,000 = 2 \times 10^8$$
$$1 \text{ million} = 1,000,000 = 10^6$$

Let $w =$ the number of gallons of drinking water that can be contaminated by 200 million gal of used motor oil.

Translate. We reword the problem.

What is | contamination caused by 1 gal | times | 200 million gal?
w | $=$ | 10^6 | $\times$ | 2×10^8

Carry out. We do the computation.

$$w = 10^6 \times (2 \times 10^8)$$
$$w = 2 \times (10^6 \times 10^8)$$
$$w = 2 \times 10^{14}$$

Check. We review the computation. Also, the answer is a very large number, as expected.

State. 200 million gal of oil can contaminate 2×10^{14} gal of drinking water.

120. $\$3.2 \times 10^{10}$

121. *Familiarize*. Express 1 billion and 2500 in scientific notation:

$$1 \text{ billion} = 1,000,000,000 = 10^9$$
$$2500 = 2.5 \times 10^3$$

Let b = the number of bytes in the network.

Translate. We reword the problem.

What is | 2500 | times | 1 gigabyte?
b | $=$ | 2.5×10^3 | $\times$ | 10^9

Carry out. We do the computation.

$$b = (2.5 \times 10^3) \times 10^9$$
$$b = 2.5 \times (10^3 \times 10^9)$$
$$b = 2.5 \times 10^{12}$$

Check. We review the computation. Also, the answer seems reasonable since it is larger than 1 billion.

State. There are 2.5×10^{12} bytes in the network.

122. $\$1.32288 \times 10^{12}$

123.
$$\frac{3}{4} - 5\left(-\frac{1}{2}\right)^2 + \frac{1}{3}$$
$$= \frac{3}{4} - 5 \cdot \frac{1}{4} + \frac{1}{3}$$
$$= \frac{3}{4} - \frac{5}{4} + \frac{1}{3}$$
$$= \frac{9}{12} - \frac{15}{12} + \frac{4}{12}$$
$$= -\frac{2}{12} = -\frac{1}{6}$$

124. $8a$

125. $-12x + (-5x) = (-12 - 5)x = -17x$

126.

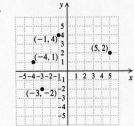

127. To plot a point with a positive first coordinate, start at the origin and move to the right. Thus, the first coordinate is positive in quadrants I and IV.

128. $t = \dfrac{cx - r}{b}$

129.

130. ◈

131.
$$\frac{(2.5 \times 10^{-8})(6.1 \times 10^{-11})}{1.28 \times 10^{-3}}$$
$$= \frac{(2.5 \cdot 6.1)}{1.28} \times \frac{(10^{-8} \cdot 10^{-11})}{10^{-3}}$$
$$= 11.9140625 \times 10^{-8+(-11)-(-3)}$$
$$= 11.9140625 \times 10^{-16}$$
$$= (1.19140625 \times 10) \times 10^{-16}$$
$$= 1.19140625 \times 10^{-15}$$

132. $4.894179894 \times 10^{26}$

133.
$$\frac{5.8 \times 10^{17}}{(4.0 \times 10^{-13})(2.3 \times 10^4)}$$
$$= \frac{5.8}{(4.0 \cdot 2.3)} \times \frac{10^{17}}{(10^{-13} \cdot 10^4)}$$
$$\approx 0.6304347826 \times 10^{17-(-13)-4}$$
$$\approx (6.304347826 \times 10^{-1}) \times 10^{26}$$
$$\approx 6.304347826 \times 10^{25}$$

134. 3.12×10^{43}

135.
$$\frac{4.2 \times 10^8[(2.5 \times 10^{-5}) \div (5.0 \times 10^{-9})]}{3.0 \times 10^{27}}$$
$$= \frac{4.2 \times 10^8[(2.5 \div 5.0) \times (10^{-5} \div 10^{-9})]}{3.0 \times 10^{27}}$$
$$= \frac{4.2 \cdot 10^8(0.5 \times 10^4)}{3.0 \times 10^{27}}$$
$$= \frac{(4.2 \times 0.5)}{3.0} \times \frac{(10^8 \cdot 10^4)}{10^{27}}$$
$$= 0.7 \times 10^{-15}$$
$$= (7 \times 10^{-1}) \times 10^{-15}$$
$$= 7 \times 10^{-16}$$

136. (a) 1.6×10^2; (b) 2.5×10^{-11}

137. $4^{-3} \cdot 8 \cdot 16 = (2^2)^{-3} \cdot 2^3 \cdot 2^4 = 2^{-6} \cdot 2^3 \cdot 2^4 = 2^1$

138. 4

139. $(5^{-12})^2 5^{25} = 5^{-24} 5^{25} = 5$

140. 7

141. $9^{23} \cdot 27^{-6} = (3^2)^{23}(3^3)^{-6} = 3^{46} \cdot 3^{-18} = 3^{28}$

142. a^n

143. False; let $x = 2$, $y = 3$, $m = 4$, and $n = 2$:
$$2^4 \cdot 3^2 = 16 \cdot 9 = 144, \text{ but}$$
$$(2 \cdot 3)^{4 \cdot 2} = 6^8 = 1,679,616$$

144. False

145. False; let $x = 5$, $y = 3$, and $m = 2$:
$$(5 - 3)^2 = 2^2 = 4, \text{ but}$$
$$5^2 - 3^2 = 25 - 9 = 16$$

Chapter 5
Polynomials and Factoring

1. Answers may vary. $10x^3 = (5x)(2x^2) = (10x^2)(x) = (-2)(-5x^3)$

2. Answers may vary. $6x^3 = (6x)(x^2) = (3x^2)(2x) = (2x^2)(3x)$

3. Answers may vary. $-6x^5 = (-2x^3)(3x^2) = (3x^4)(-2x) = (-x)(6x^4)$

4. Answers may vary. $-15x^6 = (-3x^2)(5x^4) = (3x^3)(-5x^3) = (-15x)(x^5)$

5. Answers may vary. $26x^5 = (2x^4)(13x) = (2x^3)(13x^2) = (-x^2)(-26x^3)$

6. Answers may vary. $25x^4 = (5x^2)(5x^2) = (x^3)(25x) = (-5x)(-5x^3)$

7. $x^2 - 6x = x \cdot x - x \cdot 6$
$$= x(x - 6)$$

8. $x(x + 8)$

9. $8x^2 + 4x = 4x \cdot 2x + 4x \cdot 1$
$$= 4x(2x + 1)$$

10. $5x(x - 2)$

11. $x^3 + 6x^2 = x^2 \cdot x + x^2 \cdot 6$
$$= x^2(x + 6)$$

12. $x^2(4x^2 + 1)$

13. $8x^4 - 24x^2 = 8x^2 \cdot x^2 - 8x^2 \cdot 3$
$$= 8x^2(x^2 - 3)$$

14. $5x^3(x^2 + 2)$

15. $2x^2 + 2x - 8 = 2 \cdot x^2 + 2 \cdot x - 2 \cdot 4$
$$= 2(x^2 + x - 4)$$

16. $3(2x^2 + x - 5)$

17. $5x^6 - 10x^3 + 8x^2$
$$= x^2 \cdot 5x^4 - x^2 \cdot 10x + x^2 \cdot 8$$
$$= x^2(5x^4 - 10x + 8)$$

18. $x^3(10x^2 + 6x - 3)$

19. $2x^8 + 4x^6 - 8x^4 + 10x^2$
$$= 2x^2 \cdot x^6 + 2x^2 \cdot 2x^4 - 2x^2 \cdot 4x^2 + 2x^2 \cdot 5$$
$$= 2x^2(x^6 + 2x^4 - 4x^2 + 5)$$

20. $5(x^4 - 3x^3 - 5x - 2)$

21. $x^5y^5 + x^4y^3 + x^3y^3 - x^2y^2$
$$= x^2y^2 \cdot x^3y^3 + x^2y^2 \cdot x^2y + x^2y^2 \cdot xy + x^2y^2(-1)$$
$$= x^2y^2(x^3y^3 + x^2y + xy - 1)$$

22. $x^3y^3(x^6y^3 - x^4y^2 + xy + 1)$

23. $\dfrac{5}{3}x^6 + \dfrac{4}{3}x^5 + \dfrac{1}{3}x^4 + \dfrac{1}{3}x^3$
$$= \dfrac{1}{3}x^3(5x^3) + \dfrac{1}{3}x^3(4x^2) + \dfrac{1}{3}x^3(x) + \dfrac{1}{3}x^3(1)$$
$$= \dfrac{1}{3}x^3(5x^3 + 4x^2 + x + 1)$$

24. $\dfrac{1}{7}x(5x^6 + 3x^4 - 6x^2 - 1)$

25. $y(y + 3) + 7(y + 3)$
$$= (y + 3)(y + 7) \quad \text{Factoring out the common binomial factor } y+3$$

26. $(b - 5)(b + 3)$

27. $x^2(x + 3) - 7(x + 3)$
$$= (x + 3)(x^2 - 7) \quad \text{Factoring out the common binomial factor } x + 3$$

28. $(2z + 9)(3z^2 + 1)$

29. $y^2(y + 8) + (y + 8) = y^2(y + 8) + 1(y + 8)$
$$= (y + 8)(y^2 + 1) \quad \text{Factoring out the common factor}$$

30. $(x - 7)(x^2 - 3)$

31. $x^3 + 3x^2 + 4x + 12$
$$= (x^3 + 3x^2) + (4x + 12)$$
$$= x^2(x + 3) + 4(x + 3) \quad \text{Factoring each binomial}$$
$$= (x + 3)(x^2 + 4) \quad \text{Factoring out the common factor } x + 3$$

32. $(2z + 1)(3z^2 + 1)$

33. $2x^3 + 6x^2 + 3x + 9$
$$= (2x^3 + 6x^2) + (3x + 9)$$
$$= 2x^2(x + 3) + 3(x + 3) \quad \text{Factoring each binomial}$$
$$= (x + 3)(2x^2 + 3)$$

34. $(3x + 2)(x^2 + 1)$

35.
$$9x^3 - 12x^2 + 3x - 4$$
$$= 3x^2(3x - 4) + 1(3x - 4)$$
$$= (3x - 4)(3x^2 + 1)$$

36. $(2x - 5)(5x^2 + 2)$

37.
$$5x^3 - 2x^2 + 5x - 2$$
$$= x^2(5x - 2) + 1(5x - 2)$$
$$= (5x - 2)(x^2 + 1)$$

38. $(6x - 7)(3x^2 + 5)$

39. $x^3 + 8x^2 - 3x - 24 = x^2(x + 8) - 3(x + 8)$
$$= (x + 8)(x^2 - 3)$$

40. $(x + 6)(2x^2 - 5)$

41. $w^3 - 7w^2 + 4w - 28 = w^2(w - 7) + 4(w - 7)$
$$= (w - 7)(w^2 + 4)$$

42. $(y + 8)(y^2 - 2)$

43. $x^3 - x^2 - 2x + 5 = x^2(x - 1) - 1(2x - 5)$, or
$x^3 - 2x - x^2 + 5 = x(x^2 - 2) - (x^2 - 5)$

This polynomial is not factorable using factoring by grouping.

44. Not factorable by grouping

45. $2x^3 - 8x^2 - 9x + 36 = 2x^2(x - 4) - 9(x - 4)$
$$= (x - 4)(2x^2 - 9)$$

46. $(5g - 1)(4g^2 - 5)$

47. Graph: $y = x - 6$

The equation is in the form $y = mx + b$, so we know the y-intercept is $(0, -6)$. We find two other pairs.

When $x = 5$, $y = 5 - 6 = -1$.

When $x = 2$, $y = 2 - 6 = -4$.

x	y
0	-6
5	-1
2	-4

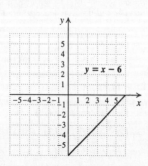

48. $\left\{ x \mid x \leq \dfrac{14}{5} \right\}$

49.
$$-13 - (-25)$$
$$= -13 + 25 \qquad \text{Adding an opposite}$$
$$= 12$$

50. $p = 2A - q$

51. $(y + 5)(y + 7) = y^2 + 7y + 5y + 35 \quad$ Using FOIL
$$= y^2 + 12y + 35$$

52. $y^2 + 14y + 49$

53. $(y + 7)(y - 7) = y^2 - 7^2 = y^2 - 49$
$$[(A + B))(A - B) = A^2 - B^2]$$

54. $y^2 - 14y + 49$

55. ◈

56. ◈

57. ◈

58. ◈

59. $4x^5 + 6x^3 + 6x^2 + 9 = 2x^3(2x^2 + 3) + 3(2x^2 + 3)$
$$= (2x^2 + 3)(2x^3 + 3)$$

60. $(x^2 + 1)(x^4 + 1)$

61. $x^{12} + x^7 + x^5 + 1 = x^7(x^5 + 1) + (x^5 + 1)$
$$= (x^5 + 1)(x^7 + 1)$$

62. Not factorable by grouping

63.
$$5x^5 - 5x^4 + x^3 - x^2 + 3x - 3$$
$$= 5x^4(x - 1) + x^2(x - 1) + 3(x - 1)$$
$$= (x - 1)(5x^4 + x^2 + 3)$$

64. $(x^2 + 2x + 3)(a + 1)$

65. ◈

66. ◈

Exercise Set 5.2

1. $x^2 + 6x + 8$

Since the constant term and the coefficient of the middle term are both positive, we look for a factorization of 8 in which both factors are positive. Their sum must be 6.

Pairs of factors	Sums of factors
1, 8	9
2, 4	6

The numbers we want are 2 and 4.

$x^2 + 6x + 8 = (x + 2)(x + 4)$.

2. $(x + 1)(x + 6)$

3. $x^2 + 9x + 8$

Since the constant term is positive and the coefficient of the middle term is positive, we look for a factorization of 8 in which both factors are positive. Their sum must be 9.

From the table in Exercise 1, we see that the numbers we want are 1 and 8.

$x^2 + 9x + 8 = (x + 1)(x + 8)$.

4. $(x + 3)(x + 4)$

5. $y^2 + 11y + 28$

Since the constant term is positive and the coefficient of the middle term is positive, we look for a factorization of 28 in which both factors are positive. Their sum must be 11.

Pairs of factors	Sums of factors
1, 28	29
2, 14	16
4, 7	11

The numbers we want are 4 and 7.

$y^2 + 11y + 28 = (y + 4)(y + 7)$

6. $(x - 3)(x - 3)$, or $(x - 3)^2$

7. $a^2 + 11a + 30$

Since the constant term is positive and the coefficient of the middle term is positive, we look for a factorization of 30 in which both factors are positive. Their sum must be 11.

Pairs of factors	Sums of factors
1, 30	31
2, 15	17
3, 10	13
5, 6	11

The numbers we want are 5 and 6.

$a^2 + 11a + 30 = (a + 5)(a + 6)$.

8. $(x + 2)(x + 7)$

9. $x^2 - 5x + 4$

Since the constant term is positive and the coefficient of the middle term is negative, we look for a factorization of 4 in which both factors are negative. Their sum must be -5.

Pairs of factors	Sums of factors
$-1, -4$	-5
$-2, -2$	-4

The numbers we want are -1 and -4.

$x^2 - 5x + 4 = (x - 1)(x - 4)$.

10. $(b + 1)(b + 4)$

11. $z^2 - 8z + 7$

Since the constant term is positive and the coefficient of the middle term is negative, we look for a factorization of 7 in which both factors are negative. Their sum must be -8. The only possible negative factors are -1 and -7. Their sum is -8, so these are the numbers we want.

$z^2 - 8z + 7 = (z - 1)(z - 7)$

12. $(a + 2)(a - 6)$

13. $x^2 - 8x + 15$

Since the constant term is positive and the coefficient of the middle term is negative, we look for a factorization of 15 in which both factors are negative. Their sum must be -8.

Pairs of factors	Sums of factors
$-1, -15$	-16
$-3, -5$	-8

The numbers we want are -3 and -5.

$x^2 - 8x + 15 = (x - 3)(x - 5)$.

14. $(d - 2)(d - 5)$

15. $x^2 - 2x - 15$

The constant term, -15, must be expressed as the product of a negative number and a positive number. Since the sum of those two numbers must be negative, the negative number must have the greater absolute value.

Pairs of factors	Sums of factors
1, -15	-14
3, -5	-2

The numbers we need are 3 and -5.

$x^2 - 2x - 15 = (x + 3)(x - 5)$.

16. $(y - 1)(y - 10)$

17. $x^2 + 2x - 15$

The constant term, -15, must be expressed as the product of a negative number and a positive number. Since the sum of those two numbers must be positive, the positive number must have the greater absolute value.

Pairs of factors	Sums of factors
-1, 15	14
-3, 5	2

The numbers we need are -3 and 5.

$x^2 + 2x - 15 = (x - 3)(x + 5)$.

18. $(x - 6)(x + 7)$

19. $3y^2 - 9y - 84 = 3(y^2 - 3y - 28)$

After factoring out the common factor, 3, we consider $y^2 - 3y + 28$. The constant term, -28, must be expressed as the product of a negative number and a positive number. Since the sum of those two numbers must be negative, the negative number must have the greater absolute value.

Pairs of factors	Sums of factors
1, -28	-29
2, -14	-12
4, -7	-3

The numbers we need are 4 and -7. The factorization of $y^2 - 3y - 28$ is $(y + 4)(y - 7)$. We must not forget the common factor, 3. Thus, $3y^2 - 9y - 84 = 3(y^2 - 3y - 28) = 3(y + 4)(y - 7)$.

20. $2(x + 2)(x - 9)$

21. $x^3 - x^2 - 42x = x(x^2 - x - 42)$

After factoring out the common factor, x, we consider $x^2 - x - 42$. The constant term, -42, must be expressed as the product of a negative number and a positive number. Since the sum of those two numbers must be negative, the negative number must have the greater absolute value.

Pairs of factors	Sums of factors
1, -42	-41
2, -21	-19
3, -14	-11
6, -7	-1

The numbers we need are 6 and -7. The factorization of $x^2 - x - 42$ is $(x + 6)(x - 7)$. We must not forget the common factor, x. Thus, $x^3 - x^2 - 42x = x(x^2 - x - 42) = x(x + 6)(x - 7)$.

22. $x(x + 2)(x - 8)$

23. $x^2 - 7x - 60$

The constant term, -60, must be expressed as the product of a negative number and a positive number. Since the sum of those two numbers must be negative, the negative number must have the greater absolute value.

Pairs of factors	Sums of factors
1, -60	-59
2, -30	-28
3, -20	-17
4, -15	-11
5, -12	-7

The numbers we need are 5 and -12.
$x^2 - 7x - 60 = (x + 5)(x - 12)$

24. $(y + 5)(y - 9)$

25. $x^2 - 72 + 6x = x^2 + 6x - 72$

The constant term, -72, must be expressed as the product of a negative number and a positive number. Since the sum of those two numbers must be positive, the positive number must have the greater absolute value.

Pairs of factors	Sums of factors
-1, 72	71
-2, 36	34
-3, 24	21
-4, 18	14
-6, 12	6

The numbers we need are -6 and 12.
$x^2 - 72 + 6x = (x - 6)(x + 12)$

26. $(x + 9)(x - 11)$

27. $5b^2 + 25b - 120 = 5(b^2 + 5b - 24)$

After factoring out the common factor, 5, we consider $b^2 + 5b - 24$. The constant term, -24, must be expressed as the product of a negative number and a positive number. Since the sum of those two numbers must be positive, the positive number must have the greater absolute value.

Pairs of factors	Sums of factors
-1, 24	23
-2, 12	10
-3, 8	5

The numbers we need are -3 and 8. The factorization of $b^2 + 5b - 24$ is $(b - 3)(b + 8)$. We must not forget the common factor. Thus, $5b^2 + 25b - 120 = 5(b^2 + 5b - 24) = 5(b - 3)(b + 8)$.

28. $c^2(c - 7)(c + 8)$

29. $x^5 + x^4 - 2x^3 = x^3(x^2 + x - 2)$

After factoring out the common factor, x^3, we consider $x^2 + x - 2$. The constant term, -2, must be expressed as the product of a negative number and a positive number. Since the sum of those two numbers must be positive, the positive number must have the greater absolute value. The only possible factors that fill these requirements are -1 and 2. These are the numbers we need. The factorization of $x^2 + x - 2$ is $(x - 1)(x + 2)$. We must not forget the common factor, x^3. Thus, $x^5 + x^4 - 2x^3 = x^3(x^2 + x - 2) = x^3(x - 1)(x + 2)$.

30. $2(a - 5)(a + 7)$

31. $x^2 + 2x + 3$

Since the constant term and the coefficient of the middle term are both positive, we look for a factorization of 3 in which both factors are positive. Their sum must be 2. The only possible pair of positive factors is 1 and 3, but their sum is not 2. Thus, this polynomial is not factorable into polynomials with integer coefficients. It is prime.

32. Prime

33. $11 - 3w + w^2 = w^2 - 3w + 11$

Since the constant term is positive and the coefficient of the middle term is negative, we look for a factorization of 11 in which both factors are negative. Their sum must be -3. The only possible pair of negative factors is -1 and -11, but their sum is not -3. Thus, this polynomial is not factorable into polynomials with integer coefficients. It is prime.

34. Prime

35. $x^2 + 20x + 99$

We look for two factors, both positive, whose product is 99 and whose sum is 20.

They are 9 and 11: $9 \cdot 11 = 99$ and $9 + 11 = 20$.

$x^2 + 20 + 99 = (x + 9)(x + 11)$

36. $(x + 10)(x + 10)$, or $(x + 10)^2$

37. $2x^3 - 40x^2 + 192x = 2x(x^2 - 20x + 96)$

After factoring out the common factor, $2x$, we consider $x^2 - 20x + 96$. We look for two factors, both negative, whose product is 96 and whose sum is -20.

They are -8 and -12: $-8(-12) = 96$ and $-8 + (-12) = -20$.

$x^2 - 20x + 96 = (x - 8)(x - 12)$, so $2x^3 - 40x^2 + 192x = 2x(x - 8)(x - 12)$.

38. $3x(x + 4)(x - 25)$

39. $4x^2 + 40x + 100 = 4(x^2 + 10x + 25)$

After factoring out the common factor, 4, we consider $x^2 + 10x + 25$. We look for two factors, both positive, whose product is 25 and whose sum is 10. They are 5 and 5.

$x^2 + 10x + 25 = (x + 5)(x + 5)$, so $4x^2 + 40x + 100 = 4(x + 5)(x + 5)$, or $4(x + 5)^2$.

40. $(x + 3)(x - 24)$

41. $y^2 - 21y + 108$

We look for two factors, both negative, whose product is 108 and whose sum is -21. They are -9 and -12.

$y^2 - 21y + 108 = (y - 9)(y - 12)$

42. $(x - 9)(x - 16)$

43. $a^6 + 9a^5 - 90a^4 = a^4(a^2 + 9a - 90)$

After factoring out the common factor, a^4, we consider $a^2 + 9a - 90$. We look for two factors, one positive and one negative, whose product is -90 and whose sum is 9. They are -6 and 15.

$a^2 + 9a - 90 = (a - 6)(a + 15)$, so $a^6 + 9a^5 - 90a^4 = a^4(a - 6)(a + 15)$.

44. $a^2(a - 11)(a + 12)$

45. $x^2 - \frac{2}{5}x + \frac{1}{25}$

We look for two factors, both negative, whose product is $\frac{1}{25}$ and whose sum is $-\frac{2}{5}$. They are $-\frac{1}{5}$ and $-\frac{1}{5}$.

$x^2 - \frac{2}{5}x + \frac{1}{25} = \left(x - \frac{1}{5}\right)\left(x - \frac{1}{5}\right)$, or $\left(x - \frac{1}{5}\right)^2$

46. $\left(t + \frac{1}{3}\right)\left(t + \frac{1}{3}\right)$, or $\left(t + \frac{1}{3}\right)^2$

47. $112 + 9y - y^2$

One way to do this is to write the polynomial in descending order and factor out -1.

$112 + 9y - y^2 = -y^2 + 9y + 112 = -(y^2 - 9y - 112)$

Now we factor $y^2 - 9y - 112$. We look for two factors, one positive and one negative, whose product is -112 and whose sum is -9. They are 7 and -16.

$y^2 - 9y - 112 = (y + 7)(y - 16)$, so $-y^2 + 9y + 112 = -(y + 7)(y - 16)$. We can also express this result as follows:

$-y^2 + 9y + 112 = -(y + 7)(y - 16) =$
$(y + 7)(-1)(y - 16) = (y + 7)(-y + 16) = (7 + y)(16 - y)$

48. $(9 - x)(12 + x)$

49. $t^2 - 0.3t - 0.10$

We look for two factors, one positive and one negative, whose product is -0.10 and whose sum is -0.3. They are 0.2 and -0.5.

$t^2 - 0.3t - 0.10 = (t + 0.2)(t - 0.5)$

50. $(y + 0.2)(y - 0.4)$

51. $p^2 + 3pq - 10q^2 = p^2 + 3qp - 10q^2$

Think of $3q$ as a "coefficient" of p. Then we look for factors of $-10q^2$ whose sum is $3q$. They are $5q$ and $-2q$.

$p^2 + 3pq - 10q^2 = (p + 5q)(p - 2q)$.

52. $(a - 3b)(a + b)$

53. $m^2 + 5mn + 5n^2 = m^2 + 5nm + 5n^2$

We look for factors of $5n^2$ whose sum is $5n$. The only reasonable possibilities are shown below.

Pairs of factors	Sums of factors
$5n, \quad n$	$6n$
$-5n, \ -n$	$-6n$

There are no factors whose sum is $5n$. Thus, the polynomial is not factorable into polynomials with integer coefficients. It is prime.

54. $(x - 8y)(x - 3y)$

55. $s^2 - 2st - 15t^2 = s^2 - 2ts - 15t^2$

We look for factors of $-15t^2$ whose sum is $-2t$. They are $-5t$ and $3t$.

$s^2 - 2st - 15t^2 = (s - 5t)(s + 3t)$

56. $(b + 10c)(b - 2c)$

57. $6a^{10} - 30a^9 - 84a^8 = 6a^8(a^2 - 5a - 14)$

After factoring out the common factor, $6a^8$, we consider $a^2 - 5a - 14$. We look for two factors, one positive and one negative, whose product is -14 and whose sum is -5. They are 2 and -7.

$a^2 - 5a - 14 = (a + 2)(a - 7)$, so $6a^{10} - 30a^9 - 84a^8 = 6a^8(a + 2)(a - 7)$.

58. $7x^7(x + 1)(x - 5)$

59. $3x - 8 = 0$

$\quad 3x = 8 \quad$ Adding 8 on both sides

$\quad x = \dfrac{8}{3} \quad$ Dividing by 3 on both sides

The solution is $\dfrac{8}{3}$.

60. $-\dfrac{7}{2}$

61. $\quad (x + 6)(3x + 4)$

$= 3x^2 + 4x + 18x + 24 \quad$ Using FOIL

$= 3x^2 + 22x + 24$

62. $49w^2 + 84w + 36$

63. *Familiarize.* Let $n =$ the number of people arrested the year before.

Translate. We reword the problem.

$$\underbrace{\text{Number arrested the year before}}_{n} \; \underbrace{\text{less } 1.2\% \text{ of}}_{- \; 1.2\% \; \cdot} \; \underbrace{\text{that number}}_{n} \; \underbrace{\text{is } 29,090.}_{= \; 29,090}$$

Carry out. We solve the equation.

$n - 1.2\% \cdot n = 29,090$

$1 \cdot n - 0.012n = 29,090$

$0.988n = 29,090$

$n \approx 29,443 \quad$ Rounding

Check. 1.2% of 29,443 is $0.012(29,443) \approx 353$ and $29,443 - 353 = 29,090$. The answer checks.

State. Approximately 29,443 people were arrested the year before.

64. $100°, 25°, 55°$

65. ◈

66. ◈

67. ◈

68. ◈

69. $y^2 + my + 50$

We look for pairs of factors whose product is 50. The sum of each pair is represented by m.

Pairs of factors whose product is −50	Sums of factors
1, 50	51
−1, −50	−51
2, 25	27
−2, −25	−27
5, 10	15
−5, −10	−15

The polynomial $y^2 + my + 50$ can be factored if m is $51, -51, 27, -27, 15,$ or -15.

70. $49, -49, 23, -23, 5, -5$

71. $x^2 + \dfrac{1}{4}x - \dfrac{1}{8}$

We look for two factors, one positive and one negative, whose product is $-\dfrac{1}{8}$ and whose sum is $\dfrac{1}{4}$. They are $\dfrac{1}{2}$ and $-\dfrac{1}{4}$.

$x^2 + \dfrac{1}{4}x - \dfrac{1}{8} = \left(x + \dfrac{1}{2}\right)\left(x - \dfrac{1}{4}\right)$

72. $\left(x + \dfrac{3}{4}\right)\left(x - \dfrac{1}{4}\right)$

73. $\dfrac{1}{3}a^3 - \dfrac{1}{3}a^2 - 2a = \dfrac{1}{3}a(a^2 - a - 6)$

After factoring out the common factor, $\dfrac{1}{3}a$, we consider $a^2 - a - 6$. We look for two factors, one positive and one negative, whose product is -6 and whose sum is -1. They are 2 and -3.

$a^2 - a - 6 = (a + 2)(a - 3)$, so $\dfrac{1}{3}a^3 - \dfrac{1}{3}a^2 - 2a = \dfrac{1}{3}a(a + 2)(a - 3)$.

74. $a^5(a - 5)\left(a + \dfrac{5}{7}\right)$

75. $x^{2m} + 11x^m + 28 = (x^m)^2 + 11x^m + 28$

We look for numbers p and q such that $x^{2m} + 11x^m + 28 = (x^m + p)(x^m + q)$. We find two factors, both positive, whose product is 28 and whose sum is 11. They are 4 and 7.

$x^{2m} + 11x^m + 28 = (x^m + 4)(x^m + 7)$

76. $(t^n - 2)(t^n - 5)$

77. $\quad (x + 1)a^2 + (x + 1)3a + (x + 1)2$

$= (x + 1)(a^2 + 3a + 2)$

After factoring out the common factor $x + 1$, we consider $a^2 + 3a + 2$. We look for two factors, whose product is 2 and whose sum is 3. they are 1 and 2.

$a^2 + 3a + 2 = (a+1)(a+2)$, so
$(x+1)a^2 + (x+1)3a + (x+1)2 =$
$(x+1)(a+1)(a+2)$.

78. $(a-5)(x+9)(x-1)$

79. We first label the drawing with additional information.

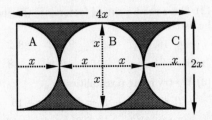

$4x$ represents the length of the rectangle and $2x$ the width. The area of the rectangle is $4x \cdot 2x$, or $8x^2$.

The area of semicircle A is $\frac{1}{2}\pi x^2$.

The area of circle B is πx^2.

The area of semicircle C is $\frac{1}{2}\pi x^2$.

$$\begin{array}{l}\text{Area of} \\ \text{shaded} \\ \text{region}\end{array} = \begin{array}{l}\text{Area of} \\ \text{rectangle}\end{array} - \begin{array}{l}\text{Area} \\ \text{of} \\ A\end{array} - \begin{array}{l}\text{Area} \\ \text{of} \\ B\end{array} - \begin{array}{l}\text{Area} \\ \text{of} \\ C\end{array}$$

$$\begin{array}{l}\text{Area of} \\ \text{shaded} \\ \text{region}\end{array} = 8x^2 - \frac{1}{2}\pi x^2 - \pi x^2 - \frac{1}{2}\pi x^2$$

$$= 8x^2 - 2\pi x^2$$

$$= 2x^2(4 - \pi)$$

The shaded area can be represented by $2x^2(4-\pi)$.

80. $x^2(\pi - 1)$

Exercise Set 5.3

1. $3x^2 + x - 4$

(1) Look for a common factor. There is none (other than 1 or -1).

(2) Because $3x^2$ can be factored as $3x \cdot x$, we have this possibility:

$$(3x + \quad)(x + \quad)$$

(3) There are 3 pairs of factors of -4 and they can be listed two ways:

$$\begin{array}{ccc} -4,1 & 4,-1 & 2,-2 \\ \text{and} \quad 1,-4 & -1,4 & -2,2 \end{array}$$

(4) Look for Outer and Inner products resulting from steps (2) and (3) for which the sum is the middle term, $-7x$. We try some possibilities:

$$(3x - 4)(x + 1) = 3x^2 - x - 4$$
$$(3x + 4)(x - 1) = 3x^2 + x - 4$$

The factorization is $(3x + 4)(x - 1)$.

2. $(2x - 1)(x + 4)$

3. $5x^2 - x - 18$

(1) There is no common factor (other than 1 or -1).

(2) Because $5x^2$ can be factored as $5x \cdot x$, we have this possibility:

$$(5x + \quad)(x + \quad)$$

(3) There are 6 pairs of factors of -18 and they can be listed two ways:

$$\begin{array}{cccc} -18,1 & 18,-1 & -9,2 & 9,-2 \\ -6,3 & 6,-3 \\ \text{and} \quad 1,-18 & -1,18 & 2,-9 & -2,9 \\ 3,-6 & -3,6 \end{array}$$

(4) Look for Outer and Inner products resulting from steps (2) and (3) for which the sum is x. We try some possibilities:

$$(5x - 18)(x + 1) = 5x^2 - 13x - 18$$
$$(5x + 18)(x - 1) = 5x^2 + 13x - 18$$
$$(5x + 9)(x - 2) = 5x^2 - x - 18$$

The factorization is $(5x + 9)(x - 2)$.

4. $(3x + 5)(x - 3)$

5. $6x^2 - 13x + 6$

(1) There is no common factor (other than 1 or -1).

(2) Because $6x^2$ can be factored as $6x \cdot x$ or $3x \cdot 2x$, we have these possibilities:

$$(6x + \quad)(x + \quad) \text{ and } (3x + \quad)(2x + \quad)$$

(3) There are 4 pairs of factors of 6 and they can be listed two ways:

$$\begin{array}{cccc} 6,1 & -6,-1 & 2,3 & -2,-3 \\ \text{and} \quad 1,6 & -1,-6 & 3,2 & -3,-2 \end{array}$$

(4) Look for Outer and Inner products resulting from steps (2) and (3) for which the sum is $-13x$.

Since the sign of the middle term is negative but the sign of the last term is positive, the two factors of 6 must both be negative. This means only four pairs from step (3) need to be considered. We can immediately reject all possibilities in which either factor has a common factor, such as $(6x + 6)$ or $(3x - 3)$, because we determined at the outset that there are no common factors. We try some possibilities:

$$(6x - 1)(x - 6) = 6x^2 - 37x + 6$$
$$(3x - 2)(2x - 3) = 6x^2 - 13x + 6$$

The factorization is $(3x - 2)(2x - 3)$.

6. $(3x - 1)(2x - 7)$

7. $3x^2 + 4x + 1$

(1) There is no common factor (other than 1 or -1).

(2) Because $3x^2$ can be factored as $3x \cdot x$, we have this possibility:

$$(3x + \quad)(x + \quad)$$

(3) There are 2 pairs of factors of 1. In this case they can be listed only one way:

$$1, 1 \quad -1, -1$$

(4) Look for Outer and Inner products resulting from steps (2) and (3) for which the sum is the middle term, $4x$. Since all coefficients are positive, we need consider only positive factors of 1. There is only one such possibility:

$$(3x + 1)(x + 1) = 3x^2 + 4x + 1$$

The factorization is $(3x + 1)(x + 1)$.

8. $(7x + 1)(x + 2)$

9. $4x^2 + 4x - 15$

(1) There is no common factor (other than 1 or -1).

(2) Because $4x^2$ can be factored as $4x \cdot x$ or $2x \cdot 2x$, we have these possibilities:

$$(4x + \quad)(x + \quad) \text{ and } (2x + \quad)(2x + \quad)$$

(3) There are 4 pairs of factors of -15 and they can be listed two ways:

$$15, -1 \quad -15, 1 \quad 5, -3 \quad -5, 3$$
$$\text{and} \quad -1, 15 \quad 1, -15 \quad -3, 5 \quad 3, -5$$

(4) We try some possibilities:

$$(4x + 15)(x - 1) = 4x^2 + 11x - 15$$
$$(2x + 15)(2x - 1) = 4x^2 + 28x - 15$$
$$(4x - 15)(x + 1) = 4x^2 - 11x - 15$$
$$(2x - 15)(2x + 1) = 4x^2 - 28x - 15$$
$$(4x + 5)(x - 3) = 4x^2 - 7x - 15$$
$$(2x + 5)(2x - 3) = 4x^2 + 4x - 15$$

The factorization is $(2x + 5)(2x - 3)$.

10. $(3x - 2)(3x + 4)$

11. $15x^2 - 19x - 10$

(1) There is no common factor (other than 1 or -1).

(2) Because $15x^2$ can be factored as $15x \cdot x$ or $5x \cdot 3x$, we have these possibilities:

$$(15x + \quad)(x + \quad) \text{ and } (5x + \quad)(3x + \quad)$$

(3) There are 4 pairs of factors of -10 and they can be listed two ways:

$$10, -1 \quad -10, 1 \quad 5, -2 \quad -5, 2$$
$$\text{and} \quad -1, 10 \quad 1, -10 \quad -2, 5 \quad 2, -5$$

(4) We can immediately reject all possibilities in which either factor has a common factor, such as $(15x + 10)$ or $(5x - 5)$, because we determined at the outset that there are no common factors. We try some possibilities:

$$(15x - 1)(x + 10) = 15x^2 + 149x - 10$$
$$(15x - 2)(x + 5) = 15x^2 + 73x - 10$$
$$(5x - 1)(3x + 10) = 15x^2 + 47x - 10$$
$$(5x + 2)(3x - 5) = 15x^2 - 19x - 10$$

The factorization is $(5x + 2)(3x - 5)$.

12. $(3x + 1)(x - 2)$

13. $9x^2 + 18x - 16$

(1) There is no common factor (other than 1 or -1).

(2) Because $9x^2$ can be factored as $9x \cdot x$ or $3x \cdot 3x$, we have these possibilities:

$$(9x + \quad)(x + \quad) \text{ and } (3x + \quad)(3x + \quad)$$

(3) There are 5 pairs of factors of -16 and they can be listed two ways:

$$16, -1 \quad -16, 1 \quad 8, -2 \quad -8, 2 \quad 4, -4$$
$$\text{and} \quad -1, 16 \quad 1, -16 \quad -2, 8 \quad 2, -8 \quad -4, 4$$

(4) We try some possibilities:

$$(9x + 16)(x - 1) = 9x^2 + 7x - 16$$
$$(3x + 16)(3x - 1) = 9x^2 + 45x - 16$$
$$(9x - 16)(x + 1) = 9x^2 - 7x - 16$$
$$(3x - 16)(3x + 1) = 9x^2 - 45x - 16$$
$$(9x + 8)(x - 2) = 9x^2 - 10x - 16$$
$$(3x + 8)(3x - 2) = 9x^2 + 18x - 16$$

The factorization is $(3x + 8)(3x - 2)$.

14. $(2x + 1)(x - 1)$

15. $2x^2 - 5x + 2$

(1) There is no common factor (other than 1 or -1).

(2) Because $2x^2$ can be factored as $2x \cdot x$, we have this possibility:

$$(2x + \quad)(x + \quad)$$

(3) There are 2 pairs of factors of 2 and they can be listed two ways:

$$2, 1 \quad -2, -1$$
$$\text{and} \quad 1, 2 \quad -1, -2$$

(4) Since the sign of the middle term is negative but the sign of the last term is positive, the two factors of 2 must both be negative. This means only two pairs from step (3) need to be considered. In addition, we can immediately reject the possibility that $(2x - 2)$ is a factor, because we determined at the outset that there are no common factors. We try the other possibility:

$$(2x - 1)(x - 2) = 2x^2 - 5x + 2$$

The factorization is $(2x - 1)(x - 2)$.

16. $(6x - 5)(3x + 2)$

17. $12x^2 - 31x + 20$

(1) There is no common factor (other than 1 or -1).

(2) Because $12x^2$ can be factored as $12x \cdot x$, $6x \cdot 2x$ or $4x \cdot 3x$, we have these possibilities:

$$(12x + \quad)(x + \quad) \text{ and } (6x + \quad)(2x + \quad) \text{ and } (4x + \quad)(3x + \quad)$$

(3) Since the sign of the middle term is negative but the sign of the last term is positive, we need to consider only negative pairs of factors of 20. There are 3 such pairs and they can be listed two ways:

$$-20, -1 \quad -10, -2 \quad -5, -4$$
$$\text{and} \quad -1, -20 \quad -2, -10 \quad -4, -5$$

(4) We can immediately reject all possibilities in which either factor has a common factor, such as $(12x - 20)$ or $(6x - 4)$, because we determined at the outset that there are no common factors. We try some of the remaining possibilities:

$$(12x - 1)(x - 20) = 12x^2 - 241x + 20$$
$$(12x - 5)(x - 4) = 12x^2 - 53x + 20$$
$$(4x - 5)(3x - 4) = 12x^2 - 31x + 20$$

The factorization is $(4x - 5)(3x - 4)$.

18. $(5x + 3)(3x + 2)$

19. $28x^2 + 38x - 6$

(1) We factor out the common factor, 2:
$$2(14x^2 + 19x - 3)$$

Then we factor the trinomial $14x^2 + 19x - 3$.

(2) Because $14x^2$ can be factored as $14x \cdot x$ or $7x \cdot 2x$, we have these possibilities:
$$(14x +)(x +) \text{ and } (7x +)(2x +)$$

(3) There are 2 pairs of factors of -3 and they can be listed two ways:

$$-3, 1 \quad 3, -1$$
$$\text{and} \quad 1, -3 \quad -1, 3$$

(4) We try some possibilities:
$$(14x - 3)(x + 1) = 14x^2 + 11x - 3$$
$$(7x + 3)(2x - 1) = 14x^2 - x - 3$$
$$(7x - 1)(2x + 3) = 14x^2 + 19x - 3$$

The factorization of $14x^2 + 19x - 3$ is $(7x - 1)(2x + 3)$. We must include the common factor in order to get a factorization of the original trinomial.
$$28x^2 + 38x - 6 = 2(7x - 1)(2x + 3)$$

20. $(7x + 4)(5x + 2)$

21. $9x^2 + 18x + 8$

(1) There is no common factor (other than 1 or -1).
(2) Because $9x^2$ can be factored as $9x \cdot x$ or $3x \cdot 3x$, we have these possibilities:
$$(9x +)(x +) \text{ and } (3x +)(3x +)$$

(3) Since all coefficients are positive, we need consider only positive pairs of factors of 8. There are 2 such pairs and they can be listed in two ways:

$$8, 1 \quad 4, 2$$
$$\text{and} \quad 1, 8 \quad 2, 4$$

(4) We try some possibilities:

$$(9x + 8)(x + 1) = 9x^2 + 17x + 8$$
$$(3x + 8)(3x + 1) = 9x^2 + 27x + 8$$
$$(9x + 4)(x + 2) = 9x^2 + 22x + 8$$
$$(3x + 4)(3x + 2) = 9x^2 + 18x + 8$$

The factorization is $(3x + 4)(3x + 2)$.

22. Prime

23. $49 + 42x + 9x^2 = 9x^2 + 42x + 49$

(1) There is no common factor (other than 1 or -1).
(2) Because $9x^2$ can be factored as $9x \cdot x$ or $3x \cdot 3x$, we have these possibilities:
$$(9x +)(x +) \text{ and } (3x +)(3x +)$$

(3) Since all coefficients are positive, we need consider only positive pairs of factors of 49. There are 2 such pairs and one pair can be listed two ways:

$$49, 1 \quad 7, 7$$
$$\text{and} \quad 1, 49$$

(4) We try some possibilities:
$$(9x + 49)(x + 1) = 9x^2 + 58x + 49$$
$$(3x + 49)(3x + 1) = 9x^2 + 150x + 49$$
$$(9x + 7)(x + 7) = 9x^2 + 70x + 49$$
$$(3x + 7)(3x + 7) = 9x^2 + 42x + 49$$

The factorization is $(3x + 7)(3x + 7)$, or $(3x + 7)^2$, or $(7 + 3x)^2$.

24. $(5x + 4)^2$

25. $24x^2 + 47x - 2$

(1) There is no common factor (other than 1 or -1).
(2) Because $24x^2$ can be factored as $24x \cdot x$, $12x \cdot 2x$, $6x \cdot 4x$, or $3x \cdot 8x$, we have these possibilities:
$$(24x +)(x +) \text{ and } (12x +)(2x +) \text{ and }$$
$$(6x +)(4x +) \text{ and } (3x +)(8x +)$$

(3) There are 2 pairs of factors of -2 and they can be listed two ways:

$$2, -1 \quad -2, 1$$
$$\text{and} \quad -1, 2 \quad 1, -2$$

(4) We can immediately reject all possibilities in which either a factor has a common factor, such as $(24x + 2)$ or $(12x - 2)$, because we determined at the outset that there are no common factors. We try some of the remaining possibilities:
$$(24x - 1)(x + 2) = 24x^2 + 47x - 2$$

The factorization is $(24x - 1)(x + 2)$.

26. $(8a + 3)(2a + 9)$

27. $35x^2 - 57x - 44$

(1) There is no common factor (other than 1 or -1).
(2) Because $35x^2$ can be factored as $35x \cdot x$ or $7x \cdot 5x$, we have these possibilities:

$(35x + \quad)(x + \quad)$ and $(7x + \quad)(5x + \quad)$

(3) There are 6 pairs of factors of -44 and they can be listed two ways:

$$1, -44 \quad -1, 44 \quad 2, -22 \quad -2, 22$$
$$4, -11 \quad -4, 11$$
and $\quad -44, 1 \quad 44, -1 \quad -22, 2 \quad 22, -2$
$$-11, 4 \quad 11, -4$$

(4) We try some possibilities:

$$(35x + 1)(x - 44) = 35x^2 - 1539x - 44$$
$$(7x + 1)(5x - 44) = 35x^2 - 303x - 44$$
$$(35x + 2)(x - 22) = 35x^2 - 768x - 44$$
$$(7x + 2)(5x - 22) = 35x^2 - 144x - 44$$
$$(35x + 4)(x - 11) = 35x^2 - 381x - 44$$
$$(7x + 4)(5x - 11) = 35x^2 - 57x - 44$$

The factorization is $(7x + 4)(5x - 11)$.

28. $2(3t - 1)(3t + 5)$

29. $2x^2 - 6x - 19$

(1) There is no common factor (other than 1 or -1).

(2) Because $2x^2$ can be factored as $2x \cdot x$ we have this possibility:

$$(2x + \quad)(x + \quad)$$

(3) There are 2 pairs of factors of -19 and they can be listed two ways:

$$19, -1 \quad -19, 1$$
and $\quad -1, 19 \quad 1, -19$

(4) We try some possibilities:

$$(2x + 19)(x - 1) = 2x^2 + 17x - 19$$
$$(2x - 1)(x + 19) = 2x^2 + 37x - 19$$

The other two possibilities will only change the sign of the middle terms in these trials.

We must conclude that $2x^2 - 6x - 19$ is prime.

30. $(2x + 5)(x - 3)$

31. $12x^2 + 28x - 24$

(1) We factor out the common factor, 4:

$$4(3x^2 + 7x - 6)$$

Then we factor the trinomial $3x^2 + 7x - 6$.

(2) Because $3x^2$ can be factored as $3x \cdot x$ we have this possibility:

$$(3x + \quad)(x + \quad)$$

(3) There are 4 pairs of factors of -6 and they can be listed two ways:

$$6, -1 \quad -6, 1 \quad 3, -2 \quad -3, 2$$
and $\quad -1, 6 \quad 1, -6 \quad -2, 3 \quad 2, -3$

(4) We can immediately reject all possibilities in which either factor has a common factor, such as $(3x + 6)$ or $(3x - 3)$, because we factored out the largest common factor at the outset. We try some of the remaining possibilities:

$$(3x - 1)(x + 6) = 3x^2 + 17x - 6$$
$$(3x - 2)(x + 3) = 3x^2 + 7x - 6$$

The factorization of $3x^2 + 7x - 6$ is $(3x - 2)(x + 3)$. We must include the common factor in order to get a factorization of the original trinomial.

$$12x^2 + 28x - 24 = 4(3x - 2)(x + 3)$$

32. $3(2x + 1)(x + 5)$

33. $30x^2 - 24x - 54$

(1) Factor out the common factor, 6:

$$6(5x^2 - 4x - 9)$$

Then we factor the trinomial $5x^2 - 4x - 9$.

(2) Because $5x^2$ can be factored as $5x \cdot x$ we have this possibility:

$$(5x + \quad)(x + \quad)$$

(3) There are 3 pairs of factors of -9 and they can be listed two ways:

$$9, -1 \quad -9, 1 \quad 3, -3$$
and $\quad -1, 9 \quad 1, -9 \quad -3, 3$

(4) We try some possibilities:

$$(5x + 9)(x - 1) = 5x^2 + 4x - 9$$
$$(5x - 9)(x + 1) = 5x^2 - 4x - 9$$

The factorization of $5x^2 - 4x - 9$ is $(5x - 9)(x + 1)$. We must include the common factor in order to get a factorization of the original trinomial.

$$30x^2 - 24x - 54 = 6(5x - 9)(x + 1)$$

34. $5(4x - 1)(x - 1)$

35. $6x^2 + 33x + 15$

(1) Factor out the common factor, 3:

$$3(2x^2 + 11x + 5)$$

Then we factor the trinomial $2x^2 + 11x + 5$.

(2) Because $2x^2$ can be factored as $2x \cdot x$ we have this possibility:

$$(2x + \quad)(x + \quad)$$

(3) Since all coefficients are positive, we need consider only positive pairs of factors of 5. There is one such pair and it can be listed two ways:

$$5, 1 \quad \text{and} \quad 1, 5$$

(4) We try some possibilities:

$$(2x + 5)(x + 1) = 2x^2 + 7x + 5$$
$$(2x + 1)(x + 5) = 2x^2 + 11x + 5$$

The factorization of $2x^2 + 11x + 5$ is $(2x + 1)(x + 5)$. We must include the common factor in order to get a factorization of the original trinomial.

$$6x^2 + 33x + 15 = 3(2x + 1)(x + 5)$$

36. $4(3x-2)(x+3)$

37. $-9+18x^2+21x=18x^2+21x-9$

(1) We factor out the common factor, 3:

$$3(6x^2+7x-3)$$

Then we factor the trinomial $6x^2+7x-3$.

(2) Because $6x^2$ can be factored as $6x \cdot x$ or $3x \cdot 2x$ we have these possibilities:

$$(6x+\ )(x+\ ) \text{ and } (3x+\ )(2x+\ )$$

(3) There are 2 pairs of factors of -3 and they can be listed two ways:

$$3,-1 \quad -3,1$$
$$\text{and} \quad -1,3 \quad 1,-3$$

(4) We can immediately reject all possibilities in which either factor has a common factor, such as $(6x+3)$ or $(3x-3)$, because we factored out the largest common factor at the outset. We try some possibilities:

$$(6x-1)(x+3)=6x^2+17x-3$$
$$(3x-1)(2x+3)=6x^2+7x-3$$

The factorization of $6x^2+7x-3$ is $(3x-1)(2x+3)$. We must include the common factor in order to get a factorization of the original trinomial.

$$-9+18x^2+21x=3(3x-1)(2x+3)$$

38. $(3x+1)(x+1)$

39. $y^2+4y-2y-8=y(y+4)-2(y+4)$
$$=(y+4)(y-2)$$

40. $(x+5)(x+2)$

41. $x^2-4x-x+4=x(x-4)-1(x-4)$
$$=(x-4)(x-1)$$

42. $(a+5)(a-2)$

43. $6x^2+4x+9x+6=2x(3x+2)+3(3x+2)$
$$=(3x+2)(2x+3)$$

44. $(3x-2)(x+1)$

45. $3x^2-4x-12x+16=x(3x-4)-4(3x-4)$
$$=(3x-4)(x-4)$$

46. $(4-3y)(6-5y)$

47. $35x^2-40x+21x-24=5x(7x-8)+3(7x-8)$
$$=(7x-8)(5x+3)$$

48. $(4x-3)(2x-7)$

49. $4x^2+6x-6x-9=2x(2x+3)-3(2x+3)$
$$=(2x+3)(2x-3)$$

50. $(x^2-3)(2x^2-5)$

51. $9x^2-42x+49$

(1) First note that there is no common factor (other than 1 or -1).

(2) Multiply the leading coefficient, 9, and the constant, 49:

$$9 \cdot 49 = 441$$

(3) Look for factors of 441 that add to -42. We need to consider only negative pairs of factors since the last term is positive and the middle term is negative.

Pairs of Factors	Sums of Factors
$-3, \quad -147$	-150
$-7, \quad -63$	-70
$-9, \quad -49$	-58
$-21 \quad -21$	-42

(4) Rewrite the middle term:

$$-42x=-21x-21x$$

(5) Factor by grouping:

$$9x^2-42x+49=9x^2-21x-21x+49$$
$$=3x(3x-7)-7(3x-7)$$
$$=(3x-7)(3x-7), \text{ or}$$
$$(3x-7)^2$$

52. $(5t+8)^2$

53. $18t^2+3t-10$

(1) First note that there is no common factor (other than 1 or -1).

(2) Multiply the leading coefficient, 18, and the constant, -10:

$$18(-10)=-180$$

(3) Look for factors of -180 that add to 3.

Pairs of Factors	Sums of Factors
$-1, \quad 180$	179
$1, \quad -180$	-179
$-2, \quad 90$	88
$2, \quad -90$	-88
$-3, \quad 60$	57
$3, \quad -60$	-57
$-4, \quad 45$	41
$4, \quad -45$	-41
$-5, \quad 36$	31
$5, \quad -36$	-31
$-6, \quad 30$	24
$6, \quad -30$	-24
$-9, \quad 20$	11
$9, \quad -20$	-11
$-10, \quad 18$	8
$10, \quad -18$	-8
$-12, \quad 15$	3

The numbers we need are -12 and 15.

(4) Rewrite the middle term:

$$3t = -12t + 15t$$

(5) Factor by grouping:

$$18t^2 + 3t - 10 = 18t^2 - 12t + 15t - 10$$
$$= 6t(3t - 2) + 5(3t - 2)$$
$$= (3t - 2)(6t + 5)$$

54. $5(x - 2)(3x + 1)$

55. $14x^2 - 35x + 14$

(1) Factor out the largest common factor, 7.

$$14x^2 - 35x + 14 = 7(2x^2 - 5x + 2)$$

(2) To factor $2x^2 - 5x + 2$ by grouping we first multiply the leading coefficient, 2, and the constant, 2:

$$2 \cdot 2 = 4$$

(3) We look for factors of 4 that add to -5. Since the last term is positive and the middle term is negative, we need consider only negative pairs of factors.

Pairs of Factors	Sums of Factors
-2, -2	-4
-1, -4	-5

The numbers we need are -1 and -4.

(4) Rewrite the middle term:

$$-5x = -x - 4x$$

(5) Factor by grouping:

$$2x^2 - 5x + 2 = 2x^2 - x - 4x + 2$$
$$= x(2x - 1) - 2(2x - 1)$$
$$= (2x - 1)(x - 2)$$

The factorization of $2x^2 - 5x + 2$ is $(2x - 1)(x - 2)$. We must include the common factor in order to get a factorization of the original trinomial:

$$14x^2 - 35x + 14 = 7(2x - 1)(x - 2)$$

56. $2(x^2 + 3x - 7)$

57. $6x^3 - 4x^2 - 10x$

(1) Factor out the largest common factor, $2x$:

$$6x^3 - 4x^2 - 10x = 2x(3x^2 - 2x - 5)$$

(2) To factor $3x^2 - 2x - 5$ by grouping we first multiply the leading coefficient, 3, and the constant, -5:

$$3(-5) = -15$$

(3) We look for factors of -15 that add to -2. The numbers we need are 3 and -5.

(4) Rewrite the middle term:

$$-2x = 3x - 5x$$

(5) Factor by grouping:

$$3x^2 - 2x - 5 = 3x^2 + 3x - 5x - 5$$
$$= 3x(x + 1) - 5(x + 1)$$
$$= (x + 1)(3x - 5)$$

The factorization of $3x^2 - 2x - 5$ is $(x + 1)(3x - 5)$. We must include the common factor in order to get a factorization of the original trinomial:

$$6x^3 - 4x^2 - 10x = 2x(x + 1)(3x - 5)$$

58. $3x(2x + 3)(3x - 1)$

59. $47 - 42y + 9y^2 = 9y^2 - 42y + 47$

(1) First note that there is no common factor (other than 1 or -1).

(2) Multiply the leading coefficient, 9, and the constant, 47:

$$9 \cdot 47 = 423$$

(3) Look for factors of 423 that add to -42. We need to consider only negative pairs of factors since the last term is positive and the middle term is negative.

Pairs of Factors	Sums of Factors
-1, -423	-424
-3, -141	-144
-9, -47	-56

There are no factors of 423 that add to -42. The polynomial is prime.

60. $(x + 1)(25x + 64)$

61. $144x^5 - 168x^4 + 48x^3$

(1) Factor out the largest common factor, $24x^3$:

$$144x^5 - 168x^4 + 48x^3 = 24x^3(6x^2 - 7x + 2)$$

(2) To factor $6x^2 + 7x + 2$ by grouping, we first multiply the leading coefficient, 6, and the constant, 2:

$$6 \cdot 2 = 12$$

(3) We look for factors of 12 that add to -7. The numbers we need are -3 and -4.

(4) Rewrite the middle term:

$$-7x = -3x - 4x$$

(5) Factor by grouping:

$$6x^2 - 7x + 2 = 6x^2 - 3x - 4x + 2$$
$$= 3x(2x - 1) - 2(2x - 1)$$
$$= (2x - 1)(3x - 2)$$

The factorization of $6x^2 - 7x + 2$ is $(2x - 1)(3x - 2)$. We must include the common factor in order to get a factorization of the original trinomial:

$$144x^5 - 168x^4 + 48x^3 = 24x^3(2x - 1)(3x - 2)$$

62. $3x(7x + 1)(8x + 1)$

63. $70a^4 - 68a^3 + 16a^2$

(1) Factor out the largest common factor, $2a^2$:

$$70a^4 - 68a^3 + 16a^2 = 2a^2(35a^2 - 34a + 8)$$

(2) To factor $35a^2 - 34a + 8$ by grouping we first multiply the leading coefficient, 35, and the constant, 8:

$$35 \cdot 8 = 280$$

(3) We look for factors of 280 that add to -34. The numbers we need are -14 and -20.

(4) Rewrite the middle term:

$$-34a = -14a - 20a$$

(5) Factor by grouping:
$$35a^2 - 34a + 8 = 35a^2 - 14a - 20a + 8$$
$$= 7a(5a - 2) - 4(5a - 2)$$
$$= (5a - 2)(7a - 4)$$

The factorization of $35a^2 - 34a + 8$ is $(5a-2)(7a-4)$. We must include the common factor in order to get a factorization of the original trinomial:
$$70a^4 - 68a^3 + 16a^2 = 2a^2(5a - 2)(7a - 4)$$

64. $t^2(2t - 3)(7t + 1)$

65. $12m^2 + mn - 20n^2$

(1) First note that there is no common factor (other than 1 or -1).

(2) Multiply the leading coefficient, 12, and the constant, -20:
$$12(-20) = -240$$

(3) Look for factors of -240 that add to 1. The numbers we need are -15 and 16.

(4) Rewrite the middle term:
$$mn = -15mn + 16mn$$

(5) Factor by grouping:
$$12m^2 + mn - 20n^2$$
$$= 12m^2 - 15mn + 16mn - 20n^2$$
$$= 3m(4m - 5n) + 4n(4m - 5n)$$
$$= (4m - 5n)(3m + 4n)$$

66. $(3a - 5b)(2a + 3b)$

67. $3p^2 - 16pq - 12q^2$

(1) First note that there is no common factor (other than 1 or -1).

(2) Multiply the leading coefficient, 3, and the constant, -12:
$$3(-12) = -36$$

(3) Look for factors of -36 that add to -16. The numbers we need are -18 and 2.

(4) Rewrite the middle term:
$$-16pq = -18pq + 2pq$$

(5) Factor by grouping:
$$3p^2 - 16pq - 12q^2 = 3p^2 - 18pq + 2pq - 12q^2$$
$$= 3p(p - 6q) + 2q(p - 6q)$$
$$= (p - 6q)(3p + 2q)$$

68. $(3a + 2b)(3a + 4b)$

69. $10s^2 + 4st - 6t^2$

(1) Factor out the largest common factor, 2:
$$10s^2 + 4st - 6t^2 = 2(5s^2 + 2st - 3t^2)$$

(2) To factor $5s^2 + 2st - 3t^2$ by grouping we first multiply the leading coefficient, 5, and the constant, -3:

$$5(-3) = -15$$

(3) We look for factors of -15 that add to 2. The numbers we need are 5 and -3.

(4) Rewrite the middle term:
$$2st = 5st - 3st$$

(5) Factor by grouping:
$$5s^2 + 2st - 3t^2 = 5s^2 + 5st - 3st - 3t^2$$
$$= 5s(s + t) - 3t(s + t)$$
$$= (s + t)(5s - 3t)$$

The factorization of $5s^2 + 2st - 3t^2$ is $(s+t)(5s-3t)$. We must include the common factor in order to get a factorization of the original trinomial:
$$10s^2 + 4st - 6t^2 = 2(s + t)(5s - 3t)$$

70. $(7p + 4q)(5p + 2q)$

71. $30a^2 + 87ab + 30b^2$

(1) Factor out the largest common factor, 3:
$$30a^2 + 87ab + 30b^2 = 3(10a^2 + 29ab + 10b^2)$$

(2) To factor $10a^2 + 29ab + 10b^2$ by grouping we first multiply the leading coefficient, 10, and the constant, 10:
$$10 \cdot 10 = 100$$

(3) We look for factors of 100 that add to 29. The numbers we need are 4 and 25.

(4) Rewrite the middle term:
$$10a^2 + 29ab + 10b^2 = 10a^2 + 4ab + 25ab + 10b^2$$
$$= 2a(5a + 2b) + 5b(5a + 2b)$$
$$= (5a + 2b)(2a + 5b)$$

The factorization of $10a^2 + 29ab + 10b^2$ is $(5a + 2b)(2a + 5b)$. We must include the common factor in order to get a factorization of the original trinomial:
$$30a^2 + 87ab + 30b^2 = 3(5a + 2b)(2a + 5b)$$

72. $6(3x - 4y)(x + y)$

73. $15a^2 - 5ab - 20b^2$

(1) Factor out the largest common factor, 5:
$$15a^2 - 5ab - 20b^2 = 5(3a^2 - ab - 4b^2)$$

(2) To factor $3a^2 - ab - 4b^2$ by grouping we first multiply the leading coefficient, 3, and the constant, -4:
$$3(-4) = -12$$

(3) We look for factors of -12 that add to -1. The numbers we need are -4 and 3.

(4) Rewrite the middle term:
$$-ab = -4ab + 3ab$$

(5) Factor by grouping:
$$3a^2 - ab - 4b^2 = 3a^2 - 4ab + 3ab - 4b^2$$
$$= a(3a - 4b) + b(3a - 4b)$$
$$= (3a - 4b)(a + b)$$

The factorization of $3a^2 - ab - 4b^2$ is $(3a - 4b)(a + b)$. We must include the common factor in order to get a factorization of the original trinomial:
$$15a^2 - 5ab - 20b^2 = 5(3a - 4b)(a + b)$$

74. $2(3a - 2b)(4a - 3b)$

75. $9x^2y^2 + 18xy^2 - 16$

Following the procedure for factoring by grouping, we can write the polynomial as $9x^2y^2 + 24xy^2 - 6xy^2 - 16$, but it is not possible to factor this by grouping, regardless of the groups chosen. Thus, the polynomial is prime.

76. $2y(2x^2 + 5x + 1)$

77. $18x^2y^2 - 3xy - 10$

(1) First note that there is no common factor (other than 1 or -1).

(2) Multiply the leading coefficient, 18, and the constant, -10:
$$18(-10) = -180$$

(3) Look for factors of -180 that add to -3. The numbers we need are 12 and -15.

(4) Rewrite the middle term:
$$-3xy = 12xy - 15xy$$

(5) Factor by grouping:
$$\begin{aligned} 18x^2y^2 - 3xy - 10 &= 18x^2y^2 + 12xy - 15xy - 10 \\ &= 6xy(3xy + 2) - 5(3xy + 2) \\ &= (3xy + 2)(6xy - 5) \end{aligned}$$

78. Prime

79. $35x^2 + 34x^3 + 8x^4 = 8x^4 + 34x^3 + 35x^2$

(1) Factor out the largest common factor, x^2:
$$x^2(8x^2 + 34x + 35)$$

(2) To factor $8x^2 + 34x + 35$ by grouping we first multiply the leading coefficient, 8, and the constant, 35:
$$8 \cdot 35 = 280$$

(3) We look for factors of 280 that add to 34. The numbers we need are 14 and 20.

(4) Rewrite the middle term:
$$34x = 14x + 20x$$

(5) Factor by grouping:
$$\begin{aligned} 8x^2 + 34x + 35 &= 8x^2 + 14x + 20x + 35 \\ &= 2x(4x + 7) + 5(4x + 7) \\ &= (4x + 7)(2x + 5) \end{aligned}$$

The factorization of $8x^2 + 34x + 35$ is $(4x+7)(2x+5)$. We must include the common factor in order to get a factorization of the original trinomial:
$$35x^2 + 34x^3 + 8x^4 = x^2(4x + 7)(2x + 5)$$

80. $x^2(2x + 3)(7x - 1)$.

81. $18a + 8 + 9a^2 = 9a^2 + 18a + 8$

(1) First note that there is no common factor (other than 1 or -1).

(2) Multiply the leading coefficient, 9, and the constant, 8:
$$9 \cdot 8 = 72$$

(3) Look for factors of 72 that add to 18. The numbers we need are 6 and 12.

(4) Rewrite the middle term:
$$18a = 6a + 12a$$

(5) Factor by grouping:
$$\begin{aligned} 9a^2 + 18a + 8 &= 9a^2 + 6a + 12a + 8 \\ &= 3a(3a + 2) + 4(3a + 2) \\ &= (3a + 2)(3a + 4) \end{aligned}$$

82. $(5a + 4)^2$

83. **Familiarize**. We will use the formula $C = 2\pi r$, where C is circumference and r is radius, to find the radius in kilometers. Then we will multiply that number by 0.62 to find the radius in miles.

Translate.
$$\underbrace{\text{Circumference}}_{40,000} = \underbrace{2}_{\approx} \cdot \underbrace{\pi \cdot \text{radius}}_{2(3.14)r}$$

Carry out. First we solve the equation.
$$\begin{aligned} 40,000 &\approx 2(3.14)r \\ 40,000 &\approx 6.28r \\ 6369 &\approx r \end{aligned}$$

Then we multiply to find the radius in miles:
$$6369(0.62) \approx 3949$$

Check. If $r = 6369$, then $2\pi r = 2(3.14)(6369) \approx 40,000$. We should also recheck the multiplication we did to find the radius in miles. Both values check.

State. The radius of the earth is about 6369 km or 3949 mi. (These values may differ slightly if a different approximation is used for π.)

84. $40°$

85. Graph: $y = \dfrac{2}{5}x - 1$

Because the equation is in the form $y = mx + b$, we know the y-intercept is $(0, -1)$. We find two other points on the line, substituting multiples of 5 for x to avoid fractions.

When $x = -5$, $y = \dfrac{2}{5}(-5) - 1 = -2 - 1 = -3$.

When $x = 5$, $y = \dfrac{2}{5}(5) - 1 = 2 - 1 = 1$.

x	y
0	-1
-5	-3
5	1

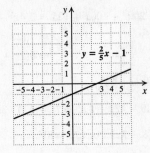

$y = \frac{2}{5}x - 1$

86. y^8

87. $(3x - 5)(3x + 5) = (3x)^2 - 5^2 = 9x^2 - 25$

88. $16a^2 - 24a + 9$

89. ◈

90. ◈

91. ◈

92. ◈

93. $16x^{10} - 8x^5 + 1 = 16(x^5)^2 - 8x^5 + 1$

(1) There is no common factor (other than 1 or -1).

(2) Because $16x^{10}$ can be factored as $16x^5 \cdot x^5$ or $8x^5 \cdot 2x^5$ or $4x^5 \cdot 4x^5$, we have these possibilities:

$$(16x^5 + \ )(x^5 + \ ) \text{ and } (8x^5 + \ )(2x^5 + \ )$$
$$\text{and } (4x^5 + \ )(4x^5 + \ )$$

(3) Since the last term is positive and the middle term is negative we need consider only negative factors of 1. The only negative pair of factors is $-1, -1$.

(4) We try some possibilities:

$$(16x^5 - 1)(x^5 - 1) = 16x^{10} - 17x^5 + 1$$
$$(8x^5 - 1)(2x^5 - 1) = 16x^{10} - 10x^5 + 1$$
$$(4x^5 - 1)(4x^5 - 1) = 16x^{10} - 8x^5 + 1$$

The factorization is $(4x^5 - 1)(4x^5 - 1)$, or $(4x^5 - 1)^2$.

94. $(3x^5 + 2)^2$

95. $20x^{2n} + 16x^n + 3 = 20(x^n)^2 + 16x^n + 3$

(1) There is no common factor (other than 1 or -1).

(2) Because $20x^{2n}$ can be factored as $20x^n \cdot x^n$, $10x^n \cdot 2x^n$, or $5x^n \cdot 4x^n$, we have these possibilities:

$$(20x^n + \ )(x^n + \ ) \text{ and } (10x^n + \ )(2x^n + \ )$$
$$\text{and } (5x^n + \ )(4x^n + \ )$$

(3) Since all the signs are positive, we need consider only the positive factor pair 3,1 when factoring 3. This pair can also be listed as 1,3.

(4) We try some possibilities:

$$(20x^n + 3)(x^n + 1) = 20x^{2n} + 23x^n + 3$$
$$(10x^n + 3)(2x^n + 1) = 20x^{2n} + 16x^n + 3$$

The factorization is $(10x^n + 3)(2x^n + 1)$.

96. $-(3x^m - 4)(5x^m - 2)$

97. $3x^{6a} - 2x^{3a} - 1 = 3(x^{3a})^2 - 2x^{3a} - 1$

(1) There is no common factor (other than 1 or -1).

(2) Because $3x^{6a}$ can be factored as $3x^{3a} \cdot x^{3a}$, we have this possibility:

$$(3x^{3a} + \ )(x^{3a} + \ )$$

(3) There is 1 pair of factors of -1 and it can be listed two ways:

$$-1, 1 \quad 1, -1$$

(4) We try these possibilities:

$$(3x^{3a} - 1)(x^{3a} + 1) = 3x^{6a} + 2x^{3a} - 1$$
$$(3x^{3a} + 1)(x^{3a} - 1) = 3x^{6a} - 2x^{3a} - 1$$

The factorization is $(3x^{3a} + 1)(x^{3a} - 1)$.

98. $x(x^n - 1)^2$

99.
$$3(a + 1)^{n+1}(a + 3)^2 - 5(a + 1)^n(a + 3)^3$$
$$= (a + 1)^n(a + 3)^2[3(a + 1) - 5(a + 3)]$$
$$\qquad\qquad\text{Removing the common factors}$$
$$= (a + 1)^n(a + 3)^2[3a + 3 - 5a - 15] \text{ Simplify-}$$
$$= (a + 1)^n(a + 3)^2(-2a - 12) \qquad \text{ing inside the}$$
$$\qquad\qquad\qquad\qquad\qquad\qquad\qquad\text{brackets}$$
$$= (a + 1)^n(a + 3)^2(-2)(a + 6) \text{ Removing the}$$
$$\qquad\qquad\qquad\qquad\qquad\qquad\text{common factor}$$
$$= -2(a + 1)^n(a + 3)^2(a + 6) \quad \text{Rearranging}$$

100. $[7(t - 3)^n - 2][(t - 3)^n + 1]$

Exercise Set 5.4

1. $x^2 - 18x + 81$

(1) Two terms, x^2 and 81, are squares.

(2) There is no minus sign before x^2 or 81.

(3) Twice the product of the square roots, $2 \cdot x \cdot 9$, is $18x$, the opposite of the remaining term, $-18x$.

Thus, $x^2 - 18x + 81$ is a perfect-square trinomial.

2. Yes

3. $x^2 + 16x - 64$

(1) Two terms, x^2 and 64, are squares.

(2) There is a minus sign before 64, so $x^2 + 16x - 64$ is not a perfect-square trinomial.

4. No

5. $x^2 - 3x + 9$

(1) Two terms, x^2 and 9, are squares.

(2) There is no minus sign before x^2 or 9.

(3) Twice the product of the square roots, $2 \cdot x \cdot 3$, is $6x$. This is neither the remaining term nor its opposite, so $x^2 - 3x + 9$ is not a perfect-square trinomial.

6. No

7. $9x^2 - 36x + 24$

(1) Only one term, $9x^2$, is a square. Thus, $9x^2 - 36x + 24$ is not a perfect-square trinomial.

8. No

9. $x^2 - 16x + 64$
$= x^2 - 2 \cdot x \cdot 8 + 8^2 = (x - 8)^2$
$\quad\uparrow \quad \uparrow \quad \uparrow \quad \uparrow \quad \uparrow$
$= A^2 - 2 \quad A \quad B + B^2 = (A - B)^2$

10. $(x - 7)^2$

11. $x^2 + 14x + 49$
$= x^2 + 2 \cdot x \cdot 7 + 7^2 = (x + 7)^2$
$\quad\uparrow \quad \uparrow \quad \uparrow \quad \uparrow \quad \uparrow$
$= A^2 + 2 \quad A \quad B + B^2 = (A + B)^2$

12. $(x + 8)^2$

13. $x^2 - 2x + 1 = x^2 - 2 \cdot x \cdot 1 + 1^2 = (x - 1)^2$

14. $5(x - 1)^2$

15. $4 + 4x + x^2 = 2^2 + 2 \cdot 2 \cdot x + x^2$
$\qquad\qquad = (2 + x)^2$

16. $(x - 2)^2$

17. $18x^2 - 12x + 2 = 2(9x^2 - 6x + 1)$
$\qquad\qquad = 2[(3x)^2 - 2 \cdot 3x \cdot 1 + 1^2]$
$\qquad\qquad = 2(3x - 1)^2$

18. $(5x + 1)^2$

19. $49 + 56y + 16y^2 = 16y^2 + 56y + 49$
$\qquad\qquad = (4y)^2 + 2 \cdot 4y \cdot 7 + 7^2$
$\qquad\qquad = (4y + 7)^2$

We could also factor as follows:
$49 + 56y + 16y^2 = 7^2 + 2 \cdot 7 \cdot 4y + (4y)^2$
$\qquad\qquad = (7 + 4y)^2$

20. $3(4m + 5)^2$

21. $x^5 - 18x^4 + 81x^3 = x^3(x^2 - 18x + 81)$
$\qquad\qquad = x^3(x^2 - 2 \cdot x \cdot 9 + 9^2)$
$\qquad\qquad = x^3(x - 9)^2$

22. $2(x - 10)^2$

23. $2x^3 - 4x^2 + 2x = 2x(x^2 - 2x + 1)$
$\qquad\qquad = 2x(x^2 - 2 \cdot x \cdot 1 + 1^2)$
$\qquad\qquad = 2x(x - 1)^2$

24. $x(x + 12)^2$

25. $20x^2 + 100x + 125 = 5(4x^2 + 20x + 25)$
$\qquad\qquad = 5[(2x)^2 + 2 \cdot 2x \cdot 5 + 5^2]$
$\qquad\qquad = 5(2x + 5)^2$

26. $3(2x + 3)^2$

27. $49 - 42x + 9x^2 = 7^2 - 2 \cdot 7 \cdot 3x + (3x)^2 = (7 - 3x)^2$

28. $(8 - 7x)^2$, or $(7x - 8)^2$

29. $5y^2 + 10y + 5 = 5(y^2 + 2y + 1)$
$\qquad\qquad = 5(y^2 + 2 \cdot y \cdot 1 + 1^2)$
$\qquad\qquad = 5(y + 1)^2$

30. $2(a + 7)^2$

31. $2 + 20x + 50x^2 = 2(1 + 10x + 25x^2)$
$\qquad\qquad = 2[1^2 + 2 \cdot 1 \cdot 5x + (5x)^2]$
$\qquad\qquad = 2(1 + 5x)^2$

32. $7(1 - a)^2$, or $7(a - 1)^2$

33. $4p^2 + 12pq + 9q^2 = (2p)^2 + 2 \cdot 2p \cdot 3q + (3q)^2$
$\qquad\qquad = (2p + 3q)^2$

34. $(5m + 2n)^2$

35. $a^2 - 14ab + 49b^2 = a^2 - 2 \cdot a \cdot 7b + (7b)^2$
$\qquad\qquad = (a - 7b)^2$

36. $(x - 3y)^2$

37. $64m^2 + 16mn + n^2 = (8m)^2 + 2 \cdot 8m \cdot n + n^2$
$\qquad\qquad = (8m + n)^2$

38. $(9p - q)^2$

39. $16s^2 - 40st + 25t^2 = (4s)^2 - 2 \cdot 4s \cdot 5t + (5t)^2$
$\qquad\qquad = (4s - 5t)^2$

40. $4(3a + 4b)^2$

41. $x^2 - 100$

(1) The first expression is a square: x^2
The second expression is a square: $100 = 10^2$
(2) The terms have different signs.
Thus, $x^2 - 100$ is a difference of squares, $x^2 - 10^2$.

42. Yes

43. $x^2 + 36$

(1) The first expression is a square: x^2
The second expression is a square: $36 = 6^2$
(2) The terms do not have different signs.
Thus, $x^2 + 36$ is not a difference of squares.

44. No

45. $x^2 - 35$

(1) The expression 35 is not a square.

Thus, $x^2 - 35$ is not a difference of squares.

46. No

47. $16x^2 - 25$

(1) The first expression is a square: $16x^2 = (4x)^2$

The second expression is a square: $25 = 5^2$

(2) The terms have different signs.

Thus, $16x^2 - 25$ is a difference of squares, $(4x)^2 - 5^2$.

48. Yes

49. $y^2 - 4 = y^2 - 2^2 = (y + 2)(y - 2)$

50. $(x + 6)(x - 6)$

51. $p^2 - 9 = p^2 - 3^2 = (p + 3)(p - 3)$

52. $(q + 1)(q - 1)$

53. $-49 + t^2 = t^2 - 49 = t^2 - 7^2 = (t + 7)(t - 7)$

54. $(m + 8)(m - 8)$

55. $a^2 - b^2 = (a + b)(a - b)$

56. $(p + q)(p - q)$

57. $m^2 n^2 - 49 = (mn)^2 - 7^2 = (mn + 7)(mn - 7)$

58. $(5 + ab)(5 - ab)$

59. $200 - 2t^2 = 2(100 - t^2) = 2(10^2 - t^2) =$
   $2(10 + t)(10 - t)$

60. $(9 + w)(9 - w)$

61. $16a^2 - 9 = (4a)^2 - 3^2 = (4a + 3)(4a - 3)$

62. $(5x + 2)(5x - 2)$

63. $4x^2 - 25y^2 = (2x)^2 - (5y)^2 = (2x + 5y)(2x - 5y)$

64. $(3a + 4b)(3a - 4b)$

65. $8x^2 - 98 = 2(4x^2 - 49) = 2[(2x)^2 - 7^2] =$
   $2(2x + 7)(2x - 7)$

66. $6(2x + 3)(2x - 3)$

67. $36x - 49x^3 = x(36 - 49x^2) = x[6^2 - (7x)^2] =$
   $x(6 + 7x)(6 - 7x)$

68. $x(4 + 9x)(4 - 9x)$

69. $49a^4 - 81 = (7a^2)^2 - 9^2 = (7a^2 + 9)(7a^2 - 9)$

70. $(5a^2 + 3)(5a^2 - 3)$

71. $5x^4 - 5$
   $= 5(x^4 - 1)$
   $= 5[(x^2)^2 - 1^2]$
   $= 5(x^2 + 1)(x^2 - 1)$
   $= 5(x^2 + 1)(x + 1)(x - 1)$ Factoring further;
                                   $x^2 - 1$ is a difference of squares

72. $(x^2 + 4)(x + 2)(x - 2)$

73. $4x^4 - 64$
   $= 4(x^4 - 16) = 4[(x^2)^2 - 4^2]$
   $= 4(x^2 + 4)(x^2 - 4)$
   $= 4(x^2 + 4)(x + 2)(x - 2)$ Factoring further;
                                     $x^2 - 4$ is a difference of squares

74. $5(x^2 + 4)(x + 2)(x - 2)$

75. $1 - y^8$
   $= 1^2 - (y^4)^2$
   $= (1 + y^4)(1 - y^4)$
   $= (1 + y^4)(1 + y^2)(1 - y^2)$ Factoring $1 - y^4$
   $= (1 + y^4)(1 + y^2)(1 + y)(1 - y)$
                                 Factoring $1 - y^2$

76. $(x^4 + 1)(x^2 + 1)(x + 1)(x - 1)$

77. $3x^3 - 24x^2 + 48x = 3x(x^2 - 8x + 16)$
   $= 3x(x^2 - 2 \cdot x \cdot 4 + 4^2)$
   $= 3x(x - 4)^2$

78. $2a^2(a - 9)^2$

79. $x^{12} - 16$
   $= (x^6)^2 - 4^2$
   $= (x^6 + 4)(x^6 - 4)$
   $= (x^6 + 4)(x^3 + 2)(x^3 - 2)$ Factoring $x^6 - 4$

80. $(x^4 + 9)(x^2 + 3)(x^2 - 3)$

81. $y^2 - \dfrac{1}{16} = y^2 - \left(\dfrac{1}{4}\right)^2$
   $= \left(y + \dfrac{1}{4}\right)\left(y - \dfrac{1}{4}\right)$

82. $\left(x + \dfrac{1}{5}\right)\left(x - \dfrac{1}{5}\right)$

83. $a^8 - 2a^7 + a^6 = a^6(a^2 - 2a + 1)$
   $= a^6(a^2 - 2 \cdot a \cdot 1 + 1^2)$
   $= a^6(a - 1)^2$

84. $x^6(x - 4)^2$

85. $16 - m^4n^4$

$= 4^2 - (m^2n^2)^2$

$= (4 + m^2n^2)(4 - m^2n^2)$

$= (4 + m^2n^2)(2 + mn)(2 - mn)$

Factoring $4 - m^2n^2$

86. $(1 + a^2b^2)(1 + ab)(1 - ab)$

87. *Familiarize*. Let $s=$ the score on the fourth test.

Translate.

$\underbrace{\text{The average score}}$ $\underbrace{\text{is at least}}$ 90.

$\downarrow$ $\downarrow$ $\downarrow$

$\dfrac{96 + 98 + 89 + s}{4}$ $\geq$ 90

Carry out. We solve the inequality.

$\dfrac{96 + 98 + 89 + s}{4} \geq 90$

$96 + 98 + 89 + s \geq 360$ Multiplying by 4

$283 + s \geq 360$

$s \geq 77$

Check. We can obtain a partial check by substituting one number greater than or equal to 77 and another number less than 77 in the inequality. This is left to the student.

State. A score of 77 or better on the last test will earn Bonnie an A in the course. In terms of the inequality we have $s \geq 77$.

88. 3.125 L

89. $(x^3y^5)(x^9y^7) = x^{3+9}y^{5+7} = x^{12}y^{12}$

90. $25a^4b^6$

91. Graph: $y = \dfrac{3}{2}x - 3$

Because the equation is in the form $y = mx + b$, we know the y-intercept is $(0, -3)$. We find two other points on the line, substituting multiples of 2 for x to avoid fractions.

When $x = -2$, $y = \dfrac{3}{2}(-2) - 3 = -3 - 3 = -6$.

When $x = 4$, $y = \dfrac{3}{2} \cdot 4 - 3 = 6 - 3 = 3$.

x	y
0	-3
-2	-6
4	3

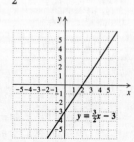

92.

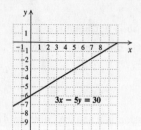

93.

94.

95.

96.

97. $x^2 - 2.25 = x^2 - (1.5)^2 = (x + 1.5)(x - 1.5)$

98. $27(3x^2 + 8)$

99. $x^2 - 5x + 25$

This is not a perfect-square trinomial, because $2 \cdot x \cdot 5 = 10x \neq -5x$. We cannot find two factors of 25 that add to -5, so this polynomial is prime.

100. $(x^4 + 2^4)(x^2 + 2^2)(x + 2)(x - 2)$, or
$(x^4 + 16)(x^2 + 4)(x + 2)(x - 2)$

101. $3x^2 - \dfrac{1}{3} = 3\left(x^2 - \dfrac{1}{9}\right)$

$= 3\left[x^2 - \left(\dfrac{1}{3}\right)^2\right]$

$= 3\left(x + \dfrac{1}{3}\right)\left(x - \dfrac{1}{3}\right)$

This exercise could also be done as follows:

$3x^2 - \dfrac{1}{3} = \dfrac{1}{3}(9x^2 - 1)$

$= \dfrac{1}{3}[(3x)^2 - 1^2]$

$= \dfrac{1}{3}(3x + 1)(3x - 1)$

102. $2x\left(3x + \dfrac{2}{5}\right)\left(3x - \dfrac{2}{5}\right)$

103. $0.49p - p^3 = p(0.49 - p^2) = p(0.7 + p)(0.7 - p)$

104. $(0.8x + 1.1)(0.8x - 1.1)$

105. $(x + 3)^4 - 18$

$= [(x + 3)^2 + 9][(x + 3)^2 - 9]$

$= [(x + 3)^2 + 9][(x + 3) + 3][(x + 3) - 3]$

$= (x^2 + 6x + 9 + 9)(x + 6)(x)$

$= x(x + 6)(x^2 + 6x + 18)$

106. $[(y - 5)^2 + z^4][y - 5 + z^2][y - 5 - z^2]$

107. $x^2 - \left(\dfrac{1}{x}\right)^2 = \left(x + \dfrac{1}{x}\right)\left(x - \dfrac{1}{x}\right)$

108. $(a^n + 7b^n)(a^n - 7b^n)$

109.
$$81 - b^{4k} = 9^2 - (b^{2k})^2$$
$$= (9 + b^{2k})(9 - b^{2k})$$
$$= (9 + b^{2k})[3^2 - (b^k)^2]$$
$$= (9 + b^{2k})(3 + b^k)(3 - b^k)$$

110. $(x + 3)(x - 3)(x^2 + 1)$

111. $9b^{2n} + 12b^n + 4 = (3b^n)^2 + 2 \cdot 3b^n \cdot 2 + 2^2 =$
$(3b^n + 2)^2$

112. $16(x^2 - 3)^2$

113.
$$(y + 3)^2 + 2(y + 3) + 1$$
$$= (y + 3)^2 + 2 \cdot (y + 3) \cdot 1 + 1^2$$
$$= [(y + 3) + 1]^2$$
$$= (y + 4)^2$$

114. $(7x + 4)^2$

115.
$$27x^3 - 63x^2 - 147x + 343$$
$$= 9x^2(3x - 7) - 49(3x - 7)$$
$$= (3x - 7)(9x^2 - 49)$$
$$= (3x - 7)(3x + 7)(3x - 7), \text{ or}$$
$$(3x - 7)^2(3x + 7)$$

116. $(x - 1)(2x^2 + x + 1)$

117.
$$a^2 + 2a + 1 - 9 = (a^2 + 2a + 1) - 9$$
$$= (a + 1)^2 - 3^2$$
$$= (a + 1 + 3)(a + 1 - 3)$$
$$= (a + 4)(a - 2)$$

118. $(y + x + 7)(y - x - 1)$

119. If $cy^2 + 6y + 1$ is the square of a binomial, then $2 \cdot a \cdot 1 = 6$ where $a^2 = c$. Then $a = 3$, so $c = a^2 = 3^2 = 9$. (The polynomial is $9y^2 + 6y + 1$.)

120. 16

121. If $x^2 + a^2 x + a^2 = (x + a)^2$, then $a^2 x = 2ax$, or $a^2 = 2a$.
We solve for a:
$$a^2 - 2a = 0$$
$$a(a - 2) = 0$$
$$a = 0 \text{ or } a = 2$$

122.
$$(x + 1)^2 - x^2$$
$$= [(x + 1) + x](x + 1 - x)$$
$$= 2x + 1$$

Exercise Set 5.5

1.
$$5x^2 - 45$$
$$= 5(x^2 - 9) \qquad \text{5 is a common factor}$$
$$= 5(x + 3)(x - 3) \quad \text{Factoring the difference of}$$
$$\text{squares}$$

2. $10(a + 8)(a - 8)$

3.
$$a^2 + 25 + 10a$$
$$= a^2 + 10a + 25 \qquad \text{Perfect-square trinomial}$$
$$= (a + 5)^2$$

4. $(y - 7)^2$

5. $2x^2 - 11x + 12$

There is no common factor (other than 1). This polynomial has three terms, but it is not a perfect-square trinomial. Multiply the leading coefficient and the constant, 2 and 12: $2 \cdot 12 = 24$. Try to factor 24 so that the sum of the factors is -11. The numbers we want are -3 and -8: $-3(-8) = 24$ and $-3 + (-8) = -11$. Split the middle term and factor by grouping.
$$2x^2 - 11x + 12 = 2x^2 - 3x - 8x + 12$$
$$= x(2x - 3) - 4(2x - 3)$$
$$= (2x - 3)(x - 4)$$

6. $(2y + 5)(4y - 1)$

7.
$$x^3 - 24x^2 + 144x$$
$$= x(x^2 - 24x + 144) \qquad x \text{ is a common factor}$$
$$= x(x^2 - 2 \cdot x \cdot 12 + 12^2) \qquad \text{Perfect-square}$$
$$\text{trinomial}$$
$$= x(x - 12)^2$$

8. $x(x - 9)^2$

9.
$$x^3 + 3x^2 - 4x - 12$$
$$= x^2(x + 3) - 4(x + 3) \quad \text{Factoring by grouping}$$
$$= (x + 3)(x^2 - 4)$$
$$= (x + 3)(x + 2)(x - 2) \quad \text{Factoring the difference}$$
$$\text{of squares}$$

10. $(x + 5)(x - 5)^2$

11.
$$24x^2 - 54$$
$$= 6(4x^2 - 9) \qquad \text{6 is a common factor}$$
$$= 6[(2x)^2 - 3^2] \qquad \text{Difference of squares}$$
$$= 6(2x + 3)(2x - 3)$$

12. $2(2x + 7)(2x - 7)$

13.
$$20x^3 - 4x^2 - 72x$$
$$= 4x(5x^2 - x - 18) \quad 4x \text{ is a common factor}$$
$$= 4x(5x + 9)(x - 2) \quad \text{Factoring the trinomial us-}$$
$$\text{ing trial and error}$$

14. $3x(x+3)(3x-5)$

15. x^2+4

The polynomial has no common factor and is not a difference of squares. It is prime.

16. Prime

17. $a^4+8a^2+8a^3+64a$

$= a(a^3+8a+8a^2+64)$ a is a common factor

$= a[a(a^2+8)+8(a^2+8)]$ Factoring by grouping

$= a(a^2+8)(a+8)$

18. $t(t^2+7)(t-3)$

19. $x^5-14x^4+49x^3$

$= x^3(x^2-14x+49)$ x^3 is a common factor

$= x^3(x^2-2\cdot x\cdot 7+7^2)$ Trinomial square

$= x^3(x-7)^2$

20. $2x^4(x+2)^2$

21. $20-6x-2x^2$

$= -2x^2-6x+20$ Rewriting

$= -2(x^2+3x-10)$ -2 is a common factor

$= -2(x+5)(x-2)$ Using trial and error

22. $-3(2x-5)(x+3)$, or $3(5-2x)(3+x)$

23. x^2+3x+1

There is no common factor (other than 1). This is not a trinomial square, because $2\cdot x\cdot 1 \neq 3x$. We try factoring by trial and error. We look for two factors whose product is 1 and whose sum is 3. There are none. The polynomial cannot be factored. It is prime.

24. Prime

25. $4x^4-64$

$= 4(x^4-16)$ 4 is a common factor

$= 4[(x^2)^2-4^2]$ Difference of squares

$= 4(x^2+4)(x^2-4)$ Difference of squares

$= 4(x^2+4)(x+2)(x-2)$

26. $5x(x^2+4)(x+2)(x-2)$

27. t^8-1

$= (t^4)^2-1^2$ Difference of squares

$= (t^4+1)(t^4-1)$ Difference of squares

$= (t^4+1)(t^2+1)(t^2-1)$ Difference of squares

$= (t^4+1)(t^2+1)(t+1)(t-1)$

28. $(1+n^4)(1+n^2)(1+n)(1-n)$

29. $x^5-4x^4+3x^3$

$= x^3(x^2-4x+3)$ x^3 is a common factor

$= x^3(x-3)(x-1)$ Factoring the trinomial using trial and error

30. $x^4(x^2-2x+7)$

31. x^2-y^2 Difference of squares

$= (x+y)(x-y)$

32. $(pq+r)(pq-r)$

33. $12n^2+24n^3 = 12n^2(1+2n)$

34. $a(x^2+y^2)$

35. $ab^2-a^2b = ab(b-a)$

36. $9mn(4-mn)$

37. $2\pi rh+2\pi r^2 = 2\pi r(h+r)$

38. $5p^2q^2(2p^2q^2+7pq+2)$

39. $(a+b)(x-3)+(a+b)(x+4)$

$= (a+b)[(x-3)+(x+4)]$ $(a+b)$ is a common factor

$= (a+b)(2x+1)$

40. $(a^3+b)(5c-1)$

41. $(x-5)(x+2)-y(x+2)$

$= (x+2)(x-5-y)$ $(x+2)$ is a common factor

42. $(n+2)(n+p)$

43. $x^2+x+xy+y$

$= x(x+1)+y(x+1)$ Factoring by grouping

$= (x+1)(x+y)$

44. $(x-2)(2x+z)$

45. $a^2-3a+ay-3y$

$= a(a-3)+y(a-3)$ Factoring by grouping

$= (a-3)(a+y)$

46. $(x-y)^2$

47. $6y^2-3y+2py-p$

$= 3y(2y-1)+p(2y-1)$ Factoring by grouping

$= (2y-1)(3y+p)$

48. $(3c+d)^2$

49. $4b^2+a^2-4ab$

$= 4b^2-4ab+a^2$

$= (2b)^2-2\cdot 2b\cdot a+a^2$ Perfect-square trinomial

$= (2b-a)^2$

This result can also be expressed as $(a-2b)^2$.

50. $7(p^2 + q^2)(p + q)(p - q)$

51. $\quad 16x^2 + 24xy + 9y^2$

$\quad = (4x)^2 + 2 \cdot 4x \cdot 3y + (3y)^2 \quad$ Perfect-square
$\qquad\qquad\qquad\qquad\qquad\qquad$ trinomial

$\quad = (4x + 3y)^2$

52. $(5z + y)^2$

53. $\quad 4x^2y^2 + 12xyz + 9z^2$

$\quad = (2xy)^2 + 2 \cdot 2xy \cdot 3z + (3z)^2 \quad$ Perfect-square
$\qquad\qquad\qquad\qquad\qquad\qquad\quad$ trinomial

$\quad = (2xy + 3z)^2$

54. $a^3(a + 5b)(a - b)$

55. $\quad a^4b^4 - 16$

$\quad = (a^2b^2)^2 - 4^2 \qquad$ Difference of squares

$\quad = (a^2b^2 + 4)(a^2b^2 - 4) \quad$ Difference of squares

$\quad = (a^2b^2 + 4)(ab + 2)(ab - 2)$

56. $(a - 2b)(a + b)$

57. $\quad 4p^2q + pq^2 + 4p^3$

$\quad = p(4pq + q^2 + 4p^2) \qquad p$ is a common factor

$\quad = p(4p^2 + 4pq + q^2) \qquad$ Rewriting

$\quad = p((2p)^2 + 2 \cdot 2p \cdot q + q^2) \quad$ Perfect-square
$\qquad\qquad\qquad\qquad\qquad\qquad$ trinomial

$\quad = p(2p + q)^2$, or

$\qquad p(q + 2p)^2$

58. $(m + 20n)(m - 18n)$

59. $\quad 3b^2 - 17ab - 6a^2$

$\quad = (3b + a)(b - 6a) \quad$ Using trial and error

60. $(mn - 8)(mn + 4)$

61. $\quad 15 + x^2y^2 + 8xy$

$\quad = x^2y^2 + 8xy + 15 \quad$ Rewriting

$\quad = (xy + 3)(xy + 5) \quad$ Using trial and error

62. $a^3(ab + 5)(ab - 2)$

63. $\quad p^2q^2 + 7pq + 6$

$\quad = (pq + 1)(pq + 6) \quad$ Using trial and error

64. $(7x + 8y)^2$

65. $\quad 4ab^5 - 32b^4 + a^2b^6$

$\quad = b^4(4ab - 32 + a^2b^2) \qquad b^4$ is a common factor

$\quad = b^4(a^2b^2 + 4ab - 32) \qquad$ Rewriting

$\quad = b^4(ab + 8)(ab - 4) \qquad$ Using trial and error

66. $2t^2(s^3 + 3t)(s^3 + 2t)$

67. $\qquad x^6 + x^5y - 2x^4y^2$

$\quad = x^4(x^2 + xy - 2y^2) \qquad x^4$ is a common factor

$\quad = x^4(x + 2y)(x - y) \qquad$ Factoring the trinomial

68. $a^2(1 + bc)^2$

69. $\quad 36a^2 - 15a + \dfrac{25}{16} = (6a)^2 - 2 \cdot 6a \cdot \dfrac{5}{4} + \left(\dfrac{5}{4}\right)^2$

$\qquad\qquad\qquad\qquad\quad = \left(6a - \dfrac{5}{4}\right)^2$

70. $\left(\dfrac{1}{9}x - \dfrac{4}{3}\right)^2$, or $\dfrac{1}{9}\left(\dfrac{1}{3}x - 4\right)^2$

71. $\quad \dfrac{1}{4}a^2 + \dfrac{1}{3}ab + \dfrac{1}{9}b^2$

$\quad = \left(\dfrac{1}{2}a\right)^2 + 2 \cdot \dfrac{1}{2}a \cdot \dfrac{1}{3}b + \left(\dfrac{1}{3}b\right)^2$

$\quad = \left(\dfrac{1}{2}a + \dfrac{1}{3}b\right)^2$

72. $(1 + 4x^6y^6)(1 + 2x^3y^3)(1 - 2x^3y^3)$

73. $\qquad b^4a - 81a^5$

$\quad = a(b^4 - 81a^4) \qquad a$ is a common factor

$\quad = a[(b^2)^2 - (9a^2)^2] \qquad$ Difference of squares

$\quad = a(b^2 + 9a^2)(b^2 - 9a^2) \quad$ Difference of squares

$\quad = a(b^2 + 9a^2)(b + 3a)(b - 3a)$

74. $(0.1x - 0.5y)^2$, or $0.01(x - 5y)^2$

75. $\qquad w^3 - 7w^2 - 4w + 28$

$\quad = w^2(w - 7) - 4(w - 7) \qquad$ Factoring by
$\qquad\qquad\qquad\qquad\qquad\qquad$ grouping

$\quad = (w - 7)(w^2 - 4)$

$\quad = (w - 7)(w + 2)(w - 2) \qquad$ Factoring a dif-
$\qquad\qquad\qquad\qquad\qquad\qquad\quad$ erence of squares

76. $(y + 8)(y + 1)(y - 1)$

77. $\quad \dfrac{y = -4x + 7}{}$

$\quad 11 \ ? \ -4(-1) + 7$

$\qquad\ \Big|\ 4 + 7$

$\quad 11 \ \Big|\ 11 \qquad\qquad$ TRUE

Since $11 = 11$ is true, $(-1, 11)$ is a solution.

$\quad \dfrac{y = -4x + 7}{}$

$\quad 7 \ ? \ -4 \cdot 0 + 7$

$\qquad \Big|\ 0 + 7$

$\quad 7 \ \Big|\ 7 \qquad\qquad$ TRUE

Since $7 = 7$ is true, $(0, 7)$ is a solution.

$\quad \dfrac{y = -4x + 7}{}$

$\ -5 \ ? \ -4 \cdot 3 + 7$

$\qquad\ \Big|\ -12 + 7$

$\ -5 \ \Big|\ -5 \qquad\qquad$ TRUE

Since $-5 = -5$ is true, $(3, -5)$ is a solution.

78. $\dfrac{7}{2}$

79. $3x + 4 = 0$

$\qquad 3x = -4$ Subtracting 4 on both sides

$\qquad x = -\dfrac{4}{3}$ Dividing by 3 on both sides

The solution is $-\dfrac{4}{3}$.

80.

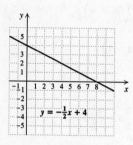

$y = -\frac{1}{2}x + 4$

81. $A = aX + bX - 7$

$\qquad A + 7 = aX + bX$

$\qquad A + 7 = X(a + b)$

$\qquad \dfrac{A + 7}{a + b} = X$

82. $\{x | x < 32\}$

83.

84. ◈

85. ◈

86. ◈

87. $-(x^5 + 7x^3 - 18x)$

$= -x(x^4 + 7x^2 - 18)$

$= -x(x^2 + 9)(x^2 - 2)$

88. $(a - 2)(a + 3)(a - 3)$

89. $3a^4 - 15a^2 + 12$

$= 3(a^4 - 5a^2 + 4)$

$= 3(a^2 - 1)(a^2 - 4)$

$= 3(a + 1)(a - 1)(a + 2)(a - 2)$

90. $(x^2 + 2)(x + 3)(x - 3)$

91. $x^3 - x^2 - 4x + 4 = x^2(x - 1) - 4(x - 1)$

$\qquad\qquad\qquad\qquad = (x - 1)(x^2 - 4)$

$\qquad\qquad\qquad\qquad = (x - 1)(x + 2)(x - 2)$

92. $(y + 1)(y - 7)(y + 3)$

93. $y^2(y - 1) - 2y(y - 1) + (y - 1)$

$= (y - 1)(y^2 - 2y + 1)$

$= (y - 1)(y - 1)^2$

$= (y - 1)^3$

94. $(2x + 3y - 2)(3x - y - 3)$

95. $\qquad (y + 4)^2 + 2x(y + 4) + x^2$

$= (y + 4)^2 + 2 \cdot (y + 4) \cdot x + x^2$ Perfect-square trinomial

$= [(y + 4) + x]^2$

96. $(2a + b + 4)(a - b + 5)$

97. $x^{2k} - 2^{2k} = x^{2 \cdot 4} - 2^{2 \cdot 4}$ Substituting 4 for k

$\qquad\qquad = x^8 - 2^8$

$\qquad\qquad = x^8 - 256$

$\qquad\qquad = (x^4 + 16)(x^4 - 16)$

$\qquad\qquad = (x^4 + 16)(x^2 + 4)(x^2 - 4)$

$\qquad\qquad = (x^4 + 16)(x^2 + 4)(x + 2)(x - 2)$

98. $11 = 11$; the factorization is probably correct.

99. $\qquad 6x^2y^2 - 23xy + 20$

$= 6(-1)^2(1)^2 - 23(-1)(1) + 20$

$= 6 \cdot 1 \cdot 1 - 23(-1)(1) + 20$

$= 6 + 23 + 20 = 49$

$\qquad (2xy - 5)(3xy - 4)$

$= [2(-1)(1) - 5][3(-1)(1) - 4]$

$= (-2 - 5)(-3 - 4)$

$= -7(-7) = 49$

Since the value of both expressions is 49, the factorization is probably correct.

Exercise Set 5.6

1. $(x + 7)(x + 6) = 0$

We use the principle of zero products.

$\qquad x + 7 = 0 \quad or \quad x + 6 = 0$

$\qquad\qquad x = -7 \quad or \qquad x = -6$

Check:

For -7:

$$\frac{(x + 7)(x + 6) = 0}{\begin{array}{c|c}(-7 + 7)(-7 + 6) \ ? \ 0 \\ 0 \cdot (-1) \\ 0 & 0 \quad \text{TRUE}\end{array}}$$

For -6:

$$\frac{(x + 7)(x + 6) = 0}{\begin{array}{c|c}(-6 + 7)(-6 + 6) \ ? \ 0 \\ 1 \cdot 0 \\ 0 & 0 \quad \text{TRUE}\end{array}}$$

The solutions are -7 and -6.

2. $-1, -2$

3. $(x-3)(x+5) = 0$

$\qquad x - 3 = 0 \quad or \quad x + 5 = 0$

$\qquad\quad x = 3 \quad or \qquad x = -5$

Check:

For 3:

$$\frac{(x-3)(x+5) = 0}{(3-3)(3+5) \ \overset{?}{|} \ 0}$$
$$\begin{array}{c|c} 0 \cdot 8 & \\ 0 & 0 \quad \text{TRUE} \end{array}$$

For -5:

$$\frac{(x-3)(x+5) = 0}{(-5-3)(-5+5) \ \overset{?}{|} \ 0}$$
$$\begin{array}{c|c} -8 \cdot 0 & \\ 0 & 0 \quad \text{TRUE} \end{array}$$

The solutions are 3 and -5.

4. $-9, 3$

5. $(2x-9)(x+4) = 0$

$\qquad 2x - 9 = 0 \quad or \quad x + 4 = 0$

$\qquad\quad 2x = 9 \quad or \qquad x = -4$

$\qquad\quad x = \dfrac{9}{2} \quad or \qquad x = -4$

The solutions are $\dfrac{9}{2}$ and -4.

6. $\dfrac{5}{3}, -1$

7. $(10x-7)(4x+9) = 0$

$\qquad 10x - 7 = 0 \quad or \quad 4x + 9 = 0$

$\qquad\quad 10x = 7 \quad or \qquad 4x = -9$

$\qquad\quad x = \dfrac{7}{10} \quad or \qquad x = -\dfrac{9}{4}$

The solutions are $\dfrac{7}{10}$ and $-\dfrac{9}{4}$.

8. $\dfrac{7}{2}, -\dfrac{4}{3}$

9. $x(x+6) = 0$

$\qquad x = 0 \quad or \quad x + 6 = 0$

$\qquad x = 0 \quad or \qquad x = -6$

The solutions are 0 and -6.

10. $0, -9$

11. $\left(\dfrac{2}{3}x - \dfrac{12}{11}\right)\left(\dfrac{7}{4}x - \dfrac{1}{12}\right) = 0$

$\qquad \dfrac{2}{3}x - \dfrac{12}{11} = 0 \qquad or \qquad \dfrac{7}{4}x - \dfrac{1}{12} = 0$

$\qquad\quad \dfrac{2}{3}x = \dfrac{12}{11} \qquad or \qquad \dfrac{7}{4}x = \dfrac{1}{12}$

$\qquad\quad x = \dfrac{3}{2} \cdot \dfrac{12}{11} \quad or \qquad x = \dfrac{4}{7} \cdot \dfrac{1}{12}$

$\qquad\quad x = \dfrac{18}{11} \qquad or \qquad x = \dfrac{1}{21}$

The solutions are $\dfrac{18}{11}$ and $\dfrac{1}{21}$.

12. $-\dfrac{1}{10}, \dfrac{1}{27}$

13. $7x(2x+9) = 0$

$\qquad 7x = 0 \quad or \quad 2x + 9 = 0$

$\qquad x = 0 \quad or \qquad 2x = -9$

$\qquad x = 0 \quad or \qquad x = -\dfrac{9}{2}$

The solutions are 0 and $-\dfrac{9}{2}$.

14. $0, -\dfrac{7}{4}$

15. $(20x - 0.4x)(7 - 0.1x) = 0$

$\qquad 20 - 0.4x = 0 \qquad or \quad 7 - 0.1x = 0$

$\qquad\quad -0.4x = -20 \quad or \qquad -0.1x = -7$

$\qquad\qquad x = 50 \quad or \qquad\quad x = 70$

The solutions are 50 and 70.

16. $\dfrac{10}{3}, 20$

17. $(x+3)(2x-5)(x-6) = 0$

$\quad x + 3 = 0 \quad or \quad 2x - 5 = 0 \quad or \quad x - 6 = 0$

$\qquad x = -3 \quad or \qquad 2x = 5 \quad or \qquad x = 6$

$\qquad x = -3 \quad or \qquad x = \dfrac{5}{2} \quad or \qquad x = 6$

The solutions are $-3, \dfrac{5}{2}$, and 6.

18. $4, -9, \dfrac{1}{3}$

19. $\qquad x^2 - 7x + 6 = 0$

$\qquad (x-6)(x-1) = 0 \quad \text{Factoring}$

$\qquad x - 6 = 0 \quad or \quad x - 1 = 0 \ \text{Using the principle}$
$\qquad\qquad\qquad\qquad\qquad\qquad\qquad \text{of zero products}$

$\qquad\quad x = 6 \quad or \qquad x = 1$

The solutions are 6 and 1.

20. $1, 5$

21. $\qquad x^2 - 4x - 21 = 0$

$\qquad (x+3)(x-7) = 0 \quad \text{Factoring}$

$\qquad x + 3 = 0 \quad or \quad x - 7 = 0 \ \text{Using the principle}$
$\qquad\qquad\qquad\qquad\qquad\qquad\qquad \text{of zero products}$

$\qquad\quad x = -3 \quad or \qquad x = 7$

The solutions are -3 and 7.

22. $-2, 9$

23. $\qquad x^2 + 9x + 14 = 0$

$\qquad (x+7)(x+2) = 0$

$\qquad x + 7 = 0 \quad or \quad x + 2 = 0$

$\qquad\quad x = -7 \quad or \qquad x = -2$

The solutions are -7 and -2.

24. $-5, -3$

25. $x^2 + 3x = 0$
$x(x + 3) = 0$
$x = 0 \quad or \quad x + 3 = 0$
$x = 0 \quad or \qquad x = -3$
The solutions are 0 and -3.

26. $-8, 0$

27. $x^2 - 9x = 0$
$x(x - 9) = 0$
$x = 0 \quad or \quad x - 9 = 0$
$x = 0 \quad or \qquad x = 9$
The solutions are 0 and 9.

28. $-4, 0$

29. $100 = x^2$
$0 = x^2 - 100$
$0 = (x + 10)(x - 10)$
$x + 10 = 0 \qquad or \quad x - 10 = 0$
$\qquad x = -10 \quad or \qquad x = 10$
The solutions are -10 and 10.

30. $-4, 4$

31. $4x^2 - 9 = 0$
$(2x + 3)(2x - 3) = 0$
$2x + 3 = \quad 0 \quad or \quad 2x - 3 = 0$
$2x = -3 \quad or \qquad 2x = 3$
$x = -\dfrac{3}{2} \quad or \qquad x = \dfrac{3}{2}$
The solutions are $-\dfrac{3}{2}$ and $\dfrac{3}{2}$.

32. $-\dfrac{2}{3}, \dfrac{2}{3}$

33. $0 = 25 + x^2 + 10x$
$0 = x^2 + 10x + 25 \quad$ Writing in descending order
$0 = (x + 5)(x + 5)$
$x + 5 = 0 \quad or \quad x + 5 = 0$
$\quad x = -5 \quad or \qquad x = -5$
The solution is -5.

34. -3

35. $1 + x^2 = 2x$
$x^2 - 2x + 1 = 0$
$(x - 1)(x - 1) = 0$
$x - 1 = 0 \quad or \quad x - 1 = 0$
$\quad x = 1 \quad or \qquad x = 1$
The solution is 1.

36. 4

37. $9x^2 = 4x$
$9x^2 - 4x = 0$
$x(9x - 4) = 0$
$x = 0 \quad or \quad 9x - 4 = 0$
$x = 0 \quad or \qquad 9x = 4$
$x = 0 \quad or \qquad x = \dfrac{4}{9}$
The solutions are 0 and $\dfrac{4}{9}$.

38. $0, \dfrac{7}{3}$

39. $3x^2 - 7x = 20$
$3x^2 - 7x - 20 = 0$
$(3x + 5)(x - 4) = 0$
$3x + 5 = 0 \qquad or \quad x - 4 = 0$
$3x = -5 \quad or \qquad x = 4$
$x = -\dfrac{5}{3} \quad or \qquad x = 4$
The solutions are $-\dfrac{5}{3}$ and 4.

40. $-1, \dfrac{5}{3}$

41. $2y^2 + 12y = -10$
$2y^2 + 12y + 10 = 0$
$2(y^2 + 6y + 5) = 0$
$2(y + 5)(y + 1) = 0$
$y + 5 = 0 \quad or \quad y + 1 = 0$
$y = -5 \quad or \qquad y = -1$
The solutions are -5 and -1.

42. $-\dfrac{1}{4}, \dfrac{2}{3}$

43. $(x - 5)(x + 1) = 16$
$x^2 - 4x - 5 = 16$
$x^2 - 4x - 21 = 0$
$(x + 3)(x - 7) = 0$
$x + 3 = 0 \quad or \quad x - 7 = 0$
$x = -3 \quad or \qquad x = 7$
The solutions are -3 and 7.

44. 2, 4

45. $y(3y + 1) = 2$
$3y^2 + y = 2$
$3y^2 + y - 2 = 0$
$(3y - 2)(y + 1) = 0$

$$3y - 2 = 0 \quad or \quad y + 1 = 0$$
$$3y = 2 \quad or \qquad y = -1$$
$$y = \frac{2}{3} \quad or \qquad y = -1$$

The solutions are $\frac{2}{3}$ and -1.

46. $-2, 7$

47.
$$81x^2 - 5 = 20$$
$$81x^2 - 25 = 0$$
$$(9x + 5)(9x - 5) = 0$$
$$9x + 5 = 0 \quad or \quad 9x - 5 = 0$$
$$9x = -5 \quad or \qquad 9x = 5$$
$$x = -\frac{5}{9} \quad or \qquad x = \frac{5}{9}$$

The solutions are $-\frac{5}{9}$ and $\frac{5}{9}$.

48. $-\frac{7}{6}, \frac{7}{6}$

49.
$$(x - 1)(5x + 4) = 2$$
$$5x^2 - x - 4 = 2$$
$$5x^2 - x - 6 = 0$$
$$(5x - 6)(x + 1) = 0$$
$$5x - 6 = 0 \quad or \quad x + 1 = 0$$
$$5x = 6 \quad or \qquad x = -1$$
$$x = \frac{6}{5} \quad or \qquad x = -1$$

The solutions are $\frac{6}{5}$ and -1.

50. $-4, -\frac{2}{3}$

51.
$$x^2 - 2x = 18 + 5x$$
$$x^2 - 7x - 18 = 0 \qquad \text{Subtracting 18 and } 5x$$
$$(x - 9)(x + 2) = 0$$
$$x - 9 = 0 \quad or \quad x + 2 = 0$$
$$x = 9 \quad or \qquad x = -2$$

The solutions are 9 and -2.

52. $-3, 1$

53.
$$(6a + 1)(a + 1) = 21$$
$$6a^2 + 7a + 1 = 21$$
$$6a^2 + 7a - 20 = 0$$
$$(3a - 4)(2a + 5) = 0$$
$$3a - 4 = 0 \quad or \quad 2a + 5 = 0$$
$$3a = 4 \quad or \qquad 2a = -5$$
$$a = \frac{4}{3} \quad or \qquad a = -\frac{5}{2}$$

The solutions are $\frac{4}{3}$ and $-\frac{5}{2}$.

54. $-\frac{3}{2}, \frac{5}{4}$

55. The solutions of the equation are the first coordinates of the x-intercepts of the graph. From the graph we see that the x-intercepts are $(-3, 0)$ and $(2, 0)$, so the solutions of the equation are -3 and 2.

56. $-1, 4$

57. We let $y = 0$ and solve for x.
$$0 = x^2 + 3x - 4$$
$$0 = (x + 4)(x - 1)$$
$$x + 4 = 0 \quad or \quad x - 1 = 0$$
$$x = -4 \quad or \qquad x = 1$$

The x-intercepts are $(-4, 0)$ and $(1, 0)$.

58. $(3, 0), (-2, 0)$

59. We let $y = 0$ and solve for x.
$$0 = x^2 - 2x - 15$$
$$0 = (x - 5)(x + 3)$$
$$x - 5 = 0 \quad or \quad x + 3 = 0$$
$$x = 5 \quad or \qquad x = -3$$

The x-intercepts are $(5, 0)$ and $(-3, 0)$.

60. $(-4, 0), (2, 0)$.

61. We let $y = 0$ and solve for x
$$0 = 2x^2 + x - 10$$
$$0 = (2x + 5)(x - 2)$$
$$2x + 5 = 0 \quad or \quad x - 2 = 0$$
$$2x = -5 \quad or \qquad x = 2$$
$$x = -\frac{5}{2} \quad or \qquad x = 2$$

The x-intercepts are $\left(-\frac{5}{2}, 0\right)$ and $(2, 0)$.

62. $\left(\frac{3}{2}, 0\right), (-3, 0)$

63. $(a + b)^2$

64. $a^2 + b^2$

65. Let x represent the smaller integer; $x + (x + 1)$

66. Let x represent the number; $2x + 5 < 19$

67. Let x represent the number; $\frac{1}{2}x - 7 > 24$

68. Let n represent the number; $n - 3 \geq 34$

69. ◈

70. ◈

71. ◈

72. ◈

73. a)
$$x = -3 \quad or \quad x = 4$$
$$x + 3 = 0 \quad or \quad x - 4 = 0$$
$(x+3)(x-4) = 0$ \quad Principle of zero products
$x^2 - x - 12 = 0$ \quad Multiplying

b)
$$x = -3 \quad or \quad x = -4$$
$$x + 3 = 0 \quad or \quad x + 4 = 0$$
$(x+3)(x+4) = 0$
$x^2 + 7x + 12 = 0$

c)
$$x = \frac{1}{2} \quad or \quad x = \frac{1}{2}$$
$$x - \frac{1}{2} = 0 \quad or \quad x - \frac{1}{2} = 0$$
$\left(x - \frac{1}{2}\right)\left(x - \frac{1}{2}\right) = 0$
$x^2 - x + \frac{1}{4} = 0, \quad$ or
$4x^2 - 4x + 1 = 0$ \quad Multiplying by 4

d)
$$x = 5 \quad or \quad x = -5$$
$$x - 5 = 0 \quad or \quad x + 5 = 0$$
$(x-5)(x+5) = 0$ \quad Principle of zero products
$x^2 - 25 = 0$ \quad Multiplying

e) $x = 0 \quad or \quad x = 0.1 \ or \quad x = \frac{1}{4}$
$x = 0 \quad or \quad x - 0.1 = 0 \quad or \quad x - \frac{1}{4} = 0$
$x = 0 \quad or \quad x - \frac{1}{10} = 0 \quad or \quad x - \frac{1}{4} = 0$
$x\left(x - \frac{1}{10}\right)\left(x - \frac{1}{4}\right) = 0$
$x\left(x^2 - \frac{7}{20}x + \frac{1}{40}\right) = 0$
$x^3 - \frac{7}{20}x^2 + \frac{1}{40}x = 0, \quad$ or
$40x^3 - 14x^2 + x = 0$ \quad Multiplying by 40

74. 4

75.
$$a(9 + a) = 4(2a + 5)$$
$$9a + a^2 = 8a + 20$$
$a^2 + a - 20 = 0$ \quad Subtracting $8a$ and 20
$(a+5)(a-4) = 0$
$a + 5 = 0 \quad or \quad a - 4 = 0$
$a = -5 \quad or \quad a = 4$
The solutions are -5 and 4.

76. 3, 5

77.
$$x^2 - \frac{9}{25} = 0$$
$$\left(x - \frac{3}{5}\right)\left(x + \frac{3}{5}\right) = 0$$
$x - \frac{3}{5} = 0 \quad or \quad x + \frac{3}{5} = 0$
$x = \frac{3}{5} \quad or \quad x = -\frac{3}{5}$
The solutions are $\frac{3}{5}$ and $-\frac{3}{5}$.

78. $-\frac{5}{6}, \frac{5}{6}$

79.
$$\frac{5}{16}x^2 = 5$$
$$\frac{5}{16}x^2 - 5 = 0$$
$5\left(\frac{1}{16}x^2 - 1\right) = 0$
$5\left(\frac{1}{4}x - 1\right)\left(\frac{1}{4}x + 1\right) = 0$
$\frac{1}{4}x - 1 = 0 \quad or \quad \frac{1}{4}x + 1 = 0$
$\frac{1}{4}x = 1 \quad or \quad \frac{1}{4}x = -1$
$x = 4 \quad or \quad x = -4$
The solutions are 4 and -4.

80. $-\frac{5}{9}, \frac{5}{9}$

81. a) $2(x^2 + 10x - 2) = 2 \cdot 0$ \quad Multiplying (a) by 2
$2x^2 + 20x - 4 = 0$
(a) and $2x^2 + 20x - 4 = 0$ are equivalent.

b) $(x-6)(x+3) = x^2 - 3x - 18$ \quad Multiplying
(b) and $x^2 - 3x - 18 = 0$ are equivalent.

c) $5x^2 - 5 = 5(x^2 - 1) = 5(x+1)(x-1) =$
$(x+1)5(x-1) = (x+1)(5x-5)$
(c) and $(x+1)(5x-5) = 0$ are equivalent.

d) $2(2x-5)(x+4) = 2 \cdot 0$ \quad Multiplying (d) by 2
$2(x+4)(2x-5) = 0$
$(2x+8)(2x-5) = 0$
(d) and $(2x+8)(2x-5) = 0$ are equivalent.

e) $4(x^2 + 2x + 9) = 4 \cdot 0$ \quad Multiplying (e) by 4
$4x^2 + 8x + 36 = 0$
(e) and $4x^2 + 8x + 36 = 0$ are equivalent.

f) $3(3x^2 - 4x + 8) = 3 \cdot 0$ \quad Multiplying (f) by 3
$9x^2 - 12x + 24 = 0$
(f) and $9x^2 - 12x + 24 = 0$ are equivalent.

82. ◈

83. ◈

84. $-3.45, 1.65$

85. $2.33, 6.77$

86. $-0.25, 0.88$

87. $-4.59, -9.15$

88. $4.55, -3.23$

89. $-3.25, -6.75$

Exercise Set 5.7

1. ***Familiarize***. Let $x =$ the number (or numbers).

Translate. We reword the problem.

$$\underbrace{\text{The square of a number}}_{x^2} \quad \underset{-}{\text{minus}} \quad \underbrace{\text{the number}}_{x} \quad \underset{=}{\text{is}} \quad \underset{2}{2}$$

Carry out. We solve the equation.

$$x^2 - x = 2$$
$$x^2 - x - 2 = 0$$
$$(x - 2)(x + 1) = 0$$

$$x - 2 = 0 \quad or \quad x + 1 = 0$$
$$x = 2 \quad or \quad x = -1$$

Check. For 2: The square of 2 is 2^2, or 4, and $4 - 2 = 2$.

For -1: The square of -1 is $(-1)^2$, or 1, and $1 - (-1) = 1 + 1 = 2$. Both numbers check.

State. The numbers are 2 and -1.

2. $-2, 3$

3. ***Familiarize***. Let $n =$ the number (or numbers).

Translate.

$$\underbrace{\text{Eight times a number}}_{8n} \quad \underset{= 15}{\text{is 15}} \quad \underset{+}{\text{plus}} \quad \underbrace{\text{the square of the number.}}_{n^2}$$

Carry out. We solve the equation.

$$8n = 15 + n^2$$
$$0 = n^2 - 8n + 15$$
$$0 = (n - 3)(n - 5)$$

$$n - 3 = 0 \; or \quad n - 5 = 0$$
$$n = 3 \; or \qquad n = 5$$

Check. For 3: Eight times 3 is $8 \cdot 3$, or 24, and $3^2 = 9$. Since $24 = 15 + 9$, the number 3 checks.

For 5: Eight times 5 is $8 \cdot 5$, or 40, and $5^2 = 25$. Since $40 = 15 + 25$, the number 5 checks also.

State. The numbers are 3 and 5.

4. $2, 4$

5. ***Familiarize***. The page numbers on facing pages are consecutive integers. Let $x =$ the smaller integer. Then $x + 1 =$ the larger integer.

Translate. We reword the problem.

$$\underbrace{\text{Smaller integer}}_{x} \quad \underset{\cdot}{\text{times}} \quad \underbrace{\text{larger integer}}_{(x + 1)} \quad \underset{=}{\text{is}} \quad \underset{110}{110.}$$

Carry out. We solve the equation.

$$x(x + 1) = 110$$
$$x^2 + x = 110$$
$$x^2 + x - 110 = 0$$
$$(x + 11)(x - 10) = 0$$

$$x + 11 = 0 \quad or \quad x - 10 = 0$$
$$x = -11 \quad or \qquad x = 10$$

Check. The solutions of the equation are -11 and 10. Since a page number cannot be negative, -11 cannot be a solution of the original problem. We only need to check 10. When $x = 10$, then $x + 1 = 11$, and $10 \cdot 11 = 110$. This checks.

State. The page numbers are 10 and 11.

6. 14 and 15

7. ***Familiarize***. Let $x =$ the smaller even integer. Then $x + 2 =$ the larger even integer.

Translate. We reword the problem.

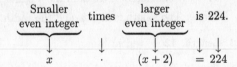

$$\underbrace{\text{Smaller even integer}}_{x} \quad \underset{\cdot}{\text{times}} \quad \underbrace{\text{larger even integer}}_{(x + 2)} \quad \underset{=}{\text{is}} \quad \underset{224}{224.}$$

Carry out.

$$x(x + 2) = 224$$
$$x^2 + 2x = 224$$
$$x^2 + 2x - 224 = 0$$
$$(x + 16)(x - 14) = 0$$

$$x + 16 = 0 \quad or \quad x - 14 = 0$$
$$x = -16 \quad or \qquad x = 14$$

Check. The solutions of the equation are -16 and 14. When x is -16, then $x + 2$ is -14 and $-16(-14) = 224$. The numbers -16 and -14 are consecutive even integers which are solutions of the problem. When x is 14, then $x + 2$ is 16 and $14 \cdot 16 = 224$. The numbers 14 and 16 are also consecutive even integers which are solutions of the problem.

State. We have two solutions, each of which consists of a pair of numbers: -16 and -14, and 14 and 16.

8. 15 and 17, -17 and -15

9. **Familiarize**. We make a drawing. Let $w =$ the width, in cm. Then $w + 5 =$ the length, in cm.

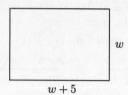

$w + 5$

Recall that the area of a rectangle is length times width.

Translate. We reword the problem.

Length times width is $\underbrace{84 \text{ cm}^2}$.

$\downarrow \qquad \downarrow \quad \downarrow \quad \downarrow \qquad \downarrow$

$(w + 5) \quad \cdot \quad w \quad = \quad 84$

Carry out. We solve the equation.

$$(w + 5)w = 84$$
$$w^2 + 5w = 84$$
$$w^2 + 5w - 84 = 0$$
$$(w + 12)(w - 7) = 0$$
$$w + 12 = 0 \quad or \quad w - 7 = 0$$
$$w = -12 \quad or \qquad w = 7$$

Check. The solutions of the equation are -12 and 7. The width of a rectangle cannot have a negative measure, so -12 cannot be a solution. Suppose the width is 7 cm. The length is 5 cm greater than the width, so the length is 12 cm and the area is $12 \cdot 7$, or 84 cm^2. The numbers check in the original problem.

State. The length is 12 cm, and the width is 7 cm.

10. Length: 12 m, width: 8 m

11. **Familiarize**. Using the labels shown on the drawing in the text, we let $b =$ the length of the base, in cm, and $b - 3 =$ the height, in cm. Recall that the formula for the area of a triangle is $\frac{1}{2} \cdot$ (base) $\cdot$ (height).

Translate.

$\frac{1}{2}$ times base times height is $\underbrace{35 \text{ cm}^2}$.

$\downarrow \quad \downarrow \qquad \downarrow \quad \downarrow \qquad \downarrow \quad \downarrow \qquad \downarrow$

$\frac{1}{2} \quad \cdot \qquad b \quad \cdot \quad (b - 3) \quad = \qquad 35$

Carry out.

$$\frac{1}{2}b(b - 3) = 35$$
$$b(b - 3) = 70 \quad \text{Multiplying by 2}$$
$$b^2 - 3b = 70$$
$$b^2 - 3b - 70 = 0$$
$$(b - 10)(b + 7) = 0$$
$$b - 10 = 0 \quad or \quad b + 7 = 0$$
$$b = 10 \quad or \qquad b = -7$$

Check. The solutions of the equation are 10 and -7. The length of the base of a triangle cannot be

negative, so -7 cannot be a solution. Suppose the length of the base is 10 cm. Then the height is $10 - 3$, or 7 cm, and the area is $\frac{1}{2} \cdot 10 \cdot 7$, or 35 cm^2. These numbers check.

State. The height is 7 cm, and the length of the base is 10 cm.

12. Height: 4 cm, base: 14 cm

13. **Familiarize**. Using the labels show on the drawing in the text, we let $x =$ the length of the foot of the sail, in ft, and $x + 5 =$ the height of the sail, in ft. Recall that the formula for the area of a triangle is $\frac{1}{2} \cdot$ (base) $\cdot$ (height).

Translate.

$\frac{1}{2}$ times base times height is $\underbrace{42 \text{ ft}^2}$.

$\downarrow \quad \downarrow \qquad \downarrow \quad \downarrow \qquad \downarrow \quad \downarrow \qquad \downarrow$

$\frac{1}{2} \quad \cdot \qquad x \quad \cdot \quad (x + 5) \quad = \qquad 42$

Carry out.

$$\frac{1}{2}x(x + 5) = 42$$
$$x(x + 5) = 84 \quad \text{Multiplying by 2}$$
$$x^2 + 5x = 84$$
$$x^2 + 5x - 84 = 0$$
$$(x + 12)(x - 7) = 0$$
$$x + 12 = 0 \quad or \quad x - 7 = 0$$
$$x = -12 \quad or \qquad x = 7$$

Check. The solutions of the equation are -12 and 7. The length of the base of a triangle cannot be negative, so -12 cannot be a solution. Suppose the length of the foot of the sail is 7 ft. Then the height is $7 + 5$, or 12 ft, and the area is $\frac{1}{2} \cdot 7 \cdot 12$, or 42 ft^2. These numbers check.

State. The length of the foot of the sail is 7 ft, and the height is 12 ft.

14. Base: 8 m, height: 16 m

15. **Familiarize**. We make a drawing. Let $l =$ the length of the cable, in ft.

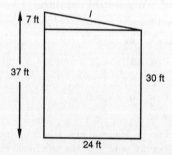

Note that we have a right triangle with hypotenuse l and legs of 24 ft and $37 - 30$, or 7 ft.

Translate. We use the Pythagorean theorem.

$$a^2 + b^2 = c^2$$
$$7^2 + 24^2 = l^2 \quad \text{Substituting}$$

Carry out.
$$7^2 + 24^2 = l^2$$
$$49 + 576 = l^2$$
$$625 = l^2$$
$$0 = l^2 - 625$$
$$0 = (l + 25)(l - 25)$$
$$l + 25 = 0 \quad \text{or} \quad l - 25 = 0$$
$$l = -25 \quad \text{or} \quad l = 25$$

Check. The integer -25 cannot be the length of the cable, because it is negative. When $l = 25$, we have $7^2 + 24^2 = 25^2$. This checks.

State. The cable is 25 ft long.

16. 8000 ft

17. *Familiarize.* We will use the formula $n^2 - n = N$.

Translate. Substitute 20 for n.
$$20^2 - 20 = N$$

Carry out. We do the computation on the left.
$$20^2 - 20 = N$$
$$400 - 20 = N$$
$$380 = N$$

Check. We can recheck the computation or we can solve $n^2 - n = 380$. The answer checks.

State. 380 games will be played.

18. 182

19. *Familiarize.* We will use the formula $n^2 - n = N$.

Translate. Substitute 132 for N.
$$n^2 - n = 132$$

Carry out.
$$n^2 - n = 132$$
$$n^2 - n - 132 = 0$$
$$(n - 12)(n + 11) = 0$$
$$n - 12 = 0 \quad \text{or} \quad n + 11 = 0$$
$$n = 12 \quad \text{or} \quad n = -11$$

Check. The solutions of the equation are 12 and -11. Since the number of teams cannot be negative, -11 cannot be a solution. But 12 checks since $12^2 - 12 = 144 - 12 = 132$.

State. There are 12 teams in the league.

20. 10

21. *Familiarize.* We make a drawing. Let $h =$ the height to which the ladder will reach, in ft.

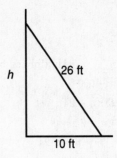

Translate. Use the Pythagorean theorem.
$$a^2 + b^2 = c^2$$
$$h^2 + 10^2 = 26^2 \quad \text{Substituting}$$

Carry out.
$$h^2 + 10^2 = 26^2$$
$$h^2 + 100 = 676$$
$$h^2 - 576 = 0$$
$$(h + 24)(h - 24) = 0$$
$$h + 24 = 0 \quad \text{or} \quad h - 24 = 0$$
$$h = -24 \quad \text{or} \quad h = 24$$

Check. The integer -24 cannot be the height because it is negative. When $h = 24$, we have $24^2 + 10^2 = 26^2$. This checks.

State. The ladder will reach to a height of 24 ft on the house.

22. 9 ft

23. *Familiarize.* We will use the formula
$$N = \frac{1}{2}(n^2 - n).$$

Translate. Substitute 12 for n.
$$N = \frac{1}{2}(12^2 - 12)$$

Carry out. We do the computation on the right.
$$N = \frac{1}{2}(12^2 - 12)$$
$$N = \frac{1}{2}(144 - 12)$$
$$N = \frac{1}{2}(132)$$
$$N = 66$$

Check. We can recheck the computation, or we can solve the equation $66 = \frac{1}{2}(n^2 - n)$. The answer checks.

State. 66 handshakes are possible.

24. 435

25. *Familiarize.* We will use the formula $N = \frac{1}{2}(n^2 - n)$, since "high fives" can be substituted for handshakes.

Translate. Substitute 300 for N.
$$300 = \frac{1}{2}(n^2 - n)$$

Carry out.

$$300 = \frac{1}{2}(n^2 - n)$$
$$600 = n^2 - n \qquad \text{Multiplying by 2}$$
$$0 = n^2 - n - 600$$
$$0 = (n - 25)(n + 24)$$
$$n - 25 = 0 \quad or \quad n + 24 = 0$$
$$n = 25 \; or \qquad n = -24$$

Check. The solutions of the equation are 25 and -24. Since the number of people cannot be negative, -24 cannot be a solution. However, 25 checks since $\frac{1}{2}(25^2 - 25) = \frac{1}{2}(625 - 25) = \frac{1}{2} \cdot 600 = 300$.

State. 25 people were on the team.

26. 20

27. *Familiarize.* We add labels to the drawing in the text. Let $h =$ the height of the antenna, in m. Then $h + 1 =$ the length of the guy wire, in m.

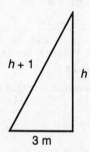

Translate. We use the Pythagorean theorem.
$$a^2 + b^2 = c^2$$
$$3^2 + h^2 = (h + 1)^2 \quad \text{Substituting}$$

Carry out.
$$3^2 + h^2 = (h + 1)^2$$
$$9 + h^2 = h^2 + 2h + 1$$
$$9 = 2h + 1$$
$$8 = 2h$$
$$4 = h$$

Check. When $h = 4$ m, then $h + 1 = 4 + 1$, or 5 m. Then we have $3^2 + 4^2 = 5^2$. This checks.

State. The antenna is 4 m tall.

28. Dining room: 12 ft by 12 ft, kitchen: 12 ft by 10 ft

29. *Familiarize.* We will use the formula $h = 48t - 16t^2$.

Translate. Substitute $\frac{1}{2}$ for t.

$$h = 48 \cdot \frac{1}{2} - 16\left(\frac{1}{2}\right)^2$$

Carry out. We do the computation on the right.

$$h = 48 \cdot \frac{1}{2} - 16\left(\frac{1}{2}\right)^2$$
$$h = 48 \cdot \frac{1}{2} - 16 \cdot \frac{1}{4}$$
$$h = 24 - 4$$
$$h = 20$$

Check. We can recheck the computation, or we can solve the equation $20 = 48t - 16t^2$. The answer checks.

State. The rocket is 20 ft high $\frac{1}{2}$ sec after it is launched.

30. 36 ft

31. *Familiarize.* We will use the formula $h = 48t - 16t^2$.

Translate. Substitute 32 for h.

$$32 = 48t - 16t^2$$

Carry out. We solve the equation.

$$32 = 48t - 16t^2$$
$$0 = -16t^2 + 48t - 32$$
$$0 = -16(t^2 - 3t + 2)$$
$$0 = -16(t - 1)(t - 2)$$
$$t - 1 = 0 \quad or \quad t - 2 = 0$$
$$t = 1 \quad or \qquad t = 2$$

Check. When $t = 1$, $h = 48 \cdot 1 - 16 \cdot 1^2 = 48 - 16 = 32$. When $t = 2$, $h = 48 \cdot 2 - 16 \cdot 2^2 = 96 - 64 = 32$. Both numbers check.

State. The rocket will be exactly 32 ft above the ground at 1 sec and at 2 sec after it is launched.

32. 3 sec after it is launched

33. Graph: $y = -\frac{2}{3}x + 1$

Since the equation is in the form $y = mx + b$, we know that the y-intercept is (0,1). We find two other solutions, using multiples of 3 for x to avoid fractions.

When $x = -3$, $y = -\frac{2}{3}(-3) + 1 = 2 + 1 = 3$.

When $x = 3$, $y = -\frac{2}{3} \cdot 3 + 1 = -2 + 1 = -1$.

x	y
0	1
-3	3
3	-1

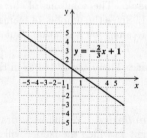

34.

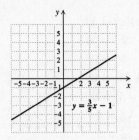

35. $7x^0 = 7 \cdot 4^0 = 7 \cdot 1 = 7$

(Any nonzero number raised to the zero power is 1.)

36. $m^6 n^6$

37. $-\dfrac{2}{3} + \dfrac{3}{4} \cdot \dfrac{1}{2} = -\dfrac{2}{3} + \dfrac{3}{8}$

$\qquad = -\dfrac{16}{24} + \dfrac{9}{24}$

$\qquad = -\dfrac{7}{24}$

38. -2

39.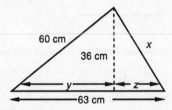

40. ◈

41. ◈

42. ◈

43. *Familiarize*. We add labels to the drawing in the text.

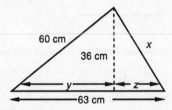

First we will use the Pythagorean theorem to find y. Then we will subtract to find z and, finally, we will use the Pythagorean theorem again to find x.

Translate. Use the Pythagorean theorem to find y.

$a^2 + b^2 = c^2$

$y^2 + 36^2 = 60^2$ Substituting

Carry out.

$\qquad y^2 + 36^2 = 60^2$

$\qquad y^2 + 1296 = 3600$

$\qquad y^2 - 2304 = 0$

$\qquad (y + 48)(y - 48) = 0$

$\quad y + 48 = 0 \qquad or \quad y - 48 = 0$

$\qquad y = -48 \quad or \qquad y = 48$

Since the length y cannot be negative, we use 48 cm. The base of the triangle measures 63 cm. We substitute to find z:

$z = 63 - 48 = 15$

Now we use the Pythagorean theorem again to find x.

$\qquad a^2 + b^2 = c^2$

$\qquad 15^2 + 36^2 = x^2 \qquad\qquad$ Substituting

$\qquad 225 + 1296 = x^2$

$\qquad\qquad 1521 = x^2$

$\qquad\qquad 0 = x^2 - 1521$

$\qquad\qquad 0 = (x + 39)(x - 39)$

$\quad x + 39 = 0 \qquad or \quad x - 39 = 0$

$\qquad x = -39 \quad or \qquad\quad x = 39$

Since the length x cannot be negative, we have $x = 39$ cm.

Check. Recheck the calculations.

State. x is 39 cm.

44. 5 ft

45. *Familiarize*. From the drawing in the text we see that the length of each half of the roof is 32 ft. Next we need to find the width w of each half of the roof, in ft. Then we will find the area of the roof. We make a drawing.

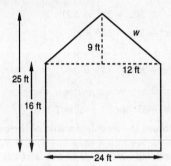

Translate. We use the Pythagorean theorem to find w.

$\qquad a^2 + b^2 = c^2$

$\qquad 9^2 + 12^2 = w^2$ Substituting

Carry out.

$\qquad 9^2 + 12^2 = w^2$

$\qquad 81 + 144 = w^2$

$\qquad\qquad 225 = w^2$

$\qquad\qquad 0 = w^2 - 225$

$\qquad\qquad 0 = (w + 15)(w - 15)$

$\quad w + 15 = 0 \qquad or \quad w - 15 = 0$

$\qquad x = -15 \quad or \qquad\quad w = 15$

Since the width of the roof cannot be negative, we use $w = 15$ ft. The roof consists of two rectangles, each of which has dimensions 15 ft by 32 ft. We find the area of the roof:

$2 \cdot 32 \cdot 15 = 960$

Check. Recheck the calculations.

State. The area of the roof is 960 ft^2.

46. 37

47. *Familiarize*. We make a drawing. Let $x =$ the length of a side of the base, in meters.

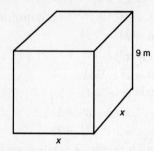

The area of each of the 4 sides of the box is $9 \cdot x$, so the total area of the sides is $4 \cdot 9 \cdot x$, or $36x$. The area of the base is $x \cdot x$, or x^2, and the area of the lid is also $x \cdot x$, or x^2. Then the total surface area is $36x + x^2 + x^2$, or $36x + 2x^2$.

***Translate*.**

$$\underbrace{\text{Total surface area}}_{\downarrow} \;\; \underbrace{\text{is}}_{\downarrow} \;\; \underbrace{350 \text{ m}^2.}_{\downarrow \quad \downarrow}$$
$$36x + 2x^2 \quad = \quad 350$$

***Carry out*.**

$$36x + 2x^2 = 350$$
$$2x^2 + 36x - 350 = 0$$
$$2(x^2 + 18x - 175) = 0$$
$$2(x + 25)(x - 7) = 0$$
$$x + 25 = 0 \quad or \quad x - 7 = 0$$
$$x = -25 \quad or \quad x = 7$$

***Check*.** Since the length of a side of the base cannot be negative, we check only 7. If $x = 7$, then the area of each side is $9 \cdot 7$, or 63 m^2, and the total area of the 4 sides is $4 \cdot 63$, or 252 m^2. The area of the base is $7 \cdot 7$, or 49 m^2, and the area of the lid is also 49 m^2. Then the total surface area is $252 + 49 + 49$, or 350 m^2. Our answer checks.

***State*.** The length of a side of the base is 7 m.

48. 30 cm by 15 cm

49. *Familiarize*. We make a drawing. Let $x =$ the depth of the gutter, in inches.

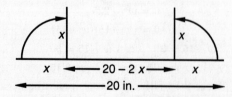

The cross-section has dimensions $20 - 2x$ by x.

***Translate*.**

$$\underbrace{\text{Area of cross-section}}_{\downarrow} \;\; \underbrace{\text{is}}_{\downarrow} \;\; \underbrace{50 \text{ in}^2.}_{\downarrow \quad \downarrow}$$
$$(20 - 2x)(x) \quad = \quad 50$$

***Carry out*.**

$$(20 - 2x)(x) = 50$$
$$20x - 2x^2 = 50$$
$$-2x^2 + 20x - 50 = 0$$
$$-2(x^2 - 10x + 25) = 0$$
$$-2(x - 5)(x - 5) = 0$$
$$x - 5 = 0 \quad or \quad x - 5 = 0$$
$$x = 5 \quad or \quad x = 5$$

***Check*.** If the depth of the gutter is 5 in., then the cross-section has dimension $20 - 2 \cdot 5$, or 10 in. by 5 in., and its area is $10 \cdot 5 = 50 \text{ in}^2$. The answer checks.

***State*.** The gutter is 5 in. deep.

50. 100 cm^2, 225 cm^2

51. *Familiarize*. First we find the length of the other leg of the right triangle. Then we find the area of the triangle, and finally we multiply by the cost per square foot of the sailcloth. We begin by making a drawing. Let $x =$ the length of the other leg of the right triangle, in feet.

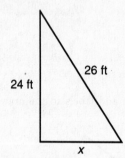

***Translate*.** We use the Pythagorean theorem to find x.

$$a^2 + b^2 = c^2$$
$$x^2 + 24^2 = 26^2 \quad \text{Substituting}$$

***Carry out*.**

$$x^2 + 24^2 = 26^2$$
$$x^2 + 576 = 676$$
$$x^2 - 100 = 0$$
$$(x + 10)(x - 10) = 0$$
$$x + 10 = 0 \quad or \quad x - 10 = 0$$
$$x = -10 \quad or \quad x = 10$$

Since the length cannot be negative we use $x = 10$. Now we find the area of the right triangle with base 10 ft and height 24 ft:

$$\frac{1}{2} \cdot \text{base} \cdot \text{height} = \frac{1}{2} \cdot 10 \cdot 24 = 120 \text{ ft}^2$$

Finally, we multiply the area, 120 ft^2, by the price per square foot of the sailcloth, \$10:

$$120 \cdot 10 = 1200$$

***Check*.** Recheck the calculations.

***State*.** A new main sail costs \$1200.

52. 4

Chapter 6

Rational Expressions and Equations

Exercise Set 6.1

1. $\dfrac{17}{-3x}$

We find the real number(s) that make the denominator 0. To do so we set the denominator equal to 0 and solve for x:

$$-3x = 0$$
$$x = 0$$

The expression is undefined for $x = 0$.

2. 0

3. $\dfrac{t-4}{t+6}$

Set the denominator equal to 0 and solve for t:

$$t - 6 = 0$$
$$t = 6$$

The expression is undefined for $t = 6$.

4. -7

5. $\dfrac{9}{2a-10}$

Set the denominator equal to 0 and solve for a:

$$2a - 10 = 0$$
$$2a = 10$$
$$a = 5$$

The expression is undefined for $a = 5$.

6. 3

7. $\dfrac{x^2+11}{x^2-3x-28}$

Set the denominator equal to 0 and solve for x:

$$x^2 - 3x - 28 = 0$$
$$(x-7)(x+4) = 0$$
$$x - 7 = 0 \quad or \quad x + 4 = 0$$
$$x = 7 \quad or \qquad x = -4$$

The expression is undefined for $x = 7$ and $x = -4$.

8. 2, 5

9. $\dfrac{m^3-2m}{m^2-25}$

Set the denominator equal to 0 and solve for m:

$$m^2 - 25 = 0$$
$$(m+5)(m-5) = 0$$
$$m + 5 = 0 \quad or \quad m - 5 = 0$$
$$m = -5 \quad or \qquad m = 5$$

The expression is undefined for $m = -5$ and $m = 5$.

10. $-7,\ 7$

11. $\dfrac{50a^2b}{40ab^3}$

$$= \dfrac{5a \cdot 10ab}{4b^2 \cdot 10ab} \qquad \text{Factoring the numerator and denominator. Note the common factor of } 10ab.$$

$$= \dfrac{5a}{4b^2} \cdot \dfrac{10ab}{10ab} \qquad \text{Rewriting as a product of two rational expressions}$$

$$= \dfrac{5a}{4b^2} \cdot 1 \qquad \dfrac{10ab}{10ab} = 1$$

$$= \dfrac{5a}{4b^2} \qquad \text{Removing the factor 1}$$

12. $\dfrac{5y}{x^2}$

13. $\dfrac{35x^2y}{14x^3y^5} = \dfrac{5 \cdot 7x^2y}{2xy^4 \cdot 7x^2y}$

$$= \dfrac{5}{2xy^4} \cdot \dfrac{7x^2y}{7x^2y}$$

$$= \dfrac{5}{2xy^4} \cdot 1$$

$$= \dfrac{5}{2xy^4}$$

14. $\dfrac{2a^2b^5}{3}$

15. $\dfrac{9x+15}{6x+10} = \dfrac{3(3x+5)}{2(3x+5)}$

$$= \dfrac{3}{2} \cdot \dfrac{3x+5}{3x+5}$$

$$= \dfrac{3}{2} \cdot 1$$

$$= \dfrac{3}{2}$$

16. $\dfrac{7}{5}$

17. $\dfrac{a^2-9}{a^2+4a+3} = \dfrac{(a+3)(a-3)}{(a+3)(a+1)}$

$$= \dfrac{a+3}{a+3} \cdot \dfrac{a-3}{a+1}$$

$$= 1 \cdot \dfrac{a-3}{a+1}$$

$$= \dfrac{a-3}{a+1}$$

18. $\dfrac{a+2}{a-3}$

19. $\dfrac{36x^6}{24x^9} = \dfrac{3 \cdot 12x^6}{2x^3 \cdot 12x^6}$

$\qquad = \dfrac{3}{2x^3} \cdot \dfrac{12x^6}{12x^6}$

$\qquad = \dfrac{3}{2x^3} \cdot 1$

$\qquad = \dfrac{3}{2x^3}$

Check: Let $x = 1$.

$\dfrac{36x^6}{24x^9} = \dfrac{36 \cdot 1^6}{24 \cdot 1^9} = \dfrac{36}{24} = \dfrac{3}{2}$

$\dfrac{3}{2x^3} = \dfrac{3}{2 \cdot 1^3} = \dfrac{3}{2}$

The answer is probably correct.

20. $\dfrac{19a^2}{6}$

21. $\dfrac{-2y + 6}{-4y} = \dfrac{-2(y - 3)}{-2 \cdot 2y}$

$\qquad = \dfrac{-2}{-2} \cdot \dfrac{y - 3}{2y}$

$\qquad = 1 \cdot \dfrac{y - 3}{2y}$

$\qquad = \dfrac{y - 3}{2y}$

Check: Let $x = 2$.

$\dfrac{-2y + 6}{-4y} = \dfrac{-2 \cdot 2 + 6}{-4 \cdot 2} = \dfrac{2}{-8} = -\dfrac{1}{4}$

$\dfrac{y - 3}{2y} = \dfrac{2 - 3}{2 \cdot 2} = \dfrac{-1}{4} = -\dfrac{1}{4}$

The answer is probably correct.

22. $\dfrac{x - 3}{x}$

23. $\dfrac{6a^2 - 3a}{7a^2 - 7a} = \dfrac{3a(2a - 1)}{7a(a - 1)}$

$\qquad = \dfrac{a}{a} \cdot \dfrac{3(2a - 1)}{7(a - 1)}$

$\qquad = 1 \cdot \dfrac{3(2a - 1)}{7(a - 1)}$

$\qquad = \dfrac{3(2a - 1)}{7(a - 1)}$

Check: Let $a = 2$.

$\dfrac{6a^2 - 3a}{7a^2 - 7a} = \dfrac{6 \cdot 2^2 - 3 \cdot 2}{7 \cdot 2^2 - 7 \cdot 2} = \dfrac{18}{14} = \dfrac{9}{7}$

$\dfrac{3(2a - 1)}{7(a - 1)} = \dfrac{3(2 \cdot 2 - 1)}{7(2 - 1)} = \dfrac{3 \cdot 3}{7 \cdot 1} = \dfrac{9}{7}$

The answer is probably correct.

24. $\dfrac{m + 1}{2m + 3}$

25. $\dfrac{t^2 - 25}{t^2 + t - 20} = \dfrac{(t + 5)(t - 5)}{(t + 5)(t - 4)}$

$\qquad = \dfrac{t + 5}{t + 5} \cdot \dfrac{t - 5}{t - 4}$

$\qquad = 1 \cdot \dfrac{t - 5}{t - 4}$

$\qquad = \dfrac{t - 5}{t - 4}$

Check: Let $t = 1$.

$\dfrac{t^2 - 25}{t^2 + t - 20} = \dfrac{1^2 - 25}{1^2 + 1 - 20} = \dfrac{-24}{-18} = \dfrac{4}{3}$

$\dfrac{t - 5}{t - 4} = \dfrac{1 - 5}{1 - 4} = \dfrac{-4}{-3} = \dfrac{4}{3}$

The answer is probably correct.

26. $\dfrac{a - 2}{a + 3}$

27. $\dfrac{3a^2 - 9a - 12}{6a^2 + 30a + 24} = \dfrac{3(a^2 - 3a - 4)}{6(a^2 + 5a + 4)}$

$\qquad = \dfrac{3(a - 4)(a + 1)}{3 \cdot 2(a + 4)(a + 1)}$

$\qquad = \dfrac{3(a + 1)}{3(a + 1)} \cdot \dfrac{a - 4}{2(a + 4)}$

$\qquad = 1 \cdot \dfrac{a - 4}{2(a + 4)}$

$\qquad = \dfrac{a - 4}{2(a + 4)}$

Check: Let $a = 1$.

$\dfrac{3a^2 - 9a - 12}{6a^2 + 30a + 24} = \dfrac{3 \cdot 1^2 - 9 \cdot 1 - 12}{6 \cdot 1^2 + 30 \cdot 1 + 24} = \dfrac{-18}{60} = -\dfrac{3}{10}$

$\dfrac{a - 4}{2(a + 4)} = \dfrac{1 - 4}{2(1 + 4)} = \dfrac{-3}{10} = -\dfrac{3}{10}$

The answer is probably correct.

28. $\dfrac{t + 2}{2(t - 4)}$

29. $\dfrac{x^2 + 8x + 16}{x^2 - 16} = \dfrac{(x + 4)(x + 4)}{(x + 4)(x - 4)}$

$\qquad = \dfrac{x + 4}{x + 4} \cdot \dfrac{x + 4}{x - 4}$

$\qquad = 1 \cdot \dfrac{x + 4}{x - 4}$

$\qquad = \dfrac{x + 4}{x - 4}$

Check: Let $x = 1$.

$\dfrac{x^2 + 8x + 16}{x^2 - 16} = \dfrac{1^2 + 8 \cdot 1 + 16}{1^2 - 16} = \dfrac{25}{-15} = -\dfrac{5}{3}$

$\dfrac{x + 4}{x - 4} = \dfrac{1 + 4}{1 - 4} = \dfrac{5}{-3} = -\dfrac{5}{3}$

The answer is probably correct.

30. $\dfrac{x + 5}{x - 5}$

31. $\dfrac{t^2 - 1}{t + 1} = \dfrac{(t+1)(t-1)}{t+1}$

$\qquad\qquad = \dfrac{t+1}{t+1} \cdot \dfrac{t-1}{1}$

$\qquad\qquad = 1 \cdot \dfrac{t-1}{1}$

$\qquad\qquad = t - 1$

Check: Let $t = 2$.

$\dfrac{t^2 - 1}{t+1} = \dfrac{2^2 - 1}{2+1} = \dfrac{3}{3} = 1$

$t - 1 = 2 - 1 = 1$

The answer is probably correct.

32. $a + 1$

33. $\dfrac{y^2 + 4}{y + 2}$ cannot be simplified.

Neither the numerator nor the denominator can be factored.

34. $\dfrac{x^2 + 1}{x + 1}$

35. $\dfrac{5x^2 - 20}{10x^2 - 40} = \dfrac{5(x^2 - 4)}{10(x^2 - 4)}$

$\qquad\qquad = \dfrac{1 \cdot \cancel{5} \cdot (\cancel{x^2 - 4})}{2 \cdot \cancel{5} \cdot (\cancel{x^2 - 4})}$

$\qquad\qquad = \dfrac{1}{2}$

Check: Let $x = 1$.

$\dfrac{5x^2 - 20}{10x^2 - 40} = \dfrac{5 \cdot 1^2 - 20}{10 \cdot 1^2 - 40} = \dfrac{-15}{-30} = \dfrac{1}{2}$

$\dfrac{1}{2} = \dfrac{1}{2}$

The answer is probably correct.

36. $\dfrac{3}{2}$

37. $\dfrac{5y + 5}{y^2 + 7y + 6} = \dfrac{5(y+1)}{(y+1)(y+6)}$

$\qquad\qquad = \dfrac{y+1}{y+1} \cdot \dfrac{5}{y+6}$

$\qquad\qquad = 1 \cdot \dfrac{5}{y+6}$

$\qquad\qquad = \dfrac{5}{y+6}$

Check: Let $x = 1$.

$\dfrac{5y + 5}{y^2 + 7y + 6} = \dfrac{5 \cdot 1 + 5}{1^2 + 7 \cdot 1 + 6} = \dfrac{10}{14} = \dfrac{5}{7}$

$\dfrac{5}{y+6} = \dfrac{5}{1+6} = \dfrac{5}{7}$

The answer is probably correct.

38. $\dfrac{6}{t - 3}$

39. $\dfrac{y^2 - 3y - 18}{y^2 - 2y - 15} = \dfrac{(y-6)(y+3)}{(y-5)(y+3)}$

$\qquad\qquad = \dfrac{y-6}{y-5} \cdot \dfrac{y+3}{y+3}$

$\qquad\qquad = \dfrac{y-6}{y-5} \cdot 1$

$\qquad\qquad = \dfrac{y-6}{y-5}$

Check: Let $y = 1$.

$\dfrac{y^2 - 3y - 18}{y^2 - 2y - 15} = \dfrac{1^2 - 3 \cdot 1 - 18}{1^2 - 2 \cdot 1 - 15} = \dfrac{-20}{-16} = \dfrac{5}{4}$

$\dfrac{y-6}{y-5} = \dfrac{1-6}{1-5} = \dfrac{-5}{-4} = \dfrac{5}{4}$

The answer is probably correct.

40. $\dfrac{a - 3}{a - 4}$

41. $\dfrac{(a-3)^2}{a^2 - 9} = \dfrac{(a-3)(a-3)}{(a+3)(a-3)}$

$\qquad\qquad = \dfrac{a-3}{a+3} \cdot \dfrac{a-3}{a-3}$

$\qquad\qquad = \dfrac{a-3}{a+3} \cdot 1$

$\qquad\qquad = \dfrac{a-3}{a+3}$

Check: Let $a = 2$.

$\dfrac{(a-3)^2}{a^2 - 9} = \dfrac{(2-3)^2}{2^2 - 9} = \dfrac{1}{-5} = -\dfrac{1}{5}$

$\dfrac{a-3}{a+3} = \dfrac{2-3}{2+3} = \dfrac{-1}{5} = -\dfrac{1}{5}$

The answer is probably correct.

42. $\dfrac{t - 2}{t + 2}$

43. $\dfrac{x-8}{8-x} = \dfrac{x-8}{-(x-8)}$

$\qquad\qquad = \dfrac{1}{-1} \cdot \dfrac{x-8}{x-8}$

$\qquad\qquad = \dfrac{1}{-1} \cdot 1$

$\qquad\qquad = -1$

Check: Let $x = 2$.

$\dfrac{x-8}{8-x} = \dfrac{2-8}{8-2} = \dfrac{-6}{6} = -1$

$-1 = -1$

The answer is probably correct.

44. -1

45. $\dfrac{q-p}{-p+q} = \dfrac{q-p}{q-p} \qquad (-p + q = q - p)$

$\qquad\qquad = 1$

Check: Let $p = 1$ and $q = 2$.

$\dfrac{q-p}{-p+q} = \dfrac{2-1}{-1+2} = \dfrac{1}{1} = 1$

$1 = 1$

The answer is probably correct.

46. -1

47. $\dfrac{5a - 15}{3 - a} = \dfrac{5(a - 3)}{3 - a}$

$\phantom{\dfrac{5a - 15}{3 - a}} = \dfrac{5(a - 3)}{-(a - 3)}$

$\phantom{\dfrac{5a - 15}{3 - a}} = \dfrac{5}{-1} \cdot \dfrac{a - 3}{a - 3}$

$\phantom{\dfrac{5a - 15}{3 - a}} = \dfrac{5}{-1} \cdot 1$

$\phantom{\dfrac{5a - 15}{3 - a}} = -5$

Check: Let $a = 1$.

$\dfrac{5a - 15}{3 - a} = \dfrac{5 \cdot 1 - 15}{3 - 1} = \dfrac{-10}{2} = -5$

$-5 = -5$

The answer is probably correct.

48. -6

49. $\dfrac{3x^2 - 3y^2}{2y^2 - 2x^2} = \dfrac{3(x^2 - y^2)}{2(y^2 - x^2)}$

$\phantom{\dfrac{3x^2 - 3y^2}{2y^2 - 2x^2}} = \dfrac{3(x^2 - y^2)}{2(-1)(x^2 - y^2)}$

$\phantom{\dfrac{3x^2 - 3y^2}{2y^2 - 2x^2}} = \dfrac{3}{2(-1)} \cdot \dfrac{x^2 - y^2}{x^2 - y^2}$

$\phantom{\dfrac{3x^2 - 3y^2}{2y^2 - 2x^2}} = \dfrac{3}{2(-1)} \cdot 1$

$\phantom{\dfrac{3x^2 - 3y^2}{2y^2 - 2x^2}} = -\dfrac{3}{2}$

Check: Let $x = 1$ and $y = 2$.

$\dfrac{3x^2 - 3y^2}{2y^2 - 2x^2} = \dfrac{3 \cdot 1^2 - 3 \cdot 2^2}{2 \cdot 2^2 - 2 \cdot 1^2} = \dfrac{-9}{6} = -\dfrac{3}{2}$

$-\dfrac{3}{2} = -\dfrac{3}{2}$

The answer is probably correct.

50. $-7(a + 1)$

51. $x^2 + 8x + 7$

The factorization is of the form $(x +)(x +)$. We look for two factors of 7 whose sum is 8. The numbers we need are 1 and 7.

$x^2 + 8x + 7 = (x + 1)(x + 7)$

52. $(x - 2)(x - 7)$

53. $5x + 2y = 20$

To find the y-intercept, solve:

$2y = 20$

$y = 10$

The y-intercept is $(0, 10)$.

To find the x-intercept, solve:

$5x = 20$

$x = 4$

The x-intercept is $(4, 0)$.

We find a third point as a check. Let $x = 2$ and solve for y.

$5 \cdot 2 + 2y = 20$

$10 + 2y = 20$

$2y = 10$

$y = 5$

The point $(2, 5)$ appears to line up with the intercepts, so we draw the graph.

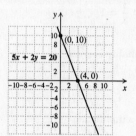

54.

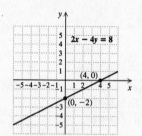

55. $\dfrac{2}{3} - \left(\dfrac{3}{4}\right)^2 = \dfrac{2}{3} - \dfrac{9}{16} = \dfrac{32}{48} - \dfrac{27}{48} = \dfrac{5}{48}$

56. $\dfrac{13}{63}$

57. ◈

58. ◈

59. ◈

60. ◈

61. $\dfrac{16y^4 - x^4}{(x^2 + 4y^2)(x - 2y)}$

$= \dfrac{(4y^2 + x^2)(4y^2 - x^2)}{(x^2 + 4y^2)(x - 2y)}$

$= \dfrac{(4y^2 + x^2)(2y + x)(2y - x)}{(x^2 + 4y^2)(x - 2y)}$

$= \dfrac{(x^2 + 4y^2)(2y + x)(-1)(x - 2y)}{(x^2 + 4y^2)(x - 2y)}$

$= \dfrac{(x^2 + 4y^2)(x - 2y)}{(x^2 + 4y^2)(x - 2y)} \cdot \dfrac{(2y + x)(-1)}{1}$

$= -2y - x, \text{ or } -x - 2y$

62. $\dfrac{a - b}{-a - b}$, or $\dfrac{b - a}{a + b}$

63.
$$\frac{(t^4-1)(t^2-9)(t-9)^2}{(t^4-81)(t^2+1)(t+1)^2}$$

$$=\frac{(t^2+1)(t+1)(t-1)(t+3)(t-3)(t-9)(t-9)}{(t^2+9)(t+3)(t-3)(t^2+1)(t+1)(t+1)}$$

$$=\frac{(\cancel{t^2+1})(\cancel{t+1})(t-1)(\cancel{t+3})(\cancel{t-3})(t-9)(t-9)}{(t^2+9)(\cancel{t+3})(\cancel{t-3})(\cancel{t^2+1})(\cancel{t+1})(t+1)}$$

$$=\frac{(t-1)(t-9)(t-9)}{(t^2+9)(t+1)},\text{ or }\frac{(t-1)(t-9)^2}{(t^2+9)(t+1)}$$

64. 1

65.
$$\frac{(x^2-y^2)(x^2-2xy+y^2)}{(x+y)^2(x^2-4xy-5y^2)}$$

$$=\frac{(x+y)(x-y)(x-y)(x-y)}{(x+y)(x+y)(x-5y)(x+y)}$$

$$=\frac{(\cancel{x+y})(x-y)(x-y)(x-y)}{(\cancel{x+y})(x+y)(x-5y)(x+y)}$$

$$=\frac{(x-y)^3}{(x+y)^2(x-5y)}$$

66. $\dfrac{1}{x-1}$

67.
$$\frac{x^5-2x^3+4x^2-8}{x^7+2x^4-4x^3-8}$$

$$=\frac{x^3(x^2-2)+4(x^2-2)}{x^4(x^3+2)-4(x^3+2)}$$

$$=\frac{(x^2-2)(x^3+4)}{(x^3+2)(x^4-4)}$$

$$=\frac{(x^2-2)(x^3+4)}{(x^3+2)(x^2+2)(x^2-2)}$$

$$=\frac{(\cancel{x^2-2})(x^3+4)}{(x^3+2)(x^2+2)(\cancel{x^2-2})}$$

$$=\frac{x^3+4}{(x^3+2)(x^2+2)}$$

68. $\dfrac{-2t^3-3}{2+3t^2}$, or $\dfrac{2t^3+3}{-2-3t^2}$

69.

Exercise Set 6.2

1. $\dfrac{7x}{5}\cdot\dfrac{x-3}{2x+1}=\dfrac{7x(x-3)}{5(2x+1)}$

2. $\dfrac{3x(5x+2)}{4(x-1)}$

3. $\dfrac{a-4}{a+6}\cdot\dfrac{a+2}{a+6}=\dfrac{(a-4)(a+2)}{(a+6)(a+6)}$

4. $\dfrac{(a+3)(a+3)}{(a+6)(a-1)}$

5. $\dfrac{2x+3}{4}\cdot\dfrac{x+1}{x-5}=\dfrac{(2x+3)(x+1)}{4(x-5)}$

6. $\dfrac{(-5)(-6)}{(3x-4)(5x+6)}$

7. $\dfrac{a-5}{a^2+1}\cdot\dfrac{a+2}{a^2-1}=\dfrac{(a-5)(a+2)}{(a^2+1)(a^2-1)}$

8. $\dfrac{(t+3)(t+3)}{(t^2-2)(t^2-2)}$

9. $\dfrac{x+1}{2+x}\cdot\dfrac{x-1}{x+1}=\dfrac{(x+1)(x-1)}{(2+x)(x+1)}$

10. $\dfrac{(m^2+5)(m^2-4)}{(m+8)(m^2-4)}$

11. $\dfrac{5a^4}{8a}\cdot\dfrac{2}{a}$

$$=\frac{5a^4\cdot2}{8a\cdot a}\qquad\text{Multiplying the numerators and the denominators}$$

$$=\frac{5\cdot a\cdot a\cdot a\cdot a\cdot2}{2\cdot4\cdot a\cdot a}\qquad\text{Factoring the numerator and the denominator}$$

$$=\frac{5\cdot\cancel{a}\cdot\cancel{a}\cdot a\cdot a\cdot\cancel{2}}{\cancel{2}\cdot4\cdot\cancel{a}\cdot\cancel{a}}\qquad\begin{array}{l}\text{Removing a factor equal}\\\text{to 1}\end{array}$$

$$=\frac{5a^2}{4}\qquad\text{Simplifying}$$

12. $\dfrac{6}{5t^6}$

13. $\dfrac{3c}{d^2}\cdot\dfrac{4d}{6c^3}$

$$=\frac{3c\cdot4d}{d^2\cdot6c^3}\qquad\begin{array}{l}\text{Multiplying the numerators and}\\\text{the denominators}\end{array}$$

$$=\frac{3\cdot c\cdot2\cdot2\cdot d}{d\cdot d\cdot3\cdot2\cdot c\cdot c\cdot c}\qquad\begin{array}{l}\text{Factoring the numera-}\\\text{tor and the denomina-}\\\text{tor}\end{array}$$

$$=\frac{\cancel{3}\cdot\cancel{c}\cdot\cancel{2}\cdot2\cdot\cancel{d}}{d\cdot d\cdot\cancel{3}\cdot\cancel{2}\cdot\cancel{c}\cdot c\cdot c}$$

$$=\frac{2}{dc^2}$$

14. $\dfrac{6x}{y^2}$

15. $\dfrac{x^2-3x-10}{(x-2)^2}\cdot\dfrac{x-2}{x-5}=\dfrac{(x^2-3x-10)(x-2)}{(x-2)^2(x-5)}$

$$=\frac{(x-5)(x+2)(x-2)}{(x-2)(x-2)(x-5)}$$

$$=\frac{(\cancel{x-5})(x+2)(\cancel{x-2})}{(\cancel{x-2})(x-2)(\cancel{x-5})}$$

$$=\frac{x+2}{x-2}$$

16. $\dfrac{t}{t+2}$

17. $\dfrac{a^2-25}{a^2-4a+3} \cdot \dfrac{2a-5}{2a+5} = \dfrac{(a^2-25)(2a-5)}{(a^2-4a+3)(2a+5)}$

$$= \dfrac{(a+5)(a-5)(2a-5)}{(a-3)(a-1)(2a+5)}$$

(No simplification is possible.)

18. $\dfrac{(x+3)(x+4)(x+1)}{(x^2+9)(x+9)}$

19. $\dfrac{a^2-9}{a^2} \cdot \dfrac{a^2-3a}{a^2+a-12} = \dfrac{(a-3)(a+3)(a)(a-3)}{a \cdot a(a+4)(a-3)}$

$$= \dfrac{(a-3)(a+3)(\cancel{a})(a-3)}{\cancel{a} \cdot a(a+4)(a-3)}$$

$$= \dfrac{(a-3)(a+3)}{a(a+4)}$$

20. 1

21. $\dfrac{4a^2}{3a^2-12a+12} \cdot \dfrac{3a-6}{2a}$

$$= \dfrac{4a^2(3a-6)}{(3a^2-12a+12)2a}$$

$$= \dfrac{2 \cdot 2 \cdot a \cdot a \cdot 3 \cdot (a-2)}{3 \cdot (a-2) \cdot (a-2) \cdot 2 \cdot a}$$

$$= \dfrac{\cancel{2} \cdot 2 \cdot \cancel{a} \cdot a \cdot \cancel{3} \cdot (\cancel{a-2})}{\cancel{3} \cdot (a-2) \cdot (\cancel{a-2}) \cdot \cancel{2} \cdot \cancel{a}}$$

$$= \dfrac{2a}{a-2}$$

22. $\dfrac{5(v-2)}{v-1}$

23. $\dfrac{t^2+2t-3}{t^2+4t-5} \cdot \dfrac{t^2-3t-10}{t^2+5t+6}$

$$= \dfrac{(t^2+2t-3)(t^2-3t-10)}{(t^2+4t-5)(t^2+5t+6)}$$

$$= \dfrac{(t+3)(t-1)(t-5)(t+2)}{(t+5)(t-1)(t+3)(t+2)}$$

$$= \dfrac{(\cancel{t+3})(\cancel{t-1})(t-5)(\cancel{t+2})}{(t+5)(\cancel{t-1})(\cancel{t+3})(\cancel{t+2})}$$

$$= \dfrac{t-5}{t+5}$$

24. $\dfrac{x+4}{x-4}$

25. $\dfrac{5a^2-180}{10a^2-10} \cdot \dfrac{20a+20}{2a-12}$

$$= \dfrac{(5a^2-180)(20a+20)}{(10a^2-10)(2a-12)}$$

$$= \dfrac{5(a+6)(a-6)(2)(10)(a+1)}{10(a+1)(a-1)(2)(a-6)}$$

$$= \dfrac{5(a+6)(\cancel{a-6})(\cancel{2})(\cancel{10})(\cancel{a+1})}{\cancel{10}(\cancel{a+1})(a-1)(\cancel{2})(\cancel{a-6})}$$

$$= \dfrac{5(a+6)}{a-1}$$

26. $\dfrac{t+7}{4(t-1)}$

27. $\dfrac{x^2-1}{x^2-9} \cdot \dfrac{(x-3)^4}{(x+1)^2}$

$$= \dfrac{(x^2-1)(x-3)^4}{(x^2-9)(x+1)^2}$$

$$= \dfrac{(x+1)(x-1)(x-3)(x-3)(x-3)(x-3)}{(x+3)(x-3)(x+1)(x+1)}$$

$$= \dfrac{(\cancel{x+1})(x-1)(\cancel{x-3})(x-3)(x-3)(x-3)}{(x+3)(\cancel{x-3})(\cancel{x+1})(x+1)}$$

$$= \dfrac{(x-1)(x-3)(x-3)(x-3)}{(x+3)(x+1)}, \text{ or}$$

$$\dfrac{(x-1)(x-3)^3}{(x+3)(x+1)}$$

28. $\dfrac{(x+2)^4(x+1)}{(x-1)^2(x+3)}$

29. $\dfrac{(t-2)^3}{(t-1)^3} \cdot \dfrac{t^2-2t+1}{t^2-4t+4}$

$$= \dfrac{(t-2)^3(t^2-2t+1)}{(t-1)^3(t^2-4t+4)}$$

$$= \dfrac{(t-2)(t-2)(t-2)(t-1)(t-1)}{(t-1)(t-1)(t-1)(t-2)(t-2)}$$

$$= \dfrac{(t-2)(t-2)(t-1)(t-1)}{(t-2)(t-2)(t-1)(t-1)} \cdot \dfrac{t-2}{t-1}$$

$$= \dfrac{t-2}{t-1}$$

30. $\dfrac{y+4}{y+2}$

31. The reciprocal of $\dfrac{x}{9}$ is $\dfrac{9}{x}$ because $\dfrac{x}{9} \cdot \dfrac{9}{x} = 1$.

32. $\dfrac{x^2+4}{3-x}$

33. The reciprocal of a^3-8a is $\dfrac{1}{a^3-8a}$ because

$$\dfrac{a^3-8a}{1} \cdot \dfrac{1}{a^3-8a} = 1.$$

34. $\dfrac{a^2-b^2}{7}$

35. The reciprocal of $\dfrac{x^2+2x-5}{x^2-4x+7}$ is $\dfrac{x^2-4x+7}{x^2+2x-5}$ because

$$\dfrac{x^2+2x-5}{x^2-4x+7} \cdot \dfrac{x^2-4x+7}{x^2+2x-5} = 1.$$

36. $\dfrac{x^2+7xy-y^2}{x^2-3xy+y^2}$

37. $\dfrac{5}{9} \div \dfrac{2}{7}$

$= \dfrac{5}{9} \cdot \dfrac{7}{2}$ Multiplying by the reciprocal of the divisor

$= \dfrac{5 \cdot 7}{9 \cdot 2}$

$= \dfrac{35}{18}$

38. $\dfrac{3}{20}$

39. $\dfrac{x}{3} \div \dfrac{x}{12}$

$= \dfrac{x}{3} \cdot \dfrac{12}{x}$ Multiplying by the reciprocal of the divisor

$= \dfrac{x \cdot 12}{3 \cdot x}$

$= \dfrac{x \cdot 3 \cdot 4}{3 \cdot x \cdot 1}$ Factoring the numerator and the denominator

$= \dfrac{3x}{3x} \cdot \dfrac{4}{1}$ Factoring the fractional expression

$= 4$ Simplifying

40. $\dfrac{x^2}{20}$

41. $\dfrac{x^5}{y^2} \div \dfrac{x^2}{y} = \dfrac{x^5}{y^2} \cdot \dfrac{y}{x^2}$

$= \dfrac{x^5 \cdot y}{y^2 \cdot x^2}$

$= \dfrac{x^2 \cdot x^3 \cdot y}{y \cdot y \cdot x^2}$

$= \dfrac{x^2 y}{x^2 y} \cdot \dfrac{x^3}{y}$

$= \dfrac{x^3}{y}$

42. $\dfrac{a^2}{b^3}$

43. $\dfrac{a+2}{a-3} \div \dfrac{a-1}{a+3} = \dfrac{a+2}{a-3} \cdot \dfrac{a+3}{a-1}$

$= \dfrac{(a+2)(a+3)}{(a-3)(a-1)}$

44. $\dfrac{y+2}{2y}$

45. $\dfrac{x^2-1}{x} \div \dfrac{x+1}{x-1} = \dfrac{x^2-1}{x} \cdot \dfrac{x-1}{x+1}$

$= \dfrac{(x^2-1)(x-1)}{x(x+1)}$

$= \dfrac{(x+1)(x-1)(x-1)}{x(x+1)}$

$= \dfrac{x+1}{x+1} \cdot \dfrac{(x-1)(x-1)}{x}$

$= \dfrac{(x-1)^2}{x}$

46. $4(y-2)$

47. $\dfrac{x+1}{6} \div \dfrac{x+1}{3} = \dfrac{x+1}{6} \cdot \dfrac{3}{x+1}$

$= \dfrac{(x+1) \cdot 3}{6(x+1)}$

$= \dfrac{3(x+1)}{2 \cdot 3(x+1)}$

$= \dfrac{3(x+1)}{3(x+1)} \cdot \dfrac{1}{2}$

$= \dfrac{1}{2}$

48. $\dfrac{a}{b}$

49. $(y^2-9) \div \dfrac{y^2-2y-3}{y^2+1} = \dfrac{(y^2-9)}{1} \cdot \dfrac{y^2+1}{y^2-2y-3}$

$= \dfrac{(y^2-9)(y^2+1)}{y^2-2y-3}$

$= \dfrac{(y+3)(y-3)(y^2+1)}{(y-3)(y+1)}$

$= \dfrac{(y+3)(y-3)(y^2+1)}{(y-3)(y+1)}$

$= \dfrac{(y+3)(y^2+1)}{y+1}$

50. $\dfrac{(x-6)(x+6)}{x-1}$

51. $\dfrac{5x-5}{16} \div \dfrac{x-1}{6} = \dfrac{5x-5}{16} \cdot \dfrac{6}{x-1}$

$= \dfrac{(5x-5) \cdot 6}{16(x-1)}$

$= \dfrac{5(x-1) \cdot 2 \cdot 3}{2 \cdot 8(x-1)}$

$= \dfrac{5(x-1) \cdot 2 \cdot 3}{2 \cdot 8(x-1)}$

$= \dfrac{15}{8}$

52. $\dfrac{1}{2}$

53. $\dfrac{-6+3x}{5} \div \dfrac{4x-8}{25} = \dfrac{-6+3x}{5} \cdot \dfrac{25}{4x-8}$

$= \dfrac{(-6+3x) \cdot 25}{5(4x-8)}$

$= \dfrac{3(x-2) \cdot 5 \cdot 5}{5 \cdot 4(x-2)}$

$= \dfrac{3(x-2) \cdot 5 \cdot 5}{5 \cdot 4(x-2)}$

$= \dfrac{15}{4}$

54. 3

55. $\dfrac{a+2}{a-1} \div \dfrac{3a+6}{a-5} = \dfrac{a+2}{a-1} \cdot \dfrac{a-5}{3a+6}$

$\qquad = \dfrac{(a+2)(a-5)}{(a-1)(3a+6)}$

$\qquad = \dfrac{(a+2)(a-5)}{(a-1)\cdot 3 \cdot(a+2)}$

$\qquad = \dfrac{(a+2)(a-5)}{(a-1)\cdot 3 \cdot(a+2)}$

$\qquad = \dfrac{a-5}{3(a-1)}$

56. $\dfrac{t+1}{4(t+2)}$

57. $\qquad (x-5) \div \dfrac{2x^2-11x+5}{4x^2-1}$

$\quad = \dfrac{x-5}{1} \cdot \dfrac{4x^2-1}{2x^2-11x+5}$

$\quad = \dfrac{(x-5)(4x^2-1)}{1\cdot(2x^2-11x+5)}$

$\quad = \dfrac{(x-5)(2x+1)(2x-1)}{1\cdot(2x-1)(x-5)}$

$\quad = \dfrac{(x-5)(2x+1)(2x-1)}{1\cdot(2x-1)(x-5)}$

$\quad = 2x+1$

58. $\dfrac{(a+7)(a+1)}{3a-7}$

59. $\dfrac{x^2-4}{x} \div \dfrac{x-2}{x+2} = \dfrac{x^2-4}{x} \cdot \dfrac{x+2}{x-2}$

$\qquad = \dfrac{(x^2-4)(x+2)}{x(x-2)}$

$\qquad = \dfrac{(x-2)(x+2)(x+2)}{x(x-2)}$

$\qquad = \dfrac{(x-2)(x+2)(x+2)}{x(x-2)}$

$\qquad = \dfrac{(x+2)^2}{x}$

60. $\dfrac{(x+y)^2}{x^2+y}$

61. $\qquad \dfrac{x^2+7x+12}{x^2-x-20} \div \dfrac{x^2+6x+9}{x^2-10x+25}$

$\quad = \dfrac{x^2+7x+12}{x^2-x-20} \cdot \dfrac{x^2-10x+25}{x^2+6x+9}$

$\quad = \dfrac{(x^2+7x+12)(x^2-10x+25)}{(x^2-x-20)(x^2+6x+9)}$

$\quad = \dfrac{(x+3)(x+4)(x-5)(x-5)}{(x+4)(x-5)(x+3)(x+3)}$

$\quad = \dfrac{(x+3)(x+4)(x-5)(x-5)}{(x+4)(x-5)(x+3)(x+3)}$

$\quad = \dfrac{x-5}{x+3}$

62. $\dfrac{a-1}{a+1}$

63. $\qquad \dfrac{t^2+4}{t^2-6t+9} \div \dfrac{t^4+16}{t^2+6t+9}$

$\quad = \dfrac{t^2+4}{t^2-6t+9} \cdot \dfrac{t^2+6t+9}{t^4+16}$

$\quad = \dfrac{(t^2+4)(t^2+6t+9)}{(t^2-6t+9)(t^4+16)}$

$\quad = \dfrac{(t^2+4)(t+3)^2}{(t-3)^2(t^4+16)}$

64. $\dfrac{(x+2)(x-2)(x-1)}{(x+1)^2(x^4+1)}$

65. $\qquad \dfrac{c^2+10c+21}{c^2-2c-15} \div (c^2+2c-35)$

$\quad = \dfrac{c^2+10c+21}{c^2-2c-25} \cdot \dfrac{1}{c^2+2c-35}$

$\quad = \dfrac{(c^2+10c+21)\cdot 1}{(c^2-2c-15)(c^2+2c-35)}$

$\quad = \dfrac{(c+7)(c+3)}{(c-5)(c+3)(c+7)(c-5)}$

$\quad = \dfrac{(c+7)(c+3)}{(c+7)(c+3)} \cdot \dfrac{1}{(c-5)(c-5)}$

$\quad = \dfrac{1}{(c-5)^2}$

66. $\dfrac{1}{1+2z-z^2}$

67. $\qquad \dfrac{(t+5)^3}{(t-5)^3} \div \dfrac{(t+5)^2}{(t-5)^2}$

$\quad = \dfrac{(t+5)^3}{(t-5)^3} \cdot \dfrac{(t-5)^2}{(t+5)^2}$

$\quad = \dfrac{(t+5)^3(t-5)^2}{(t-5)^3(t+5)^2}$

$\quad = \dfrac{(t+5)^2(t-5)^2}{(t+5)^2(t-5)^2} \cdot \dfrac{t+5}{t-5}$

$\quad = \dfrac{t+5}{t-5}$

68. $\dfrac{y-3}{y+3}$

69. **Familiarize.** Let $x =$ the number.

Translate.

$\underline{\text{Sixteen}}$	$\underset{\text{than}}{\text{more}}$	$\overbrace{\begin{array}{c}\text{the square}\\ \text{of a}\\ \text{number}\end{array}}$	is	$\overbrace{\begin{array}{c}\text{eight times}\\ \text{the number.}\end{array}}$
$\downarrow$	$\downarrow$	$\downarrow$	$\downarrow$	$\downarrow$
16	+	x^2	=	$8x$

Carry out.
$$16 + x^2 = 8x$$
$$x^2 - 8x + 16 = 0$$
$$(x-4)(x-4) = 0$$

$$x - 4 = 0 \quad \text{or} \quad x - 4 = 0$$
$$x = 4 \quad \text{or} \quad x = 4$$

Check. The square of 4, which is 16, plus 16 is 32, and eight times 4 is 32. The number checks.

State. The number is 4.

70. $2x^2 + 16$

71. $(8x^3 - 3x^2 + 7) - (8x^2 + 3x - 5) =$
$8x^3 - 3x^2 + 7 - 8x^2 - 3x + 5 =$
$8x^3 - 11x^2 - 3x + 12$

72. $0.06y^3 - 0.09y^2 + 0.01y - 1$

73. $\dfrac{2}{5} - \left(\dfrac{3}{2}\right)^2 = \dfrac{2}{5} - \dfrac{9}{4} = \dfrac{8}{20} - \dfrac{45}{20} = -\dfrac{37}{20}$

74. $\dfrac{49}{45}$

75. ◈

76. ◈

77. ◈

78. ◈

79. $\dfrac{2a^2 - 5ab}{c - 3d} \div (4a^2 - 25b^2)$

$= \dfrac{2a^2 - 5ab}{c - 3d} \cdot \dfrac{1}{4a^2 - 25b^2}$

$= \dfrac{a(2a - 5b)}{(c - 3d)(2a + 5b)(2a - 5b)}$

$= \dfrac{2a - 5b}{2a - 5b} \cdot \dfrac{a}{(c - 3d)(2a + 5b)}$

$= \dfrac{a}{(c - 3d)(2a + 5b)}$

80. 1

81. $\dfrac{3a^2 - 5ab - 12b^2}{3ab + 4b^2} \div (3b^2 - ab)^2$

$= \dfrac{3a^2 - 5ab - 12b^2}{3ab + 4b^2} \cdot \dfrac{1}{(3b^2 - ab)^2}$

$= \dfrac{(3a + 4b)(a - 3b)}{b(3a + 4b) \cdot [b(3b - a)]^2}$

$= \dfrac{(3a + 4b)(-1)(3b - a)}{b(3a + 4b)(b^2)(3b - a)(3b - a)}$

$= \dfrac{(3a + 4b)(-1)(3b - a)}{b(3a + 4b)(b^2)(3b - a)(3b - a)}$

$= -\dfrac{1}{b^3(3b - a)}, \ \text{or} \ \dfrac{1}{b^3(a - 3b)}$

82. $\dfrac{1}{(x + y)^3(3x + y)}$

83. $\dfrac{y^2 - 4xy}{y - x} \div \dfrac{16x^2y^2 - y^4}{4x^2 - 3xy - y^2} \div \dfrac{4}{x^3y^3}$

$= \dfrac{y^2 - 4xy}{y - x} \cdot \dfrac{4x^2 - 3xy - y^2}{16x^2y^2 - y^4} \cdot \dfrac{x^3y^3}{4}$

$= \dfrac{y(y - 4x)(4x + y)(x - y)(x^3y^3)}{(y - x)(y^2)(4x + y)(4x - y)(4)}$

$= \dfrac{y(-1)(4x - y)(4x + y)(x - y) \cdot x^3 \cdot y \cdot y^2}{-(x - y)(y^2)(4x + y)(4x - y)(4)}$

$= \dfrac{(4x - y)(4x + y)(x - y)(y^2)}{(4x - y)(4x + y)(x - y)(y^2)} \cdot \dfrac{y(-1) \cdot x^3 \cdot y}{(-1)4}$

$= \dfrac{x^3y^2}{4}$

84. $\dfrac{(z + 4)^3}{3(z - 4)^2}$

85. $\dfrac{x^2 - x + xy - y}{x^2 + 6x - 7} \div \dfrac{x^2 + 2xy + y^2}{4x + 4y}$

$= \dfrac{x^2 - x + xy - y}{x^2 + 6x - 7} \cdot \dfrac{4x + 4y}{x^2 + 2xy + y^2}$

$= \dfrac{x(x - 1) + y(x - 1)}{x^2 + 6x - 7} \cdot \dfrac{4x + 4y}{x^2 + 2xy + y^2}$

$= \dfrac{(x - 1)(x + y) \cdot 4(x + y)}{(x + 7)(x - 1)(x + y)(x + y)}$

$= \dfrac{(x - 1)(x + y)(x + y)}{(x - 1)(x + y)(x + y)} \cdot \dfrac{4}{x + 7}$

$= \dfrac{4}{x + 7}$

86. $\dfrac{x(x^2 + 1)}{3(x + y - 1)}$

87. $\dfrac{t^4 - 1}{t^4 - 81} \cdot \dfrac{t^2 - 9}{t^2 + 1} \div \dfrac{(t + 1)^2}{(t - 9)^2}$

$= \dfrac{t^4 - 1}{t^4 - 81} \cdot \dfrac{t^2 - 9}{t^2 + 1} \cdot \dfrac{(t - 9)^2}{(t + 1)^2}$

$= \dfrac{(t^4 - 1)(t^2 - 9)(t - 9)^2}{(t^4 - 81)(t^2 + 1)(t + 1)^2}$

$= \dfrac{(t^2 + 1)(t + 1)(t - 1)(t + 3)(t - 3)(t - 9)(t - 9)}{(t^2 + 9)(t + 3)(t - 3)(t^2 + 1)(t + 1)(t + 1)}$

$= \dfrac{(t^2 + 1)(t + 1)(t - 1)(t + 3)(t - 3)(t - 9)(t - 9)}{(t^2 + 9)(t + 3)(t - 3)(t^2 + 1)(t + 1)(t + 1)}$

$= \dfrac{(t - 1)(t - 9)(t - 9)}{(t^2 + 9)(t + 1)}, \ \text{or}$

$\dfrac{(t - 1)(t - 9)^2}{(t^2 + 9)(t + 1)}$

88. 1

89.
$$\left(\frac{y^2+5y+6}{y^2}\cdot\frac{3y^3+6y^2}{y^2-y-12}\right)\div\frac{y^2-y}{y^2-2y-8}$$

$$=\frac{y^2+5y+6}{y^2}\cdot\frac{3y^3+6y^2}{y^2-y-12}\cdot\frac{y^2-2y-8}{y^2-y}$$

$$=\frac{(y+3)(y+2)(3y^2)(y+2)(y-4)(y+2)}{y^2(y-4)(y+3)(y)(y-1)}$$

$$=\frac{y^2(y-4)(y+3)}{y^2(y-4)(y+3)}\cdot\frac{3(y+2)(y+2)(y+2)}{y(y-1)}$$

$$=\frac{3(y+2)^3}{y(y-1)}$$

90. $\dfrac{a-3b}{c}$

Exercise Set 6.3

1. $\dfrac{4}{x}+\dfrac{9}{x}=\dfrac{13}{x}$ Adding numerators

2. $\dfrac{13}{a^2}$

3. $\dfrac{x}{15}+\dfrac{2x+1}{15}=\dfrac{3x+1}{15}$ Adding numerators

4. $\dfrac{4a-4}{7}$

5. $\dfrac{5}{a+3}+\dfrac{1}{a+3}=\dfrac{6}{a+3}$

6. $\dfrac{13}{x+2}$

7. $\dfrac{9}{a+2}-\dfrac{3}{a+2}=\dfrac{6}{a+2}$ Subtracting numerators

8. $\dfrac{6}{x+7}$

9.
$$\frac{3y+9}{2y}-\frac{y+1}{2y}$$

$$=\frac{3y+9-(y+1)}{2y}$$

$$=\frac{3y+9-y-1}{2y}\quad\text{Removing parentheses}$$

$$=\frac{2y+8}{2y}$$

$$=\frac{2(y+4)}{2y}$$

$$=\frac{2(y+4)}{2\cdot y}$$

$$=\frac{y+4}{y}$$

10. $\dfrac{t+4}{4t}$

11.
$$\frac{9x+8}{x+1}+\frac{2x+3}{x+1}$$

$$=\frac{11x+11}{x+1}\qquad\text{Adding numerators}$$

$$=\frac{11(x+1)}{x+1}\qquad\text{Factoring}$$

$$=\frac{11(x+1)}{x+1}\qquad\text{Removing a factor equal to 1}$$

$$=11$$

12. 5

13.
$$\frac{9x+8}{x+1}-\frac{2x+3}{x+1}=\frac{9x+8-(2x+3)}{x+1}$$

$$=\frac{9x+8-2x-3}{x+1}$$

$$=\frac{7x+5}{x+1}$$

14. $\dfrac{a+6}{a+4}$

15.
$$\frac{a^2}{a-4}+\frac{a-20}{a-4}=\frac{a^2+a-20}{a-4}$$

$$=\frac{(a+5)(a-4)}{a-4}$$

$$=\frac{(a+5)(a-4)}{a-4}$$

$$=a+5$$

16. $x+2$

17.
$$\frac{x^2}{x-2}-\frac{6x-8}{x-2}=\frac{x^2-(6x-8)}{x-2}$$

$$=\frac{x^2-6x+8}{x-2}$$

$$=\frac{(x-4)(x-2)}{x-2}$$

$$=\frac{(x-4)(x-2)}{x-2}$$

$$=x-4$$

18. $a-5$

19.
$$\frac{t^2+4t}{t-1}+\frac{2t-7}{t-1}=\frac{t^2+6t-7}{t-1}$$

$$=\frac{(t+7)(t-1)}{t-1}$$

$$=\frac{(t+7)(t-1)}{t-1}$$

$$=t+7$$

20. $y+6$

21. $\dfrac{x+1}{x^2+5x+6} + \dfrac{2}{x^2+5x+6} = \dfrac{x+3}{x^2+5x+6}$

$= \dfrac{x+3}{(x+3)(x+2)}$

$= \dfrac{\cancel{x+3}}{\cancel{(x+3)}(x+2)}$

$= \dfrac{1}{x+2}$

22. $\dfrac{1}{x-1}$

23. $\dfrac{a^2+5}{a^2+5a-6} - \dfrac{6}{a^2+5a-6} = \dfrac{a^2-1}{a^2+5a-6}$

$= \dfrac{(a+1)(a-1)}{(a+6)(a-1)}$

$= \dfrac{(a+1)\cancel{(a-1)}}{(a+6)\cancel{(a-1)}}$

$= \dfrac{a+1}{a+6}$

24. $\dfrac{a+3}{a-4}$

25. $\dfrac{t^2-3t}{t^2+6t+9} + \dfrac{2t-12}{t^2+6t+9} = \dfrac{t^2-t-12}{t^2+6t+9}$

$= \dfrac{(t-4)(t+3)}{(t+3)^2}$

$= \dfrac{(t-4)\cancel{(t+3)}}{(t+3)\cancel{(t+3)}}$

$= \dfrac{t-4}{t+3}$

26. $\dfrac{y-5}{y+4}$

27. $\dfrac{2x^2+x}{x^2-8x+12} - \dfrac{x^2-2x+10}{x^2-8x+12}$

$= \dfrac{2x^2+x-(x^2-2x+10)}{x^2-8x+12}$

$= \dfrac{2x^2+x-x^2+2x-10}{x^2-8x+12}$

$= \dfrac{x^2+3x-10}{x^2-8x+12}$

$= \dfrac{(x+5)(x-2)}{(x-6)(x-2)}$

$= \dfrac{(x+5)\cancel{(x-2)}}{(x-6)\cancel{(x-2)}}$

$= \dfrac{x+5}{x-6}$

28. $\dfrac{x+6}{x-5}$

29. $\dfrac{7-2x}{x^2-6x+8} + \dfrac{3-3x}{x^2-6x+8}$

$= \dfrac{10-5x}{x^2-6x+8}$

$= \dfrac{5(2-x)}{(x-4)(x-2)}$

$= \dfrac{5(-1)(x-2)}{(x-4)(x-2)}$

$= \dfrac{5(-1)\cancel{(x-2)}}{(x-4)\cancel{(x-2)}}$

$= \dfrac{-5}{x-4}$

30. $\dfrac{-5}{t-4}$

31. $\dfrac{x-7}{x^2+3x-4} - \dfrac{2x-3}{x^2+3x-4}$

$= \dfrac{x-7-(2x-3)}{x^2+3x-4}$

$= \dfrac{x-7-2x+3}{x^2+3x-4}$

$= \dfrac{-x-4}{x^2+3x-4}$

$= \dfrac{-(x+4)}{(x+4)(x-1)}$

$= \dfrac{-1\cancel{(x+4)}}{\cancel{(x+4)}(x-1)}$

$= \dfrac{-1}{x-1}$

32. $\dfrac{-4}{x-1}$

33. $15 = 3 \cdot 5$
$27 = 3 \cdot 3 \cdot 3$
$LCM = 3 \cdot 3 \cdot 3 \cdot 5$, or 135

34. 30

35. $8 = 2 \cdot 2 \cdot 2$
$9 = 3 \cdot 3$
$LCM = 2 \cdot 2 \cdot 2 \cdot 3 \cdot 3$, or 72

36. 60

37. $6 = 2 \cdot 3$
$9 = 3 \cdot 3$
$21 = 3 \cdot 7$
$LCM = 2 \cdot 3 \cdot 3 \cdot 7$, or 126

38. 360

39. $6x^2 = 2 \cdot 3 \cdot x \cdot x$
$12x^3 = 2 \cdot 2 \cdot 3 \cdot x \cdot x \cdot x$
$LCM = 2 \cdot 2 \cdot 3 \cdot x \cdot x \cdot x$, or $12x^3$

40. $8a^2b^2$

41. $15a^4b^7 = 3 \cdot 5 \cdot a \cdot a \cdot a \cdot a \cdot b \cdot b \cdot b \cdot b \cdot b \cdot b \cdot b$

$10a^2b^8 = 2 \cdot 5 \cdot a \cdot a \cdot b \cdot b \cdot b \cdot b \cdot b \cdot b \cdot b \cdot b$

$\text{LCM} = 2 \cdot 3 \cdot 5 \cdot a \cdot a \cdot a \cdot a \cdot b \cdot b \cdot b \cdot b \cdot b \cdot b \cdot b \cdot b,$

$\quad\quad \text{or } 30a^4b^8$

42. $18a^5b^7$

43. $2(y-3) = 2 \cdot (y-3)$

$6(y-3) = 2 \cdot 3 \cdot (y-3)$

$\text{LCM} = 2 \cdot 3 \cdot (y-3), \text{ or } 6(y-3)$

44. $8(x-1)$

45. $x^2 - 4 = (x+2)(x-2)$

$x^2 + 5x + 6 = (x+3)(x+2)$

$\text{LCM} = (x+2)(x-2)(x+3)$

46. $(x+2)(x+1)(x-2)$

47. $t^3 + 4t^2 + 4t = t(t^2 + 4t + 4) = t(t+2)(t+2)$

$t^2 - 4t = t(t-4)$

$\text{LCM} = t(t+2)(t+2)(t-4) = t(t+2)^2(t-4)$

48. $y^2(y+1)(y-1)$

49. $10x^2y = 2 \cdot 5 \cdot x \cdot x \cdot y$

$6y^2z = 2 \cdot 3 \cdot y \cdot y \cdot z$

$5xz^3 = 5 \cdot x \cdot z \cdot z \cdot z$

$\text{LCM} = 2 \cdot 3 \cdot 5 \cdot x \cdot x \cdot y \cdot y \cdot z \cdot z \cdot z = 30x^2y^2z^3$

50. $24x^3y^5z^2$

51. $a + 1 = a + 1$

$(a-1)^2 = (a-1)(a-1)$

$a^2 - 1 = (a+1)(a-1)$

$\text{LCM} = (a+1)(a-1)(a-1) = (a+1)(a-1)^2$

52. $2(x+y)^2(x-y)$

53. $m^2 - 5m + 6 = (m-3)(m-2)$

$m^2 - 4m + 4 = (m-2)(m-2)$

$\text{LCM} = (m-3)(m-2)(m-2) = (m-3)(m-2)^2$

54. $(2x+1)(x+2)(x-1)$

55. $10v^2 + 30v = 10v(v+3) = 2 \cdot 5 \cdot v(v+3)$

$5v^2 + 35v + 60 = 5(v^2 + 7v + 12)$

$\quad\quad\quad\quad\quad\quad = 5(v+4)(v+3)$

$\text{LCM} = 2 \cdot 5 \cdot v(v+3)(v+4) = 10v(v+3)(v+4)$

56. $12a(a+2)(a+3)$

57. $9x^3 - 9x^2 - 18x = 9x(x^2 - x - 2)$

$\quad\quad\quad\quad\quad\quad = 3 \cdot 3 \cdot x(x-2)(x+1)$

$6x^5 - 24x^4 + 24x^3 = 6x^3(x^2 - 4x + 4)$

$\quad\quad\quad\quad\quad\quad = 2 \cdot 3 \cdot x \cdot x \cdot x(x-2)(x-2)$

$\text{LCM} = 2 \cdot 3 \cdot 3 \cdot x \cdot x \cdot x(x-2)(x-2)(x+1) =$

$18x^3(x-2)^2(x+1)$

58. $x^3(x+2)^2(x-2)$

59. $6x^5 = 2 \cdot 3 \cdot x \cdot x \cdot x \cdot x \cdot x$

$12x^3 = 2 \cdot 2 \cdot 3 \cdot x \cdot x \cdot x$

The LCD is $2 \cdot 2 \cdot 3 \cdot x \cdot x \cdot x \cdot x \cdot x$, or $12x^5$.

The factor of the LCD that is missing from the first denominator is 2. We multiply by 1 using 2/2:

$$\frac{13}{6x^5} \cdot \frac{2}{2} = \frac{26}{12x^5}$$

The second denominator is missing two factors of x, or x^2. We multiply by 1 using x^2/x^2:

$$\frac{y}{12x^3} \cdot \frac{x^2}{x^2} = \frac{x^2y}{12x^5}$$

60. $\dfrac{3a^3}{10a^6}, \ \dfrac{2b}{10a^6}$

61. $2a^2b = 2 \cdot a \cdot a \cdot b$

$8ab^2 = 2 \cdot 2 \cdot 2 \cdot a \cdot b \cdot b$

The LCD is $2 \cdot 2 \cdot 2 \cdot a \cdot a \cdot b \cdot b$, or $8a^2b^2$.

We multiply the first expression by $\dfrac{4b}{4b}$ to obtain the LCD:

$$\frac{3}{2a^2b} \cdot \frac{4b}{4b} = \frac{12b}{8a^2b^2}$$

We multiply the second expression by a/a to obtain the LCD:

$$\frac{5}{8ab^2} \cdot \frac{a}{a} = \frac{5a}{8a^2b^2}$$

62. $\dfrac{21y}{9x^4y^3}, \ \dfrac{4x^3}{9x^4y^3}$

63. The LCD is $(x+2)(x-2)(x+3)$. (See Exercise 45.)

$$\frac{x+1}{x^2-4} = \frac{x+1}{(x+2)(x-2)} \cdot \frac{x+3}{x+3}$$

$$= \frac{(x+1)(x+3)}{(x+2)(x-2)(x+3)}$$

$$\frac{x-2}{x^2+5x+6} = \frac{x-2}{(x+3)(x+2)} \cdot \frac{x-2}{x-2}$$

$$= \frac{(x-2)^2}{(x+3)(x+2)(x-2)}$$

64. $\dfrac{(x-4)(x+8)}{(x+3)(x-3)(x+8)}, \ \dfrac{(x+2)(x-3)}{(x+3)(x+8)(x-3)}$

65. $x^2 - 19x + 60$

Since the last term is positive and the middle term is negative, we look for a pair of negative factors of 60 whose sum is -19. The numbers we need are -4 and -15.

$$x^2 - 19x + 60 = (x - 4)(x - 15)$$

66. $(x + 12)(x - 3)$

67. The shaded area has dimensions $x - 6$ by $x - 3$. Then the area is $(x - 6)(x - 3)$, or $x^2 - 9x + 18$.

68. $s^2 - \pi r^2$

69. $-\dfrac{7}{24} + \dfrac{5}{18} = -\dfrac{21}{72} + \dfrac{20}{72} = -\dfrac{1}{72}$

70. $-\dfrac{13}{60}$

71. ◈

72. ◈

73. ◈

74. ◈

75.
$$\frac{x+y}{x^2-y^2} + \frac{x-y}{x^2-y^2} - \frac{2x}{x^2-y^2}$$
$$= \frac{x+y+x-y-2x}{x^2-y^2}$$
$$= \frac{0}{x^2-y^2}$$
$$= 0$$

76. $\dfrac{18x+5}{x-1}$

77.
$$\frac{2x+11}{x-3} \cdot \frac{3}{x+4} + \frac{-1}{4+x} \cdot \frac{6x+3}{x-3}$$
$$= \frac{3(2x+11)}{(x-3)(x+4)} + \frac{-1(6x+3)}{(4+x)(x-3)}$$
$$= \frac{3(2x+11) - 1(6x+3)}{(x-3)(x+4)}$$
$$= \frac{6x+33 - 6x - 3}{(x-3)(x+4)}$$
$$= \frac{30}{(x-3)(x+4)}$$

78. $\dfrac{x}{3x+1}$

79. $72 = 2 \cdot 2 \cdot 2 \cdot 3 \cdot 3$

$90 = 2 \cdot 3 \cdot 3 \cdot 5$

$96 = 2 \cdot 2 \cdot 2 \cdot 2 \cdot 2 \cdot 3$

LCM $= 2 \cdot 2 \cdot 2 \cdot 2 \cdot 2 \cdot 3 \cdot 3 \cdot 5$, or 1440

80. $120(x + 1)(x - 1)^2$, or $120(x+1)(x-1)(1-x)$

81. The number of minutes after 5:00 A.M. when the shuttles will first leave at the same time again is the LCM of their departure intervals, 25 minutes and 35 minutes.

$25 = 5 \cdot 5$

$35 = 5 \cdot 7$

LCM $= 5 \cdot 5 \cdot 7$, or 175

Thus, the shuttles will leave at the same time 175 minutes after 5:00 A.M., or at 7:55 A.M.

82. 24 min

83. ◈

84. ◈

Exercise Set 6.4

1. $\dfrac{3}{x} + \dfrac{4}{x^2} = \dfrac{3}{x} + \dfrac{4}{x \cdot x}$ LCD $= x \cdot x$, or x^2

$$= \frac{3}{x} \cdot \frac{x}{x} + \frac{4}{x \cdot x}$$
$$= \frac{3x+4}{x^2}$$

2. $\dfrac{5x+6}{x^2}$

3. $\left.\begin{array}{l} 6r = 2 \cdot 3 \cdot r \\ 8r = 2 \cdot 2 \cdot 2 \cdot r \end{array}\right\}$LCD $= 2 \cdot 2 \cdot 2 \cdot 3 \cdot r$, or $24r$

$$\frac{5}{6r} - \frac{5}{8r} = \frac{5}{6r} \cdot \frac{4}{4} - \frac{5}{8r} \cdot \frac{3}{3}$$
$$= \frac{20-15}{24r}$$
$$= \frac{5}{24r}$$

4. $\dfrac{-29}{18t}$

5. $\left.\begin{array}{l} xy^2 = x \cdot y \cdot y \\ x^2y = x \cdot x \cdot y \end{array}\right\}$LCD $= x \cdot x \cdot y \cdot y$, or x^2y^2

$$\frac{7}{xy^2} + \frac{3}{x^2y} = \frac{7}{xy^2} \cdot \frac{x}{x} + \frac{3}{x^2y} \cdot \frac{y}{y}$$
$$= \frac{7x+3y}{x^2y^2}$$

6. $\dfrac{2d^2+7c}{c^2d^3}$

7. $\left.\begin{array}{l} 9t^3 = 3 \cdot 3 \cdot t \cdot t \cdot t \\ 6t^2 = 2 \cdot 3 \cdot t \cdot t \end{array}\right\}$LCD $= 2 \cdot 3 \cdot 3 \cdot t \cdot t \cdot t$, or $18t^3$

$$\frac{8}{9t^3} - \frac{5}{6t^2} = \frac{8}{9t^3} \cdot \frac{2}{2} - \frac{5}{6t^2} \cdot \frac{3t}{3t}$$
$$= \frac{16-15t}{18t^3}$$

8. $\dfrac{-2xy - 18}{3x^2y^3}$

9. LCD = 24   (See Example 1.)

$$\dfrac{x+5}{8} + \dfrac{x-3}{12} = \dfrac{x+5}{8} \cdot \dfrac{3}{3} + \dfrac{x-3}{12} \cdot \dfrac{2}{2}$$

$$= \dfrac{3(x+5)}{24} + \dfrac{2(x-3)}{24}$$

$$= \dfrac{3x+15}{24} + \dfrac{2x-6}{24}$$

$$= \dfrac{5x+9}{24} \quad \text{Adding numerators}$$

10. $\dfrac{5x+7}{18}$

11. $\left.\begin{array}{l} 2 = 2 \\ 4 = 2 \cdot 2 \end{array}\right\}$ LCD = 4

$$\dfrac{a+2}{2} - \dfrac{a-4}{4} = \dfrac{a+2}{2} \cdot \dfrac{2}{2} - \dfrac{a-4}{4}$$

$$= \dfrac{2a+4}{4} - \dfrac{a-4}{4}$$

$$= \dfrac{2a+4-(a-4)}{4}$$

$$= \dfrac{2a+4-a+4}{4}$$

$$= \dfrac{a+8}{4}$$

12. $\dfrac{-x-4}{6}$

13. $\left.\begin{array}{l} 3a^2 = 3 \cdot a \cdot a \\ 9a = 3 \cdot 3 \cdot a \end{array}\right\}$ LCD = $3 \cdot 3 \cdot a \cdot a$, or $9a^2$

$$\dfrac{2a-1}{3a^2} + \dfrac{5a+1}{9a} = \dfrac{2a-1}{3a^2} \cdot \dfrac{3}{3} + \dfrac{5a+1}{9a} \cdot \dfrac{a}{a}$$

$$= \dfrac{6a-3}{9a^2} + \dfrac{5a^2+a}{9a^2}$$

$$= \dfrac{5a^2+7a-3}{9a^2}$$

14. $\dfrac{a^2+16a+16}{16a^2}$

15. $\left.\begin{array}{l} 4x = 4 \cdot x \\ x = x \end{array}\right\}$ LCD = $4x$

$$\dfrac{x-1}{4x} - \dfrac{2x+3}{x} = \dfrac{x-1}{4x} - \dfrac{2x+3}{x} \cdot \dfrac{4}{4}$$

$$= \dfrac{x-1}{4x} - \dfrac{8x+12}{4x}$$

$$= \dfrac{x-1-(8x+12)}{4x}$$

$$= \dfrac{x-1-8x-12}{4x}$$

$$= \dfrac{-7x-13}{4x}$$

16. $\dfrac{7z-12}{12z}$

17. $\left.\begin{array}{l} c^2d = c \cdot c \cdot d \\ cd^2 = c \cdot d \cdot d \end{array}\right\}$ LCD = $c \cdot c \cdot d \cdot d$, or c^2d^2

$$\dfrac{2c-d}{c^2d} + \dfrac{c+d}{cd^2} = \dfrac{2c-d}{c^2d} \cdot \dfrac{d}{d} + \dfrac{c+d}{cd^2} \cdot \dfrac{c}{c}$$

$$= \dfrac{d(2c-d) + c(c+d)}{c^2d^2}$$

$$= \dfrac{2cd - d^2 + c^2 + cd}{c^2d^2}$$

$$= \dfrac{c^2 + 3cd - d^2}{c^2d^2}$$

18. $\dfrac{x^2 + 4xy + y^2}{x^2y^2}$

19. $\left.\begin{array}{l} 2x^2y = 2 \cdot x \cdot x \cdot y \\ xy^2 = x \cdot y \cdot y \end{array}\right\}$ LCD = $2 \cdot x \cdot x \cdot y \cdot y$, or $2x^2y^2$

$$\dfrac{5x+3y}{2x^2y} - \dfrac{3x+4y}{xy^2} = \dfrac{5x+3y}{2x^2y} \cdot \dfrac{y}{y} - \dfrac{3x+4y}{xy^2} \cdot \dfrac{2x}{2x}$$

$$= \dfrac{5xy + 3y^2}{2x^2y^2} - \dfrac{6x^2 + 8xy}{2x^2y^2}$$

$$= \dfrac{5xy + 3y^2 - (6x^2 + 8xy)}{2x^2y^2}$$

$$= \dfrac{5xy + 3y^2 - 6x^2 - 8xy}{2x^2y^2}$$

$$= \dfrac{3y^2 - 3xy - 6x^2}{2x^2y^2}$$

(Although $3y^2 - 3xy - 6x^2$ can be factored, doing so will not enable us to simplify the result further.)

20. $\dfrac{4x^2 - 13xt + 9t^2}{3x^2t^2}$

21. The denominators cannot be factored, so the LCD is their product, $(x-1)(x+1)$.

$$\dfrac{2}{x-1} + \dfrac{2}{x+1} = \dfrac{2}{x-1} \cdot \dfrac{x+1}{x+1} + \dfrac{2}{x+1} \cdot \dfrac{x-1}{x-1}$$

$$= \dfrac{2(x+1) + 2(x-1)}{(x-1)(x+1)}$$

$$= \dfrac{2x+2 + 2x - 2}{(x-1)(x+1)}$$

$$= \dfrac{4x}{(x-1)(x+1)}$$

22. $\dfrac{6x}{(x-2)(x+2)}$

23. The denominators cannot be factored, so the LCD is their product, $(z-1)(z+1)$.

$$\frac{2z}{z-1} - \frac{3z}{z+1} = \frac{2z}{z-1} \cdot \frac{z+1}{z+1} - \frac{3z}{z+1} \cdot \frac{z-1}{z-1}$$

$$= \frac{2z^2+2z}{(z-1)(z+1)} - \frac{3z^2-3z}{(z-1)(z+1)}$$

$$= \frac{2z^2+2z-(3z^2-3z)}{(z-1)(z+1)}$$

$$= \frac{2z^2+2z-3z^2+3z}{(z-1)(z+1)}$$

$$= \frac{-z^2+5z}{(z-1)(z+1)}$$

(Although $-z^2+5z$ can be factored, doing so will not enable us to simplify the result further.)

24. $\dfrac{2x-40}{(x+5)(x-5)}$

25. $\left.\begin{array}{l} x+5 = x+5 \\ 4x = 4 \cdot x \end{array}\right\}$ LCD $= 4x(x+5)$

$$\frac{2}{x+5} + \frac{3}{4x} = \frac{2}{x+5} \cdot \frac{4x}{4x} + \frac{3}{4x} \cdot \frac{x+5}{x+5}$$

$$= \frac{2 \cdot 4x + 3(x+5)}{4x(x+5)}$$

$$= \frac{8x + 3x + 15}{4x(x+5)}$$

$$= \frac{11x+15}{4x(x+5)}$$

26. $\dfrac{11x+2}{3x(x+1)}$

27. $\dfrac{8}{x^2-4} - \dfrac{3}{x+2}$

$$= \frac{8}{(x+2)(x-2)} - \frac{3}{x+2}, \text{ LCD } = (x+2)(x-2)$$

$$= \frac{8}{(x+2)(x-2)} - \frac{3}{x+2} \cdot \frac{x-2}{x-2}$$

$$= \frac{8 - 3(x-2)}{(x+2)(x-2)}$$

$$= \frac{8 - 3x + 6}{(x+2)(x-2)}$$

$$= \frac{14 - 3x}{(x+2)(x-2)}$$

28. $\dfrac{3-5t}{2t(t-1)}$

29. $\dfrac{4x}{x^2-25} + \dfrac{x}{x+5}$

$$= \frac{4x}{(x+5)(x-5)} + \frac{x}{x+5} \quad \text{LCD} = (x+5)(x-5)$$

$$= \frac{4x + x(x-5)}{(x+5)(x-5)}$$

$$= \frac{4x + x^2 - 5x}{(x+5)(x-5)}$$

$$= \frac{x^2 - x}{(x+5)(x-5)}$$

(Although x^2-x can be factored, doing so will not enable us to simplify the result further.)

30. $\dfrac{x^2+6x}{(x+4)(x-4)}$

31. $\dfrac{t}{t-3} - \dfrac{5}{4t-12}$

$$= \frac{t}{t-3} - \frac{5}{4(t-3)} \quad \text{LCD } = 4(t-3)$$

$$= \frac{t}{t-3} \cdot \frac{4}{4} - \frac{5}{4(t-3)}$$

$$= \frac{4t-5}{4(t-3)}$$

32. $\dfrac{16}{3(z+4)}$

33. $\dfrac{2}{x+3} + \dfrac{4}{(x+3)^2} \quad \text{LCD } (x+3)^2$

$$= \frac{2}{x+3} \cdot \frac{x+3}{x+3} + \frac{4}{(x+3)^2}$$

$$= \frac{2(x+3)+4}{(x+3)^2}$$

$$= \frac{2x+6+4}{(x+3)^2}$$

$$= \frac{2x+10}{(x+3)^2}$$

(Although $2x+10$ can be factored, doing so will not enable us to simplify the result further.)

34. $\dfrac{3x-1}{(x-1)^2}$

35. $\dfrac{2}{5x^2+5x} - \dfrac{4}{3x+3}$

$$= \frac{2}{5x(x+1)} - \frac{4}{3(x+1)} \quad \text{LCD } = 15x(x+1)$$

$$= \frac{2}{5x(x+1)} \cdot \frac{3}{3} - \frac{4}{3(x+1)} \cdot \frac{5x}{5x}$$

$$= \frac{6-20x}{15x(x+1)}$$

(Although $6-20x$ can be factored, doing so will not enable us to simplify the result further.)

36. $\dfrac{-t-9}{(t+3)(t-3)}$

37.
$$\frac{3a}{4a-20}+\frac{9a}{6a-30}$$

$$=\frac{3a}{2\cdot 2(a-5)}+\frac{9a}{2\cdot 3(a-5)}$$
$$\text{LCD}=2\cdot 2\cdot 3(a-5)$$

$$=\frac{3a}{2\cdot 2(a-5)}\cdot\frac{3}{3}+\frac{9a}{2\cdot 3(a-5)}\cdot\frac{2}{2}$$

$$=\frac{9a+18a}{2\cdot 2\cdot 3(a-5)}$$

$$=\frac{27a}{2\cdot 2\cdot 3(a-5)}$$

$$=\frac{\cancel{3}\cdot 9\cdot a}{2\cdot 3\cdot \cancel{3}(a-5)}$$

$$=\frac{9a}{4(a-5)}$$

38. $\dfrac{11a}{10(a-2)}$

39. $\text{LCD}=(y-t)(y+t)$

$$\frac{t}{y-t}-\frac{y}{y+t}=\frac{t}{y-t}\cdot\frac{y+t}{y+t}-\frac{y}{y+t}\cdot\frac{y-t}{y-t}$$

$$=\frac{t(y+t)-y(y-t)}{(y-t)(y+t)}$$

$$=\frac{ty+t^2-y^2+ty}{(y-t)(y+t)}$$

$$=\frac{t^2+2ty-y^2}{(y-t)(y+t)}$$

40. $\dfrac{-2a^2}{(x+a)(x-a)}$

41. $\text{LCD}=x(x-5)$

$$\frac{x}{x-5}+\frac{x-5}{x}=\frac{x}{x-5}\cdot\frac{x}{x}+\frac{x-5}{x}\cdot\frac{x-5}{x-5}$$

$$=\frac{x^2+(x-5)^2}{x(x-5)}$$

$$=\frac{x^2+x^2-10x+25}{x(x-5)}$$

$$=\frac{2x^2-10x+25}{x(x-5)}$$

42. $\dfrac{2x^2+8x+16}{x(x+4)}$

43.
$$\frac{x}{x^2+5x+6}-\frac{2}{x^2+3x+2}$$

$$=\frac{x}{(x+3)(x+2)}-\frac{2}{(x+2)(x+1)}$$
$$\text{LCD}=(x+3)(x+2)(x+1)$$

$$=\frac{x}{(x+3)(x+2)}\cdot\frac{x+1}{x+1}-\frac{2}{(x+2)(x+1)}\cdot\frac{x+3}{x+3}$$

$$=\frac{x^2+x}{(x+3)(x+2)(x+1)}-\frac{2x+6}{(x+3)(x+2)(x+1)}$$

$$=\frac{x^2+x-(2x+6)}{(x+3)(x+2)(x+1)}$$

$$=\frac{x^2+x-2x-6}{(x+3)(x+2)(x+1)}$$

$$=\frac{x^2-x-6}{(x+3)(x+2)(x+1)}$$

$$=\frac{(x-3)(x+2)}{(x+3)(x+2)(x+1)}$$

$$=\frac{(x-3)\cancel{(x+2)}}{(x+3)\cancel{(x+2)}(x+1)}$$

$$=\frac{x-3}{(x+3)(x+1)}$$

44. $\dfrac{x-5}{(x+3)(x+5)}$

45.
$$\frac{x}{x^2+2x+1}+\frac{1}{x^2+5x+4}$$

$$=\frac{x}{(x+1)(x+1)}+\frac{1}{(x+1)(x+4)}$$
$$\text{LCD}=(x+1)^2(x+4)$$

$$=\frac{x}{(x+1)(x+1)}\cdot\frac{x+4}{x+4}+\frac{1}{(x+1)(x+4)}\cdot\frac{x+1}{x+1}$$

$$=\frac{x(x+4)+1\cdot(x+1)}{(x+1)^2(x+4)}=\frac{x^2+4x+x+1}{(x+1)^2(x+4)}$$

$$=\frac{x^2+5x+1}{(x+1)^2(x+4)}$$

46. $\dfrac{12a-11}{(a+2)(a-1)(a-3)}$

47.
$$\frac{x}{x^2+15x+56}-\frac{6}{x^2+13x+42}$$

$$=\frac{x}{(x+7)(x+8)}-\frac{6}{(x+6)(x+7)}$$
$$\text{LCD}=(x+7)(x+8)(x+6)$$

$$=\frac{x}{(x+7)(x+8)}\cdot\frac{x+6}{x+6}-\frac{6}{(x+6)(x+7)}\cdot\frac{x+8}{x+8}$$

$$=\frac{x^2+6x}{(x+7)(x+8)(x+6)}-\frac{6x+48}{(x+7)(x+8)(x+6)}$$

$$=\frac{x^2+6x-(6x+48)}{(x+7)(x+8)(x+6)}$$

$$=\frac{x^2+6x-6x-48}{(x+7)(x+8)(x+6)}$$

$$=\frac{x^2-48}{(x+7)(x+8)(x+6)}$$

48. $\dfrac{-8x - 88}{(x+1)(x+16)(x+8)}$

49. $\dfrac{3}{x^2 - 9} + \dfrac{2}{x^2 - x - 6}$

$= \dfrac{3}{(x+3)(x-3)} + \dfrac{2}{(x+2)(x-3)}$

$\quad\quad\quad\quad\quad\quad \text{LCD} = (x+3)(x-3)(x+2)$

$= \dfrac{3}{(x+3)(x-3)} \cdot \dfrac{x+2}{x+2} + \dfrac{2}{(x+2)(x-3)} \cdot \dfrac{x+3}{x+3}$

$= \dfrac{3(x+2) + 2(x+3)}{(x+3)(x-3)(x+2)}$

$= \dfrac{3x + 6 + 2x + 6}{(x+3)(x-3)(x+2)}$

$= \dfrac{5x + 12}{(x+3)(x-3)(x+2)}$

50. $\dfrac{3z^2 + 19z - 20}{(z-2)^2(z+3)}$

51. $\dfrac{5}{x-1} - \dfrac{6}{1-x} = \dfrac{5}{x-1} - \dfrac{6}{1-x} \cdot \dfrac{-1}{-1}$

$= \dfrac{5}{x-1} - \dfrac{-6}{-1+x}$

$= \dfrac{5}{x-1} - \dfrac{-6}{x-1}$

$= \dfrac{5 - (-6)}{x-1}$

$= \dfrac{5 + 6}{x-1}$

$= \dfrac{11}{x-1}$

52. $\dfrac{4x - 5}{4}$

53. $\dfrac{t^2}{t-2} + \dfrac{4}{2-t} = \dfrac{t^2}{t-2} + \dfrac{4}{2-t} \cdot \dfrac{-1}{-1}$

$= \dfrac{t^2}{t-2} + \dfrac{-4}{-2+t}$

$= \dfrac{t^2 - 4}{t-2}$

$= \dfrac{(t+2)(t-2)}{t-2}$

$= \dfrac{(t+2)\cancel{(t-2)}}{\cancel{t-2}}$

$= t + 2$

54. $y + 3$

55. $\dfrac{a-3}{a^2 - 25} + \dfrac{a-3}{25 - a^2} = \dfrac{a-3}{a^2 - 25} + \dfrac{a-3}{25 - a^2} \cdot \dfrac{-1}{-1}$

$= \dfrac{a-3}{a^2 - 25} + \dfrac{-a+3}{a^2 - 25}$

$= \dfrac{a - 3 - a + 3}{a^2 - 25}$

$= \dfrac{0}{a^2 - 25}$

$= 0$

56. $\dfrac{2b - 14}{b^2 - 16}$

57. $\dfrac{4-p}{25 - p^2} + \dfrac{p+1}{p-5}$

$= \dfrac{4-p}{(5+p)(5-p)} + \dfrac{p+1}{p-5}$

$= \dfrac{4-p}{(5+p)(5-p)} \cdot \dfrac{-1}{-1} + \dfrac{p+1}{p-5}$

$= \dfrac{p-4}{(p+5)(p-5)} + \dfrac{p+1}{p-5} \quad \text{LCD} = (p+5)(p-5)$

$= \dfrac{p-4}{(p+5)(p-5)} + \dfrac{p+1}{p-5} \cdot \dfrac{p+5}{p+5}$

$= \dfrac{p - 4 + p^2 + 6p + 5}{(p+5)(p-5)}$

$= \dfrac{p^2 + 7p + 1}{(p+5)(p-5)}$

58. $\dfrac{y^2 + 10y + 11}{(y+7)(y-7)}$

59. $\dfrac{5x}{x^2 - 9} - \dfrac{4}{3 - x}$

$= \dfrac{5x}{(x+3)(x-3)} - \dfrac{4}{3-x}$

$= \dfrac{5x}{(x+3)(x-3)} - \dfrac{4}{3-x} \cdot \dfrac{-1}{-1}$

$= \dfrac{5x}{(x+3)(x-3)} - \dfrac{-4}{x-3} \quad \text{LCD} = (x+3)(x-3)$

$= \dfrac{5x}{(x+3)(x-3)} - \dfrac{-4}{x-3} \cdot \dfrac{x+3}{x+3}$

$= \dfrac{5x - (-4)(x+3)}{(x+3)(x-3)}$

$= \dfrac{5x + 4x + 12}{(x+3)(x-3)}$

$= \dfrac{9x + 12}{(x+3)(x-3)}$

(Although $9x + 12$ can be factored, doing so will not enable us to simplify the result further.)

60. $\dfrac{13x + 20}{(4+x)(4-x)}$, or $\dfrac{-13x - 20}{(x+4)(x-4)}$

61. $\dfrac{3x + 2}{3x + 6} + \dfrac{x}{4 - x^2}$

$= \dfrac{3x + 2}{3(x+2)} + \dfrac{x}{(2+x)(2-x)}$

$\quad\quad\quad\quad \text{LCD} = 3(x+2)(2-x)$

$= \dfrac{3x + 2}{3(x+2)} \cdot \dfrac{2-x}{2-x} + \dfrac{x}{(2+x)(2-x)} \cdot \dfrac{3}{3}$

$= \dfrac{(3x+2)(2-x) + x \cdot 3}{3(x+2)(2-x)}$

$= \dfrac{-3x^2 + 4x + 4 + 3x}{3(x+2)(2-x)}$

$= \dfrac{-3x^2 + 7x + 4}{3(x+2)(2-x)}$, or

$\dfrac{3x^2 - 7x - 4}{3(x+2)(x-2)}$

62. $\dfrac{-a-2}{(a+1)(a-1)}$

63. $\dfrac{4-a^2}{a^2-9} - \dfrac{a-2}{3-a}$

$= \dfrac{4-a^2}{(a+3)(a-3)} - \dfrac{a-2}{3-a}$

$= \dfrac{4-a^2}{(a+3)(a-3)} - \dfrac{a-2}{3-a} \cdot \dfrac{-1}{-1}$

$= \dfrac{4-a^2}{(a+3)(a-3)} - \dfrac{2-a}{a-3}$　　LCD $= (a+3)(a-3)$

$= \dfrac{4-a^2}{(a+3)(a-3)} - \dfrac{2-a}{a-3} \cdot \dfrac{a+3}{a+3}$

$= \dfrac{4-a^2-(2a+6-a^2-3a)}{(a+3)(a-3)}$

$= \dfrac{4-a^2-2a-6+a^2+3a}{(a+3)(a-3)}$

$= \dfrac{a-2}{(a+3)(a-3)}$

64. $\dfrac{10x+6y}{(x+y)(x-y)}$

65. $\dfrac{4y}{y^2-1} - \dfrac{2}{y} - \dfrac{2}{y+1}$

$= \dfrac{4y}{(y+1)(y-1)} - \dfrac{2}{y} - \dfrac{2}{y+1}$

　　　LCD $= y(y+1)(y-1)$

$= \dfrac{4y}{(y+1)(y-1)} \cdot \dfrac{y}{y} - \dfrac{2}{y} \cdot \dfrac{(y+1)(y-1)}{(y+1)(y-1)} -$

$\qquad\qquad \dfrac{2}{y+1} \cdot \dfrac{y(y-1)}{y(y-1)}$

$= \dfrac{4y^2-(2y^2-2)-(2y^2-2y)}{y(y+1)(y-1)}$

$= \dfrac{4y^2-2y^2+2-2y^2+2y}{y(y+1)(y-1)}$

$= \dfrac{2y+2}{y(y+1)(y-1)}$

$= \dfrac{2(y+1)}{y(y+1)(y-1)}$

$= \dfrac{2}{y(y-1)}$

66. $\dfrac{2x-3}{2-x}$

67. $\dfrac{2z}{1-2z} + \dfrac{3z}{2z+1} - \dfrac{3}{4z^2-1}$

$= \dfrac{2z}{1-2z} + \dfrac{3z}{2z+1} - \dfrac{3}{(2z+1)(2z-1)}$

$= \dfrac{2z}{1-2z} \cdot \dfrac{-1}{-1} + \dfrac{3z}{2z+1} - \dfrac{3}{(2z+1)(2z-1)}$

$= \dfrac{-2z}{2z-1} + \dfrac{3z}{2z+1} - \dfrac{3}{(2z-1)(2z+1)}$

　　　LCD $= (2z-1)(2z+1)$

$= \dfrac{-2z}{2z-1} \cdot \dfrac{2z+1}{2z+1} + \dfrac{3z}{2z+1} \cdot \dfrac{2z-1}{2z-1} - \dfrac{3}{(2z-1)(2z+1)}$

$= \dfrac{(-4z^2-2z)+(6z^2-3z)-3}{(2z-1)(2z+1)}$

$= \dfrac{2z^2-5z-3}{(2z-1)(2z+1)}$

$= \dfrac{(z-3)(2z+1)}{(2z-1)(2z+1)}$

$= \dfrac{z-3}{2z-1}$

68. 0

69. $\dfrac{2r}{r^2-s^2} + \dfrac{1}{r+s} - \dfrac{1}{r-s}$

$= \dfrac{2r}{(r+s)(r-s)} + \dfrac{1}{r+s} - \dfrac{1}{r-s}$

　　　LCD $= (r+s)(r-s)$

$= \dfrac{2r}{(r+s)(r-s)} + \dfrac{1}{r+s} \cdot \dfrac{r-s}{r-s} - \dfrac{1}{r-s} \cdot \dfrac{r+s}{r+s}$

$= \dfrac{2r+(r-s)-(r+s)}{(r+s)(r-s)}$

$= \dfrac{2r-2s}{(r+s)(r-s)}$

$= \dfrac{2(r-s)}{(r+s)(r-s)}$

$= \dfrac{2}{r+s}$

70. $\dfrac{1}{2c-1}$

71. Graph: $y = \dfrac{1}{2}x - 5$

Since the equation is in the form $y = mx + b$, we know the y−intercept is $(0, -5)$. We find two other solutions, substituting multiples of 2 for x to avoid fractions.

When $x = 2$, $y = \dfrac{1}{2} \cdot 2 - 5 = 1 - 5 = -4$.

When $x = 4$, $y = \dfrac{1}{2} \cdot 4 - 5 = 2 - 5 = -3$.

x	y
0	-5
2	-4
4	-3

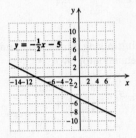

72.

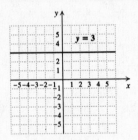

73. Graph: $y = 3$

All solutions are of the form $(x, 3)$. The graph is a line parallel to the x−axis with y−intercept $(0, 3)$.

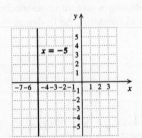

74.

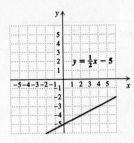

75.
$$3x - 7 = 5x + 9$$
$$-2x - 7 = 9 \qquad \text{Adding } -5x \text{ on both sides}$$
$$-2x = 16 \qquad \text{Adding } 7 \text{ on both sides}$$
$$x = -8 \qquad \text{Dividing by } -2 \text{ on both sides}$$

The solution is -8.

76. $\dfrac{5}{6}$

77. $x^2 - 8x + 15 = 0$
$$(x - 3)(x - 5) = 0$$
$$x - 3 = 0 \ \text{ or } \ x - 5 = 0 \quad \text{Principle of zero products}$$
$$x = 3 \ \text{ or } \qquad x = 5$$

The solutions are 3 and 5.

78. $-2, 9$

79. ◈

80. ◈

81. ◈

82. ◈

83. $P = 2l + 2w$
$$= 2\left(\frac{x}{x+4}\right) + 2\left(\frac{x}{x+5}\right)$$
$$= \frac{2x}{x+4} + \frac{2x}{x+5} \qquad \text{LCD} = (x+4)(x+5)$$
$$= \frac{2x}{x+4} \cdot \frac{x+5}{x+5} + \frac{2x}{x+5} \cdot \frac{x+4}{x+4}$$
$$= \frac{2x^2 + 10x + 2x^2 + 8x}{(x+4)(x+5)}$$
$$= \frac{4x^2 + 18x}{(x+4)(x+5)}$$

(Although $4x^2 + 18x$ can be factored, doing so will not enable us to simplify the result further.)

$$A = lw$$
$$= \left(\frac{x}{x+4}\right)\left(\frac{x}{x+5}\right)$$
$$= \frac{x^2}{(x+4)(x+5)}$$

84. Perimeter: $\dfrac{10x - 14}{(x+4)(x-5)}$; area: $\dfrac{6}{(x+4)(x-5)}$

85.
$$\frac{2x+11}{x-3} \cdot \frac{3}{x+4} + \frac{2x+1}{4+x} \cdot \frac{3}{3-x}$$
$$= \frac{6x+33}{(x-3)(x+4)} + \frac{6x+3}{(4+x)(3-x)}$$
$$= \frac{6x+33}{(x-3)(x+4)} + \frac{6x+3}{(4+x)(3-x)} \cdot \frac{-1}{-1}$$
$$= \frac{6x+33}{(x-3)(x+4)} + \frac{-6x-3}{(x+4)(x-3)}$$
$$= \frac{6x+33-6x-3}{(x-3)(x+4)}$$
$$= \frac{30}{(x-3)(x+4)}$$

86. $\dfrac{x}{3x+1}$

176

Chapter 6: Rational Expressions and Equations

87.

$$\frac{x}{x^4-y^4}-\left(\frac{1}{x+y}\right)^2$$

$$=\frac{x}{(x^2+y^2)(x+y)(x-y)}-\frac{1}{(x+y)^2}$$

$$\text{LCD}=(x^2+y^2)(x+y)^2(x-y)$$

$$=\frac{x}{(x^2+y^2)(x+y)(x-y)}\cdot\frac{x+y}{x+y}-$$
$$\frac{1}{(x+y)^2}\cdot\frac{(x^2+y^2)(x-y)}{(x^2+y^2)(x-y)}$$

$$=\frac{x(x+y)-(x^2+y^2)(x-y)}{(x^2+y^2)(x+y)^2(x-y)}$$

$$=\frac{x^2+xy-(x^3-x^2y+xy^2-y^3)}{(x^2+y^2)(x+y)^2(x-y)}$$

$$=\frac{x^2+xy-x^3+x^2y-xy^2+y^3}{(x^2+y^2)(x+y)^2(x-y)}$$

88. $\dfrac{-3xy-3a+6x}{(y-3)^2(a+2x)(a-2x)}$

89.

$$\frac{2x^2+5x-3}{2x^2-9x+9}+\frac{x+1}{3-2x}+\frac{4x^2+8x+3}{x-3}\cdot\frac{x+3}{9-4x^2}$$

$$=\frac{2x^2+5x-3}{(2x-3)(x-3)}+\frac{x+1}{3-2x}+$$
$$\frac{(4x^2+8x+3)(x+3)}{(x-3)(3+2x)(3-2x)}$$

$$=\frac{2x^2+5x-3}{(2x-3)(x-3)}\cdot\frac{-1}{-1}+\frac{x+1}{3-2x}+$$
$$\frac{4x^3+20x^2+27x+9}{(x-3)(3+2x)(3-2x)}$$

$$=\frac{-2x^2-5x+3}{(3-2x)(x-3)}+\frac{x+1}{3-2x}+\frac{4x^3+20x^2+27x+9}{(x-3)(3+2x)(3-2x)}$$

$$\text{LCD}=(x-3)(3+2x)(3-2x)$$

$$=\frac{-2x^2-5x+3}{(3-2x)(x-3)}\cdot\frac{3+2x}{3+2x}+\frac{x+1}{3-2x}\cdot\frac{(x-3)(3+2x)}{(x-3)(3+2x)}+$$
$$\frac{4x^3+20x^2+27x+9}{(x-3)(3+2x)(3-2x)}$$

$$=[(-4x^3-16x^2-9x+9+2x^3-x^2-12x-9+$$
$$4x^3+20x^2+27x+9)]/$$
$$[(x-3)(3+2x)(3-2x)]$$

$$=\frac{2x^3+3x^2+6x+9}{(x-3)(3+2x)(3-2x)}$$

$$=\frac{x^2(2x+3)+3(2x+3)}{(x-3)(3+2x)(3-2x)}$$

$$=\frac{(2x+3)(x^2+3)}{(x-3)(3+2x)(3-2x)}$$

$$=\frac{x^2+3}{(x-3)(3-2x)},\text{ or }\frac{-x^2-3}{(x-3)(2x-3)}$$

90. $\dfrac{5(a^2+2ab-b^2)}{(a-b)(3a+b)(3a-b)}$

91. Answers may vary. $\dfrac{a}{a-b}+\dfrac{3b}{b-a}$

Exercise Set 6.5

1.

$$\frac{3+\frac{1}{2}}{9-\frac{1}{4}}\qquad\text{LCD is }4$$

$$=\frac{3+\frac{1}{2}}{9-\frac{1}{4}}\cdot\frac{4}{4}\qquad\text{Multiplying by }\frac{4}{4}$$

$$=\frac{\left(3+\frac{1}{2}\right)4}{\left(9-\frac{1}{4}\right)4}\qquad\begin{array}{l}\text{Multiplying numerator and}\\\text{denominator by }4\end{array}$$

$$=\frac{3\cdot4+\frac{1}{2}\cdot4}{9\cdot4-\frac{1}{4}\cdot4}$$

$$=\frac{12+2}{36-1}=\frac{14}{35}$$

$$=\frac{2}{5}$$

2. $\dfrac{4}{25}$

3.

$$\frac{1+\frac{1}{3}}{5-\frac{5}{27}}$$

$$=\frac{1\cdot\frac{3}{3}+\frac{1}{3}}{5\cdot\frac{27}{27}-\frac{5}{27}}\qquad\begin{array}{l}\text{Getting a common denominator}\\\text{in numerator and in denominator}\end{array}$$

$$=\frac{\frac{3}{3}+\frac{1}{3}}{\frac{135}{27}-\frac{5}{27}}$$

$$=\frac{\frac{4}{3}}{\frac{130}{27}}\qquad\begin{array}{l}\text{Adding in the numerator; sub-}\\\text{tracting in the denominator}\end{array}$$

$$=\frac{4}{3}\cdot\frac{27}{130}\qquad\begin{array}{l}\text{Multiplying by the reciprocal of}\\\text{the divisor}\end{array}$$

$$=\frac{2\cdot2\cdot3\cdot9}{3\cdot2\cdot65}$$

$$=\frac{2\cdot2\cdot3\cdot9}{3\cdot2\cdot65}$$

$$=\frac{18}{65}$$

4. 3

5. $\dfrac{\dfrac{s}{3}+s}{\dfrac{3}{s}+s}$ LCD is $3s$

$= \dfrac{\dfrac{s}{3}+s}{\dfrac{3}{s}+s} \cdot \dfrac{3s}{3s}$

$= \dfrac{\left(\dfrac{s}{3}+s\right)(3s)}{\left(\dfrac{3}{s}+s\right)(3s)}$

$= \dfrac{\dfrac{5}{3}\cdot 3s + s\cdot 3s}{\dfrac{3}{s}\cdot 3s + s\cdot 3s}$

$= \dfrac{s^2+3s^2}{9+3s^2}$

$= \dfrac{4s^2}{9+3s^2}$

6. $\dfrac{1-5x}{1+3x}$

7. $\dfrac{\dfrac{3}{x}}{\dfrac{2}{x}+\dfrac{1}{x^2}}$ LCD is x^2

$= \dfrac{\dfrac{3}{x}}{\dfrac{2}{x}+\dfrac{1}{x^2}} \cdot \dfrac{x^2}{x^2}$

$= \dfrac{\dfrac{3}{x}\cdot x^2}{\left(\dfrac{2}{x}+\dfrac{1}{x^2}\right)x^2}$

$= \dfrac{3x}{\dfrac{2}{x}\cdot x^2 + \dfrac{1}{x^2}\cdot x^2}$

$= \dfrac{3x}{2x+1}$

8. $\dfrac{4x-1}{2x}$

9. $\dfrac{\dfrac{2a-5}{3a}}{\dfrac{a-1}{6a}}$

$= \dfrac{2a-5}{3a}\cdot \dfrac{6a}{a-1}$ Multiplying by the reciprocal of the divisor

$= \dfrac{(2a-5)\cdot 2\cdot 3a}{3a\cdot(a-1)}$

$= \dfrac{(2a-5)\cdot 2\cdot 3a}{3a\cdot(a-1)}$

$= \dfrac{2(2a-5)}{a-1}$

$= \dfrac{4a-10}{a-1}$

10. $\dfrac{3a+12}{a^2-2a}$

11. $\dfrac{\dfrac{x}{4}-\dfrac{4}{x}}{\dfrac{1}{4}+\dfrac{1}{x}}$ LCD is $4x$

$= \dfrac{\dfrac{x}{4}-\dfrac{4}{x}}{\dfrac{1}{4}+\dfrac{1}{x}} \cdot \dfrac{4x}{4x}$

$= \dfrac{\dfrac{x}{4}\cdot 4x - \dfrac{4}{x}\cdot 4x}{\dfrac{1}{4}\cdot 4x + \dfrac{1}{x}\cdot 4x}$

$= \dfrac{x^2-16}{x+4}$

$= \dfrac{(x+4)(x-4)}{x+4}$

$= \dfrac{(x+4)(x-4)}{(x+4)\cdot 1}$

$= x-4$

12. $\dfrac{24+3x}{x^2-24}$

13. $\dfrac{\dfrac{1}{3}+\dfrac{1}{x}}{\dfrac{3+x}{3}}$ LCD is $3x$

$= \dfrac{\dfrac{1}{3}+\dfrac{1}{x}}{\dfrac{3+x}{3}} \cdot \dfrac{3x}{3x}$

$= \dfrac{\dfrac{1}{3}\cdot 3x + \dfrac{1}{x}\cdot 3x}{\left(\dfrac{3+x}{3}\right)(3x)}$

$= \dfrac{x+3}{x(3+x)}$

$= \dfrac{(x+3)\cdot 1}{x(3+x)}$ $(3+x=x+3)$

$= \dfrac{1}{x}$

14. $-\dfrac{1}{a}$

15. $\dfrac{\dfrac{1}{t^2}+1}{\dfrac{1}{t}-1}$ LCD is t^2

$= \dfrac{\dfrac{1}{t^2}+1}{\dfrac{1}{t}-1} \cdot \dfrac{t^2}{t^2}$

$= \dfrac{\dfrac{1}{t^2}\cdot t^2 + 1\cdot t^2}{\dfrac{1}{t}\cdot t^2 - 1\cdot t^2}$

$= \dfrac{1+t^2}{t-t^2}$

(Although the denominator can be factored, doing so will not enable us to simplify further.)

16. $\dfrac{2x^2 + x}{2x^2 - 1}$

17. $\dfrac{\dfrac{x^2}{x^2 - y^2}}{\dfrac{x}{x + y}}$

$= \dfrac{x^2}{x^2 - y^2} \cdot \dfrac{x + y}{x}$ Multiplying by the reciprocal of the divisor

$= \dfrac{x^2(x + y)}{(x^2 - y^2)(x)}$

$= \dfrac{x \cdot x \cdot (x + y)}{(x + y)(x - y)(x)}$

$= \dfrac{\cancel{x} \cdot x \cdot \cancel{(x + y)}}{\cancel{(x + y)}(x - y)(\cancel{x})}$

$= \dfrac{x}{x - y}$

18. $\dfrac{a^2 + 3a}{2}$

19. $\dfrac{\dfrac{3}{a} - \dfrac{4}{a^2}}{\dfrac{2}{a^3} + \dfrac{3}{a}}$ LCD is a^3

$= \dfrac{\dfrac{3}{a} - \dfrac{4}{a^2}}{\dfrac{2}{a^3} + \dfrac{3}{a}} \cdot \dfrac{a^3}{a^3}$

$= \dfrac{\dfrac{3}{a} \cdot a^3 - \dfrac{4}{a^2} \cdot a^3}{\dfrac{2}{a^3} \cdot a^3 + \dfrac{3}{a} \cdot a^3}$

$= \dfrac{3a^2 - 4a}{2 + 3a^2}$

(Although the numerator can be factored, doing so will not enable us to simplify further.)

20. $\dfrac{5 + x}{2x^2 - 3x}$

21. $\dfrac{\dfrac{2}{7a^4} - \dfrac{1}{14a}}{\dfrac{3}{5a^2} + \dfrac{2}{15a}} = \dfrac{\dfrac{2}{7a^4} \cdot \dfrac{2}{2} - \dfrac{1}{14a} \cdot \dfrac{a^3}{a^3}}{\dfrac{3}{5a^2} \cdot \dfrac{3}{3} + \dfrac{2}{15a} \cdot \dfrac{a}{a}}$

$= \dfrac{\dfrac{4 - a^3}{14a^4}}{\dfrac{9 + 2a}{15a^2}}$

$= \dfrac{4 - a^3}{14a^4} \cdot \dfrac{15a^2}{9 + 2a}$

$= \dfrac{15 \cdot \cancel{a^2}(4 - a^3)}{14a^2 \cdot \cancel{a^2}(9 + 2a)}$

$= \dfrac{15(4 - a^3)}{14a^2(9 + 2a)}$, or $\dfrac{60 - 15a^3}{126a^2 + 28a^3}$

22. $\dfrac{10 - 3x^2}{12x^2 + 6}$

23. $\dfrac{\dfrac{x}{5y^3} - \dfrac{3}{10y}}{\dfrac{x}{10y} + \dfrac{3}{y^4}} = \dfrac{\dfrac{x}{5y^3} - \dfrac{3}{10y}}{\dfrac{x}{10y} + \dfrac{3}{y^4}} \cdot \dfrac{10y^4}{10y^4}$

$= \dfrac{\dfrac{x}{5y^3} \cdot 10y^4 - \dfrac{3}{10y} \cdot 10y^4}{\dfrac{x}{10y} \cdot 10y^4 + \dfrac{3}{y^4} \cdot 10y^4}$

$= \dfrac{2xy - 3y^3}{xy^3 + 30}$

(Although the numerator can be factored, doing so will not enable us to simplify further.)

24. $\dfrac{3a + 8b}{15b^2 - 2}$

25. $\dfrac{\dfrac{5}{ab^4} + \dfrac{2}{a^3b}}{\dfrac{5}{a^3b} - \dfrac{3}{ab}} = \dfrac{\dfrac{5}{ab^4} \cdot \dfrac{a^2}{a^2} + \dfrac{2}{a^3b} \cdot \dfrac{b^3}{b^3}}{\dfrac{5}{a^3b} - \dfrac{3}{ab} \cdot \dfrac{a^2}{a^2}}$

$= \dfrac{\dfrac{5a^2 + 2b^3}{a^3b^4}}{\dfrac{5 - 3a^2}{a^3b}}$

$= \dfrac{5a^2 + 2b^3}{a^3b^4} \cdot \dfrac{a^3b}{5 - 3a^2}$

$= \dfrac{\cancel{a^3}\cancel{b}(5a^2 + 2b^3)}{\cancel{a^3}\cancel{b} \cdot b^3(5 - 3a^2)}$

$= \dfrac{5a^2 + 2b^3}{b^3(5 - 3a^2)}$, or $\dfrac{5a^2 + 2b^3}{5b^3 - 3a^2b^3}$

26. $\dfrac{2y^2 + 3xy}{2x + y^2}$

27. $\dfrac{2 - \dfrac{3}{x^2}}{2 + \dfrac{3}{x^4}} = \dfrac{2 - \dfrac{3}{x^2}}{2 + \dfrac{3}{x^4}} \cdot \dfrac{x^4}{x^4}$

$= \dfrac{2 \cdot x^4 - \dfrac{3}{x^2} \cdot x^4}{2 \cdot x^4 + \dfrac{3}{x^4} \cdot x^4}$

$= \dfrac{2x^4 - 3x^2}{2x^4 + 3}$

28. $\dfrac{3a^4 - 2}{2a^4 + 3a}$

29.
$$\cfrac{t - \cfrac{2}{t}}{t + \cfrac{5}{t}} = \cfrac{t \cdot \cfrac{t}{t} - \cfrac{2}{t}}{t \cdot \cfrac{t}{t} + \cfrac{5}{t}}$$

$$= \cfrac{\cfrac{t^2 - 2}{t}}{\cfrac{t^2 + 5}{t}}$$

$$= \frac{t^2 - 2}{t} \cdot \frac{t}{t^2 + 5}$$

$$= \frac{\cancel{t}(t^2 - 2)}{\cancel{t}(t^2 + 5)}$$

$$= \frac{t^2 - 2}{t^2 + 5}$$

30. $\dfrac{x^2 + 3}{x^2 - 2}$

31.
$$\cfrac{7 - \cfrac{5}{ab^3}}{\cfrac{4 + a}{a^2 b}} = \cfrac{7 - \cfrac{5}{ab^3}}{\cfrac{4 + a}{a^2 b}} \cdot \frac{a^2 b^3}{a^2 b^3}$$

$$= \frac{7 \cdot a^2 b^3 - \dfrac{5}{ab^3} \cdot a^2 b^3}{\dfrac{4 + a}{a^2 b} \cdot a^2 b^3}$$

$$= \frac{7a^2 b^3 - 5a}{b^2(4 + a)}, \text{ or } \frac{7a^2 b^3 - 5a}{4b^2 + ab^2}$$

32. $\dfrac{5x^3 y + 3x}{3 + x}$

33.
$$\cfrac{\cfrac{a - 7}{a^3}}{\cfrac{3}{a^2} + \cfrac{2}{a}} = \cfrac{\cfrac{a - 7}{a^3}}{\cfrac{3}{a^2} + \cfrac{2}{a} \cdot \cfrac{a}{a}}$$

$$= \cfrac{\cfrac{a - 7}{a^3}}{\cfrac{3 + 2a}{a^2}}$$

$$= \frac{a - 7}{a^3} \cdot \frac{a^2}{3 + 2a}$$

$$= \frac{\cancel{a^2}(a - 7)}{a \cdot \cancel{a^2}(3 + 2a)}$$

$$= \frac{a - 7}{a(3 + 2a)}, \text{ or } \frac{a - 7}{3a + 2a^2}$$

34. $\dfrac{x + 5}{2x - 3}$

35.
$$\cfrac{1 + \cfrac{a}{5b^2}}{\cfrac{a}{10b} - 1} = \cfrac{1 + \cfrac{a}{5b^2}}{\cfrac{a}{10b} - 1} \cdot \frac{10b^2}{10b^2}$$

$$= \frac{1 \cdot 10b^2 + \dfrac{a}{5b^2} \cdot 10b^2}{\dfrac{a}{10b} \cdot 10b^2 - 1 \cdot 10b^2}$$

$$= \frac{10b^2 + 2a}{ab - 10b^2}$$

(Although both the numerator and denominator can be factored, doing so will not enable us to simplify further.)

36. $\dfrac{x - 2}{x - 3}$

37.
$$\cfrac{x - 2 - \cfrac{x}{3}}{x + 7 - \cfrac{4}{5x}} = \cfrac{x - 2 + \cfrac{x}{3}}{x + 7 - \cfrac{4}{5x}} \cdot \frac{15x}{15x}$$

$$= \frac{x \cdot 15x - 2 \cdot 15x + \dfrac{x}{3} \cdot 15x}{x \cdot 15x + 7 \cdot 15x - \dfrac{4}{5x} \cdot 15x}$$

$$= \frac{15x^2 - 30x + 5x^2}{15x^2 + 105x - 12}$$

$$= \frac{20x^2 - 30x}{15x^2 + 105x - 12}$$

38. $\dfrac{a^2 + 5a - 3}{a^2 - 3a + 5}$

39.
$$3x - 5 + 2(4x - 1) = 12x - 3$$
$$3x - 5 + 8x - 2 = 12x - 3$$
$$11x - 7 = 12x - 3$$
$$-7 = x - 3$$
$$-4 = x$$

The solution is -4.

40. -4

41. $(5x^4 - 6x^3 + 23x^2 - 79x + 24) - (-18x^4 - 56x^3 + 84x - 17) = 5x^4 - 6x^3 + 23x^2 - 79x + 24 + 18x^4 + 56x^3 - 84x + 17 = 23x^4 + 50x^3 + 23x^2 - 163x + 41$

42. 14 yd

43. ◈

44. ◈

45. ◈

46. ◈

47.
$$\frac{1}{\dfrac{2}{x-1} - \dfrac{1}{3x-2}}$$

$$= \frac{1}{\dfrac{2}{x-1} - \dfrac{1}{3x-2}} \cdot \frac{(x-1)(3x-2)}{(x-1)(3x-2)}$$

$$= \frac{(x-1)(3x-2)}{\left(\dfrac{2}{x-1} - \dfrac{1}{3x-2}\right)(x-1)(3x-2)}$$

$$= \frac{(x-1)(3x-2)}{\dfrac{2}{x-1}(x-1)(3x-2) - \dfrac{1}{3x-2}(x-1)(3x-2)}$$

$$= \frac{(x-1)(3x-2)}{2(3x-2) - (x-1)}$$

$$= \frac{(x-1)(3x-2)}{6x-4-x+1}$$

$$= \frac{(x-1)(3x-2)}{5x-3}$$

48. $-\dfrac{ac}{bd}$

49.
$$\frac{\dfrac{a}{b} + \dfrac{c}{d}}{\dfrac{b}{a} + \dfrac{d}{c}} = \frac{\dfrac{a}{b} \cdot \dfrac{d}{d} + \dfrac{c}{d} \cdot \dfrac{b}{b}}{\dfrac{b}{a} \cdot \dfrac{c}{c} + \dfrac{d}{c} \cdot \dfrac{a}{a}}$$

$$= \frac{\dfrac{ad+bc}{bd}}{\dfrac{bc+ad}{ac}}$$

$$= \frac{ad+bc}{bd} \cdot \frac{ac}{bc+ad}$$

$$= \frac{(ad+bc)(ac)}{bd(bc+ad)}$$

$$= \frac{ac}{bd}$$

50. x^5

51.
$$1 + \frac{1}{1 + \dfrac{1}{1 + \dfrac{1}{x}}} = 1 + \frac{1}{1 + \dfrac{1}{\dfrac{x+1}{x}}}$$

$$= 1 + \frac{1}{1 + \dfrac{x}{x+1}}$$

$$= 1 + \frac{1}{\dfrac{x+1+x}{x+1}}$$

$$= 1 + \frac{1}{\dfrac{2x+1}{x+1}}$$

$$= 1 + \frac{x+1}{2x+1}$$

$$= \frac{2x+2+x+1}{2x+1}$$

$$= \frac{3x+2}{2x+1}$$

52. $\dfrac{-2z(5z-2)}{(2+z)(-13z+6)}$

53. ◈

54. ◼

Exercise Set 6.6

1. $\dfrac{4}{5} - \dfrac{2}{3} = \dfrac{x}{9}$, LCD = 45

$$45\left(\frac{4}{5} - \frac{2}{3}\right) = 45 \cdot \frac{x}{9}$$

$$45 \cdot \frac{4}{5} - 45 \cdot \frac{2}{3} = 45 \cdot \frac{x}{9}$$

$$36 - 30 = 5x$$

$$6 = 5x$$

$$\frac{6}{5} = x$$

Check:
$$\frac{\dfrac{4}{5} - \dfrac{2}{3} = \dfrac{x}{9}}{\dfrac{4}{5} - \dfrac{2}{3} \;?\; \dfrac{\dfrac{6}{5}}{9}}$$

$$\frac{12}{15} - \frac{10}{15} \;\bigg|\; \frac{6}{5} \cdot \frac{1}{9}$$

$$\frac{2}{15} \;\bigg|\; \frac{2}{15} \qquad \text{TRUE}$$

This checks, so the solution is $\dfrac{6}{5}$.

2. $-\dfrac{17}{2}$

3. $\dfrac{3}{5} + \dfrac{1}{8} = \dfrac{1}{x}$, LCD = 40x

$$40x\left(\frac{3}{5} + \frac{1}{8}\right) = 40x \cdot \frac{1}{x}$$

$$40x \cdot \frac{3}{5} + 40x \cdot \frac{1}{8} = 40x \cdot \frac{1}{x}$$

$$24x + 5x = 40$$

$$29x = 40$$

$$x = \frac{40}{29}$$

Check:
$$\frac{\dfrac{3}{5} + \dfrac{1}{8} = \dfrac{1}{x}}{\dfrac{3}{5} + \dfrac{1}{8} \;?\; \dfrac{1}{\dfrac{40}{29}}}$$

$$\frac{24}{40} + \frac{5}{40} \;\bigg|\; 1 \cdot \frac{29}{40}$$

$$\frac{29}{40} \;\bigg|\; \frac{29}{40} \qquad \text{TRUE}$$

This checks, so the solution is $\dfrac{40}{29}$.

4. $\frac{2}{3}$

5.
$$\frac{1}{8} + \frac{1}{10} = \frac{1}{t}, \text{ LCD} = 40t$$
$$40t\left(\frac{1}{8} + \frac{1}{10}\right) = 40t \cdot \frac{1}{t}$$
$$40t \cdot \frac{1}{8} + 40t \cdot \frac{1}{10} = 40t \cdot \frac{1}{t}$$
$$5t + 4t = 40$$
$$9t = 40$$
$$t = \frac{40}{9}$$

Check:

$$\frac{\frac{1}{8} + \frac{1}{10} = \frac{1}{t}}{\begin{array}{c|c} \frac{1}{8} + \frac{1}{10} & \frac{1}{\frac{40}{9}} \\ \frac{5}{40} + \frac{4}{40} & 1 \cdot \frac{9}{40} \\ \frac{9}{40} & \frac{9}{40} \quad \text{TRUE} \end{array}}$$

This checks, so the solution is $\frac{40}{9}$.

6. $\frac{24}{7}$

7.
$$x + \frac{3}{x} = -4, \text{ LCD} = x$$
$$x\left(x + \frac{3}{x}\right) = -4 \cdot x$$
$$x \cdot x + x \cdot \frac{3}{x} = -4 \cdot x$$
$$x^2 + 3 = -4x$$
$$x^2 + 4x + 3 = 0$$
$$(x+3)(x+1) = 0$$
$$x + 3 = 0 \quad or \quad x + 1 = 0$$
$$x = -3 \quad or \quad x = -1$$

Check:

$$\frac{x + \frac{3}{x} = -4}{\begin{array}{c|c} -3 + \frac{3}{-3} & -4 \\ -3 - 1 & \\ -4 & -4 \text{ TRUE} \end{array}} \qquad \frac{x + \frac{3}{x} = -4}{\begin{array}{c|c} -1 + \frac{3}{-1} & -4 \\ -1 - 3 & \\ -4 & -4 \text{ TRUE} \end{array}}$$

Both of these check, so the two solutions are -3 and -1.

8. $-4, -1$

9.
$$\frac{x}{7} - \frac{7}{x} = 0, \text{ LCD} = 7x$$
$$7x\left(\frac{x}{7} - \frac{7}{x}\right) = 7x \cdot 0$$
$$7x \cdot \frac{x}{7} - 7x \cdot \frac{7}{x} = 7x \cdot 0$$
$$x^2 - 49 = 0$$
$$(x+7)(x-7) = 0$$
$$x + 7 = 0 \quad or \quad x - 7 = 0$$
$$x = -7 \quad or \quad x = 7$$

Check:

$$\frac{\frac{x}{7} - \frac{7}{x} = 0}{\begin{array}{c|c} \frac{-7}{7} - \frac{7}{-7} & 0 \\ -1 - (-1) & \\ -1 + 1 & \\ 0 & 0 \text{ TRUE} \end{array}} \qquad \frac{\frac{x}{7} - \frac{7}{x} = 0}{\begin{array}{c|c} \frac{7}{7} - \frac{7}{7} & 0 \\ 1 - 1 & \\ 0 & 0 \text{ TRUE} \end{array}}$$

Both of these check, so the two solutions are -7 and 7.

10. $-6, 6$

11.
$$\frac{4}{x} = \frac{5}{x} - \frac{1}{2}$$
$$2x\left(\frac{4}{x}\right) = 2x\left(\frac{5}{x} - \frac{1}{2}\right)$$
$$2x\left(\frac{4}{x}\right) = 2x \cdot \frac{5}{x} - 2x \cdot \frac{1}{2}$$
$$8 = 10 - x$$
$$x = 2$$

Check:

$$\frac{\frac{4}{x} = \frac{5}{x} - \frac{1}{2}}{\begin{array}{c|c} \frac{4}{2} & \frac{5}{2} - \frac{1}{2} \\ 2 & \frac{4}{2} \\ 2 & 2 \quad \text{TRUE} \end{array}}$$

This checks, so the solution is 2.

12. 3

13.
$$\frac{3}{4x} + \frac{5}{x} = 1, \text{ LCD} = 4x$$
$$4x\left(\frac{3}{4x} + \frac{5}{x}\right) = 4x \cdot 1$$
$$4x \cdot \frac{3}{4x} + 4x \cdot \frac{5}{x} = 4x \cdot 1$$
$$3 + 20 = 4x$$
$$23 = 4x$$
$$\frac{23}{4} = x$$

Check:

$$\frac{3}{4x} + \frac{5}{x} = 1$$

$$\frac{\frac{3}{4 \cdot \frac{23}{4}} + \frac{5}{\frac{23}{4}} \;\bigg|\; ?\; 1}{}$$

$$\frac{3}{23} + 5 \cdot \frac{4}{23}$$

$$\frac{3}{23} + \frac{20}{23}$$

$$\frac{23}{23}$$

$$1 \quad\bigg|\quad 1 \quad \text{TRUE}$$

This checks, so the solution is $\frac{23}{4}$.

14. $\frac{14}{3}$

15.
$$\frac{a-2}{a+3} = \frac{3}{8}, \text{LCD} = 8(a+3)$$

$$8(a+3) \cdot \frac{a-2}{a+3} = 8(a+3) \cdot \frac{3}{8}$$

$$8(a-2) = 3(a+3)$$

$$8a - 16 = 3a + 9$$

$$5a = 25$$

$$a = 5$$

Check:

$$\frac{a-2}{a+3} = \frac{3}{8}$$

$$\frac{\frac{5-2}{5+3} \;\bigg|\; ?\; \frac{3}{8}}{}$$

$$\frac{3}{8} \quad\bigg|\quad \frac{3}{8} \quad \text{TRUE}$$

This checks, so the solution is 5.

16. 10

17.
$$\frac{5}{x-1} = \frac{3}{x+2},$$
$$\text{LCD} = (x-1)(x+2)$$

$$(x-1)(x+2) \cdot \frac{5}{x-1} = (x-1)(x+2) \cdot \frac{3}{x+2}$$

$$5(x+2) = 3(x-1)$$

$$5x + 10 = 3x - 3$$

$$2x = -13$$

$$x = -\frac{13}{2}$$

This checks, so the solution is $-\frac{13}{2}$.

18. 5

19.
$$\frac{x}{8} - \frac{x}{12} = \frac{1}{8}, \text{LCD} = 24$$

$$24\left(\frac{x}{8} - \frac{x}{12}\right) = 24 \cdot \frac{1}{8}$$

$$24 \cdot \frac{x}{8} - 24 \cdot \frac{x}{12} = 24 \cdot \frac{1}{8}$$

$$3x - 2x = 3$$

$$x = 3$$

This checks, so the solution is 3.

20. $\frac{5}{2}$

21.
$$\frac{x+2}{5} - 1 = \frac{x-2}{4}, \text{LCD} = 20$$

$$20\left(\frac{x+2}{5} - 1\right) = 20 \cdot \frac{x-2}{4}$$

$$20 \cdot \frac{x+2}{5} - 20 \cdot 1 = 20 \cdot \frac{x-2}{4}$$

$$4(x+2) - 20 = 5(x-2)$$

$$4x + 8 - 20 = 5x - 10$$

$$4x - 12 = 5x - 10$$

$$-2 = x$$

This checks, so the solution is -2.

22. -1

23.
$$\frac{x-7}{x-9} = \frac{2}{x-9}$$

$$(x-9) \cdot \frac{x-7}{x-9} = (x-9) \cdot \frac{2}{x-9}$$

$$x - 7 = 2$$

$$x = 9$$

Check:

$$\frac{x-7}{x-9} = \frac{2}{x-9}$$

$$\frac{\frac{9-7}{9-9} \;\bigg|\; ?\; \frac{2}{9-9}}{}$$

$$\frac{2}{0} \quad\bigg|\quad \frac{2}{0} \quad \text{UNDEFINED}$$

The number 9 is not a solution of the original equation because it results in division by 0. The equation has no solution.

24. No solution

25.
$$\frac{3}{x+4} = \frac{5}{x}, \text{LCD} = x(x+4)$$

$$x(x+4) \cdot \frac{3}{x+4} = x(x+4) \cdot \frac{5}{x}$$

$$3x = 5(x+4)$$

$$3x = 5x + 20$$

$$-2x = 20$$

$$x = -10$$

This checks, so the solution is -10.

26. $-\dfrac{21}{5}$

27.
$$\frac{2b-3}{3b+2} = \frac{2b+1}{3b-2},$$
$$\text{LCD} = (3b+2)(3b-2)$$
$$(3b+2)(3b-2)\cdot\frac{2b-3}{3b+2} = (3b+2)(3b-2)\cdot\frac{2b+1}{3b-2}$$
$$(3b-2)(2b-3) = (3b+2)(2b+1)$$
$$6b^2 - 13b + 6 = 6b^2 + 7b + 2$$
$$-13b + 6 = 7b + 2$$
$$4 = 20b$$
$$\frac{1}{5} = b$$

This checks, so the solution is $\dfrac{1}{5}$.

28. $\dfrac{5}{3}$

29.
$$\frac{4}{x-3} + \frac{2x}{x^2-9} = \frac{1}{x+3},$$
$$\text{LCD} = (x-3)(x+3)$$
$$(x-3)(x+3)\left(\frac{4}{x-3} + \frac{2x}{(x+3)(x-3)}\right) =$$
$$(x-3)(x+3)\cdot\frac{1}{x+3}$$
$$4(x+3) + 2x = x - 3$$
$$4x + 12 + 2x = x - 3$$
$$6x + 12 = x - 3$$
$$5x = -15$$
$$x = -3$$

The number -3 is not a solution of the original equation because it results in division by 0. There is no solution.

30. No solution

31.
$$\frac{5}{y-3} - \frac{30}{y^2-9} = 1$$
$$\frac{5}{y-3} - \frac{30}{(y+3)(y-3)} = 1,$$
$$\text{LCD} = (y-3)(y+3)$$
$$(y-3)(y+3)\left(\frac{5}{y-3} - \frac{30}{(y+3)(y-3)}\right) =$$
$$(y-3)(y+3)\cdot 1$$
$$5(y+3) - 30 = (y+3)(y-3)$$
$$5y + 15 - 30 = y^2 - 9$$
$$0 = y^2 - 5y + 6$$
$$0 = (y-3)(y-2)$$
$$y - 3 = 0 \ \text{ or } \ y - 2 = 0$$
$$y = 3 \ \text{ or } \quad y = 2$$

Only 2 checks, so the solution is 2.

32. $\dfrac{1}{2}$

33.
$$\frac{4}{8-a} = \frac{4-a}{a-8}$$
$$\frac{-1}{-1}\cdot\frac{4}{8-a} = \frac{4-a}{a-8}$$
$$\frac{-4}{a-8} = \frac{4-a}{a-8}, \ \text{LCD} = a-8$$
$$(a-8)\cdot\frac{-4}{a-8} = (a-8)\cdot\frac{4-a}{a-8}$$
$$-4 = 4 - a$$
$$-8 = -a$$
$$8 = a$$

The number 8 is not a solution of the original equation because it results in division by 0. There is no solution.

34. -13

35.
$$\left(\frac{3}{5}\right)^{-2} = \frac{3^{-2}}{5^{-2}}$$
$$= \frac{5^2}{3^2}$$
$$= \frac{25}{9}$$

36. $-\dfrac{1}{64}$

37. $(a^2b^5)^{-3} = a^{-6}b^{-15}$, or $\dfrac{1}{a^6b^{15}}$

38. $x^8 y^{12}$

39.
$$\frac{5}{x^2-9} - \frac{2}{x^2-5x+6}$$
$$= \frac{5}{(x+3)(x-3)} - \frac{2}{(x-2)(x-3)},$$
$$\text{LCD} = (x+3)(x-3)(x-2)$$
$$= \frac{5}{(x+3)(x-3)}\cdot\frac{x-2}{x-2} - \frac{2}{(x-2)(x-3)}\cdot\frac{x+3}{x+3}$$
$$= \frac{5(x-2) - 2(x+3)}{(x+3)(x-3)(x-2)}$$
$$= \frac{5x - 10 - 2x - 6}{(x+3)(x-3)(x-2)}$$
$$= \frac{3x - 16}{(x+3)(x-3)(x-2)}$$

40. $\dfrac{2x+23}{3(x-4)(x+4)}$

41. ◈

42. ◈

43.

44. ◈

45.
$$x + 1 + \frac{x-1}{x-3} = \frac{2}{x-3},$$
$$\text{LCD} = x - 3$$
$$(x-3)\left(x + 1 + \frac{x-1}{x-3}\right) = \frac{2}{x-3} \cdot (x-3)$$
$$x(x-3) + 1 \cdot (x-3) + x - 1 = 2$$
$$x^2 - 3x + x - 3 + x - 1 = 2$$
$$x^2 - x - 4 = 2$$
$$x^2 - x - 6 = 0$$
$$(x-3)(x+2) = 0$$
$$x - 3 = 0 \quad or \quad x + 2 = 0$$
$$x = 3 \quad or \qquad x = -2$$
Only -2 checks. The solution is -2.

46. 7

47.
$$\frac{x}{x^2 + 3x - 4} + \frac{x+1}{x^2 + 6x + 8} =$$
$$\frac{2x}{x^2 + x - 2}$$
$$\frac{x}{(x+4)(x-1)} + \frac{x+1}{(x+4)(x+2)} =$$
$$\frac{2x}{(x+2)(x-1)},$$
$$\text{LCD} = (x+4)(x-1)(x+2)$$
$$(x+4)(x-1)(x+2)\left(\frac{x}{(x+4)(x-1)} + \frac{x+1}{(x+4)(x+2)}\right) =$$
$$(x+4)(x-1)(x+2) \cdot \frac{2x}{(x+2)(x-1)}$$
$$x(x+2) + (x+1)(x-1) = 2x(x+4)$$
$$x^2 + 2x + x^2 - 1 = 2x^2 + 8x$$
$$2x^2 + 2x - 1 = 2x^2 + 8x$$
$$2x - 1 = 8x$$
$$-1 = 6x$$
$$-\frac{1}{6} = x$$
This checks, so the solution is $-\frac{1}{6}$.

48. 3

49.
$$\frac{x^2}{x^2 - 4} = \frac{x}{x+2} - \frac{2x}{2-x}$$
$$\frac{x^2}{x^2 - 4} = \frac{x}{x+2} - \frac{2x}{2-x} \cdot \frac{-1}{-1}$$
$$\frac{x^2}{(x+2)(x-2)} = \frac{x}{x+2} - \frac{-2x}{x-2},$$
$$\text{LCD} = (x+2)(x-2)$$
$$(x+2)(x-2) \cdot \frac{x^2}{(x+2)(x-2)} =$$
$$(x+2)(x-2)\left(\frac{x}{x+2} - \frac{-2x}{x-2}\right)$$
$$x^2 = x(x-2) - (-2x)(x+2)$$
$$x^2 = x^2 - 2x + 2x^2 + 4x$$
$$x^2 = 3x^2 + 2x$$
$$0 = 2x^2 + 2x$$
$$0 = 2x(x+1)$$
$$2x = 0 \quad or \quad x + 1 = 0$$
$$x = 0 \quad or \qquad x = -1$$
Both of these check, so the solutions are -1 and 0.

50. 4

51.
$$\frac{1}{x-1} + x + 1 = \frac{5x-4}{x-1}, \quad \text{LCD} = x - 1$$
$$(x-1)\left(\frac{1}{x-1} + x + 1\right) = (x-1) \cdot \frac{5x-4}{x-1}$$
$$1 + x(x-1) + x - 1 = 5x - 4$$
$$1 + x^2 - x + x - 1 = 5x - 4$$
$$x^2 = 5x - 4$$
$$x^2 - 5x + 4 = 0$$
$$(x-1)(x-4) = 0$$
$$x - 1 = 0 \quad or \quad x - 4 = 0$$
$$x = 1 \quad or \qquad x = 4$$
Only 4 checks, so the solution is 4.

52. -6

53. ▨

54. ▨

Exercise Set 6.7

1. *Familiarize.* Let x = the number.

Translate.

$$\underbrace{\text{A number}}_{} \quad \text{minus} \quad \underbrace{\text{five times its reciprocal}}_{} \quad \text{is} \quad 4.$$

$$\downarrow \qquad\qquad \downarrow \qquad\qquad\quad \downarrow \qquad\qquad \downarrow \quad \downarrow$$

$$x \qquad\qquad - \qquad\qquad 5 \cdot \frac{1}{x} \qquad\quad = \quad 4$$

Carry out.

$$x - \frac{5}{x} = 4$$

$$x\left(x - \frac{5}{x}\right) = x \cdot 4 \quad \text{Multiplying by the LCD}$$

$$x^2 - 5 = 4x$$

$$x^2 - 4x - 5 = 0$$

$$(x - 5)(x + 1) = 0$$

$$x - 5 = 0 \quad or \quad x + 1 = 0$$

$$x = 5 \quad or \qquad x = -1$$

Check. Five times the reciprocal of 5 is $5 \cdot \frac{1}{5}$, or 1. Since $5 - 1 = 4$, the number 5 is a solution. Five times the reciprocal of -1 is $5(-1)$, or -5. Since $-1 - (-5) = -1 + 5$, or 4, the number -1 is a solution.

State. The solutions are 5 and -1.

2. $-1, 4$

3. *Familiarize*. Let $x =$ the number.

Translate. We reword the problem.

$$\underbrace{\text{A number}}_{\downarrow} \quad \underset{\downarrow}{\text{plus}} \quad \underbrace{\text{its reciprocal}}_{\downarrow} \quad \underset{\downarrow}{\text{is}} \quad \underset{\downarrow}{2}.$$

$$x \quad + \quad \frac{1}{x} \quad = \quad 2$$

Carry out. We solve the equation.

$$x + \frac{1}{x} = 2$$

$$x\left(x + \frac{1}{x}\right) = x \cdot 2 \quad \text{Multiplying by the LCD}$$

$$x^2 + 1 = 2x$$

$$x^2 - 2x + 1 = 0$$

$$(x - 1)(x - 1) = 0$$

$$x - 1 = 0 \quad or \quad x - 1 = 0$$

$$x = 1 \quad or \qquad x = 1$$

Check. The reciprocal of 1 is 1. Since $1 + 1 = 2$, the number 1 is a solution.

State. The solution is 1.

4. 1, 5

5. *Familiarize*. The job takes Juanita 12 hours working alone and Antoine 16 hours working alone. Then in 1 hour Juanita does $\frac{1}{12}$ of the job and Antoine does $\frac{1}{16}$ of the job. Working together, they can do $\frac{1}{12} + \frac{1}{16}$, or $\frac{7}{48}$ of the job in 1 hour. In four hours, Juanita does $4 \cdot \frac{1}{12}$ of the job and Antoine does $4 \cdot \frac{1}{16}$ of the job. Working together they can do $4 \cdot \frac{1}{12} + 4 \cdot \frac{1}{16}$, or $\frac{7}{12}$ of the job in 4 hours. In 7

hours they can do $7 \cdot \frac{1}{12} + 7 \cdot \frac{1}{16}$, or $\frac{49}{48}$ or $1\frac{1}{48}$ of the job which is more of the job than needs to be done. The answer is somewhere between 4 hr and 7 hr.

Translate. If they work together t hours, then Juanita does $t\left(\frac{1}{12}\right)$ of the job and Antoine does $t\left(\frac{1}{16}\right)$ of the job. We want some number t such that

$$t\left(\frac{1}{12}\right) + t\left(\frac{1}{16}\right) = 1.$$

Carry out. We solve the equation.

$$\frac{t}{12} + \frac{t}{16} = 1, \text{ LCD} = 48$$

$$48\left(\frac{t}{12} + \frac{t}{16}\right) = 48 \cdot 1$$

$$4t + 3t = 48$$

$$7t = 48$$

$$t = \frac{48}{7}, \text{ or } 6\frac{6}{7}$$

Check. The check can be done by repeating the computations. We also have a partial check in that we expected from our familiarization step that the answer would be between 4 hr and 7 hr.

State. Working together, it takes them $6\frac{6}{7}$ hr to build the shed.

6. $2\frac{2}{9}$ hr

7. *Familiarize*. The job takes Vern 45 min working alone and Nina 60 min working alone. Then in 1 minute Vern does $\frac{1}{45}$ of the job and Nina does $\frac{1}{60}$ of the job. Working together, they can do $\frac{1}{45} + \frac{1}{60}$, or $\frac{7}{180}$ of the job in 1 minute. In 20 minutes, Vern does $\frac{20}{45}$ of the job and Nina does $\frac{20}{60}$ of the job. Working together, they can do $\frac{20}{45} + \frac{20}{60}$, or $\frac{7}{9}$ of the job. In 30 minutes, they can do $\frac{30}{45} + \frac{30}{60}$, or $\frac{7}{6}$ of the job which is more of the job than needs to be done. The answer is somewhere between 20 minutes and 30 minutes.

Translate. If they work together t minutes, then Vern does $t\left(\frac{1}{45}\right)$ of the job and Nina does $t\left(\frac{1}{60}\right)$ of the job. We want some number t such that

$$t\left(\frac{1}{45}\right) + t\left(\frac{1}{60}\right) = 1.$$

Carry out. We solve the equation.

$$\frac{t}{45} + \frac{t}{60} = 1, \text{ LCD} = 180$$

$$180\left(\frac{t}{45} + \frac{t}{60}\right) = 180 \cdot 1$$

$$4t + 3t = 180$$

$$7t = 180$$

$$t = \frac{180}{7}, \text{ or } 25\frac{5}{7}$$

Check. The check can be done by repeating the computations. We also have a partial check in that we expected from our familiarization step that the answer would be between 20 minutes and 30 minutes.

State. It would take them $25\frac{5}{7}$ minutes to complete the job working together.

8. $1\frac{5}{7}$ hr

9. ***Familiarize***. The job takes Raul 48 hours working along and Mira 36 hours working alone. Then in 1 hour Raul does $\frac{1}{48}$ of the job and Mira does $\frac{1}{36}$ of the job. Working together they can do $\frac{1}{48} + \frac{1}{36}$, or $\frac{7}{144}$ of the job in 1 hour. In two hours, Raul does $2\left(\frac{1}{48}\right)$ of the job and Mira does $2\left(\frac{1}{36}\right)$ of the job. Working together they can do $2\left(\frac{1}{48}\right) + 2\left(\frac{1}{36}\right)$, or $\frac{7}{72}$ of the job in two hours. In 10 hours, they can do $\frac{35}{72}$ of the job. In 25 hours, they can do $\frac{175}{144}$, or $1\frac{31}{144}$ of the job which is more of the job than needs to be done. The answer is somewhere between 10 hr and 25 hr.

Translate. If they work together t hours, then Raul does $t\left(\frac{1}{48}\right)$ of the job and Mira does $t\left(\frac{1}{36}\right)$ of the job. We want some number t such that

$$t\left(\frac{1}{48}\right) + t\left(\frac{1}{36}\right) = 1.$$

Carry out. We solve the equation.

$$\frac{t}{48} + \frac{t}{36} = 1, \text{ LCD} = 144$$

$$144\left(\frac{t}{48} + \frac{t}{36}\right) = 144 \cdot 1$$

$$3t + 4t = 144$$

$$7t = 144$$

$$t = \frac{144}{7}, \text{ or } 20\frac{4}{7}$$

Check. The check can be done by repeating the computations. We also have a partial check in that we expected from our familiarization step that the answer would be between 10 hr and 25 hr.

State. Working together, it takes them $20\frac{4}{7}$ hr to complete the job.

10. $5\frac{1}{7}$ hr

11. ***Familiarize***. Let $t =$ the number of minutes it takes Bobbi and Blanche to pick a quart of raspberries, working together.

Translate. We use the work principle.

$$\frac{t}{20} + \frac{t}{25} = 1$$

Carry out. We solve the equation.

$$\frac{t}{20} + \frac{t}{25} = 1, \text{ LCD} = 100$$

$$100\left(\frac{t}{20} + \frac{t}{25}\right) = 100 \cdot 1$$

$$5t + 4t = 100$$

$$9t = 100$$

$$t = \frac{100}{9}, \text{ or } 11\frac{1}{9}$$

Check. Note that in $\frac{100}{9}$ min, the portion of the job done is $\frac{1}{20} \cdot \frac{100}{9} + \frac{1}{25} \cdot \frac{100}{9} = \frac{5}{9} + \frac{4}{9} = 1$ job.

State. It would take $11\frac{1}{9}$ min for Bobbi and Blanche to pick a quart of raspberries, working together.

12. $3\frac{3}{7}$ hr

13. ***Familiarize***. Let $t =$ the number of minutes it would take to fax the report if the two fax machines worked together.

Translate. We use the work principle.

$$\frac{t}{10} + \frac{t}{8} = 1$$

Carry out. We solve the equation.

$$\frac{t}{10} + \frac{t}{8} = 1, \text{ LCD} = 40$$

$$40\left(\frac{t}{10} + \frac{t}{8}\right) = 40 \cdot 1$$

$$4t + 5t = 40$$

$$9t = 40$$

$$t = \frac{40}{9}, \text{ or } 4\frac{4}{9}$$

Check. Note that in $\frac{40}{9}$ min, the portion of the job done is $\frac{1}{10} \cdot \frac{40}{9} + \frac{1}{8} \cdot \frac{40}{9} = \frac{4}{9} + \frac{5}{9} = 1$ job.

State. It would take $4\frac{4}{9}$ min for the two machines to fax the report, working together.

14. $22\frac{2}{9}$ min

15. *Familiarize*. We complete the table shown in the text.

$$d \quad = \quad r \quad \cdot \quad t$$

	Distance	Speed	Time
Train	300	r	t
Police car	700	$r + 40$	t

***Translate*.** We can replace the t's in the table above using the formula $t = d/r$.

	Distance	Speed	Time
Train	300	r	$\dfrac{300}{r}$
Police car	700	$r + 40$	$\dfrac{700}{r+40}$

Since times are the same for both vehicles, we have the equation

$$\frac{300}{r} = \frac{700}{r + 40}.$$

***Carry out*.** We multiply by the LCD, $r(r + 40)$.

$$r(r+40) \cdot \frac{300}{r} = r(r+40) \cdot \frac{700}{r+40}$$

$$300(r + 40) = 700r$$

$$300r + 12{,}000 = 700r$$

$$12{,}000 = 400r$$

$$30 = r$$

If r is 30 mph, then $r + 40$ is 70 mph.

***Check*.** If the train travels 30 mph and the police car travels 70 mph, then the police car is traveling 40 mph faster than the train. The time for the train is 300/30, or 10 hr. The time for the police car is 700/70, or 10 hr. The times are the same. The values check.

***State*.** The speed of the train is 30 mph, and the speed of the police car is 70 mph.

16. Bill: 50 mph, Hillary: 80 mph

17. *Familiarize*. We complete the table shown in the text.

$$d \quad = \quad r \quad \cdot \quad t$$

	Distance	Speed	Time
Freight	330	$r - 14$	t
Passenger	400	r	t

***Translate*.** We can replace the t's in the table above using the formula $t = d/r$.

	Distance	Speed	Time
Freight	330	$r - 14$	$\dfrac{330}{r-14}$
Passenger	400	r	$\dfrac{400}{r}$

Since the times are the same for both trains, we have the equation

$$\frac{330}{r - 14} = \frac{400}{r}.$$

***Carry out*.** We multiply by the LCD, $r(r - 14)$.

$$r(r - 14) \cdot \frac{330}{r - 14} = r(r - 14) \cdot \frac{400}{r}$$

$$330r = 400(r - 14)$$

$$330r = 400r - 5600$$

$$-70r = -5600$$

$$r = 80$$

If $r = 80$, then $r - 14 = 66$.

***Check*.** If the passenger train travels 80 km/h and the freight train travels 66 km/h, the freight train's speed is 14 km/h slower than the passenger train's speed. The freight train's time is 330/66, or 5 hr. The passenger train's time is 400/80, or 5 hr. Since the times are the same the answer checks.

***State*.** The speed of the passenger train is 80 km/h. The speed of the freight train is 66 km/h.

18. Passenger: 80 km/h, freight: 65 km/h

19. *Familiarize*. Let r = Ted's speed and t = the time the bicyclists travel, in hours. Organize the information in a table.

	Distance	Speed	Time
Ted	42	r	t
Joni	57	$r + 5$	t

***Translate*.** We can replace the t's in the table above using the formula $r = d/t$.

	Distance	Speed	Time
Ted	42	r	$\dfrac{42}{r}$
Joni	57	$r + 5$	$\dfrac{57}{r+5}$

Since the times are the same for both bicyclists, we have the equation

$$\frac{42}{r} = \frac{57}{r + 5}.$$

***Carry out*.** We multiply by the LCD, $r(r + 5)$.

$$r(r + 5) \cdot \frac{42}{r} = r(r + 5) \cdot \frac{57}{r + 5}$$

$$42(r + 5) = 57r$$

$$42r + 210 = 57r$$

$$210 = 15r$$

$$14 = r$$

If $r = 14$, then $r + 5 = 19$.

***Check*.** If Ted's speed is 14 km/h and Joni's speed is 19 km/h, then Joni bicycles 5 km/h faster than Ted. Ted's time is 42/14, or 3 hr. Joni's time is 57/19, or 3 hr. Since the times are the same, the answer checks.

State. Ted travels at 14 km/h, and Joni travels at 19 km/h.

20. Mark's: 35 km/h, Ellie's: 40 km/h

21. Familiarize. Let t = the time it takes Caledonia to drive to town and organize the given information in a table.

	Distance	Speed	Time
Caledonia	15	r	t
Manley	20	r	$t+1$

Translate. We can replace the r's in the table above using the formula $r = d/t$.

	Distance	Speed	Time
Caledonia	15	$\frac{15}{t}$	t
Manley	20	$\frac{20}{t+1}$	$t+1$

Since the speeds are the same for both riders, we have the equation

$$\frac{15}{t} = \frac{20}{t+1}.$$

Carry out. We multiply by the LCD, $t(t+1)$.

$$t(t+1) \cdot \frac{15}{t} = t(t+1) \cdot \frac{20}{t+1}$$
$$15(t+1) = 20t$$
$$15t + 15 = 20t$$
$$15 = 5t$$
$$3 = t$$

If $t = 3$, then $t + 1 = 3 + 1$, or 4.

Check. If Caledonia's time is 3 hr and Manley's time is 4 hr, then Manley's time is 1 hr more than Caledonia's. Caledonia's speed is 15/3, or 5 mph. Manley's speed is 20/4, or 5 mph. Since the speeds are the same, the answer checks.

State. It takes Caledonia 3 hr to drive to town.

22. $1\frac{1}{3}$ hr

23. $\frac{750 \text{ students}}{50 \text{ faculty}} = 15$ students per faculty member

24. $16\frac{\text{mi}}{\text{gal}}$

25. $\frac{5440 \text{ ft}}{5 \text{ sec}} = 1088$ ft/sec

26. 2.3 km/h

27.
$$\frac{a}{b} = \frac{c}{d}$$
$$\frac{8}{5} = \frac{6}{d}$$
$$5d \cdot \frac{8}{5} = 5d \cdot \frac{6}{d}$$
$$8d = 30$$
$$d = \frac{30}{8} = \frac{15}{4} \text{ cm}$$

28. $\frac{90}{7}$ cm

29. Let $c = b + 2$ and $d = b - 2$.
$$\frac{a}{b} = \frac{c}{d}$$
$$\frac{15}{b} = \frac{b+2}{b-2}$$
$$b(b-2) \cdot \frac{15}{b} = b(b-2) \cdot \frac{b+2}{b-2}$$
$$15(b-2) = b(b+2)$$
$$15b - 30 = b^2 + 2b$$
$$0 = b^2 - 13b + 30$$
$$0 = (b-3)(b-10)$$
$$b - 3 = 0 \ \text{ or } \ b - 10 = 0$$
$$b = 3 \ \text{ or } \qquad b = 10$$

If $b = 3$ m, then $c = 3 + 2$, or 5 m and $d = 3 - 2$, or 1 m.

If $b = 10$ m, then $c = 10 + 2$, or 12 m and $d = 10 - 2$, or 8 m.

30. If $b = 3$ m, $c = 6$ m, $d = 1$ m; if $b = 12$ m, $c = 15$ m, $d = 10$ m

31. Familiarize. The coffee beans from 14 trees are required to produce 7.7 kilograms of coffee, and we wish to find how many trees are required to produce 320 kilograms of coffee. We can set up ratios:
$$\frac{T}{320} \qquad \frac{14}{7.7}$$

Translate. Assuming the two ratios are the same, we can translate to a proportion.

$$\begin{array}{c} \text{Trees} \rightarrow \\ \text{Kilograms} \rightarrow \end{array} \frac{T}{320} = \frac{14}{7.7} \begin{array}{c} \leftarrow \text{Trees} \\ \leftarrow \text{Kilograms} \end{array}$$

Carry out. We solve the proportion.
$$320 \cdot \frac{T}{320} = 320 \cdot \frac{14}{7.7}$$
$$T = \frac{4480}{7.7}, \text{ or } \frac{4480}{\frac{77}{10}}$$
$$T = 581\frac{9}{11}$$

Check. $\dfrac{581\frac{9}{11}}{320} = \dfrac{\frac{6400}{11}}{320} = \dfrac{6400}{11} \cdot \dfrac{1}{320} = \dfrac{320 \cdot 20}{11 \cdot 320} =$

$\dfrac{20}{11}$ and $\dfrac{14}{7.7} = \dfrac{14}{\frac{77}{10}} = \dfrac{14}{1} \cdot \dfrac{10}{77} = \dfrac{7 \cdot 2 \cdot 10}{7 \cdot 11} = \dfrac{20}{11}$

The ratios are the same.

State. 582 trees are required to produce 320 kg of coffee. (We round to the nearest whole number.)

32. 702 km

33. *Familiarize.* The ratio of milk to flour is $\dfrac{3}{13}$, so 3 cups of milk are used when 13 cups of flour are used. We wish to find how many cups of flour are used when 5 cups of milk are used. We can set up ratios:

$$\dfrac{5}{F} \qquad \dfrac{3}{13}$$

Translate. Assuming the two ratios are the same we can translate to a proportion.

$$\text{Milk} \to \dfrac{5}{F} = \dfrac{3}{13} \gets \text{Milk}$$
$$\text{Flour} \to \qquad \qquad \gets \text{Flour}$$

Carry out. We solve the proportion.

$$13F \cdot \dfrac{5}{F} = 13F \cdot \dfrac{3}{13}$$
$$65 = 3F$$
$$\dfrac{65}{3} = F, \text{ or}$$
$$21\dfrac{2}{3} = F$$

Check. $\dfrac{5}{21\frac{2}{3}} = \dfrac{5}{\frac{65}{3}} = 5 \cdot \dfrac{3}{65} = \dfrac{3}{13}$. The ratios are the same.

State. $21\dfrac{2}{3}$ cups of flour are used if 5 cups of milk are used.

34. 1.92 grams

35. We write a proportion and then solve it.

$$\dfrac{b}{6} = \dfrac{7}{4}$$
$$b = \dfrac{7}{4} \cdot 6$$
$$b = \dfrac{42}{4}, \text{ or } 10.5$$

$\left(\text{Note that the proportions } \dfrac{6}{b} = \dfrac{4}{7}, \dfrac{b}{7} = \dfrac{6}{4}, \text{ or } \dfrac{7}{b} = \dfrac{4}{6} \right.$
could also be used.$\Big)$

36. 6.75

37. We write a proportion and then solve it.

$$\dfrac{4}{f} = \dfrac{6}{4}$$
$$4f \cdot \dfrac{4}{f} = 4f \cdot \dfrac{6}{4}$$
$$16 = 6f$$
$$\dfrac{8}{3} = f \qquad \text{Simplifying}$$

$\Big($One of the following proportions could also be used:
$\dfrac{f}{4} = \dfrac{4}{6}, \dfrac{4}{f} = \dfrac{9}{6}, \dfrac{f}{4} = \dfrac{6}{9}, \dfrac{4}{9} = \dfrac{f}{6}, \dfrac{9}{4} = \dfrac{6}{f}\Big)$

38. 7.5

39. We write a proportion and then solve it.

$$\dfrac{4}{10} = \dfrac{6}{l}$$
$$10l \cdot \dfrac{4}{10} = 10l \cdot \dfrac{6}{l}$$
$$4l = 60$$
$$l = 15 \text{ ft}$$

$\Big($One of the following proportions could also be used:
$\dfrac{4}{6} = \dfrac{10}{l}, \dfrac{10}{4} = \dfrac{l}{6}, \text{ or } \dfrac{6}{4} = \dfrac{l}{10}\Big)$

40. 4.5 ft

41. *Familiarize.* The ratio of trout tagged to the total number of trout in the lake, T, is $\dfrac{112}{T}$. Of the 82 trout caught later, there were 32 tagged trout. The ratio of tagged trout to trout caught is $\dfrac{32}{82}$.

Translate. Assuming the two ratios are the same, we can translate to a proportion.

$$\begin{array}{c} \text{Trout tagged} \\ \text{originally} \\ \text{Trout} \\ \text{in lake} \end{array} \longrightarrow \dfrac{112}{T} = \dfrac{32}{82} \begin{array}{c} \longleftarrow \text{Tagged trout} \\ \text{caught later} \\ \longleftarrow \text{Trout} \\ \text{caught later} \end{array}$$

Carry out. We solve the proportion.

$$82T \cdot \dfrac{112}{T} = 82T \cdot \dfrac{32}{82}$$
$$82 \cdot 112 = T \cdot 32$$
$$9184 = 32T$$
$$\dfrac{9184}{32} = T$$
$$287 = T$$

Check.
$$\dfrac{112}{287} \approx 0.39 \qquad \dfrac{32}{82} \approx 0.39$$
The ratios are the same.

State. We estimate that there are 287 trout in the lake.

42. 954

43. *Familiarize.* The ratio of moose tagged to the total number of moose in the park, M, is $\dfrac{69}{M}$. Of the 40 moose caught later, 15 are tagged. The ratio of tagged moose to moose caught is $\dfrac{15}{40}$.

Translate. We translate to a proportion.

$$\begin{array}{ll} \text{Moose originally} & \text{Tagged moose}\\ \text{tagged} \to \dfrac{69}{M} = \dfrac{15}{40} \leftarrow & \text{caught later}\\ \text{Moose} \to & \text{Moose}\\ \text{in forest} & \text{caught later} \end{array}$$

Carry out. We solve the proportion. We multiply by the LCD, $40M$.

$$40M \cdot \frac{69}{M} = 40M \cdot \frac{15}{40}$$
$$40 \cdot 69 = M \cdot 15$$
$$2760 = 15M$$
$$\frac{2760}{15} = M$$
$$184 = M$$

Check.
$$\frac{69}{184} = 0.375 \qquad \frac{15}{40} = 0.375$$
The ratios are the same.

State. We estimate that there are 184 moose in the park.

44. 42

45. *Familiarize*. Let M = the number of miles Emmanuel will drive in 4 years if he continues to drive at the current rate. We set up two ratios:

$$\frac{16,000}{1\frac{1}{2}} \qquad \frac{M}{4}$$

Translate. We write a proportion.

$$\begin{array}{l} \text{Miles} \to \dfrac{16,000}{1.5} = \dfrac{M}{4} \leftarrow \text{Miles}\\ \text{Years} \to \qquad\qquad\quad \leftarrow \text{Years} \end{array}$$

Carry out. We solve the proportion.

$$1.5(4) \cdot \frac{16,000}{1.5} = 1.5(4) \cdot \frac{M}{4}$$
$$64,000 = 1.5M$$
$$42,666.\overline{6} = M$$

If this possible answer is correct, Emmanuel will not exceed the 45,000 miles allowed for four years.

Check.
$$\frac{16,000}{1.5} = 10,666.\overline{6} \qquad \frac{42,666.\overline{6}}{4} = 10,666.\overline{6}$$
The ratios are the same.

State. At this rate, Emmanuel will not exceed the mileage allowed for four years.

46. 20

47. *Familiarize*. The ratio of deer tagged to the total number of deer in the park, D, is $\dfrac{25}{D}$. Of the 36 deer caught later, 4 had tags. The ratio of tagged deer to deer caught is $\dfrac{4}{36}$.

Translate. Assuming the two ratios are the same, we can translate to a proportion.

$$\begin{array}{ll} \text{Deer tagged} & \text{Tagged deer}\\ \text{originally} \to \dfrac{25}{D} = \dfrac{4}{36} \leftarrow & \text{caught later}\\ \text{Deer} \to & \text{Deer}\\ \text{in park} & \text{caught later} \end{array}$$

Carry out. We solve the proportion.
$$36D \cdot \frac{25}{D} = 36D \cdot \frac{4}{36}$$
$$900 = 4D$$
$$225 = D$$

Check.
$$\frac{25}{225} = \frac{1}{9} \qquad \frac{4}{36} = \frac{1}{9}$$
The ratios are the same.

State. We estimate that there are 225 deer in the park.

48. (a) 1.92 tons; (b) 28.8 lb

49. *Familiarize*. The ratio of the weight of an object on Mars to the weight of an object on the earth is 0.4 to 1.

a) We wish to find how much a 12-ton rocket would weigh on Mars.

b) We wish to find how much a 120-lb astronaut would weigh on Mars.

We can set up ratios.
$$\frac{0.4}{1} \qquad \frac{T}{12} \qquad \frac{P}{120}$$

Translate. Assuming the ratios are the same, we can translate to proportions.

a)
$$\begin{array}{ll} \text{Weight} & \text{Weight}\\ \text{on Mars} \to \dfrac{0.4}{1} = \dfrac{T}{12} \leftarrow & \text{on Mars}\\ \text{Weight} \to & \text{Weight}\\ \text{on earth} & \text{on earth} \end{array}$$

b)
$$\begin{array}{ll} \text{Weight} & \text{Weight}\\ \text{on Mars} \to \dfrac{0.4}{1} = \dfrac{P}{120} \leftarrow & \text{on Mars}\\ \text{Weight} \to & \text{Weight}\\ \text{on earth} & \text{on earth} \end{array}$$

Carry out. We solve each proportion.

a) $\dfrac{0.4}{1} = \dfrac{T}{12}$ b) $\dfrac{0.4}{1} = \dfrac{P}{120}$

$\quad 12(0.4) = T \qquad\qquad 120(0.4) = P$

$\qquad\quad 4.8 = T \qquad\qquad\qquad 48 = P$

Check. $\dfrac{0.4}{1} = 0.4$, $\dfrac{4.8}{12} = 0.4$, and $\dfrac{48}{120} = 0.4$.
The ratios are the same.

State.

a) A 12-ton rocket would weigh 4.8 tons on Mars.

b) A 120-lb astronaut would weigh 48 lb on Mars.

50. $\dfrac{36}{68}$

51. $(x+2) - (x+1) = x + 2 - x - 1 = 1$

52. $x^2 - 1$

53.
$$(4y^3 - 5y^2 + 7y - 24) - (-9y^3 + 9y^2 - 5y + 49)$$
$$= 4y^3 - 5y^2 + 7y - 24 + 9y^3 - 9y^2 + 5y - 49$$
$$= 13y^3 - 14y^2 + 12y - 73$$

54. $25,704 \text{ ft}^2$

55.
$$ar + st = n$$
$$st = n - ar$$
$$t = \frac{n - ar}{s}$$

56. $v = \dfrac{m}{u - a}$

57. ◈

58. ◈

59. ◈

60. ◈

61. *Familiarize.* Let t = the number of hours it would take Ann to sew the quilt, working alone. Then $t + 6$ = the number of hours it would take Betty, working alone. In 4 hr Ann does $4 \cdot \dfrac{1}{t}$ of the job and Betty does $4 \cdot \dfrac{1}{t+6}$ of the job, and together they do 1 complete job.

Translate. We use the information above to write an equation.
$$\frac{4}{t} + \frac{4}{t+6} = 1$$

Carry out. We solve the equation.
$$\frac{4}{t} + \frac{4}{t+6} = 1, \text{ LCD} = t(t+6)$$
$$t(t+6)\left(\frac{4}{t} + \frac{4}{t+6}\right) = t(t+6) \cdot 1$$
$$4(t+6) + 4t = t(t+6)$$
$$4t + 24 + 4t = t^2 + 6t$$
$$0 = t^2 - 2t - 24$$
$$0 = (t-6)(t+4)$$
$$t - 6 = 0 \quad or \quad t + 4 = 0$$
$$t = 6 \quad or \qquad t = -4$$

Check. Time cannot be negative in this application, so we check only 6. If Ann can sew the quilt working alone in 6 hr, then it would take Betty $6+6$, or 12 hr, working alone. In 4 hr they would do
$$\frac{4}{6} + \frac{4}{12} = \frac{2}{3} + \frac{1}{3} = 1 \text{ complete job.}$$
The answer checks.

State. It would take Ann 6 hr to sew the quilt working alone, and it would take Betty 12 hr working alone.

62. 2 mph

63. *Familiarize.* Let t = the number of minutes after 5:00 at which the hands will first be together. When the minute hand moves through t minutes, the hour hand moves through $t/12$ minutes. At 5:00 the hour hand is on the 25-minute mark, so at t minutes after 5:00 it is at $25 + t/12$.

Translate. We equate the positions of the minute hand and the hour hand.
$$t = 25 + \frac{t}{12}$$

Carry out. We solve the equation.
$$t = 25 + \frac{t}{12}$$
$$12 \cdot t = 12\left(25 + \frac{t}{12}\right)$$
$$12t = 300 + t$$
$$11t = 300$$
$$t = \frac{300}{11}, \text{ or } 27\frac{3}{11}$$

Check. When the minute hand is at $27\frac{3}{11}$ minutes after 5:00, the hour hand is at $25 + \dfrac{\frac{300}{11}}{12} =$
$$25 + \frac{300}{11} \cdot \frac{1}{12} = 25 + \frac{25}{11} = 25 + 2\frac{3}{11} = 27\frac{3}{11} \text{ minutes}$$
after 5:00 also. The answer checks.

State. After 5:00 the hands on a clock will first be together in $27\frac{3}{11}$ minutes or at $27\frac{3}{11}$ minutes after 5:00.

64. $\dfrac{D}{B} = \dfrac{C}{A}; \dfrac{A}{C} = \dfrac{B}{D}; \dfrac{D}{C} = \dfrac{B}{A}$

65. *Familiarize.* The job takes Rosina 8 days working alone and Ng 10 days working alone. Let x represent the number of days it would take Oscar working alone. Then in 1 day Rosina does $\dfrac{1}{8}$ of the job, Ng does $\dfrac{1}{10}$ of the job, and Oscar does $\dfrac{1}{x}$ of the job. In 1 day they would complete $\dfrac{1}{8} + \dfrac{1}{10} + \dfrac{1}{x}$ of the job, and in 3 days they would complete $3\left(\dfrac{1}{8} + \dfrac{1}{10} + \dfrac{1}{x}\right)$, or $\dfrac{3}{8} + \dfrac{3}{10} + \dfrac{3}{x}$.

Translate. The amount done in 3 days is one entire job, so we have
$$\frac{3}{8} + \frac{3}{10} + \frac{3}{x} = 1.$$

Carry out. We solve the equation.

$$\frac{3}{8} + \frac{3}{10} + \frac{3}{x} = 1, \quad \text{LCD} = 40x$$

$$40x\left(\frac{3}{8} + \frac{3}{10} + \frac{3}{x}\right) = 40x \cdot 1$$

$$40x \cdot \frac{3}{8} + 40x \cdot \frac{3}{10} + 40x \cdot \frac{3}{x} = 40x$$

$$15x + 12x + 120 = 40x$$

$$120 = 13x$$

$$\frac{120}{13} = x$$

Check. If it takes Oscar $\frac{120}{13}$, or $9\frac{3}{13}$ days, to complete the job, then in one day Oscar does $\dfrac{1}{\frac{120}{13}}$, or $\frac{13}{120}$, of the job, and in 3 days he does $3\left(\frac{13}{120}\right)$, or $\frac{13}{40}$, of the job. The portion of the job done by Rosina, Ng, and Oscar in 3 days is

$$\frac{3}{8} + \frac{3}{10} + \frac{13}{40} = \frac{15}{40} + \frac{12}{40} + \frac{13}{40} = \frac{40}{40} = 1 \text{ entire job.}$$

The answer checks.

State. It will take Oscar $9\frac{3}{13}$ days to write the program working alone.

66. Michelle: 6 hr, Sal: 3 hr, Kristen: 4 hr

67. *Familiarize*. Let $t =$ the number of hours it takes to wire one house, working together. We want to find the number of hours it takes to wire two houses, working together.

Translate. We write an equation.

$$\frac{t}{28} + \frac{t}{34} = 2$$

Carry out. We solve the equation.

$$\frac{t}{28} + \frac{t}{34} = 2, \quad \text{LCD} = 476$$

$$476\left(\frac{t}{28} + \frac{t}{34}\right) = 476 \cdot 2$$

$$17t + 14t = 952$$

$$31t = 952$$

$$t = \frac{952}{31}, \text{ or } 32\frac{22}{31}$$

Check. If $30\frac{22}{31}$ hr, Janet does $\frac{952}{31} \cdot \frac{1}{28} = \frac{34}{31}$ of one complete job and Linus does $\frac{952}{31} \cdot \frac{1}{34}$, or $\frac{28}{31}$ of one complete job. Together they do $\frac{34}{31} + \frac{28}{31}$, or $\frac{62}{31}$, or 2 complete jobs. The answer checks.

State. It will take Janet and Linus $30\frac{22}{31}$ hr to wire two houses, working together.

68. $66\frac{2}{3}$ ft

69. *Familiarize*. Let $M =$ the number of votes Melanie received. Then her opponent received $450 - M$ votes. We can write two proportions.

$$\frac{3}{2} \qquad \frac{M}{450 - M}$$

Translate. We write a proportion.

$$\frac{3}{2} = \frac{M}{450 - M}$$

Carry out. We solve the proportion.

$$2(450 - M) \cdot \frac{3}{2} = 2(450 - M) \cdot \frac{M}{450 - M}$$

$$3(450 - M) = 2M$$

$$1350 - 3M = 2M$$

$$1350 = 5M$$

$$270 = M$$

Check.

$$\frac{3}{2} = 1.5 \qquad \frac{270}{450 - 270} = \frac{270}{180} = 1.5$$

The ratios are the same.

State. Melanie received 270 votes.

70. 45 mph

71. ◈

72. ◈

Exercise Set 6.8

1. $S = 2\pi r h$

$$\frac{S}{2\pi r} = h \qquad \text{Multiplying by } \frac{1}{2\pi r}$$

2. $t = \dfrac{A - P}{Pr}$

3. $A = \frac{1}{2}bh$

$$2A = bh \quad \text{Multiplying by 2}$$

$$\frac{2A}{b} = h \quad \text{Multiplying by } \frac{1}{b}$$

4. $g = \dfrac{2s}{t^2}$

5. $\dfrac{1}{180} = \dfrac{n - 2}{s}$

$$\frac{s}{180} = n - 2 \quad \text{Multiplying by } s$$

$$\frac{s}{180} + 2 = n, \qquad \text{Adding 2}$$

$$\text{or } \frac{s + 360}{180} = n$$

6. $a = \dfrac{2S - nl}{n}$

7. $$V = \frac{1}{3}k(B+b-2n)$$

$$3V = k(B+b-2n) \quad \text{Multiplying by 3}$$

$$3V = kB+kb-2kn \quad \text{Removing parentheses}$$

$$3V-kB+2kn = kb \quad \text{Adding } -kb+2kn$$

$$\frac{3V-kB+2kn}{k} = b \quad \text{Multiplying by } \frac{1}{k}$$

8. $P = \dfrac{A}{1+rt}$

9. $rl - rS = L$

$r(l - S) = L \quad \text{Factoring out } r$

$r = \dfrac{L}{l - S} \quad \text{Dividing by } l - S$

10. $m = \dfrac{T}{g - f}$

11. $$A = \frac{1}{2}h(b_1 + b_2)$$

$$2A = h(b_1 + b_2) \quad \text{Multiplying by 2}$$

$$\frac{2A}{b_1 + b_2} = h \quad \text{Multiplying by } \frac{1}{b_1 + b_2}$$

12. $h = \dfrac{S - 2\pi r^2}{2\pi r}$

13. $ab = ac + d$

$ab - ac = d \quad \text{Subtracting } ac \text{ to get all terms involving } a \text{ on one side}$

$a(b - c) = d \quad \text{Factoring out } a$

$a = \dfrac{d}{b - c} \quad \text{Dividing by } b - c$

14. $n = \dfrac{p}{p - m}$

15. $\dfrac{m}{r} = s$

$m = rs \quad \text{Multiplying by } r$

$\dfrac{m}{s} = r \quad \text{Dividing by } s$

16. $m = \dfrac{V}{d}$

17. $a + b = \dfrac{c}{d}$

$d(a + b) = c \quad \text{Multiplying by } d$

$d = \dfrac{c}{a + b} \quad \text{Dividing by } a + b$

18. $n = \dfrac{m}{p - q}$

19. $I = \dfrac{V_1 - V_2}{R}$

$IR = V_1 - V_2 \quad \text{Multiplying by } R$

$R = \dfrac{V_1 - V_2}{I} \quad \text{Dividing by } I$

20. $t = \dfrac{M - g}{r + s}$

21. $$\frac{1}{p} + \frac{1}{q} = \frac{1}{f}$$

$$pqf\left(\frac{1}{p} + \frac{1}{q}\right) = pqf \cdot \frac{1}{f} \quad \text{Multiplying by the LCD, } pqf, \text{ to clear fractions}$$

$$pqf \cdot \frac{1}{p} + pqf \cdot \frac{1}{q} = pq$$

$$qf + pf = pq$$

$$f(q + p) = pq$$

$$f = \frac{pq}{q + p}$$

22. $R = \dfrac{r_1 r_2}{r_2 + r_1}$

23. $a = \dfrac{v_2 - v_1}{t}$

$at = v_2 - v_1 \quad \text{Multiplying by } t$

$t = \dfrac{v_2 - v_1}{a} \quad \text{Dividing by } a$

24. $p = \dfrac{ar}{v^2 L}$

25. $P = 2(l + w)$

$P = 2l + 2w \quad \text{Removing parentheses}$

$P - 2w = 2l \quad \text{Subtracting } 2w$

$\dfrac{P - 2w}{2} = l, \quad \text{Dividing by 2}$

or $\dfrac{P}{2} - w = l$

26. $n = \dfrac{a}{c(1 + b)}$

27. $ab - ac = \dfrac{Q}{M}$

$a(b - c) = \dfrac{Q}{M} \quad \text{Factoring out } a$

$a = \dfrac{Q}{M(b - c)} \quad \text{Multiplying by } \dfrac{1}{b - c}$

28. $b = \dfrac{a}{3S - 2}$

29. $C = \dfrac{Ka - b}{a}$

$aC = Ka - b \quad \text{Multiplying by } a$

$b = Ka - aC \quad \text{Adding } -aC \text{ and } b$

$b = a(K - C) \quad \text{Factoring out } a$

$\dfrac{b}{K - C} = a \quad \text{Multiplying by } \dfrac{1}{K - C}$

30. $F = \dfrac{9C + 160}{5}$, or $F = \dfrac{9}{5}C + 32$

31. $V = \dfrac{4}{3}\pi r^3$

$3V = 4\pi r^3$ Multiplying by 3

$\dfrac{3V}{4\pi} = r^3$ Multiplying by $\dfrac{1}{4\pi}$

32. $g = \dfrac{fm + t}{m}$

33. $S = \dfrac{rl - a}{r - l}$

$(r - l) \cdot S = (r - l) \cdot \dfrac{rl - a}{r - l}$ Multiplying by $r-l$

$Sr - Sl = rl - a$

$Sr - rl = Sl - a$ Adding Sl and $-rl$

$r(S - l) = Sl - a$ Factoring out r

$r = \dfrac{Sl - a}{S - l}$ Multiplying by $\dfrac{1}{S - l}$

34. $m = \dfrac{-t}{f - g}$, or $m = \dfrac{t}{g - f}$

35. Graph: $y = \dfrac{4}{5}x + 1$

x	y
-5	-3
0	1
5	5

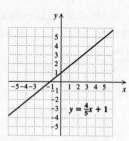

36.

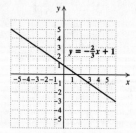

37. $-\dfrac{3}{5}x = \dfrac{9}{20}$

$-\dfrac{5}{3}\left(-\dfrac{3}{5}x\right) = -\dfrac{5}{3} \cdot \dfrac{9}{20}$

$x = -\dfrac{45}{60}$, or $-\dfrac{3}{4}$

The solution is $-\dfrac{3}{4}$.

38.

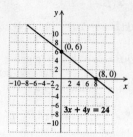

39. $x^2 - 13x - 30$

We look for two factors of -30 whose sum is -13. The numbers we need are -15 and 2.

$x^2 - 13x - 30 = (x - 15)(x + 2)$

40. $-3x^3 - 5x^2 + 5$

41. ◈

42. ◈

43. ◈

44. ◈

45. $\dfrac{n_1}{p_1} + \dfrac{n_2}{p_2} = \dfrac{n_2 - n_1}{R}$

$p_1 p_2 R\left(\dfrac{n_1}{p_1} + \dfrac{n_2}{p_2}\right) = p_1 p_2 R \cdot \dfrac{n_2 - n_1}{R}$

$n_1 p_2 R + n_2 p_1 R = p_1 p_2 n_2 - p_1 p_2 n_1$

$n_1 p_2 R + p_1 p_2 n_1 = p_1 p_2 n_2 - n_2 p_1 R$

$n_1 p_2 R + p_1 p_2 n_1 = n_2(p_1 p_2 - p_1 R)$

$\dfrac{n_1 p_2 R + p_1 p_2 n_1}{p_1 p_2 - p_1 R} = n_2$

46. $T = \dfrac{FP}{u + EF}$

47. $N = \dfrac{(b + d)f_1 - v}{(b - v)f_2}$

$N = \dfrac{bf_1 + df_1 - v}{bf_2 - vf_2}$

$N(bf_2 - vf_2) = bf_1 + df_1 - v$

$Nbf_2 - Nvf_2 = bf_1 + df_1 - v$

$v - Nvf_2 = bf_1 + df_1 - Nbf_2$

$v(1 - Nf_2) = bf_1 + df_1 - Nbf_2$

$v = \dfrac{bf_1 + df_1 - Nbf_2}{1 - Nf_2}$, or

$v = \dfrac{Nbf_2 - bf_1 - df_1}{Nf_2 - 1}$

48. $-40°$

Chapter 7
More With Graphing

Exercise Set 7.1

1. Slope $\frac{3}{5}$; y-intercept $(0, 2)$

We plot $(0, 2)$ and from there move up 3 units and right 5 units. This locates the point $(5, 5)$. We plot $(5, 5)$ and draw a line passing through $(0, 2)$ and $(5, 5)$.

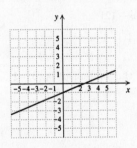

2.

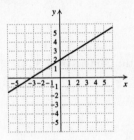

3. Slope $\frac{5}{3}$; y-intercept $(0, -3)$

We plot $(0, -3)$ and from there move up 5 units and right 3 units. This locates the point $(3, 2)$. We plot $(3, 2)$ and draw a line passing through $(0, -3)$ and $(3, 2)$.

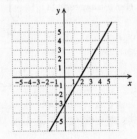

4.

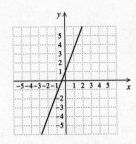

5. Slope $-\frac{3}{4}$; y-intercept $(0, 5)$

We plot $(0, 5)$. We can think of the slope as $\frac{-3}{4}$, so from $(0, 5)$ we move down 3 units and right 4 units. This locates the point $(4, 2)$. We plot $(4, 2)$ and draw a line passing through $(0, 5)$ and $(4, 2)$.

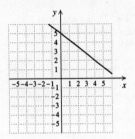

6.

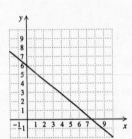

7. Slope 2; y-intercept $(0, -4)$

We plot $(0, -4)$. We can think of the slope as $\frac{2}{1}$, so from $(0, -4)$ we move up 2 units and right 1 unit. This locates the point $(1, -2)$. We plot $(1, -2)$ and draw a line passing through $(0, -4)$ and $(1, -2)$.

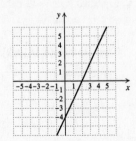

8.

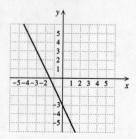

9. Slope -3; y-intercept $(0, 2)$

We plot $(0, 2)$. We can think of the slope as $\dfrac{-3}{1}$, so from $(0, 2)$ we move down 3 units and right 1 unit. This locates the point $(1, -1)$. We plot $(1, -1)$ and draw a line passing through $(0, 2)$ and $(1, -1)$.

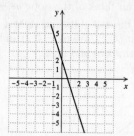

10.

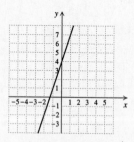

11. We read the slope and y-intercept from the equation.
$$y = \frac{3}{7}x + 6$$

The slope is $\dfrac{3}{7}$.   The y-intercept is $(0, 6)$.

12. $-\dfrac{3}{8}$, $(0, 7)$

13. We read the slope and y-intercept from the equation.
$$y = -\frac{5}{6}x + 2$$

The slope is $-\dfrac{5}{6}$.   The y-intercept is $(0, 2)$.

14. $\dfrac{7}{2}$, $(0, 4)$

15. $y = \dfrac{9}{4}x - 7$

$y = \dfrac{9}{4}x + (-7)$

The slope is $\dfrac{9}{4}$, and the y-intercept is $(0, -7)$.

16. $\dfrac{2}{9}$, $(0, -1)$

17. $y = -\dfrac{2}{5}x$

$y = -\dfrac{2}{5}x + 0$

The slope is $-\dfrac{2}{5}$, and the y-intercept is $(0, 0)$.

18. $\dfrac{4}{3}$, $(0, 0)$

19. We solve for y to rewrite the equation in the form $y = mx + b$.
$$-2x + y = 4$$
$$y = 2x + 4$$

The slope is 2, and the y-intercept is $(0, 4)$.

20. 5, $(0, 5)$

21. $3x - 4y = 12$
$$-4y = -3x + 12$$
$$y = -\frac{1}{4}(-3x + 12)$$
$$y = \frac{3}{4}x - 3$$

The slope is $\dfrac{3}{4}$, and the y-intercept is $(0, -3)$.

22. $\dfrac{3}{2}$, $(0, -9)$

23. $x - 5y = -8$
$$-5y = -x - 8$$
$$y = -\frac{1}{5}(-x - 8)$$
$$y = \frac{1}{5}x + \frac{8}{5}$$

The slope is $\dfrac{1}{5}$, and the y-intercept is $\left(0, \dfrac{8}{5}\right)$.

24. $\dfrac{1}{6}$, $\left(0, -\dfrac{3}{2}\right)$

25. $y = 4$
$$y = 0x + 4$$

The slope is 0, and the y-intercept is $(0, 4)$.

26. 0, $(0, 8)$

27. We use the slope-intercept equation, substituting 3 for m and 7 for b:
$$y = mx + b$$
$$y = 3x + 7$$

28. $y = -4x - 2$

29. We use the slope-intercept equation, substituting $\frac{7}{8}$ for m and -1 for b:

$$y = mx + b$$
$$y = \frac{7}{8}x - 1$$

30. $y = \frac{5}{7}x + 4$

31. We use the slope-intercept equation, substituting $-\frac{5}{3}$ for m and -8 for b:

$$y = mx + b$$
$$y = -\frac{5}{3}x - 8$$

32. $y = \frac{3}{4}x + 23$

33. We use the slope-intercept equation, substituting -2 for m and 3 for b:

$$y = mx + b$$
$$y = -2x + 3$$

34. $y = 7x - 6$

35. $y = \frac{3}{5}x + 2$

First we plot the y-intercept $(0,2)$. We can start at the y-intercept and use the slope, $\frac{3}{5}$, to find another point. We move up 3 units and right 5 units to get a new point $(5,5)$. Thinking of the slope as $\frac{-3}{-5}$ we can start at $(0,2)$ and move down 3 units and left 5 units to get another point $(-5,-1)$.

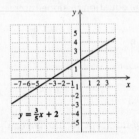

36.

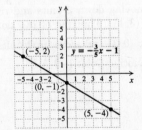

37. $y = -\frac{3}{5}x + 1$

First we plot the y-intercept $(0,1)$. We can start at the y-intercept and, thinking of the slope as $\frac{-3}{5}$, find another point by moving down 3 units and right 5

units to the point $(5,-2)$. Thinking of the slope as $\frac{3}{-5}$ we can start at $(0,1)$ and move up 3 units and left 5 units to get another point $(-5,4)$.

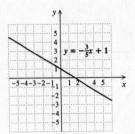

38.

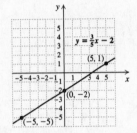

39. $y = \frac{5}{3}x + 3$

First we plot the y-intercept $(0,3)$. We can start at the y-intercept and use the slope, $\frac{5}{3}$, to find another point. We move up 5 units and right 3 units to get a new point $(3,8)$. Thinking of the slope as $\frac{-5}{-3}$ we can start at $(0,3)$ and move down 5 units and left 3 units to get another point $(-3,-2)$.

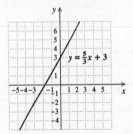

40.

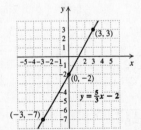

41. $y = -\frac{3}{2}x - 2$

First we plot the y-intercept $(0,-2)$. We can start at the y-intercept and, thinking of the slope as $\frac{-3}{2}$, find another point by moving down 3 units and right 2 units to the point $(2,-5)$. Thinking of the slope as $\dfrac{3}{-2}$ we can start at $(0, -2)$ and move up 3 units and left 2 units to get another point $(-2, 1)$.

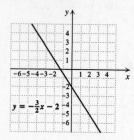

42.

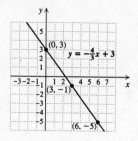

43. We first rewrite the equation in slope-intercept form.

$$2x + y = 1$$
$$y = -2x + 1$$

Now we plot the y-intercept $(0, 1)$. We can start at the y-intercept and, thinking of the slope as $\dfrac{-2}{1}$, find another point by moving down 2 units and right 1 unit to the point $(1, -1)$. In a similar manner, we can move from the point $(1, -1)$ to find a third point $(2, -3)$.

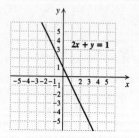

44. $3x + y = 2$

$$y = -3x + 2$$

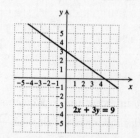

45. We first rewrite the equation in slope-intercept form.

$$3x - y = 4$$
$$-y = -3x + 4$$
$$y = 3x - 4 \quad \text{Multiplying by } -1$$

Now we plot the y-intercept $(0, -4)$. We can start at the y-intercept and, thinking of the slope as $\dfrac{3}{1}$, find another point by moving up 3 units and right 1 unit to the point $(1, -1)$. In a similar manner, we can move from the point $(1, -1)$ to find a third point $(2, 2)$.

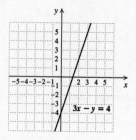

46. $2x - y = 5$

$$y = 2x - 5$$

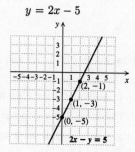

47. We first rewrite the equation in slope-intercept form.

$$2x + 3y = 9$$
$$3y = -2x + 9$$
$$y = \dfrac{1}{3}(-2x + 9)$$
$$y = -\dfrac{2}{3}x + 3$$

Now we plot the y-intercept $(0, 3)$. We can start at the y-intercept and, thinking of the slope as $\dfrac{-2}{3}$, find another point by moving down 2 units and right 3 units to the point $(3, 1)$. Thinking of the slope as $\dfrac{2}{-3}$ we can start at $(0, 3)$ and move up 2 units and left 3 units to get another point $(-3, 5)$.

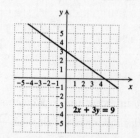

48.

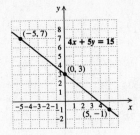

49. We first rewrite the equation in slope-intercept form.

$$x - 4y = 12$$
$$-4y = -x + 12$$
$$y = -\frac{1}{4}(-x + 12)$$
$$y = \frac{1}{4}x - 3$$

Now we plot the y-intercept $(0, -3)$. We can start at the y-intercept and use the slope, $\frac{1}{4}$, to find another point. We move up 1 unit and right 4 units to the point $(4, -2)$. Thinking of the slope as $\frac{-1}{-4}$ we can start at $(0, -3)$ and move down 1 unit and left 4 units to get another point $(-4, -4)$.

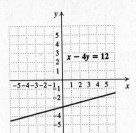

50.

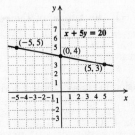

51. Two points on the graph are $(0, 9)$ and $(7, 16)$, so we see that the y-intercept will be $(0, 9)$. Now we find the slope:

$$m = \frac{16 - 9}{7 - 0} = \frac{7}{7} = 1$$

Then the equation is $y = x + 9$.

To graph the equation we first plot $(0, 9)$. We can think of the slope as $\frac{1}{1}$, so from the y-intercept we move up 1 unit and right 1 unit to the point $(1, 10)$. We plot $(1, 10)$ and draw the line passing through $(0, 9)$ and $(1, 10)$.

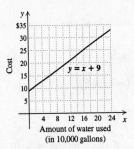

Since the slope is 1, the rate is \$1 per 10,000 gallons.

52.

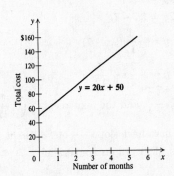

\$20 per month

53. Two points on the graph are $(0, 16)$ and $(1, 16 + 1.5)$, or $(1, 17.5)$, so the y-intercept will be $(0, 16)$. Now we find the slope:

$$m = \frac{17.5 - 16}{1 - 0} = \frac{1.5}{1} = 1.5$$

Then the equation is $y = 1.5x + 16$.

54. $y = 2.5x + 4.95$

55. $y = \frac{2}{3}x + 7$: The slope is $\frac{2}{3}$, and the y-intercept is $(0, 7)$.

$y = \frac{2}{3}x - 5$: The slope is $\frac{2}{3}$, and the y-intercept is $(0, -5)$.

Since both lines have slope $\frac{2}{3}$ but different y-intercepts, their graphs are parallel.

56. No

57. The equation $y = 2x - 5$ represents a line with slope 2 and y-intercept $(0, -5)$. We rewrite the second equation in slope-intercept form.

$$4x + 2y = 9$$
$$2y = -4x + 9$$
$$y = \frac{1}{2}(-4x + 9)$$
$$y = -2x + \frac{9}{2}$$

The slope is -2 and the y-intercept is $\left(0, \frac{9}{2}\right)$. Since the lines have different slopes, their graphs are not parallel.

58. Yes

59. Rewrite each equation in slope-intercept form.

$$3x + 4y = 8$$
$$4y = -3x + 8$$
$$y = \frac{1}{4}(-3x + 8)$$
$$y = -\frac{3}{4}x + 2$$

The slope is $-\frac{3}{4}$, and the y-intercept is $(0, 2)$.

$$7 - 12y = 9x$$
$$-12y = 9x - 7$$
$$y = -\frac{1}{12}(9x - 7)$$
$$y = -\frac{3}{4}x + \frac{7}{12}$$

The slope is $-\frac{3}{4}$, and the y-intercept is $\left(0, \frac{7}{12}\right)$.

Since both lines have slope $-\frac{3}{4}$ but different y-intercepts, their graphs are parallel.

60. No

61.
$$2x^2 + 6x = 0$$
$$2x(x + 3) = 0$$
$$x = 0 \quad or \quad x + 3 = 0$$
$$x = 0 \quad or \quad\quad x = -3$$

The solutions are 0 and -3.

62. $x(x + 7)(x - 2)$

63. *Familiarize*. Let $y = $ the smaller odd integer. Then $y + 2 = $ the larger odd integer.

Translate. We reword the problem.

$$\underbrace{\text{Smaller odd integer}}_{y} \quad \underset{\cdot}{\text{times}} \quad \underbrace{\text{larger odd integer}}_{(y+2)} \quad \underset{=}{\text{is 195.}} \quad 195$$

Carry out. We solve the equation.

$$y(y + 2) = 195$$
$$y^2 + 2y = 195$$
$$y^2 + 2y - 195 = 0$$
$$(y + 15)(y - 13) = 0$$
$$y + 15 = 0 \quad or \quad y - 13 = 0$$
$$y = -15 \quad or \quad\quad y = 13$$

Check. If $y = -15$, then $y + 2 = -13$. These are consecutive odd integers and their product is $(-15)(-13)$, or 195. This pair checks. If $y = 13$ then $y + 2 = 15$. These are consecutive odd integers and their product is $13 \cdot 15$, or 195. This pair checks also.

State. The integers are -15 and -13 or 13 and 15.

64. -1, 11

65.
$$3x - 4(9 - x) = 17$$
$$3x - 36 + 4x = 17$$
$$7x - 36 = 17$$
$$7x = 53$$
$$x = \frac{53}{7}$$

The solution is $\frac{53}{7}$.

66. $\frac{3}{8}$

67. ◈

68. ◈

69. ◈

70. ▨

71. Rewrite each equation in slope-intercept form.

$$3y = 5x - 3$$
$$y = \frac{1}{3}(5x - 3)$$
$$y = \frac{5}{3}x - 1$$

The slope is $\frac{5}{3}$.

$$3x + 5y = 10$$
$$5y = -3x + 10$$
$$y = \frac{1}{5}(-3x + 10)$$
$$y = -\frac{3}{5}x + 2$$

The slope is $-\frac{3}{5}$.

Since $\frac{5}{3}\left(-\frac{3}{5}\right) = -1$, the graphs of the equations are perpendicular.

72. Yes

73. Rewrite each equation in slope-intercept form.

$$3x + 5y = 10$$
$$5y = -3x + 10$$
$$y = \frac{1}{5}(-3x + 10)$$
$$y = -\frac{3}{5}x + 2$$

The slope is $-\frac{3}{5}$.

$$15x + 9y = 18$$
$$9y = -15x + 18$$
$$y = \frac{1}{9}(-15x + 18)$$
$$y = -\frac{5}{3}x + 2$$

The slope is $-\frac{5}{3}$.

Since $-\frac{3}{5}\left(-\frac{5}{3}\right) = 1 \neq -1$, the graphs of the equations are not perpendicular.

74. Yes

75. Since $x = 5$ represents a vertical line and $y = \frac{1}{2}$ represents a horizontal line, the graphs of the equations are perpendicular.

76. No

77. See the answer section in the text.

78. $y = \frac{3}{2}x - 2$

79. Rewrite $2x + 5y = 6$ in slope-intercept form.

$$2x + 5y = 6$$
$$5y = -2x + 6$$
$$y = \frac{1}{5}(-2x + 6)$$
$$y = -\frac{2}{5}x + \frac{6}{5}$$

The slope is $-\frac{2}{5}$.

The slope of a line perpendicular to this line is a number m such that

$$-\frac{2}{5}m = -1, \text{ or}$$
$$m = \frac{5}{2}.$$

We graph the line whose equation we want to find. First we plot the given point $(2, 6)$. Now think of the slope as $\frac{-5}{-2}$. From the point $(2, 6)$ go down 5 units and left 2 units to the point $(0, 1)$. Plot this point and draw the graph.

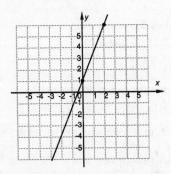

We see that the y-intercept is $(0, 1)$, so the desired equation is $y = \frac{5}{2}x + 1$.

80. ◈

Exercise Set 7.2

1. $y - y_1 = m(x - x_1)$

We substitute 5 for m, 3 for x_1, and 7 for y_1.

$$y - 7 = 5(x - 3)$$

2. $y - 0 = -2(x - (-3))$

3. $y - y_1 = m(x - x_1)$

We substitute $\frac{3}{4}$ for m, 2 for x_1, and 4 for y_1.

$$y - 4 = \frac{3}{4}(x - 2)$$

4. $y - 2 = -1\left(x - \frac{1}{2}\right)$

5. $y - y_1 = m(x - x_1)$

We substitute 1 for m, 2 for x_1, and -6 for y_1.

$$y - (-6) = 1(x - 2)$$

6. $y - (-2) = 6(x - 4)$

7. $y - y_1 = m(x - x_1)$

We substitute -3 for m, -4 for x_1, and 0 for y_1.

$$y - 0 = -3(x - (-4))$$

8. $y - 3 = -3(x - 0)$

9. $y - y_1 = m(x - x_1)$

We substitute $\frac{2}{3}$ for m, 5 for x_1, and 6 for y_1.

$$y - 6 = \frac{2}{3}(x - 5)$$

10. $y - 7 = \frac{5}{6}(x - 2)$

11. First we write the equation in point-slope form.

$$y - y_1 = m(x - x_1)$$
$$y - 4 = 2(x - 1) \quad \text{Substituting}$$

Next we find an equivalent equation of the form $y = mx + b$.

$$y - 4 = 2(x - 1)$$
$$y - 4 = 2x - 2$$
$$y = 2x + 2$$

12. $y = 4x + 1$

13. We first write the equation in point-slope form and then find a equivalent equation of the form $y = mx + b$.

$$y - y_1 = m(x - x_1)$$
$$y - 5 = -1(x - 3) \quad \text{Substituting}$$
$$y - 5 = -x + 3$$
$$y = -x + 8$$

14. $y = x - 5$

15. We first write the equation in point-slope form and then find a equivalent equation of the form $y = mx + b$.

$$y - y_1 = m(x - x_1)$$
$$y - 3 = \frac{1}{2}(x - (-2))$$
$$y - 3 = \frac{1}{2}(x + 2)$$
$$y - 3 = \frac{1}{2}x + 1$$
$$y = \frac{1}{2}x + 4$$

16. $y = -\frac{1}{3}x - 7$

17. We first write the equation in point-slope form and then find a equivalent equation of the form $y = mx + b$.

$$y - y_1 = m(x - x_1)$$
$$y - (-5) = -\frac{1}{3}(x - (-6))$$
$$y + 5 = -\frac{1}{3}(x + 6)$$
$$y + 5 = -\frac{1}{3}x - 2$$
$$y = -\frac{1}{3}x - 7$$

18. $y = \frac{1}{5}x + 8$

19. We first write the equation in point-slope form and then find a equivalent equation of the form $y = mx + b$.

$$y - (-3) = \frac{5}{4}(x - 4)$$
$$y + 3 = \frac{5}{4}x - 5$$
$$y = \frac{5}{4}x - 8$$

20. $y = \frac{4}{3}x + 12$

21. $(-6, 1)$ and $(2, 3)$

First we find the slope.
$$m = \frac{1 - 3}{-6 - 2} = \frac{-2}{-8} = \frac{1}{4}$$
Then we use the point-slope equation.
$$y - y_1 = m(x - x_1)$$
We substitute $\frac{1}{4}$ for m, -6 for x_1, and 1 for y_1.

$$y - 1 = \frac{1}{4}(x - (-6))$$
$$y - 1 = \frac{1}{4}(x + 6)$$
$$y - 1 = \frac{1}{4}x + \frac{3}{2}$$
$$y = \frac{1}{4}x + \frac{5}{2}$$

We also could substitute $\frac{1}{4}$ for m, 2 for x_1, and 3 for y_1.
$$y - 3 = \frac{1}{4}(x - 2)$$
$$y - 3 = \frac{1}{4}x - \frac{1}{2}$$
$$y = \frac{1}{4}x + \frac{5}{2}$$

22. $y = x + 4$

23. $(0, 4)$ and $(4, 2)$

First we find the slope.
$$m = \frac{4 - 2}{0 - 4} = \frac{2}{-4} = -\frac{1}{2}$$
Then we use the point-slope equation.
$$y - y_1 = m(x - x_1)$$
We substitute $-\frac{1}{2}$ for m, 0 for x_1, and 4 for y_1.
$$y - 4 = -\frac{1}{2}(x - 0)$$
$$y - 4 = -\frac{1}{2}x$$
$$y = -\frac{1}{2}x + 4$$

24. $y = \frac{1}{2}x$

25. $(3, 2)$ and $(1, 5)$

First we find the slope.
$$m = \frac{2 - 5}{3 - 1} = \frac{-3}{2} = -\frac{3}{2}$$
Then we use the point-slope equation.
$$y - y_1 = m(x - x_1)$$
We substitute $-\frac{3}{2}$ for m, 3 for x_1, and 2 for y_1.
$$y - 2 = -\frac{3}{2}(x - 3)$$
$$y - 2 = -\frac{3}{2}x + \frac{9}{2}$$
$$y = -\frac{3}{2}x + \frac{13}{2}$$

26. $y = x + 5$

27. $(5, 0)$ and $(0, -2)$

First we find the slope.
$$m = \frac{0 - (-2)}{5 - 0} = \frac{2}{5}$$

Then we use the point-slope equation.

$$y - y_1 = m(x - x_1)$$

We substitute $\frac{2}{5}$ for m, 5 for x_1, and 0 for y_1.

$$y - 0 = \frac{2}{5}(x - 5)$$

$$y = \frac{2}{5}x - 2$$

28. $y = \frac{5}{3}x + \frac{4}{3}$

29. $(-2, -4)$ and $(2, -1)$

First we find the slope.

$$m = \frac{-4 - (-1)}{-2 - 2} = \frac{-4 + 1}{-2 - 2} = \frac{-3}{-4} = \frac{3}{4}$$

Then we use the point-slope equation.

$$y - y_1 = m(x - x_1)$$

We substitute $\frac{3}{4}$ for m, -2 for x_1, and -4 for y_1.

$$y - (-4) = \frac{3}{4}(x - (-2))$$

$$y + 4 = \frac{3}{4}(x + 2)$$

$$y + 4 = \frac{3}{4}x + \frac{3}{2}$$

$$y = \frac{3}{4}x - \frac{5}{2}$$

30. $y = -4x - 7$

31. $y - 4 = \frac{1}{2}(x - 3)$ Point-slope form

The line has slope $\frac{1}{2}$ and passes through $(3, 4)$. We plot $(3, 4)$ and then find a second point by moving up 1 unit and right 2 units to $(5, 5)$. We draw the line through these points.

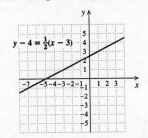

32.

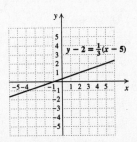

33. $y - 2 = -\frac{1}{2}(x - 5)$ Point-slope form

The line has slope $-\frac{1}{2}$, or $\frac{1}{-2}$ passes through $(5, 2)$. We plot $(5, 2)$ and then find a second point by moving up 1 unit and left 2 units to $(3, 3)$. We draw the line through these points.

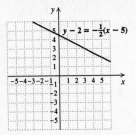

34.

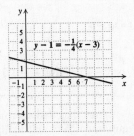

35. $y + 1 = \frac{1}{2}(x - 3)$, or $y - (-1) = \frac{1}{2}(x - 3)$

The line has slope $\frac{1}{2}$ and passes through $(3, -1)$. We plot $(3, -1)$ and then find a second point by moving up 1 unit and right 2 units to $(5, 0)$. We draw the line through these points.

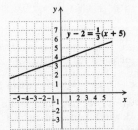

36.

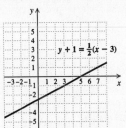

37. $y + 2 = 3(x + 1)$, or $y - (-2) = 3(x - (-1))$

The line has slope 3, or $\frac{3}{1}$, and passes through $(-1, -2)$. We plot $(-1, -2)$ and then find a second point by moving up 3 units and right 1 unit to $(0, 1)$. We draw the line through these points.

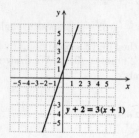

38.

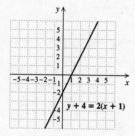

39. $y - 4 = -2(x + 1)$, or $y - 4 = -2(x - (-1))$

The line has slope -2, or $\dfrac{-2}{1}$, and passes through $(-1, 4)$. We plot $(-1, 4)$ and then find a second point by moving down 2 units and right 1 unit to $(0, 2)$. We draw the line through these points.

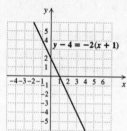

40.

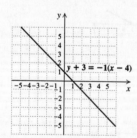

41. $y + 3 = -(x + 2)$, or $y - (-3) = -1(x - (-2))$

The line has slope -1, or $\dfrac{-1}{1}$, and passes through $(-2, -3)$. We plot $(-2, -3)$ and then find a second point by moving down 1 unit and right 1 unit to $(-1, -4)$. We draw the line through these points.

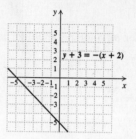

42.

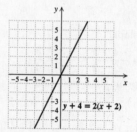

43. Graph $5x - 2y = 5$.

Find the y-intercept.
$$-2y = 5$$
$$y = -\frac{5}{2}$$

The y-intercept is $\left(0, -\dfrac{5}{2}\right)$.

Find the x-intercept.
$$5x = 5$$
$$x = 1$$

The x-intercept is $(1, 0)$.

Find a third point on the line. Let $x = 3$ and solve for y.
$$5 \cdot 3 - 2y = 5$$
$$15 - 2y = 5$$
$$-2y = -10$$
$$y = 5$$

A third point is $(3, 5)$.

Plot these points and draw the graph.

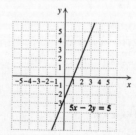

44.

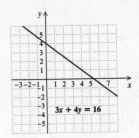

45. Graph $y = -\frac{4}{3}x - 6$.

x	y
-6	2
-3	-2
0	-6

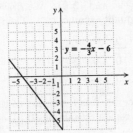

46.

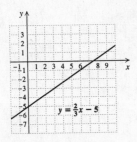

47. $3x - 5 \le 7x + 2$

$$-5 \le 4x + 2$$
$$-7 \le 4x$$
$$-\frac{7}{4} \le x$$

The solution set is $\left\{ x \Big| x \ge -\frac{7}{4} \right\}$.

48. $\left\{ x \Big| x > \frac{17}{7} \right\}$

49. ◈

50. ◈

51. ◈

52. ◈

53. First find the slope of $3x - y + 4 = 0$.

$$3x - y + 4 = 0$$
$$3x + 4 = y$$

The slope is 3.

Then find an equation of the line containing $(-5, 2)$ and having slope 3.

$$y - y_1 = m(x - x_1)$$

We substitute 3 for m, -5 for x_1, and 2 for y_1.

$$y - 2 = 3(x - (-5))$$
$$y - 2 = 3(x + 5)$$
$$y - 2 = 3x + 15$$
$$y = 3x + 17$$

54. $y = \frac{1}{5}x - 2$

55. First find the slope of the line passing through $(2, 7)$ and $(-1, -3)$.

$$m = \frac{-3 - 7}{-1 - 2} = \frac{-10}{-3} = \frac{10}{3}$$

Now find an equation of the line containing the point $(-1, 5)$ and having slope $\frac{10}{3}$.

$$y - 5 = \frac{10}{3}(x - (-1))$$
$$y - 5 = \frac{10}{3}(x + 1)$$
$$y - 5 = \frac{10}{3}x + \frac{10}{3}$$
$$y = \frac{10}{3}x + \frac{25}{3}$$

56. $y = \frac{1}{2}x + 1$

57. ◈ ▨

Exercise Set 7.3

1. We use alphabetical order of variables. We replace x by -3 and y by -5.

$$\frac{x + 3y < -18}{}$$

$$\frac{-3 + 3(-5) \ ? \ -18}{}$$
$$-3 - 15 \ \Big|$$
$$-18 \ \Big| \ -18 \ \text{FALSE}$$

Since $-18 < -18$ is false, $(-3, -5)$ is not a solution.

2. Yes

3. We use alphabetical order of variables. We replace x by -1 and y by $\frac{1}{2}$.

$$\frac{6y - 5x \ge 7}{}$$

$$\frac{6 \cdot \frac{1}{2} - 5(-1) \ ? \ 7}{}$$
$$3 + 5 \ \Big|$$
$$8 \ \Big| \ 7 \ \text{TRUE}$$

Since $8 \ge 7$ is true, $\left(-1, \frac{1}{2} \right)$ is a solution.

4. Yes

5. Graph $y \leq x + 3$.

First graph the line $y = x + 3$. The intercepts are $(0,3)$ and $(-3,0)$. We draw a solid line since the inequality symbol is $\leq$. Then we pick a test point that is not on the line We try $(0,0)$.

$$\frac{y \leq x + 3}{0 \; ? \; 0 + 3}$$
$$0 \; \mid \; 3 \qquad \text{TRUE}$$

We see that $(0,0)$ is a solution of the inequality, so we shade the region that contains $(0,0)$.

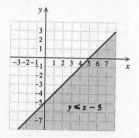

6.

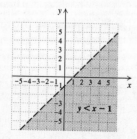

7. Graph $y < x - 1$.

First graph the line $y = x - 1$. The intercepts are $(0,-1)$ and $(1,0)$. We draw a dashed line since the inequality symbol is $<$. Then we pick a test point that is not on the line. We try $(0,0)$.

$$\frac{y < x - 1}{0 \; ? \; 0 - 1}$$
$$0 \; \mid \; -1 \qquad \text{FALSE}$$

Since $(0,0)$ is not a solution of the inequality, we shade the region that does not contain $(0,0)$.

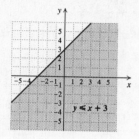

8.

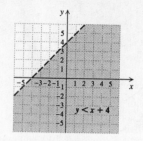

9. Graph $y \geq x - 2$.

First graph the line $y = x - 2$. The intercepts are $(0,-2)$ and $(2,0)$. We draw a solid line since the inequality symbol is $\geq$. Then we test the point $(0,0)$.

$$\frac{y \geq x - 2}{0 \; ? \; 0 - 2}$$
$$0 \; \mid \; -2 \qquad \text{TRUE}$$

Since $(0,0)$ is a solution of the inequality, we shade the region containing $(0,0)$.

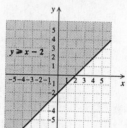

10.

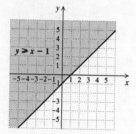

11. Graph $y \leq 2x - 1$.

First graph the line $y = 2x - 1$. The intercepts are $(0,-1)$ and $\left(\frac{1}{2},0\right)$. We draw a solid line since the inequality symbol is $\leq$. Then we test the point $(0,0)$.

$$\frac{y \leq 2x - 1}{0 \; ? \; 2 \cdot 0 - 1}$$
$$0 \; \mid \; -1 \qquad \text{FALSE}$$

Since $(0,0)$ is not a solution of the inequality, we shade the region that does not contain $(0,0)$.

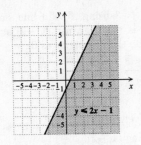

$y \le 2x - 1$

12.

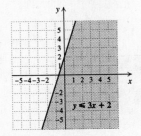

$y \le 3x + 2$

13. Graph $x - y \le 3$.

First graph the line $x - y = 3$. The intercepts are $(0, -3)$ and $(3, 0)$. We draw a solid line since the inequality symbol is $\le$. Then we test the point $(0, 0)$.

$$\begin{array}{c|c} x - y \le 3 \\ \hline 0 + 0 \ ? \ 3 \\ 0 & 3 \ \ \text{TRUE} \end{array}$$

Since $(0, 0)$ is a solution of the inequality, we shade the region that contains $(0, 0)$.

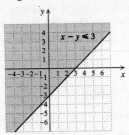

$x - y \le 3$

14.

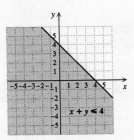

$x + y \le 4$

15. Graph $x + y > 7$.

First graph the line $x + y = 7$. The intercepts are $(0, 7)$ and $(7, 0)$. We draw a dashed line since the inequality symbol is $>$. Then we test the point $(0, 0)$.

$$\begin{array}{c|c} x + y > 7 \\ \hline 0 + 0 \ ? \ 7 \\ 0 & 7 \ \ \text{FALSE} \end{array}$$

Since $(0, 0)$ is not a solution of the inequality, we shade the region that does not contain $(0, 0)$.

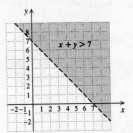

$x + y > 7$

16.

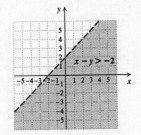

$x - y > -2$

17. Graph $y \ge 1 - 2x$.

First graph the line $y = 1 - 2x$. The intercepts are $(0, 1)$ and $\left(\frac{1}{2}, 0\right)$. We draw a solid line since the inequality symbol is $\ge$. Then we test the point $(0, 0)$.

$$\begin{array}{c|c} y \ge 1 - 2x \\ \hline 0 \ ? \ 1 - 2 \cdot 0 \\ 0 & 1 \hspace{1cm} \text{FALSE} \end{array}$$

Since $(0, 0)$ is not a solution of the inequality, we shade the region that does not contain $(0, 0)$.

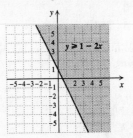

$y \ge 1 - 2x$

18.

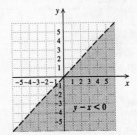

$y - x < 0$

19. Graph $y - 3x > 0$.

First graph the line $y - 3x = 0$, or $y = 3x$. Two points on the line are $(0, 0)$ and $(1, 3)$. We draw a dashed line, since the inequality symbol is $>$. Then we test the point $(2, 1)$, which is not a point on the line.

$$\frac{y - 3x > 0}{1 - 3 \cdot 2 \ ? \ 0}$$

$$1 - 6 \quad \Big|$$

$$-5 \quad \Big| \quad 0 \quad \text{FALSE}$$

Since $(2, 1)$ is not a solution of the inequality, we shade the region that does not contain $(2, 1)$.

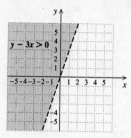

20.

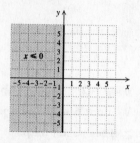

21. Graph $x \geq 3$.

First graph the line $x = 3$ using a solid line since the inequality symbol is $\geq$. Then use $(4, -3)$ as a test point. We can write the inequality as $x + 0y \geq 3$.

$$\frac{x + 0y \geq 3}{4 + 0(-3) \ ? \ 3}$$

$$4 \quad \Big| \quad 3 \quad \text{TRUE}$$

Since $(4, -3)$ is a solution of the inequality, we shade the region containing $(4, -3)$.

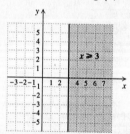

22.

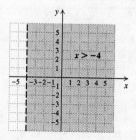

23. Graph $y \leq 3$.

Graph the line $y = 3$ using a solid line since the inequality symbol is $\leq$. Then use $(1, -2)$ as a test point. We can write the inequality as $0x + y \leq 3$.

$$\frac{x + 0y \leq 3}{0 \cdot 1 + (-2) \ ? \ 3}$$

$$-2 \quad \Big| \quad 3 \quad \text{TRUE}$$

Since $(1, -2)$ is a solution of the inequality, we shade the region containing $(1, -2)$.

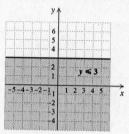

24.

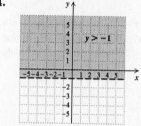

25. Graph $y \geq -5$.

Graph the line $y = -5$ using a solid line since the inequality symbol is $\geq$. Then use $(2, 3)$ as a test point. We can write the inequality as $0x + y \geq -5$.

$$\frac{0x + y \geq -5}{0 \cdot 2 + 3 \ ? \ -5}$$

$$3 \quad \Big| \quad -5 \quad \text{TRUE}$$

Since $(2, 3)$ is a solution of the inequality, we shade the region containing $(2, 3)$.

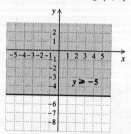

26.

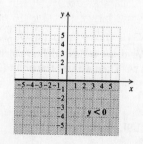

27. Graph $x < 4$.

Graph the line $x = 4$ using a dashed line since the inequality symbol is $<$. Then use $(-1, 2)$ as a test point. We can write the inequality as $x + 0y < 4$.

$$\begin{array}{c|c} x + 0y < 4 \\ \hline -1 + 0 \cdot 2 \;?\; 4 \\ -1 & 4 \quad \text{TRUE} \end{array}$$

Since $(-1, 2)$ is a solution of the inequality, we shade the region containing $(-1, 2)$.

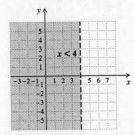

28.

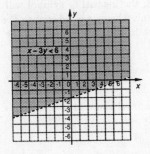

29. Graph $x - y < -10$.

Graph the line $x - y = -10$. The intercepts are $(0, 10)$ and $(-10, 0)$. We draw a dashed line since the inequality symbol is $<$. Then we test the point $(0, 0)$.

$$\begin{array}{c|c} x - y < -10 \\ \hline 0 - 0 \;?\; -10 \\ 0 & -10 \quad \text{FALSE} \end{array}$$

Since $(0, 0)$ is not a solution of the inequality, we shade the region that does not contain $(0, 0)$.

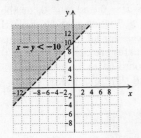

30.

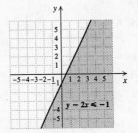

31. Graph $2x + 3y \le 12$.

First graph the line $2x + 3y = 12$. The intercepts are $(0, 4)$ and $(6, 0)$. We draw a solid line since the inequality symbol is $\le$. Then we test the point $(0, 0)$.

$$\begin{array}{c|c} 2x + 3y \le 12 \\ \hline 2 \cdot 0 + 3 \cdot 0 \;?\; 12 \\ 0 & 12 \quad \text{TRUE} \end{array}$$

Since $(0, 0)$ is a solution of the inequality, we shade the region containing $(0, 0)$.

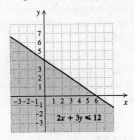

32.

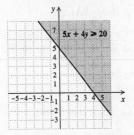

33. We multiply using columns.

$$\begin{array}{r} 2x^2 - x - 1 \\ x + 3 \\ \hline 6x^2 - 3x - 3 \quad \text{Multiplying by 3} \\ 2x^3 - x^2 - x \phantom{{}-3} \quad \text{Multiplying by } x \\ \hline 2x^3 + 5x^2 - 4x - 3 \quad \text{Adding} \end{array}$$

34. $9x^2 - 25$

35. $\quad 3a^3 + 18a^2 - 4a - 24$

$\quad = 3a^2(a + 6) - 4(a + 6) \qquad$ Factoring by grouping

$\quad = (a + 6)(3a^2 - 4)$

36. $\dfrac{x - 7}{x + 5}$

37. $2(y-5) - 3y = 8$

$\quad 2y - 10 - 3y = 8$

$\qquad\quad -y - 10 = 8$

$\qquad\qquad\quad -y = 18$

$\qquad\qquad\quad\; y = -18$

The solution is -18.

38. $\dfrac{43}{7}$

39. ◈

40. ◈

41. ◈

42. ◈

43. The c children weigh $35c$ kg, and the a adults weigh $75a$ kg. Together, the children and adults weigh $35c + 75a$ kg. When this total is more than 1000 kg the elevator is overloaded, so we have $35c + 75a > 1000$. (Of course, c and a would also have to be non-negative, so we show only the portion of the graph that is in the first quadrant.)

To graph $35c + 75a > 1000$, we first graph $35c + 75a = 1000$ using a dashed line. Two points on the line are $(4, 20)$ and $(5, 11)$. (We are using alphabetical order of variables.) Then we test the point $(0,0)$.

$$\frac{35c + 75a > 1000}{35\cdot 0 + 75\cdot 0\;?\;1000}$$

$$\begin{array}{c|c} 0 & 1000 \quad \text{FALSE} \end{array}$$

Since $(0,0)$ is not a solution of the inequality, we shade the region that does not contain $(0,0)$.

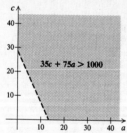

44.

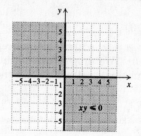

45. The sum of the riser and tread is given by $r + t$. This should not be less than 17 inches, so we write the inequality $r + t \ge 17$. To graph $r+t \ge 17$, we first graph $r + t = 17$ using a solid line. The intercepts are $(0, 17)$ and $(17, 0)$. Then we test the point $(0,0)$.

$$\frac{r + t \ge 17}{0 + 0\;?\;17}$$

$$\begin{array}{c|c} 0 & 17 \quad \text{FALSE} \end{array}$$

Since $(0,0)$ is not a solution of the inequality, we shade the region that does not contain $(0,0)$.

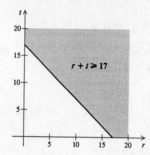

46. $y > x - 2$

47. The boundary line is a vertical line that passes through $(-2, 0)$, so its equation is $x = -2$. Since the line is solid, the inequality symbol will be $\le$ or $\ge$. To determine which, we substitute the coordinates of a point in the shaded region. We will use $(0,0)$.

$$\frac{x\quad -2}{0\;?\;-2}$$

Since $0 \ge -2$ is true, the correct symbol is $\ge$. The inequality is $x \ge -2$.

48.

49. Graph $xy \ge 0$.

From the principle of zero products, we know that $xy = 0$ when $x = 0$ or $y = 0$. Therefore, the graph contains the lines $x = 0$ and $y = 0$, or the y- and x-axes. Also, $xy > 0$ when x and y have the same sign. This is the case for all points in the first quadrant (both coordinates are positive) and in the third quadrant (both coordinates are negative). Thus, we shade the first and third quadrants.

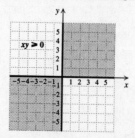

Exercise Set 7.4

1. We substitute to find k.

$y = kx$ y varies directly as x.

$28 = k \cdot 7$ Substituting 28 for y and 7 for x

$\dfrac{28}{7} = k$

$4 = k$ k is the constant of variation.

The equation of variation is $y = 4x$.

2. $y = 3.75x$

3. We substitute to find k.

$y = kx$ y varies directly as x.

$0.7 = k \cdot 0.4$ Substituting 0.7 for y and 0.4 for x

$\dfrac{0.7}{0.4} = k$

$\dfrac{7}{4} = k$, or $k = 1.75$

The equation of variation is $y = 1.75x$.

4. $y = 1.6x$

5. We substitute to find k.

$y = kx$

$400 = k \cdot 125$ Substituting 400 for y and 125 for x

$\dfrac{400}{125} = k$

$\dfrac{16}{5} = k$, or $k = 3.2$

The equation of variation is $y = 3.2x$.

6. $y = 3.6x$

7. We substitute to find k.

$y = kx$

$200 = k \cdot 300$ Substituting 200 for y and 300 for x

$\dfrac{200}{300} = k$

$\dfrac{2}{3} = k$

The equation of variation is $y = \dfrac{2}{3}x$.

8. $y = \dfrac{25}{3}x$

9. We substitute to find k.

$y = \dfrac{k}{x}$

$45 = \dfrac{k}{2}$ Substituting 45 for y and 2 for x

$90 = k$

The equation of variation is $y = \dfrac{90}{x}$.

10. $y = \dfrac{80}{x}$

11. We substitute to find k.

$y = \dfrac{k}{x}$

$7 = \dfrac{k}{10}$ Substituting 7 for y and 10 for x

$70 = k$

The equation of variation is $y = \dfrac{70}{x}$.

12. $y = \dfrac{1}{x}$

13. We substitute to find k.

$y = \dfrac{k}{x}$

$6.25 = \dfrac{k}{0.16}$ Substituting 6.25 for y and 0.16 for x

$1 = k$

The equation of variation is $y = \dfrac{1}{x}$.

14. $y = \dfrac{1050}{x}$

15. We substitute to find k.

$y = \dfrac{k}{x}$

$42 = \dfrac{k}{50}$ Substituting 42 for y and 50 for x

$2100 = k$

The equation of variation is $y = \dfrac{2100}{x}$.

16. $y = \dfrac{0.06}{x}$

17. *Familiarize and Translate*. The problem states that we have direct variation between the variables P and H. Thus, an equation $P = kH$ applies.

Carry out. First find an equation of variation.

$P = kH$

$78.75 = k \cdot 15$ Substituting 78.75 for P and 15 for H

$\dfrac{78.75}{15} = k$

$5.25 = k$

The equation of variation is $P = 5.25H$. When $H = 35$, we have:

$P = 5.25H$

$P = 5.25(35)$ Substituting 35 for H

$P = 183.75$

Check. This check might be done by repeating the computations. We might also do some reasoning about the answer. The paycheck increased from \$78.75 to \$183.75. Similarly, the hours increased from 15 to 35. The ratios 15/78.75 and 35/183.75 are the same value: about 0.19.

State. For 35 hours work, the paycheck is $183.75.

18. 16,445

19. **Familiarize and Translate**. This problem states that we have direct variation between S and W. Thus, an equation $S = kW$ applies.

Carry out. First find an equation of variation.

$$S = kW$$

$$40 = k \cdot 14 \quad \text{Substituting 40 for } S \text{ and} \\ 14 \text{ for } W$$

$$\frac{40}{14} = k$$

$$\frac{20}{7} = k$$

The equation of variation is $S = \frac{20}{7}W$. When $W = 8$, we have:

$$S = \frac{20}{7}W$$

$$S = \frac{20}{7} \cdot 8 \qquad \text{Substituting 8 for } W$$

$$S = \frac{160}{7}, \text{ or } 22\frac{6}{7}$$

Check. A check might be done by repeating the computation. We can also do some reasoning about the answer. The number of servings decreased from 40 to $22\frac{6}{7}$. Similarly, the weight decreased from 14 kg to 8 kg. The ratios 14/40 and $8/22\frac{6}{7}$ are the same value: 0.35.

State. $22\frac{6}{7}$ servings can be obtained from an 8-kg turkey.

20. 320 cm^3

21. **Familiarize and Translate**. The problem states that we have inverse variation between I and R. Thus, an equation $I = \frac{k}{R}$ applies.

Carry out. First find an equation of variation.

$$I = \frac{k}{R}$$

$$2 = \frac{k}{960} \quad \text{Substituting 2 for } I \text{ and} \\ 960 \text{ for } R$$

$$1920 = k$$

The equation of variation is $I = \frac{1920}{R}$. When $R = 540$, we have:

$$I = \frac{k}{R}$$

$$I = \frac{1920}{540} \quad \text{Substituting 540 for } R$$

$$I = \frac{32}{9}, \text{ or } 3\frac{5}{9}$$

Check. A check might be done by repeating the computations. We can also do some reasoning about the answer. Note that as the resistance decreases, the current increases as expected. Also note that $2 \cdot 960$ and $\frac{32}{9} \cdot 540$ are both 1920.

State. The current is $3\frac{5}{9}$ amperes when the resistance is 540 ohms.

22. $18\frac{1}{3}$ lb

23. **Familiarize and Translate**. The problem states that we have direct variation between the variables M and E. Thus, an equation $M = kE$ applies.

Carry out. First find an equation of variation.

$$M = kE$$

$$79.42 = k \cdot 209 \quad \text{Substituting 79.42 for } M \\ \text{and 209 for } E$$

$$\frac{79.42}{209} = k$$

$$0.38 = k$$

The equation of variation is $M = 0.38E$. When $E = 176$, we have:

$$M = 0.38E$$

$$M = 0.38(176) \quad \text{Substituting 176 for } E$$

$$M = 66.88$$

Check. A check might be done by repeating the computations. We can also do some reasoning about the answer. Note that the weight on Earth decreases and the weight on Mars also decreases as expected. Also note that the ratios 79.42/209 and 66.88/176 are both 0.38.

State. A person who weighs 176 lb on Earth will weigh 66.88 lb on Mars.

24. 2.4 ft

25. **Familiarize and Translate**. The problem states that we have inverse variation between t and r. Thus, an equation $t = \frac{k}{r}$ applies.

Carry out. First find an equation of variation.

$$t = \frac{k}{r}$$

$$90 = \frac{k}{1200} \quad \text{Substituting 90 for } t \\ \text{and 1200 for } r$$

$$90 \cdot 1200 = k$$

$$108,000 = k$$

The equation of variation is $t = \frac{108,000}{r}$. When $r = 2000$, we have:

$$t = \frac{108,000}{r}$$

$$t = \frac{108,000}{2000}$$

$$t = 54$$

Check. A check might be done by repeating the computations. We can also do some reasoning about the answer. Note that as the rate increases the time decreases as expected. Also note that $90 \cdot 1200$ and $54 \cdot 2000$ are both 108,000.

State. It will take 54 min to empty the tank at the rate of 2000 L/min.

26. 46.7¢, 1.9¢

27. Familiarize and Translate. The problem states that we have inverse variation between the variables m and n. Thus, an equation $m = \dfrac{k}{n}$ applies.

Carry out. First we find an equation of variation.

$$m = \frac{k}{n}$$

$$2.5 = \frac{k}{16} \quad \begin{array}{l}\text{Substituting 2.5 for } m \\ \text{and 16 for } n\end{array}$$

$$40 = k$$

The equation of variation is $m = \dfrac{40}{n}$. When $m = 4$,

$$m = \frac{40}{n}$$

$$4 = \frac{40}{n}$$

$$n = 10 \quad \text{Multiplying by } \frac{n}{4}$$

Check. We can repeat the computations. We also note that the products $(2.5)(16)$ and $4 \cdot 10$ are both 40. Also note that as the number of minutes increases the number of questions decreases as expected.

State. 10 questions would appear on a quiz in which students have 4 minutes per question.

28. 5

29. $2a - 3b = 4$

$$-3b = -2a + 4 \qquad \text{Adding } -2a$$

$$b = -\frac{1}{3}(-2a + 4) \quad \text{Multiplying by } -\frac{1}{3}$$

$$b = \frac{2}{3}a - \frac{4}{3}, \text{ or}$$

$$b = \frac{2a - 4}{3}$$

30. $m = \dfrac{p - 2n}{4}$

31. $(5x + 9y) - (2x - 3y)$

$= 5x + 9y - 2x + 3y$

$= 3x + 12y$

32. $-3a + b$

33. $3x - 5y = 6$

$$-5y = -3x + 6$$

$$y = -\frac{1}{5}(-3x + 6)$$

$$y = \frac{3}{5}x - \frac{6}{5}$$

x	y
-3	-3
2	0
7	3

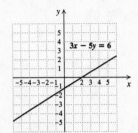

34.

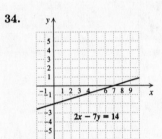

35. ◈

36. ◈

37. ◈

38. ◈

39. $P = kS$

40. $C = kr$

$k = 2\pi$

41. $B = kN$

42. $A = kr^2$

$k = \pi$

43. $S = kv^6$

44. $p^2 = kt$

45. $I = \dfrac{k}{d^2}$

46. $D = \dfrac{k}{V}$

47. $V = kr^3$

$k = \dfrac{4}{3}\pi$

48. $P = kv^3$

49. ◈

Chapter 8

Systems Of Equations And Problem Solving

Exercise Set 8.1

1. We check by substituting alphabetically 3 for x and 2 for y.

$$
\begin{array}{c|c}
2x + 3y = 12 & x - 4y = -5 \\
\hline
2 \cdot 3 + 3 \cdot 2 \ ? \ 12 & 3 - 4 \cdot 2 \ ? \ -5 \\
6 + 6 & 3 - 8 \\
12 \ \big|\ 12 \ \ \text{TRUE} & -5 \ \big|\ -5 \ \ \text{TRUE}
\end{array}
$$

The ordered pair $(3, 2)$ is a solution of each equation. Therefore it is a solution of the system of equations.

2. Yes

3. We check by substituting alphabetically 3 for a and 2 for b.

$$
\begin{array}{c|c}
3b - 2a = 0 & b + 2a = 15 \\
\hline
3 \cdot 2 - 2 \cdot 3 \ ? \ 0 & 2 + 2 \cdot 3 \ ? \ 15 \\
6 - 6 & 2 + 6 \\
0 \ \big|\ 0 \ \ \text{TRUE} & 8 \ \big|\ 15 \ \ \text{FALSE}
\end{array}
$$

The ordered pair $(3, 2)$ is not a solution of $b + 2a = 15$. Therefore it is not a solution of the system of equations.

4. Yes

5. We check by substituting alphabetically 15 for x and 20 for y.

$$
\begin{array}{c|c}
3x - 2y = 5 \\
\hline
3 \cdot 15 - 2 \cdot 20 \ ? \ 5 \\
45 - 40 \\
5 \ \big|\ 5 \ \ \text{TRUE}
\end{array}
$$

$$
\begin{array}{c|c}
6x - 5y = -10 \\
\hline
6 \cdot 15 - 5 \cdot 20 \ ? \ -10 \\
90 - 100 \\
-10 \ \big|\ -10 \ \ \text{TRUE}
\end{array}
$$

The ordered pair $(15, 20)$ is a solution of each equation. Therefore it is a solution of the system of equations.

6. Yes

7. We graph the equations.

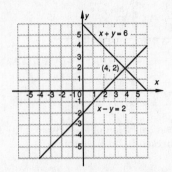

The "apparent" solution of the system, $(4, 2)$ should be checked in both equations.

Check:

$$
\begin{array}{c|c}
x + y = 6 & x - y = 2 \\
\hline
4 + 2 \ ? \ 6 & 4 - 2 \ ? \ 2 \\
6 \ \big|\ 6 \ \ \text{TRUE} & 2 \ \big|\ 2 \ \ \text{TRUE}
\end{array}
$$

The solution is $(4, 2)$.

8. $(2, 1)$

9. We graph the equations.

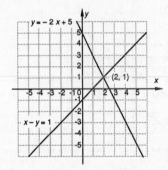

The "apparent" solution of the system, $(2, 1)$ should be checked in both equations.

Check:

$$
\begin{array}{c|c}
y = -2x + 5 & x - y = 1 \\
\hline
1 \ ? \ -2 \cdot 2 + 5 & 2 - 1 \ ? \ 1 \\
-4 + 5 & 1 \ \big|\ 1 \ \ \text{TRUE} \\
1 \ \big|\ 1 \ \ \ \ \text{TRUE} &
\end{array}
$$

The solution is $(2, 1)$.

10. $(3, 1)$

11. We graph the equations.

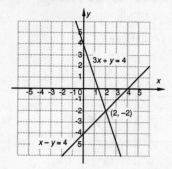

The "apparent" solution of the system, $(2, -2)$ should be checked in both equations.

Check:

$3x + y = 4$	$x - y = 4$
$3 \cdot 2 + (-2)$? 4	$2 - (-2)$? 4
$6 - 2$	$2 + 2$
$4 \mid 4$ TRUE	$4 \mid 4$ TRUE

The solution is $(2, -2)$.

12. $(1, 1)$

13. We graph the equations.

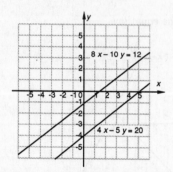

The lines are parallel. There is no solution.

14. Infinite number of solutions

15. We graph the equations.

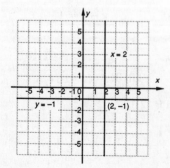

The "apparent" solution of the system, $(2, -1)$, should be checked in both equations.

Check:

$x = 2$	$y = -1$
2 ? 2 TRUE	-1 ? -1 TRUE

The solution is $(2, -1)$.

16. $\left(\dfrac{1}{3}, 1\right)$

17. We graph the equations.

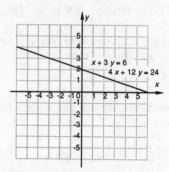

The equations represent the same line. There is an infinite number of solutions.

18. $(-4, 2)$

19. We graph the equations.

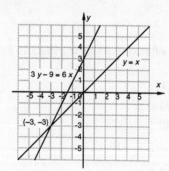

The "apparent" solution of the system, $(-3, -3)$ should be checked in both equations.

Check:

$3y - 9 = 6x$	$y = x$
$3(-3) - 9$? $6(-3)$	-3 ? -3 TRUE
$-9 - 9 \mid -18$	
$-18 \mid -18$ TRUE	

The solution is $(-3, -3)$.

20. $(-6, -2)$

21. We graph the equations.

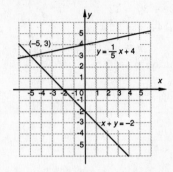

The "apparent" solution of the system, $(-5, 3)$, should be checked in both equations.

Check:

$$y = \frac{1}{5}x + 4$$

$$3 \ ? \ \frac{1}{5}(-5) + 4$$

$$-1 + 4$$

$$3 \mid 3 \quad \text{TRUE}$$

$$x + y = -2$$

$$-5 + 3 \ ? \ -2$$

$$-2 \mid -2 \quad \text{TRUE}$$

The solution is $(-5, 3)$.

22. No solution

23. We graph the equations.

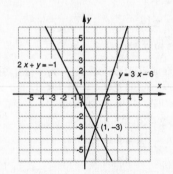

The "apparent" solution of the system, $(1, -3)$, should be checked in both equations.

Check:

$$2x + y = -1$$

$$2 \cdot 1 + (-3) \ ? \ -1$$

$$2 - 3$$

$$-1 \mid -1 \quad \text{TRUE}$$

$$y = 3x - 6$$

$$-3 \ ? \ 3 \cdot 1 - 6$$

$$3 - 6$$

$$-3 \mid -3 \quad \text{TRUE}$$

The solution is $(1, -3)$.

24. $(2, 2)$

25. We graph the equations.

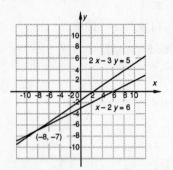

The "apparent" solution of the system, $(-8, -7)$, should be checked in both equations.

Check:

$$2x - 3y = 5$$

$$2(-8) - 3(-7) \ ? \ 5$$

$$-16 + 21$$

$$5 \mid 5 \quad \text{TRUE}$$

$$x - 2y = 6$$

$$-8 - 2(-7) \ ? \ 6$$

$$-8 + 14$$

$$6 \mid 6 \quad \text{TRUE}$$

The solution is $(-8, -7)$.

26. $(-12, 11)$

27. We graph the equations.

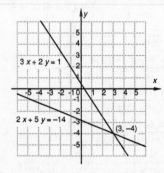

The "apparent" solution of the system, $(3, -4)$, should be checked in both equations.

Check:

$$3x + 2y = 1$$

$$3 \cdot 3 + 2(-4) \ ? \ 1$$

$$9 - 8$$

$$1 \mid 1 \quad \text{TRUE}$$

$$2x + 5y = -14$$
$$\begin{array}{c|c} 2 \cdot 3 + 5(-4) \ ? \ -14 & \\ 6 - 20 & \\ \hline -14 & -14 \ \ \text{TRUE} \end{array}$$

The solution is $(3, -4)$.

28. $(-3, 5)$

29. We graph the equations.

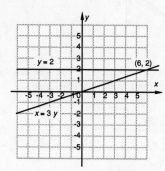

The "apparent" solution of the system, $(6, 2)$, should be checked in both equations.

Check:

$$\begin{array}{c|c} x = 3y & y = 2 \\ \hline 6 \ ? \ 3 \cdot 2 & 2 \ ? \ 2 \ \ \text{TRUE} \\ 6 \ | \ 6 \quad \text{TRUE} & \end{array}$$

The solution is $(6, 2)$.

30. $(3, 6)$

31. $\dfrac{x+2}{x-4} - \dfrac{x+1}{x+4}$, LCD is $(x-4)(x+4)$

$$= \frac{x+2}{x-4} \cdot \frac{x+4}{x+4} - \frac{x+1}{x+4} \cdot \frac{x-4}{x-4}$$

$$= \frac{(x+2)(x+4) - (x+1)(x-4)}{(x-4)(x+4)}$$

$$= \frac{x^2 + 6x + 8 - (x^2 - 3x - 4)}{(x-4)(x+4)}$$

$$= \frac{x^2 + 6x + 8 - x^2 + 3x + 4}{(x-4)(x+4)}$$

$$= \frac{9x + 12}{(x-4)(x+4)}$$

(Although $9x + 12$ can be factored as $3(3x+4)$, doing so will not enable us to simplify further.)

32. $\dfrac{2x+5}{x+3}$

33. The polynomial has three terms, so it is a trinomial.

34. Binomial

35. $2(7 - y) + 3y = 12$
$$14 - 2y + 3y = 12$$
$$14 + y = 12$$
$$y = -2$$

The solution is -2.

36. $\dfrac{26}{7}$

37. ◈

38. ◈

39. ◈

40. ◈

41. Systems in which the graphs of the equations coincide contain dependent equations. This is the case in Exercises 14 and 17.

42. Exercises 7-12, 14-21, 23-30

43. Systems in which the graphs of the equations are parallel are inconsistent. This is the case in Exercises 13 and 22.

44. Exercises 7-13, 15, 16, 18-30

45. Answers may vary. Any two independent equations with $(2, -4)$ as a solution will do.
$$2x - y = 8$$
$$x + 3y = -10$$

46. Answers may vary. $x + y = 1$

47. $(2, -3)$ is a solution of $Ax - 3y = 13$. Substitute 2 for x and -3 for y and solve for A.
$$Ax - 3y = 13$$
$$A \cdot 2 - 3(-3) = 13$$
$$2A + 9 = 13$$
$$2A = 4$$
$$A = 2$$

$(2, -3)$ is a solution of $x - By = 8$. Substitute 2 for x and -3 for y and solve for B.
$$x - By = 8$$
$$2 - B(-3) = 8$$
$$2 + 3B = 8$$
$$3B = 6$$
$$B = 2$$

48. $(1.25, 1.5)$

49. a) Let $x =$ the number of months the water heater operates and $y =$ the cost. The electric water heater costs \$100 per month to operate, so we have $y = 100x$. The gas water heater costs \$25 per month plus the \$250 purchase price plus the \$150 installation fee, so we have $y = 25x + 250 + 150$, or $y = 25x + 400$.

b)

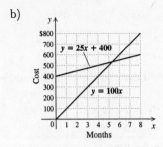

c) The first coordinate of the points of intersection of the graphs appears to be $5\frac{1}{3}$, so Elizabeth will break even after the gas water heater has operated for $5\frac{1}{3}$ months.

50. a) Copy card: $y = 18$, per page: $y = 0.08x$

b)

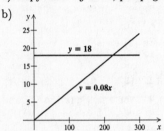

c) More than 225

51. $(41.5, 17.1)$

Exercise Set 8.2

1. $x + y = -2,$ (1)
     $y = x - 6$ (2)

Substitute $x - 6$ for y in Equation (1) and solve for x.

$$x + y = -2 \quad (1)$$
$$x + (x - 6) = -2 \quad \text{Substituting}$$
$$2x - 6 = -2$$
$$2x = 4$$
$$x = 2$$

Next we substitute 2 for x in either equation of the original system and solve for y.

$$y = x - 6 \quad (2)$$
$$y = 2 - 6 \quad \text{Substituting}$$
$$y = -4$$

We check the ordered pair $(2, -4)$.

$x + y = -2$		$y = x - 6$	
$2 + (-4)$? -2		-4 ? $2 - 6$	
-2 $\mid$ -2 TRUE		-4 $\mid$ -4 TRUE	

Since $(2, -4)$ checks in both equations, it is the solution.

2. $(9, 1)$

3. $x = y + 1,$ (1)
     $x + 2y = 4$ (2)

Substitute $y + 1$ for x in Equation (1) and solve for y.

$$x + 2y = 4 \quad (2)$$
$$(y + 1) + 2y = 4 \quad \text{Substituting}$$
$$3y + 1 = 4$$
$$3y = 3$$
$$y = 1$$

Next we substitute 1 for y in either equation of the original system and solve for x.

$$x = y + 1 \quad (1)$$
$$x = 1 + 1 \quad \text{Substituting}$$
$$x = 2$$

We check the ordered pair $(2, 1)$.

$x = y + 1$		$x + 2y = 4$	
2 ? $1 + 1$		$2 + 2 \cdot 1$? 4	
2 $\mid$ 2 TRUE		$2 + 2$	
		4 $\mid$ 4 TRUE	

Since $(2, 1)$ checks in both equations, it is the solution.

4. $(2, -1)$

5. $y = 2x - 5,$ (1)
     $3y - x = 5$ (2)

Substitute $2x - 5$ for y in Equation (2) and solve for x.

$$3y - x = 5 \quad (2)$$
$$3(2x - 5) - x = 5 \quad \text{Substituting}$$
$$6x - 15 - x = 5$$
$$5x - 15 = 5$$
$$5x = 20$$
$$x = 4$$

Next we substitute 4 for x in either equation of the original system and solve for y.

$$y = 2x - 5 \quad (1)$$
$$y = 2 \cdot 4 - 5 \quad \text{Substituting}$$
$$y = 8 - 5$$
$$y = 3$$

We check the ordered pair $(4, 3)$.

$y = 2x - 5$		$3y - x = 5$	
3 ? $2 \cdot 4 - 5$		$3 \cdot 3 - 4$? 5	
	$8 - 5$	$9 - 4$	
3 $\mid$ 3 TRUE		5 $\mid$ 5 TRUE	

Since $(4, 3)$ checks in both equations, it is the solution.

6. $(1,3)$

7. $4x + y = 2,$ (1)
$\quad\quad y = -2x$ (2)

Substitute $-2x$ for y in Equation (1) and solve for x.

$$4x + y = 2 \quad (1)$$
$$4x + (-2x) = 2 \quad \text{Substituting}$$
$$2x = 2$$
$$x = 1$$

Next we substitute 1 for x in either equation of the original system and solve for y.

$$y = -2x \quad (2)$$
$$y = -2 \cdot 1 \quad \text{Substituting}$$
$$y = -2$$

We check the ordered pair $(1, -2)$.

$4x + y = 2$		$y = -2x$	
$4 \cdot 1 + (-2)$? 2		-2 ? $-2 \cdot 1$	
$4 - 2$		-2	-2 TRUE
2	2 TRUE		

Since $(1, -2)$ checks in both equations, it is the solution.

8. $(-30, 10)$

9. $2x + 3y = 8,$ (1)
$\quad\quad x = y - 6$ (2)

Substitute $y - 6$ for x in Equation (1) and solve for y.

$$2x + 3y = 8 \quad (1)$$
$$2(y - 6) + 3y = 8 \quad \text{Substituting}$$
$$2y - 12 + 3y = 8$$
$$5y - 12 = 8$$
$$5y = 20$$
$$y = 4$$

Next we substitute 4 for y in either equation of the original system and solve for x.

$$x = y - 6 \quad (2)$$
$$x = 4 - 6 \quad \text{Substituting}$$
$$x = -2$$

We check the ordered pair $(-2, 4)$.

$2x + 3y = 8$		$x = y - 6$	
$2(-2) + 3 \cdot 4$? 8		-2 ? $4 - 6$	
$-4 + 12$		-2	-2 TRUE
8	8 TRUE		

Since $(-2, 4)$ checks in both equations, it is the solution.

10. $(-3, 5)$

11. $x = 2y + 1,$ (1)
$\quad 3x - 6y = 2$ (2)

We substitute $2y + 1$ for x in Equation (2) and solve for y.

$$3x - 6y = 2 \quad (2)$$
$$3(2y + 1) - 6y = 2$$
$$6y + 3 - 6y = 2$$
$$3 = 2$$

We obtain a false equation, so the system has no solution.

12. Infinite number of solutions

13. $s + t = -4,$ (1)
$\quad s - t = 2$ (2)

We solve Equation (2) for s.

$$s - t = 2 \quad (2)$$
$$s = t + 2 \quad (3)$$

We substitute $t + 2$ for s in Equation (1) and solve for t.

$$s + t = -4 \quad (1)$$
$$(t + 2) + t = -4 \quad \text{Substituting}$$
$$2t + 2 = -4$$
$$2t = -6$$
$$t = -3$$

Now we substitute -3 for t in either of the original equations or in Equation (3) and solve for s. It is easiest to use (3).

$$s = t + 2 = -3 + 2 = -1$$

We check the ordered pair $(-1, -3)$.

$s + t = -4$	
$-1 + (-3)$? -4	
-4	-4 TRUE

$s - t = 2$	
$-1 - (-3)$? 2	
$-1 + 3$	
2	2 TRUE

Since $(-1, -3)$ checks in both equations, it is the solution.

14. $(2, -4)$

15. $x - y = 5,$ (1)
$\quad x + 2y = 7$ (2)

Solve Equation (1) for x.

$$x - y = 5 \quad (1)$$
$$x = y + 5 \quad (3)$$

Substitute $y + 5$ for x in Equation (2) and solve for y.

$$x + 2y = 7 \quad (2)$$
$$(y + 5) + 2y = 7 \quad \text{Substituting}$$
$$3y + 5 = 7$$
$$3y = 2$$
$$y = \frac{2}{3}$$

Substitute $\frac{2}{3}$ for y in Equation (3) and compute x.

$$x = y + 5 = \frac{2}{3} + 5 = \frac{2}{3} + \frac{15}{3} = \frac{17}{3}$$

The ordered pair $\left(\frac{17}{3}, \frac{2}{3}\right)$ checks in both equations. It is the solution.

16. $\left(\dfrac{17}{3}, \dfrac{16}{3}\right)$

17. $x - 2y = 7, \quad (1)$
$3x - 21 = 6y \quad (2)$

Solve Equation (1) for x.

$$x - 2y = 7$$
$$x = 2y + 7$$

Substitute $2y + 7$ for x in Equation (2) and solve for y.

$$3x - 21 = 6y \quad (2)$$
$$3(2y + 7) - 21 = 6y \quad \text{Substituting}$$
$$6y + 21 - 21 = 6y$$
$$6y = 6y$$

The last equation is true for any choice of y, so there is an infinite number of solutions.

18. Infinite number of solutions

19. $y = -2x + 3, \quad (1)$
$2y = -4x + 6 \quad (2)$

Substitute $-2x + 3$ for y in Equation (2) and solve for x.

$$2y = -4x + 6 \quad (2)$$
$$2(-2x + 3) = -4x + 6 \quad \text{Substituting}$$
$$-4x + 6 = -4x + 6$$

The last equation is true for any choice of x, so there is an infinite number of solutions.

20. No solution

21. $x + 2y = 10, \quad (1)$
$3x + 4y = 8 \quad (2)$

Solve Equation (1) for x.

$$x + 2y = 10 \quad (1)$$
$$x = -2y + 10 \quad (3)$$

Substitute $-2y + 10$ for x in Equation (2) and solve for y.

$$3x + 4y = 8 \quad (2)$$
$$3(-2y + 10) + 4y = 8 \quad \text{Substituting}$$
$$-6y + 30 + 4y = 8$$
$$-2y + 30 = 8$$
$$-2y = -22$$
$$y = 11$$

Substitute 11 for y in Equation (3) and compute x.

$$x = -2y + 10 = -2 \cdot 11 + 10 = -22 + 10 = -12$$

The ordered pair $(-12, 11)$ checks in both equations. It is the solution.

22. $\left(\dfrac{25}{8}, -\dfrac{11}{4}\right)$

23. $3a + 2b = 2, \quad (1)$
$-2a + b = 8 \quad (2)$

Solve Equation (2) for b.

$$-2a + b = 8 \quad (2)$$
$$b = 2a + 8 \quad (3)$$

Substitute $2a + 8$ for b in Equation (1) and solve for a.

$$3a + 2b = 2 \quad (1)$$
$$3a + 2(2a + 8) = 2$$
$$3a + 4a + 16 = 2$$
$$7a + 16 = 2$$
$$7a = -14$$
$$a = -2$$

Substitute -2 for a in Equation (3) and compute b.

$$b = 2(-2) + 8 = -4 + 8 = 4$$

The ordered pair $(-2, 4)$ checks in both equations. It is the solution.

24. $(-3, 0)$

25. $y - 2x = 0, \quad (1)$
$3x + 7y = 17 \quad (2)$

Solve Equation (1) for y.

$$y - 2x = 0 \quad (1)$$
$$y = 2x \quad (3)$$

Substitute $2x$ for y in Equation (2) and solve for x.

$$3x + 7y = 17 \quad (2)$$
$$3x + 7(2x) = 17 \quad \text{Substituting}$$
$$3x + 14x = 17$$
$$17x = 17$$
$$x = 1$$

Substitute 1 for x in Equation (3) and compute y.

$$y = 2x = 2 \cdot 1 = 2$$

The ordered pair $(1, 2)$ checks in both equations. It is the solution.

26. $(6, 3)$

27. $8x + 2y = 6$,   (1)
   $4x = 3 - y$ (2)

Solve Equation (2) for y.

   $4x = 3 - y$ (2)

   $y = 3 - 4x$ (3)

Substitute $3 - 4x$ for y in Equation (1) and solve for x.

$$8x + 2y = 6 \quad (1)$$
$$8x + 2(3 - 4x) = 6 \quad \text{Substituting}$$
$$8x + 6 - 8x = 6$$
$$6 = 6$$

The last equation is true for any choice of x, so there is an infinite number of solutions.

28. No solution

29. $x - 3y = -1$,   (1)
   $5y - 2x = 4$ (2)

Solve Equation (1) for x.

   $x - 3y = -1$ (1)

   $x = 3y - 1$ (3)

Substitute $3y - 1$ for x in Equation (2) and solve for y.

$$5y - 2x = 4 \quad\quad (2)$$
$$5y - 2(3y - 1) = 4 \quad\quad \text{Substituting}$$
$$5y - 6y + 2 = 4$$
$$-y + 2 = 4$$
$$-y = 2$$
$$y = -2$$

Now substitute -2 for y in Equation (3) and compute x.

$$x = 3(-2) - 1 = -6 - 1 = -7$$

The ordered pair $(-7, -2)$ checks in both equations. It is the solution.

30. $(-3, -4)$

31. $5x - y = 0$,   (1)
   $5x - y = -2$ (2)

Solve Equation (1) for y.

   $5x - y = 0$ (1)

   $5x = y$ (3)

Substitute $5x$ for y in Equation (2) and solve for x.

$$5x - y = -2 \quad (2)$$
$$5x - 5x = -2 \quad \text{Substituting}$$
$$0 = -2$$

We obtain a false equation, so the system has no solution.

32. No solution

33. *Familiarize*. We let $x =$ the larger number and $y =$ the smaller number.

Translate.

$\underbrace{\text{The sum of two numbers}}$ is 49.
$\quad\quad\quad\quad\downarrow \quad\quad\quad\quad\quad \downarrow \;\downarrow$
$\quad\quad\; x + y \quad\quad\quad\quad = 49$

$\underbrace{\text{One number}}$ is 3 $\underbrace{\text{more than}}$ $\underbrace{\text{the other}}$.
$\quad\downarrow \quad\quad \downarrow\;\downarrow \quad\quad \downarrow \quad\quad\quad \downarrow$
$\quad x \quad\quad = 3 \quad + \quad\quad y$

The resulting system is

   $x + y = 49$,   (1)

   $x = 3 + y$.   (2)

Carry out. We solve the system of equations. We substitute $3 + y$ for x in Equation (1) and solve for y.

$$x + y = 49 \quad (1)$$
$$(3 + y) + y = 49 \quad \text{Substituting}$$
$$3 + 2y = 49$$
$$2y = 46$$
$$y = 23$$

Next we substitute 23 for y in either equation of the original system and solve for x.

$$x + y = 49 \quad (1)$$
$$x + 23 = 49 \quad \text{Substituting}$$
$$x = 26$$

Check. The sum of 23 and 26 is 49. The number 26 is 3 more than 23. These numbers check.

State. The numbers are 26 and 23.

34. 29 and 27

35. *Familiarize*. Let $x =$ the larger number and $y =$ the smaller number.

Translate.

$\underbrace{\text{The sum of two numbers}}$ is 58.
$\quad\quad\quad\quad\downarrow \quad\quad\quad\quad\quad \downarrow\;\downarrow$
$\quad\quad\; x + y \quad\quad\quad\quad = 58$

$\underbrace{\text{The difference of two numbers}}$ is 14.
$\quad\quad\quad\quad\downarrow \quad\quad\quad\quad\quad\quad \downarrow\;\downarrow$
$\quad\quad\; x - y \quad\quad\quad\quad\quad = 14$

The resulting system is

   $x + y = 58$,   (1)

   $x - y = 14$.   (2)

Carry out. We solve the system.

We solve Equation (2) for x.

   $x - y = 14$ (2)

   $x = y + 14$ (3)

We substitute $y + 14$ for x in Equation (1) and solve for y.

$$x + y = 58 \quad (1)$$
$$(y + 14) + y = 58 \quad \text{Substituting}$$
$$2y + 14 = 58$$
$$2y = 44$$
$$y = 22$$

Now we substitute 22 for y in Equation (3) and compute x.

$$x = y + 14 = 22 + 14 = 36$$

Check. The sum of 36 and 22 is 58. The difference between 36 and 22, $36 - 22$, is 14. The numbers check.

State. The numbers are 36 and 22.

36. 39 and 27

37. Familiarize. Let $x =$ the larger number and $y =$ the smaller number.

Translate.

The difference between two numbers is 16.
$$x - y \qquad\qquad\qquad = 16$$

Three times the larger number is seven times the smaller number.
$$3x \qquad = \qquad 7y$$

The resulting system is

$$x - y = 16, \quad (1)$$
$$3x = 7y. \quad (2)$$

Carry out. We solve the system.

We solve Equation (1) for x.

$$x - y = 16 \qquad (1)$$
$$x = y + 16 \qquad (3)$$

We substitute $y + 16$ for x in Equation (2) and solve for y.

$$3x = 7y \quad (2)$$
$$3(y + 16) = 7y \quad \text{Substituting}$$
$$3y + 48 = 7y$$
$$48 = 4y$$
$$12 = y$$

Next we substitute 12 for y in Equation (3) and compute x.

$$x = y + 16 = 12 + 16 = 28$$

Check. The difference between 28 and 12, $28 - 12$, is 16. Three times the larger, $3 \cdot 28$ or 84, is seven times the smaller, $7 \cdot 12 = 84$. The numbers check.

State. The numbers are 28 and 12.

38. 22 and 4

39. Familiarize. Let $x =$ one angle and $y =$ the other angle.

Translate. Since the angles are supplementary, we have one equation.

$$x + y = 180$$

The second sentence can be translated as follows:

One angle is 30° less than twice the other.
$$x \qquad = \qquad 2y - 30$$

The resulting system is

$$x + y = 180, \quad (1)$$
$$x = 2y - 30. \quad (2)$$

Carry out. We solve the system.

We substitute $2y - 30$ for x in Equation (1) and solve for y.

$$x + y = 180 \quad (1)$$
$$2y - 30 + y = 180$$
$$3y - 30 = 180$$
$$3y = 210$$
$$y = 70$$

Next we substitute 70 for y in Equation (2) and solve for x.

$$x = 2y - 30 = 2 \cdot 70 - 30 = 140 - 30 = 110$$

Check. The sum of the angles is $70° + 110°$, or $180°$, so the angles are supplementary. If $30°$ is subtracted from twice $70°$, we have $2 \cdot 70° - 30°$, or $110°$, which is the other angle. The answer checks.

State. One angle is $70°$, and the other is $110°$.

40. $133°$, $47°$

41. Familiarize. We let $x =$ the larger angle and $y =$ the smaller angle.

Translate. Since the angles are complementary, we have one equation.

$$x + y = 90$$

We reword and translate the second statement.

The difference of two angles is 34°.
$$x - y \qquad = 34$$

The resulting system is

$$x + y = 90, \quad (1)$$
$$x - y = 34. \quad (2)$$

Carry out. We solve the system.

We first solve Equation (2) for x.

$$x - y = 34 \qquad (2)$$
$$x = y + 34 \qquad (3)$$

Substitute $y + 34$ for x in Equation (1) and solve for y.

$$x + y = 90 \quad (1)$$
$$y + 34 + y = 90$$
$$2y + 34 = 90$$
$$2y = 56$$
$$y = 28$$

Next we substitute 28 for y in Equation (3) and solve for x.

$$x = y + 34 = 28 + 34 = 62$$

Check. The sum of the angles is $62° + 28°$, or $90°$, so the angles are complementary. The difference of the angles is $62° - 28°$, or $34°$. These numbers check.

State. The angles are $62°$ and $28°$.

42. $32°, 58°$

43. *Familiarize*. Recall that the perimeter of a rectangle with length l and width w is given by $2l + 2w$.

Translate.

The perimeter is 1300 mi.

$$2l + 2w = 1300$$

The length is 110 mi more than the width.

$$l = w + 110$$

The resulting system is

$$2l + 2w = 1300, \quad (1)$$
$$l = w + 110. \quad (2)$$

Carry out. We solve the system.

Substitute $w + 110$ for l in Equation (1) and solve for w.

$$2l + 2w = 1300 \quad (1)$$
$$2(w + 110) + 2w = 1300 \quad \text{Substituting}$$
$$2w + 220 + 2w = 1300$$
$$4w + 220 = 1300$$
$$4w = 1080$$
$$w = 270$$

Now substitute 270 for w in Equation (2).

$$l = w + 110 \quad (2)$$
$$l = 270 + 110 \quad \text{Substituting}$$
$$l = 380$$

Check. If the length is 380 mi and the width is 270 mi, the perimeter would be $2 \cdot 380 + 2 \cdot 270$, or $760 + 540$, or 1300 mi. Also, the length is 110 mi more than the width. These numbers check.

State. The length is 380 mi, and the width is 270 mi.

44. 365 mi, 275 mi

45. *Familiarize*. Recall that the perimeter of a rectangle with length l and width w is given by $2l + 2w$.

Translate.

The perimeter is 120 ft.

$$2l + 2w = 120$$

The length is twice the width.

$$l = 2w$$

The resulting system is

$$2l + 2w = 120, \quad (1)$$
$$l = 2w. \quad (2)$$

Carry out. We solve the system.

Substitute $2w$ for l in Equation (1) and solve for w.

$$2 \cdot 2w + 2w = 120 \quad (1)$$
$$4w + 2w = 120$$
$$6w = 120$$
$$w = 20$$

Now substitute 20 for w in Equation (2).

$$l = 2w \quad (2)$$
$$l = 2 \cdot 20 \quad \text{Substituting}$$
$$l = 40$$

Check. If the length is 40 ft and the width is 20 ft, the perimeter would be $2 \cdot 40 + 2 \cdot 20$, or $80 + 40$, or 120 ft. Also, the length is twice the width. These numbers check.

State. The length is 40 ft, and the width is 20 ft.

46. Height of wall: 20 ft, width of service zone: 5 ft

47. *Familiarize*. Let $l =$ the length and $w =$ the width. The perimeter is $l + l + w + w$, or $2l + 2w$.

Translate.

The perimeter is 340 yd.

$$2l + 2w = 340$$

The length is 50 yd more than the width.

$$l = 50 + w$$

The resulting system is

$$2l + 2w = 340, \quad (1)$$
$$l = 50 + w. \quad (2)$$

Carry out. We solve the system. We substitute $50 + w$ for l in Equation (1) and solve for w.

$$2l + 2w = 340 \quad (1)$$
$$2(50 + w) + 2w = 340$$
$$100 + 2w + 2w = 340$$
$$100 + 4w = 340$$
$$4w = 240$$
$$w = 60$$

Next we substitute 60 for w in Equation (2) and solve for l.

$$l = 50 + w = 50 + 60 = 110$$

Check. The perimeter is $2 \cdot 110 + 2 \cdot 60$, or 340 yd. Also 50 yd more than the width is $50 + 60$, or 110 yd. The answer checks.

State. The length is 110 yd, and the width is 60 yd.

48. 120 yd, $53\frac{1}{3}$ yd

49. $6x^2 - 13x + 6$

The possibilities are $(x+\ \ )(6x+\ \ )$ and $(2x+\ \ )(3x+\ \ )$. We look for a pair of factors of the last term, 6, which produces the correct middle term. Since the last term is positive and the middle term is negative, we need only consider negative pairs. The factorization is $(2x - 3)(3x - 2)$.

50. $(4p + 3)(p - 1)$

51. $4x^2 + 3x + 2$

The possibilities are $(x+\ \ )(4x+\ \ )$ and $(2x+\ \ )(2x+\ \ )$. We look for a pair of factors of the last term, 2, which produce the correct middle term. Since the last term and the middle term are both positive, we need only consider positive pairs. We find that there is no possibility that works. The trinomial is prime.

52. $(3a + 5)(3a - 5)$

53. $\quad 2(5x - 3y) - 5(2x + y)$
$= 10x - 6y - 10x - 5y$
$= -11y$

54. $23x$

55. ◈

56. ◈

57. ◈

58. ◈

59. $\quad y - 2.35x = -5.97, \quad (1)$
$\qquad 2.14y - x = 4.88 \qquad (2)$

Solve Equation (1) for y.

$$y - 2.35x = -5.97 \qquad (1)$$
$$y = 2.35x - 5.97 \quad (3)$$

Substitute $2.35x - 5.97$ for y in Equation (2) and solve for x.

$$2.14(2.35x - 5.97) - x = 4.88$$
$$5.029x - 12.7758 - x = 4.88$$
$$4.029x = 17.6558$$
$$x \approx 4.382$$

Substitute 4.382 for x in Equation (3) and solve for y.

$$y = 2.35x - 5.97 = 2.35(4.382) - 5.97 \approx 4.328$$

The solution is $(4.382, 4.328)$.

60. $(7, -1)$

61. $\quad \dfrac{x}{4} + \dfrac{3y}{4} = 1, \quad (1)$
$\qquad \dfrac{x}{5} - \dfrac{y}{2} = 3 \qquad (2)$

Clear the fractions.

$x + 3y = 4, \quad (1a) \qquad$ Multiplying Equation (1) by 4

$2x - 5y = 30 \quad (2a) \qquad$ Multiplying Equation (2) by 10

Solve Equation (1a) for x.

$$x + 3y = 4 \qquad (1a)$$
$$x = -3y + 4 \quad (3)$$

Substitute $-3y + 4$ for x in Equation (2a) and solve for y.

$$2(-3y + 4) - 5y = 30$$
$$-6y + 8 - 5y = 30$$
$$-11y + 8 = 30$$
$$-11y = 22$$
$$y = -2$$

Substitute -2 for y in Equation (3) and solve for x.

$$x = -3y + 4 = -3(-2) + 4 = 6 + 4 = 10$$

The ordered pair $(10, -2)$ checks in both equations, so it is the solution.

62. $(2, -1)$

63. $\quad x + y + z = 4, \quad (1)$
$\qquad x - 2y - z = 1, \quad (2)$
$\qquad y = -1 \qquad\qquad (3)$

Substitute -1 for y in Equations (1) and (2).

$x + y + z = 4 \quad (1) \qquad\qquad x - 2y - z = 1 \quad (2)$
$x + (-1) + z = 4 \qquad\qquad x - 2(-1) - z = 1$
$x + z = 5 \qquad\qquad\qquad\quad x + 2 - z = 1$
$\qquad\qquad\qquad\qquad\qquad\qquad\quad x - z = -1$

We now have a system of two equations in two variables.

$$x + z = 5, \quad (4)$$
$$x - z = -1 \quad (5)$$

We solve Equation (5) for x.

$$x - z = -1 \quad (5)$$
$$x = z - 1 \quad (6)$$

We substitute $z - 1$ for x in Equation (4) and solve for z.

$$x + z = 5 \quad (4)$$
$$(z - 1) + z = 5 \quad \text{Substituting}$$
$$2z - 1 = 5$$
$$2z = 6$$
$$z = 3$$

Next we substitute 3 for z in Equation (6) and compute x.

$$x = z - 1 = 3 - 1 = 2$$

We check the ordered triple $(2, -1, 3)$.

$$\begin{array}{c|c} x+y+z=4 \\ \hline 2+(-1)+3 \;?\; 4 \\ 4 \;\big|\; 4 \text{ TRUE} \end{array} \qquad \begin{array}{c|c} x-2y-z=1 \\ \hline 2-2(-1)-3 \;?\; 1 \\ 2+2-3 \\ 1 \;\big|\; 1 \text{ TRUE} \end{array}$$

$$\begin{array}{c|c} y=-1 \\ \hline -1 \;?\; -1 \text{ TRUE} \end{array}$$

Since $(2, -1, 3)$ checks in all three equations, it is the solution.

64. $(30, 50, 100)$

65. ◈

66. Answers may vary.
$$2x + 3y = 5,$$
$$5x + 4y = 2$$

Exercise Set 8.3

1.
$$\begin{array}{rl} x + y = 12 & (1) \\ x - y = 6 & (2) \\ \hline 2x \quad\;\; = 18 & \text{Adding} \\ x = 9 \end{array}$$

Substitute 9 for x in one of the original equations and solve for y.

$$x + y = 12 \quad (1)$$
$$9 + y = 12 \quad \text{Substituting}$$
$$y = 3$$

Check:
$$\begin{array}{c|c} x+y=12 \\ \hline 9+3 \;?\; 12 \\ 12 \;\big|\; 12 \text{ TRUE} \end{array} \qquad \begin{array}{c|c} x-y=6 \\ \hline 9-3 \;?\; 6 \\ 6 \;\big|\; 6 \text{ TRUE} \end{array}$$

Since $(9, 3)$ checks, it is the solution.

2. $(5, -2)$

3.
$$\begin{array}{rl} x + y = 6 & (1) \\ -x + 2y = 15 & (2) \\ \hline 3y = 21 & \text{Adding} \\ y = 7 \end{array}$$

Substitute 7 for y in one of the original equations and solve for x.

$$x + y = 6 \quad (1)$$
$$x + 7 = 6 \quad \text{Substituting}$$
$$x = -1$$

Check:
$$\begin{array}{c|c} x+y=6 \\ \hline -1+7 \;?\; 6 \\ 6 \;\big|\; 6 \text{ TRUE} \end{array}$$

$$\begin{array}{c|c} -x+2y=15 \\ \hline -(-1)+2\cdot 7 \;?\; 15 \\ 1+14 \\ 15 \;\big|\; 15 \text{ TRUE} \end{array}$$

Since $(-1, 7)$ checks, it is the solution.

4. $(5, 1)$

5.
$$\begin{array}{rl} 3x - y = 9 & (1) \\ 2x + y = 6 & (2) \\ \hline 5x \quad\;\; = 15 & \text{Adding} \\ x = 3 \end{array}$$

Substitute 3 for x in one of the original equations and solve for y.

$$2x + y = 6 \quad (2)$$
$$2 \cdot 3 + y = 6 \quad \text{Substituting}$$
$$6 + y = 6$$
$$y = 0$$

Check:
$$\begin{array}{c|c} 3x-y=9 \\ \hline 3\cdot 3-0 \;?\; 9 \\ 9-0 \\ 9 \;\big|\; 9 \text{ TRUE} \end{array} \qquad \begin{array}{c|c} 2x+y=6 \\ \hline 2\cdot 3+0 \;?\; 6 \\ 6+0 \\ 6 \;\big|\; 6 \text{ TRUE} \end{array}$$

Since $(3, 0)$ checks, it is the solution.

6. $(2, 7)$

7.
$$\begin{array}{rl} 2a + 3b = 7 & (1) \\ -2a + b = 5 & (2) \\ \hline 4b = 12 \\ b = 3 \end{array}$$

Substitute 3 for b in one of the original equations and solve for a.

$$2a + 3b = 7 \quad (1)$$
$$2a + 3 \cdot 3 = 7$$
$$2a + 9 = 7$$
$$2a = -2$$
$$a = -1$$

Check:

$$\begin{array}{c|c} 2a + 3b = 7 & \\ \hline 2(-1) + 3 \cdot 3 \ ? \ 7 & \\ -2 + 9 & \\ 7 & 7 \quad \text{TRUE} \end{array}$$

$$\begin{array}{c|c} -2a + b = 5 & \\ \hline -2(-1) + 3 \ ? \ 5 & \\ 2 + 3 & \\ 5 & 5 \quad \text{TRUE} \end{array}$$

Since $(-1, 3)$ checks, it is the solution.

8. $\left(2, \dfrac{4}{5}\right)$

9.
$$8x - 5y = -9 \quad (1)$$
$$\underline{3x + 5y = -2} \quad (2)$$
$$11x \quad\ = -11 \quad \text{Adding}$$
$$x = -1$$

Substitute -1 for x in either of the original equations and solve for y.

$$3x + 5y = -2 \quad \text{Equation (2)}$$
$$3(-1) + 5y = -2 \quad \text{Substituting}$$
$$-3 + 5y = -2$$
$$5y = 1$$
$$y = \frac{1}{5}$$

Check:

$$\begin{array}{c|c} 8x - 5y = -9 & \\ \hline 8(-1) - 5\left(\frac{1}{5}\right) \ ? \ -9 & \\ -8 - 1 & \\ -9 & -9 \quad \text{TRUE} \end{array}$$

$$\begin{array}{c|c} 3x + 5y = -2 & \\ \hline 3(-1) + 5\left(\frac{1}{5}\right) \ ? \ -2 & \\ -3 + 1 & \\ -2 & -2 \quad \text{TRUE} \end{array}$$

Since $\left(-1, \dfrac{1}{5}\right)$ checks, it is the solution.

10. $(-2, 3)$

11.
$$7a - 6b = \ \ 8,$$
$$\underline{-7a + 6b = -8}$$
$$0 = \ \ 0 \quad \text{Adding}$$

The equation $0 = 0$ is always true, so the system has an infinite number of solutions.

12. Infinite number of solutions

13.
$$-x - y = 8, \quad (1)$$
$$2x - y = -1 \quad (2)$$

We multiply by -1 on both sides of Equation (1) and then add.

$$x + y = -8 \quad \text{Multiplying by } -1$$
$$\underline{2x - y = -1}$$
$$3x \quad\ = -9 \quad \text{Adding}$$
$$x = -3$$

Substitute -3 for x in one of the original equations and solve for y.

$$2x - y = -1 \quad (2)$$
$$2(-3) - y = -1 \quad \text{Substituting}$$
$$-6 - y = -1$$
$$-y = 5$$
$$y = -5$$

Check:

$$\begin{array}{c|c} -x - y = 8 & \\ \hline -(-3) - (-5) \ ? \ 8 & \\ 3 + 5 & \\ 8 & 8 \quad \text{TRUE} \end{array}$$

$$\begin{array}{c|c} 2x - y = -1 & \\ \hline 2(-3) - (-5) \ ? \ -1 & \\ -6 + 5 & \\ -1 & -1 \quad \text{TRUE} \end{array}$$

Since $(-3, -5)$ checks, it is the solution.

14. $(-1, -6)$

15.
$$x + 3y = 19,$$
$$x - y = -1$$

We multiply by -1 on both sides of Equation (2) and then add.

$$x + 3y = 19$$
$$\underline{-x + y = \ \ 1} \quad \text{Multiplying by } -1$$
$$4y = 20 \quad \text{Adding}$$
$$y = \ \ 5$$

Substitute 5 for y in one of the original equations and solve for x.

$$x - y = -1 \quad (2)$$
$$x - 5 = -1 \quad \text{Substituting}$$
$$x = 4$$

Check:

$$\begin{array}{c|c}
x + 3y = 19 \\
\hline
4 + 3 \cdot 5 \ ? \ 19 \\
4 + 15 \\
19 \ \big| \ 19 \ \text{TRUE}
\end{array}
\qquad
\begin{array}{c|c}
x - y = -1 \\
\hline
4 - 5 \ ? \ -1 \\
-1 \ \big| \ -1 \ \text{TRUE}
\end{array}$$

Since $(4, 5)$ checks, it is the solution.

16. $(3, 1)$

17. $x + y = 5, \quad (1)$
        $5x - 3y = 17 \quad (2)$

We multiply by 3 on both sides of Equation (1) and then add.

$$\begin{array}{ll}
3x + 3y = 15 & \text{Multiplying by 3} \\
\underline{5x - 3y = 17} & \\
8x \quad\;\; = 32 & \\
x \;\;= 4 &
\end{array}$$

Substitute 4 for x in one of the original equations and solve for y.

$$\begin{array}{ll}
x + y = 5 & (1) \\
4 + y = 5 & \text{Substituting} \\
y = 1 &
\end{array}$$

Check:

$$\begin{array}{c|c}
x + y = 5 \\
\hline
4 + 1 \ ? \ 5 \\
5 \ \big| \ 5 \ \text{TRUE}
\end{array}
\qquad
\begin{array}{c|c}
5x - 3y = 17 \\
\hline
5 \cdot 4 - 3 \cdot 1 \ ? \ -1 \\
20 - 3 \\
17 \ \big| \ 17 \ \text{TRUE}
\end{array}$$

Since $(4, 1)$ checks, it is the solution.

18. $(10, 3)$

19. $2w - 3z = -1, \quad (1)$
        $3w + 4z = \;\; 24 \quad (2)$

We use the multiplication principle with both equations and then add.

$$\begin{array}{ll}
8w - 12z = -4 & \text{Multiplying (1) by 4} \\
\underline{9w + 12z = 72} & \text{Multiplying (2) by 3} \\
17w \qquad\;\; = 68 & \text{Adding} \\
w \;\; = 4 &
\end{array}$$

Substitute 4 for w in one of the original equations and solve for z.

$$\begin{array}{ll}
3w + 4z = 24 & \text{Equation (2)} \\
3 \cdot 4 + 4z = 24 & \text{Substituting} \\
12 + 4z = 24 & \\
4z = 12 & \\
z = 3 &
\end{array}$$

Check:

$$\begin{array}{c|c}
2w - 3z = -1 \\
\hline
2 \cdot 4 - 3 \cdot 3 \ ? \ -1 \\
8 - 9 \\
-1 \ \big| \ -1 \qquad \text{TRUE}
\end{array}$$

$$\begin{array}{c|c}
3w + 4z = 24 \\
\hline
3 \cdot 4 + 4 \cdot 3 \ ? \ 24 \\
12 + 12 \\
24 \ \big| \ 24 \qquad \text{TRUE}
\end{array}$$

Since $(4, 3)$ checks, it is the solution.

20. $(1, -1)$

21. $2a + 3b = -1, \quad (1)$
        $3a + 5b = -2 \quad (2)$

We use the multiplication principle with both equations and then add.

$$\begin{array}{ll}
-10a - 15b = \;\;\; 5 & \text{Multiplying (1) by } -5 \\
\underline{9a + 15b = -6} & \text{Multiplying (2) by 3} \\
-a \qquad\quad\; = -1 & \text{Adding} \\
a = \;\; 1 &
\end{array}$$

Substitute 1 for a in one of the original equations and solve for b.

$$\begin{array}{ll}
2a + 3b = -1 & \text{Equation (1)} \\
2 \cdot 1 + 3b = -1 & \text{Substituting} \\
2 + 3b = -1 & \\
3b = -3 & \\
b = -1 &
\end{array}$$

Check:

$$\begin{array}{c|c}
2a + 3b = -1 \\
\hline
2 \cdot 1 + 3(-1) \ ? \ -1 \\
2 - 3 \\
-1 \ \big| \ -1 \qquad \text{TRUE}
\end{array}$$

$$\begin{array}{c|c}
3a + 5b = -2 \\
\hline
3 \cdot 1 + 5(-1) \ ? \ -2 \\
3 - 5 \\
-2 \ \big| \ -2 \qquad \text{TRUE}
\end{array}$$

Since $(1, -1)$ checks, it is the solution.

22. $(4, -1)$

23. $x = 3y, \quad (1)$
        $5x + 14 = y \quad (2)$

We first get each equation in the form $Ax + By = C$.

$$\begin{array}{lll}
x - 3y = 0, & (1a) & \text{Adding } -3y \\
5x - y = -14 & (2a) & \text{Adding } -y - 14
\end{array}$$

We multiply by -5 on both sides of Equation (1a) and add.

$$\begin{array}{ll}
-5x + 15y = \quad 0 & \text{Multiplying by } -5 \\
\underline{5x - \quad y = -14} & \\
14y = -14 & \text{Adding} \\
y = \;\; -1 &
\end{array}$$

Substitute -1 for y in Equation (1) and solve for x.

$$\begin{array}{l}
x - 3y = 0 \\
x - 3(-1) = 0 \qquad \text{Substituting} \\
x + 3 = 0 \\
x = -3
\end{array}$$

Check:

$$x = 3y$$

$$-3 \ ? \ 3(-1)$$

$$-3 \mid -3 \qquad \text{TRUE}$$

$$5x + 14 = y$$

$$5(-3) + 14 \ ? \ -1$$

$$-15 + 14$$

$$-1 \mid -1 \qquad \text{TRUE}$$

Since $(-3, -1)$ checks, it is the solution.

24. $(2,5)$

25. $4x - 10y = 13, \quad (1)$
$-2x + 5y = 8 \quad (2)$

We multiply by 2 on both sides of Equation (2) and then add.

$$4x - 10y = 13$$
$$\underline{-4x + 10y = 16} \qquad \text{Multiplying by 2}$$
$$0 = 29$$

The equation $0 = 29$ is false for any pair (x,y), so there is no solution.

26. $(2, 1)$

27. $3x = 8y + 11,$
$x + 6y - 8 = 0$

We first get each equation in the form $Ax + By = C$.

$$3x - 8y = 11, \quad (1) \quad \text{Adding } -8y$$
$$x + 6y = 8 \quad (2) \quad \text{Adding } 8$$

We multiply by -3 on both sides of Equation (2) and add.

$$3x - 8y = 11$$
$$\underline{-3x - 18y = -24} \qquad \text{Multiplying by } -3$$
$$-26y = -13 \qquad \text{Adding}$$
$$y = \frac{1}{2}$$

Substitute $\frac{1}{2}$ for y in Equation (1) and solve for x.

$$3x - 8y = 11$$
$$3x - 8 \cdot \frac{1}{2} = 11 \qquad \text{Substituting}$$
$$3x - 4 = 11$$
$$3x = 15$$
$$x = 5$$

Check:

$$3x = 8y + 11$$

$$3 \cdot 5 \ ? \ 8 \cdot \frac{1}{2} + 11$$

$$15 \mid 4 + 11$$

$$15 \mid 15 \qquad \text{TRUE}$$

$$x + 6y - 8 = 0$$

$$5 + 6 \cdot \frac{1}{2} - 8 \ ? \ 0$$

$$5 + 3 - 8$$

$$0 \mid 0 \quad \text{TRUE}$$

Since $\left(5, \frac{1}{2}\right)$ checks, it is the solution.

28. $(50, 18)$

29. $3x + 5y = 4, \quad (1)$
$-2x + 3y = 10 \quad (2)$

We use the multiplication principle with both equations and then add.

$$6x + 10y = 8 \qquad \text{Multiplying (1) by 2}$$
$$\underline{-6x + 9y = 30} \qquad \text{Multiplying (2) by 3}$$
$$19y = 38 \qquad \text{Adding}$$
$$y = 2$$

Substitute 2 for y in one of the original equations and solve for x.

$$3x + 5y = 4 \quad (1)$$
$$3x + 5 \cdot 2 = 4$$
$$3x + 10 = 4$$
$$3x = -6$$
$$x = -2$$

Check:

$$3x + 5y = 4$$

$$3(-2) + 5 \cdot 2 \ ? \ 4$$

$$-6 + 10$$

$$4 \mid 4 \qquad \text{TRUE}$$

$$-2x + 3y = 10$$

$$-2(-2) + 3 \cdot 2 \ ? \ 10$$

$$4 + 6$$

$$10 \mid 10 \qquad \text{TRUE}$$

Since $(-2, 2)$ checks, it is the solution

30. No solution

31. $0.06x + 0.05y = 0.07,$
$0.04x - 0.03y = 0.11$

We first multiply each equation by 100 to clear the decimals.

$$6x + 5y = 7, \quad (1)$$
$$4x - 3y = 11 \quad (2)$$

We use the multiplication principle with both equations of the resulting system.

$$18x + 15y = 21 \qquad \text{Multiplying (1) by 3}$$
$$\underline{20x - 15y = 55} \qquad \text{Multiplying (2) by 5}$$
$$38x = 76 \qquad \text{Adding}$$
$$x = 2$$

Substitute 2 for x in Equation (1) and solve for y.

$$6x + 5y = 7$$
$$6 \cdot 2 + 5y = 7$$
$$12 + 5y = 7$$
$$5y = -5$$
$$y = -1$$

Check:

$$
\begin{array}{c|c}
0.06x + 0.05y = 0.07 & \\
\hline
0.06(2) + 0.05(-1) \; ? \; 0.07 & \\
0.12 - 0.05 & \\
0.07 & 0.07 \quad \text{TRUE}
\end{array}
$$

$$
\begin{array}{c|c}
0.04x - 0.03y = 0.11 & \\
\hline
0.04(2) - 0.03(-1) \; ? \; 0.11 & \\
0.08 + 0.03 & \\
0.11 & 0.11 \quad \text{TRUE}
\end{array}
$$

Since $(2, -1)$ checks, it is the solution.

32. $(10, -2)$

33. $x + \dfrac{9}{2}y = \dfrac{15}{4},$

   $\dfrac{9}{10}x - y = \dfrac{9}{20}$

First we clear fractions. We multiply both sides of the first equation by 4 and both sides of the second equation by 20.

$$4\left(x + \frac{9}{2}y\right) = 4 \cdot \frac{15}{4}$$
$$4x + 4 \cdot \frac{9}{2}y = 15$$
$$4x + 18 = 15$$

$$20\left(\frac{9}{10}x - y\right) = 20 \cdot \frac{9}{20}$$
$$20 \cdot \frac{9}{10}x - 20y = 9$$
$$18x - 20y = 9$$

The resulting system is

   $4x + 18y = 15,$ (1)
   $18x - 20y = 9.$ (2)

We use the multiplication principle with both equations.

$$
\begin{aligned}
72x + 324y &= 270 \qquad \text{Multiplying (1) by 18} \\
-72x + 80y &= -36 \qquad \text{Multiplying (2) by } -4 \\
\hline
404y &= 234 \\
y &= \frac{234}{404}, \text{ or } \frac{117}{202}
\end{aligned}
$$

Substitute $\dfrac{117}{202}$ for y in (1) and solve for x.

$$4x + 18\left(\frac{117}{202}\right) = 15$$
$$4x + \frac{1053}{101} = 15$$
$$4x = \frac{462}{101}$$
$$x = \frac{1}{4} \cdot \frac{462}{101}$$
$$x = \frac{231}{202}$$

The ordered pair $\left(\dfrac{231}{202}, \dfrac{117}{202}\right)$ checks in both equations. It is the solution.

34. $\left(\dfrac{231}{202}, \dfrac{117}{202}\right)$

35. *Familiarize.* We let m = the number of miles driven and c = the total cost of the car rental.

Translate. We reword and translate the first statement, using $0.30 for 30¢.

$$
\begin{array}{ccccccc}
\$53.95 & \text{plus} & 30\text{¢} & \text{times} & \text{the number of miles driven} & \text{is} & \text{cost.} \\
\downarrow & \downarrow & \downarrow & \downarrow & \downarrow & \downarrow & \downarrow \\
53.95 & + & 0.30 & \cdot & m & = & c
\end{array}
$$

We reword and translate the second statement using $0.20 for 20¢.

$$
\begin{array}{ccccccc}
\$54.95 & \text{plus} & 20\text{¢} & \text{times} & \text{the number of miles driven} & \text{is} & \text{cost.} \\
\downarrow & \downarrow & \downarrow & \downarrow & \downarrow & \downarrow & \downarrow \\
54.95 & + & 0.20 & \cdot & m & = & c
\end{array}
$$

We have a system of equations:

   $53.95 + 0.30m = c,$
   $54.95 + 0.20m = c$

Carry out. We solve the system of equations. We clear the decimals by multiplying both sides of each equation by 100.

   $5395 + 30m = 100c,$ (1)
   $5495 + 20m = 100c$ (2)

We multiply (1) by -1 and then add.

$$
\begin{aligned}
-5395 - 30m &= -100c \\
5495 + 20m &= 100c \\
\hline
100 - 10m &= 0 \qquad \text{Adding} \\
100 &= 10m \\
10 &= m
\end{aligned}
$$

Check. For 10 mi, the cost of the Avis car is $53.95 + 0.30(10)$, or $53.95 + 3$, or $56.95. For 10 mi, the cost of the other car is $54.95 + 0.20(10)$, or $54.95 + 2$, or $56.95, so the costs are the same when the mileage is 10.

State. When the cars are driven 10 miles, the cost will be the same.

36. 5 miles

37. *Familiarize*. We let x = the larger angle and y = the smaller angle.

Translate. We reword and translate the first statement.

The sum of two angles is 90°.

$$x + y \qquad = \qquad 90$$

We reword and translate the second statement.

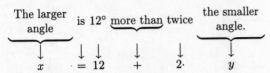

We have a system of equations:

$$x + y = 90,$$
$$x = 12 + 2y$$

Carry out. We solve the system. We will use the elimination method, although we could also easily use the substitution method. First we get the second equation in the form $Ax + By = C$.

$$x + y = 90 \quad (1)$$
$$x - 2y = 12 \quad (2) \quad \text{Adding } -2y$$

Now we multiply Equation (2) by 2 and add.

$$2x + 2y = 180$$
$$\underline{x - 2y = 12}$$
$$3x = 192$$
$$x = 64$$

Then we substitute 64 for x in Equation (1) and solve for y.

$$x + y = 90 \quad (1)$$
$$64 + y = 90 \quad \text{Substituting}$$
$$y = 26$$

Check. The sum of the angles is $64° + 26°$, or $90°$, so the angles are complementary. The larger angle, $64°$, is $12°$ more than twice the smaller angle, $26°$. These numbers check.

State. The angles are $64°$ and $26°$.

38. 58°, 32°

39. *Familiarize*. Let x = the smaller angle and y = the larger angle.

Translate. We reword the problem.

The smaller angle plus the larger angle is 180°.

$$x + y = 180$$

The larger angle is 5° more than 4 times the smaller angle.

$$y = 5 + 4 \cdot x$$

The resulting system is

$$x + y = 180,$$
$$y = 5 + 4x.$$

Carry out. We solve the system. We will use the elimination method although we could also easily use the substitution method. First we get the second equation in the form $Ax + By = C$.

$$x + y = 180 \quad (1)$$
$$-4x + y = 5 \quad (2) \text{ Adding } -4x$$

Now we multiply Equation (2) by -1 and add.

$$x + y = 180$$
$$\underline{4x - y = -5}$$
$$5x = 175$$
$$x = 35$$

Then we substitute 35 for x in Equation (1) and solve for y.

$$x + y = 180 \quad (1)$$
$$35 + y = 180 \quad \text{Substituting}$$
$$y = 145$$

Check. The sum of the angles is $35° + 145°$, or $180°$, so the angles are supplementary. Also, $5°$ more than four times the $35°$ angle is $5° + 4 \cdot 35°$, or $5° + 140°$, or $145°$, the other angle. These numbers check.

State. The angles are $35°$ and $145°$.

40. 105°, 75°

41. *Familiarize*. We let x = the number of acres of hay that should be planted and y = the number of acres of oats that should be planted.

Translate. We reword and translate the first statement.

Total number of acres is 31.

$$x + y \qquad = \quad 31$$

Now we reword and translate the second statement.

Number of acres of hay is 9 acres more than number of acres of oats.

$$x = 9 + y$$

The resulting system is

$$x + y = 31,$$
$$x = 9 + y$$

Carry out. We solve the system. We will use the elimination method, although we could also easily use the substitution method. First we get the second equation in the form $Ax + By = C$. Then we add the equations.

$$x + y = 31 \quad (1)$$
$$\underline{x - y = 9} \quad (2) \quad \text{Subtracting } y$$
$$2x = 40 \quad \text{Adding}$$
$$x = 20$$

Now we substitute 20 for x in Equation (1) and solve for y.

$$x + y = 31 \quad (1)$$
$$20 + y = 31 \quad \text{Substituting}$$
$$y = 11$$

Check. The total number of acres is $20 + 11$, or 31. Also, the number of acres of hay is 9 more than the number of acres of oats. These numbers check.

State. The owners should plant 20 acres of hay and 11 acres of oats.

42. Chardonnay: 480 acres, Riesling: 340 acres

43. *Familiarize*. Let $l = $ the length of the frame and $w = $ the width.

Translate.

The perimeter is 12 ft.
$$2l + 2w = 12$$

The length is twice the width.
$$l = 2w$$

The resulting system is
$$2l + 2w = 12,$$
$$l = 2w.$$

Carry out. We solve the system. We will use the elimination method, although we could also easily use the substitution method. First we get the second equation in the form $Al + Bw = C$. Then we add the equations.

$$2l + 2w = 12 \quad (1)$$
$$\underline{l - 2w = 0} \quad (2)$$
$$3l \qquad = 12$$
$$l = 4$$

Substitute 4 for l in Equation (2) and solve for w.

$$l - 2w = 0$$
$$4 - 2w = 0$$
$$4 = 2w$$
$$2 = w$$

Check. The perimeter is $2 \cdot 4 + 2 \cdot 2$, or 12 ft. Twice the width is $2 \cdot 2$, or 4 ft, which is the length. These numbers check.

State. The length of the frame is 4 ft, and the width is 2 ft.

44. 9 yd by 6 yd

45. $\dfrac{(a^2 b^{-3})^4}{a^5 b^{-6}} = \dfrac{a^{2(4)} b^{-3(4)}}{a^5 b^{-6}} = \dfrac{a^8 b^{-12}}{a^5 b^{-6}} =$

$a^{8-5} b^{-12-(-6)} = a^3 b^{-6}$, or $\dfrac{a^3}{b^6}$

46. $\dfrac{9a^4}{4b^6}$

47. $\quad 4x^2 + 20x + 25$
$$= (2x)^2 + 2 \cdot 2x \cdot 5 + 5^2 \quad \text{Perfect square}$$
$$\text{trinomial}$$
$$= (2x + 5)^2$$

48. $(3a - 4)^2$

49. $3.7(5) + 2.9(4) = 18.5 + 11.6 = 30.1$

50. 6.9

51. ◈

52. ◈

53. ◈

54. ◈

55. $\quad y = 3x + 4, \qquad (1)$
$$3 + y = 2(y - x) \quad (2)$$

Substitute $3x + 4$ for y in Equation (2) and solve for x.

$$3 + y = 2(y - x) \qquad (1)$$
$$3 + (3x + 4) = 2((3x + 4) - x) \quad \text{Substituting}$$
$$3x + 7 = 2(2x + 4)$$
$$3x + 7 = 4x + 8$$
$$7 = x + 8$$
$$-1 = x$$

Now substitute -1 for x in Equation (1) and compute y.

$$y = 3(-1) + 4 = -3 + 4 = 1$$

The ordered pair $(-1, 1)$ checks, so it is the solution.

56. $(2, 5)$

57. $\quad 0.05x + y = 4,$
$$\dfrac{x}{2} + \dfrac{y}{3} = 1\dfrac{1}{3}$$

Multiply the first equation by 100 to clear the decimal. Also, multiply the second equation by 6 to clear the fractions.

$$5x + 100y = 400, \quad (1)$$
$$3x + 2y = 8 \qquad (2)$$

Multiply Equation (2) by -50 and add.

$$5x + 100y = 400$$
$$\underline{-150x - 100y = -400}$$
$$-145x \qquad = 0$$
$$x = 0$$

Now substitute 0 for x in Equation (1) or (2) and solve for y.

$$3x + 2y = 8 \quad (2)$$
$$3 \cdot 0 + 2y = 8$$
$$2y = 8$$
$$y = 4$$

The ordered pair $(0, 4)$ checks, so it is the solution.

58. $\left(\dfrac{1}{2}, -\dfrac{1}{2}\right)$

59. $y = ax + b,$ (1)
    $y = x + c$ (2)

Substitute $x + c$ for y in Equation (1) and solve for x.

$$y = ax + b$$
$$x + c = ax + b \quad \text{Substituting}$$
$$x - ax = b - c$$
$$(1 - a)x = b - c$$
$$x = \frac{b - c}{1 - a}$$

Substitute $\dfrac{b - c}{1 - a}$ for x in Equation (2) and simplify to find y.

$$y = x + c$$
$$y = \frac{b - c}{1 - a} + c$$
$$y = \frac{b - c}{1 - a} + c \cdot \frac{1 - a}{1 - a}$$
$$y = \frac{b - c + c - ac}{1 - a}$$
$$y = \frac{b - ac}{1 - a}$$

The ordered pair $\left(\dfrac{b - c}{1 - a}, \dfrac{b - ac}{1 - a}\right)$ checks and is the solution.

60. $\left(\dfrac{-b - c}{a}, 1\right)$

61. Familiarize. Let x represent the number of rabbits and y the number of pheasants in the cage. Each rabbit has one head and four feet. Thus, there are x rabbit heads and $4x$ rabbit feet in the cage. Each pheasant has one head and two feet. Thus, there y pheasant heads and $2y$ pheasant feet in the cage.

Translate. We reword the problem.

Rabbit heads plus pheasant heads is 35.
$$x + y = 35$$

Rabbit feet plus pheasant feet is 94.
$$4x + 2y = 94$$

The resulting system is
$$x + y = 35, \quad (1)$$
$$4x + 2y = 94. \quad (2)$$

Carry out. We solve the system of equations. We multiply Equation (1) by -2 and then add.

$$\begin{array}{r} -2x - 2y = -70 \\ 4x + 2y = 94 \\ \hline 2x = 24 \quad \text{Adding} \\ x = 12 \end{array}$$

Substitute 12 for x in one of the original equations and solve for y.

$$x + y = 35 \quad (1)$$
$$12 + y = 35 \quad \text{Substituting}$$
$$y = 23$$

Check. If there are 12 rabbits and 23 pheasants, the total number of heads in the cage is $12 + 23$, or 35. The total number of feet in the cage is $4 \cdot 12 + 2 \cdot 23$, or $48 + 46$, or 94. The numbers check.

State. There are 12 rabbits and 23 pheasants.

62. Patrick: 6, mother: 30

63. Familiarize. Let $x =$ the man's age and $y =$ his daughter's age. Five years ago their ages were $x - 5$ and $y - 5$.

Translate.

Dividing the sum of a man's age and 5 by 5 yields his daughter's age.
$$\frac{x + 5}{5} = y$$

The man's age 5 years ago was 8 times his daughter's age 5 years ago.
$$x - 5 = 8(y - 5)$$

We have a system of equations:
$$\frac{x + 5}{5} = y$$
$$x - 5 = 8(y - 5)$$

Carry out. Solve the system.

Multiply the first Equation by 5 to clear the fraction.
$$x + 5 = 5y$$
$$x - 5y = -5$$

Simplify the second equation.
$$x - 5 = 8(y - 5)$$
$$x - 5 = 8y - 40$$
$$x - 8y = -35$$

The resulting system is
$$x - 5y = -5, \quad (1)$$
$$x - 8y = -35. \quad (2)$$

Multiply Equation (2) by -1 and add.

$$\begin{array}{r} x - 5y = -5 \\ -x + 8y = 35 \quad \text{Multiplying by } -1 \\ \hline 3y = 30 \quad \text{Adding} \\ y = 10 \end{array}$$

Substitute 10 for y in Equation (1) and solve for x.

$$x - 5y = -5$$
$$x - 5 \cdot 10 = -5 \quad \text{Substituting}$$
$$x - 50 = -5$$
$$x = 45$$

Possible solution: Man is 45, daughter is 10.

Check. If 5 is added to the man's age, $5 + 45$, the result is 50. If 50 is divided by 5, the result is 10, the daughter's age. Five years ago the father and daughter were 40 and 5, respectively, and $40 = 8 \cdot 5$. The numbers check.

State. The man is 45 years old; his daughter is 10 years old.

64. Base: 9 ft, height: 6 ft

Exercise Set 8.4

1. Familiarize. Let $x =$ the number of cars and $y =$ the number of trucks.

Translate. We reword the problem.

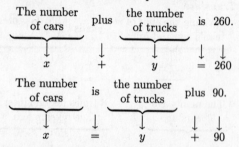

Carry out. We solve the system of equations.

$$x + y = 260 \quad (1)$$
$$x = y + 90 \quad (2)$$

We substitute $y + 90$ for x in Equation (1) and solve for y.

$$x + y = 260$$
$$(y + 90) + y = 260 \quad \text{Substituting}$$
$$2y + 90 = 260$$
$$2y = 170$$
$$y = 85$$

Next we substitute 85 for y in one of the original equations and solve for x.

$$x = y + 90 \quad (2)$$
$$x = 85 + 90 \quad \text{Substituting}$$
$$x = 175$$

Check. If there are 175 cars and 85 trucks, then the total number of vehicles is $175 + 85$, or 260. Since the number of trucks plus 90 is $85 + 90$, or 175, we know that it is true that the number of cars is 90 more than the number of trucks.

State. The firm should have 175 cars and 85 trucks.

2. 11 km

3. Familiarize. Let $x =$ the cost of one slice of pizza and $y =$ the cost of one soda.

Translate. We reword the problem.

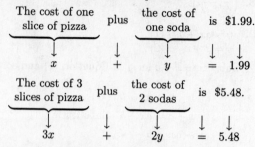

We have translated to a system of equations:

$$x + y = 1.99,$$
$$3x + 2y = 5.48$$

Carry out. We will use elimination. We first multiply both equations by 100 to clear decimals.

$$100x + 100y = 199 \quad (1)$$
$$300x + 200y = 548 \quad (2)$$

We solve using elimination.

$$\begin{array}{ll} -200x - 200y = -398 & \text{Multiplying (1) by } -2 \\ 300x + 200y = 548 & (2) \\ \hline 100x = 150 & \text{Adding} \\ x = 1.5 & \text{Dividing by 100} \end{array}$$

We go back to Equation (1) and substitute 1.5 for x.

$$100x + 100y = 199$$
$$100(1.5) + 100y = 199$$
$$150 + 100y = 199$$
$$100y = 49$$
$$y = 0.49$$

Check. If a slice of pizza costs $1.50 and a soda costs $0.49, then together they cost $1.99. Also, 3 slices of pizza and 2 sodas cost $3(\$1.50) + 2(\$0.49)$, or $5.48. These numbers check.

State. One soda costs $0.49, and one slice of pizza costs $1.50.

4. $0.99

5. Familiarize. Let h and s represent the number of bags of trash generated each month by the Hendersons and the Savickis, respectively.

Translate. Since the Hendersons generate two and a half times as much trash as the Savickis, we have

$$h = 2.5s.$$

To find a second equation, we reword some information.

We have a system of equations:

$$h = 2.5s, \quad (1)$$
$$h + s = 14 \quad (2)$$

Carry out. We solve by substitution. Substitute $2.5s$ for h in Equation (2) and solve for s.

$$h + s = 14$$
$$2.5s + s = 14$$
$$3.5s = 14$$
$$s = 4$$

Substitute 4 for s in Equation (1) and solve for h.

$$h = 2.5s = 2.5(4) = 10$$

Check. Two and a half times 4 is 10, and $10 + 4$ is 14. The numbers check.

State. Each month the Hendersons produce 10 bags of trash, and the Savickis produce 4 bags.

6. Mazza: 28 rolls, Kranepool: 8 rolls

7. *Familiarize*. Let $t =$ the number of two-pointers made and $f =$ the number of foul shots made.

Translate. To find one equation we rephrase the information about the points scored.

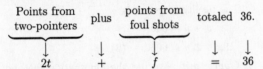

Points from two-pointers	plus	points from foul shots	totaled	36.
$2t$	$+$	f	$=$	36

Since a total of 22 shots were made, we have a second equation.

$$t + f = 22$$

The resulting system is

$$2t + f = 36, \quad (1)$$
$$t + f = 22 \quad (2)$$

Carry out. We solve using elimination.

$$2t + f = 36 \quad (1)$$
$$\underline{-t - f = -22} \quad \text{Multiplying (2) by } -1$$
$$t = 14 \quad \text{Adding}$$

Substitute 14 for t in Equation (2) and solve for f.

$$t + f = 22 \quad (2)$$
$$14 + f = 22$$
$$f = 8$$

Check. If O'Neil made 14 two-pointers and 8 foul shots, he made $14 + 8$, or 22 shots for a total of $2 \cdot 14 + 8$, or $28 + 8$, or 36 points. The numbers check.

State. O'Neil made 8 foul shots.

8. Two-point: 36, foul: 28

9. *Familiarize*. Let $x =$ the number of 5-cent bottles or cans collected and $y =$ the number of 10-cent bottles or cans collected. We organize the given information in a table.

	$0.05	$0.10	Total
Number	x	y	436
Total Value	$0.05x$	$0.10y$	26.60

Translate. A system of two equations can be formed using the rows of the table.

$$x + y = 436,$$
$$0.05x + 0.10y = 26.60$$

Carry out. First we multiply on both sides of the second equation by 100 to clear the decimals.

$$x + y = 436, \quad (1)$$
$$5x + 10y = 2660 \quad (2)$$

Now multiply Equation (1) by -5 and add.

$$-5x - 5y = -2180$$
$$\underline{5x + 10y = 2660}$$
$$5y = 480$$
$$y = 96$$

Substitute 96 for y in Equation (1) and solve for x.

$$x + y = 436$$
$$x + 96 = 436$$
$$x = 340$$

Check. If 340 5-cent bottles and cans and 96 10-cent bottles and cans were collected, then a total of $340 + 96$, or 436 bottles and cans were collected. Their total value is $\$0.05(340) + \$0.10(96)$, or $\$17 + \$9.60$, or $\$26.60$. These numbers check.

State. 340 5-cent bottles and cans and 96 10-cent bottles and cans were collected.

10. 33 soft-serve, 42 hard-pack

11. *Familiarize*. Let $x =$ the number of full-price admissions and $y =$ the number of half-price admissions. We organize the information in a table.

	Full-price	Half-price	Total
Admission	$3.00	$1.50	
Number	x	y	435
Money Taken In	$3x$	$1.50y$	967.50

Translate. We use the last two rows of the table to form a system of equations.

$$x + y = 435,$$
$$3x + 1.50y = 967.50$$

Carry out. First multiply the second equation by 10 to clear the decimals.

$$x + y = 435, \quad (1)$$
$$30x + 15y = 9675 \quad (2)$$

Now multiply Equation (1) by -15 and add.

$$-15x - 15y = -6525$$
$$\underline{30x + 15y = 9675}$$
$$15x = 3150$$
$$x = 210$$

Substitute 210 for x in Equation (1) and solve for y.

$$x + y = 435$$
$$210 + y = 435$$
$$y = 225$$

Check.   If there were 210 full-price admissions and 225 half-price admissions, there was a total of $210 + 225$, or 435 admissions. The total amount collected was $\$3(210) + \$1.50(225)$, or $\$630 + \$337.50$, or $\$967.50$. These numbers check.

State. There were 210 full-price admissions and 225 half-price admissions.

12. 175 student tickets, 315 other tickets

13. Familiarize.   Let x = the number of cardholders tickets that were sold and y = the number of non-cardholders tickets. We arrange the information in a table.

	Card-holders	Non-card-holders	Total
Price	$1.25	$2	
Number sold	x	y	203
Money taken in	1.25x	2y	$310

Translate.   The last two rows of the table give us two equations. The total number of tickets sold was 203, so we have

$$x + y = 203.$$

The total amount of money collected was $310, so we have

$$1.25x + 2y = 310.$$

We can multiply the second equation on both sides by 100 to clear decimals. The resulting system is

$$x + y = 203, \qquad (1)$$
$$125x + 200y = 31,000. \qquad (2)$$

Carry out.   We use the elimination method. We multiply on both sides of Equation (1) by -125 and then add.

$$-125x - 125y = -25,375 \quad \text{Multiplying by } -125$$
$$\underline{125x + 200y = 31,000}$$
$$75y = 5625$$
$$y = 75$$

We go back to Equation (1) and substitute 75 for y.

$$x + y = 203$$
$$x + 75 = 203$$
$$x = 128$$

Check.   The number of tickets sold was $128 + 75$, or 203. The money collected was $\$1.25(128) + \$2(75)$, or $\$160 + \$150$, or $\$310$. These numbers check.

State. 128 cardholders tickets and 75 non-cardholders tickets were sold.

14. 70 student tickets, 130 adult tickets

15. Familiarize.   Let x = the number of kg of Brazilian coffee to be used and y = the number of kg of Turkish

coffee to be used. Organize the given information in a table.

Type of coffee	Brazilian	Turkish	Mixture
Cost of coffee	$19	$22	$20
Amount (in kg)	x	y	300
Value	19x	22y	$20(300); or $6000

Translate.   The second and third rows of the table give us two equations. Since the total amount of the mixture is 300 lb, we have

$$x + y = 300.$$

The value of the Brazilian coffee is $19x$ (x lb at $19 per pound), the value of the Turkish coffee is $22y$ (y lb at $22 per pound), and the value of the mixture is $\$20(300)$ or $\$6000$. Thus we have

$$19x + 22y = 6000.$$

The resulting system is

$$x + y = 300, \qquad (1)$$
$$19x + 22y = 6000. \qquad (2)$$

Carry out.   We use the elimination method. We multiply on both sides of Equation (1) by -19 and then add.

$$-19x - 19y = -5700$$
$$\underline{19x + 22y = 6000}$$
$$3y = 300$$
$$y = 100$$

Now substitute 100 for y in Equation (1) and solve for x.

$$x + y = 300$$
$$x + 100 = 300$$
$$x = 200$$

Check.   The sum of 100 and 200 is 300. The value of the mixture is $\$19(200) + \$22(100)$, or $\$3800 + \$2200$, or $\$6000$. These numbers check.

State. 200 kg of Brazilian coffee and 100 kg of Turkish coffee should be used.

16. 30 lb of sunflower seed, 20 lb of rolled oats

17. Familiarize.   Let x and y represent the number of pounds of peanuts and Brazil nuts to be used, respectively. We organize the given information in a table.

Type of nut	Peanuts	Brazil nuts	Mixture
Cost per pound	$2.52	$3.80	$3.44
Amount	x	y	480
Value	2.52x	3.80y	3.44(480); or 1651.20

Translate. The last two rows of the table form a system of equations.

$$x + y = 480,$$
$$2.52x + 3.80y = 1651.20$$

Carry out. First we multiply the second equation by 100 to clear decimals.

$$x + y = 480, \quad (1)$$
$$252x + 380y = 165,120 \quad (2)$$

Now multiply Equation (1) by -252 and add.

$$-252x - 252y = -120,960$$
$$252x + 380y = 165,120$$
$$\overline{\qquad\qquad 128y = 44,160}$$
$$y = 345$$

Substitute 345 for y in Equation (1) and solve for x.

$$x + y = 480$$
$$x + 345 = 480$$
$$x = 135$$

Check. The sum of 135 and 345 is 480. The value of the mixture is $2.52(135) + 3.80(345)$, or $340.20 + 1311$, or $1651.20. These numbers check.

State. 135 lb of peanuts and 345 lb of Brazil nuts should be used.

18. 6 kg of cashews, 4 kg of pecans

19. *Familiarize*. We complete the table in the text. Note that x represents the number of milliliters of solution A to be used and y represents the number of milliliters of solution B.

Type of solution	50%-acid	80%-acid	68%-acid
Amount of solution	x	y	200
Percent acid	50%	80%	68%
Amount of acid in solution	$0.5x$	$0.8y$	0.68×200, or 136

Translate. Since the total amount of solution is 200 mL, we have

$$x + y = 200.$$

The amount of acid in the mixture is to be 68% of 200 mL, or 136 mL. The amounts of acid from the two solutions are $50\%x$ and $80\%y$. Thus

$$50\%x + 80\%y = 136,$$
$$\text{or} \quad 0.5x + 0.8y = 136,$$
$$\text{or} \quad 5x + 8y = 1360 \quad \text{Clearing decimals}$$

Carry out. We use the elimination method.

$$x + y = 200, \quad (1)$$
$$5x + 8y = 1360 \quad (2)$$

We multiply Equation (1) by -5 and then add.

$$-5x - 5y = -1000$$
$$5x + 8y = 1360$$
$$\overline{\qquad 3y = 360}$$
$$y = 120$$

Next we substitute 120 for y in one of the original equations and solve for x.

$$x + y = 200 \quad (1)$$
$$x + 120 = 200 \quad \text{Substituting}$$
$$x = 80$$

Check. The sum of 80 and 120 is 200. Now 50% of 80 is 40 and 80% of 120 is 96. These add up to 136. The numbers check.

State. 80 mL of the 50%-acid solution and 120 mL of the 80%-acid solution should be used.

20. 50 oz of Clear Shine, 40 oz of Sunstream

21. *Familiarize*. Let x and y represent the number of liters of 28%-fungicide solution and 40%-fungicide solution to be used in the mixture, respectively. We organize the given information in a table.

Type of solution	28%	40%	36%
Amount of solution	x	y	300
Percent fungicide	28%	40%	36%
Amount of fungicide in solution	$0.28x$	$0.4y$	$0.36(300)$, or 108

Translate. We get a system of equations from the first and third rows of the table.

$$x + y = 300,$$
$$0.28x + 0.4y = 108$$

Clearing decimals we have

$$x + y = 300, \quad (1)$$
$$28x + 40y = 10,800 \quad (2)$$

Carry out. We use the elimination method. Multiply Equation (1) by -28 and add.

$$-28x - 28y = -8400$$
$$28x + 40y = 10,800$$
$$\overline{\qquad 12y = 2400}$$
$$y = 200$$

Now substitute 200 for y in Equation (1) and solve for x.

$$x + y = 300$$
$$x + 200 = 300$$
$$x = 100$$

Check. The sum of 100 and 200 is 300. The amount of fungicide in the mixture is $0.28(100) + 0.4(200)$, or $28 + 80$, or 108 L. These numbers check.

State. 100 L of the 28%-fungicide solution and 200 L of the 40%-fungicide solution should be used in the mixture.

22. 128 L of 80%, 72 L of 30%

23. *Familiarize*. Let x and y represent the number of gallons of 87-octane gas and 93-octane gas to be blended, respectively. We organize the given information in a table.

Type of gasoline	87-octane	93-octane	91-octane
Amount of gas	x	y	12
Octane rating	87	93	91
Mixture	$87x$	$93y$	$91 \cdot 12$, or 1092

Translate. We get a system of equations from the first and third rows of the table.

$$x + \quad y = 12, \qquad (1)$$
$$87x + 93y = 1092 \qquad (2)$$

Carry out. We use the elimination method. First we multiply Equation (1) by -87 and add.

$$
\begin{array}{r}
-87x - 87y = -1044 \\
87x + 93y = 1092 \\
\hline
6y = 48 \\
y = 8
\end{array}
$$

Now substitute 8 for y in Equation (1) and solve for x.

$$x + y = 12$$
$$x + 8 = 12$$
$$x = 4$$

Check. The sum of 4 and 8 is 12. The mixture is $87 \cdot 4 + 93 \cdot 8$, or $348 + 744$, or 1092. These numbers check.

State. 4 gal of 87-octane gas and 8 gal of 93-octane gas should be blended.

24. 12 type A, 4 type B

25. *Familiarize*. Let x and y represent the number of ounces of three-fourths pure gold alloy and five-twelfths pure gold alloy to be used, respectively. We organize the given information in a table.

Type of alloy	$\frac{3}{4}$ gold	$\frac{5}{12}$ gold	$\frac{2}{3}$ gold
Amount of alloy	x	y	60
Fraction of pure gold	$\frac{3}{4}$	$\frac{5}{12}$	$\frac{2}{3}$
Amount of pure gold	$\frac{3}{4}x$	$\frac{5}{12}y$	$\frac{2}{3} \cdot 60$, or 40

Translate. We get a system of equations from the first and third rows of the table.

$$x + \quad y = 60,$$
$$\frac{3}{4}x + \frac{5}{12}y = 40$$

Carry out. First we multiply the second equation by 12 to clear the fractions.

$$x + \quad y = 60, \quad (1)$$
$$9x + 5y = 480 \quad (2)$$

Now multiply Equation (1) by -5 and add.

$$
\begin{array}{r}
-5x - 5y = -300 \\
9x + 5y = 480 \\
\hline
4x = 180 \\
x = 45
\end{array}
$$

Substitute 45 for x in Equation (1) and solve for y.

$$x + y = 60$$
$$45 + y = 60$$
$$y = 15$$

Check. The sum of 45 and 15 is 60. The amount of pure gold in the mixture is $\frac{3}{4} \cdot 45 + \frac{5}{12} \cdot 15$, or $\frac{135}{4} + \frac{25}{4}$, or 40 oz. These numbers check.

State. 45 oz of the three-fourths pure gold alloy and 15 oz of the five-twelfths pure gold alloy should be melted and mixed.

26. 70 dimes, 33 quarters

27. *Familiarize*. Let n represent the number of nickels and q represent the number of quarters. Then $5n$ represents the value of the nickels in cents and $25q$ represents the value of the quarters in cents. The total value of the collection is \$1.25, or 125¢. The total number of coins is 13.

Translate.

Number of nickels	plus	number of quarters	is	13.
$\downarrow$	$\downarrow$	$\downarrow$	$\downarrow$	$\downarrow$
n	$+$	q	$=$	13

Value of nickels	plus	value of quarters	is	125¢.
$\downarrow$	$\downarrow$	$\downarrow$	$\downarrow$	$\downarrow$
$5n$	$+$	$25q$	$=$	125

We have a system of equations.

$$n + \quad q = 13, \quad (1)$$
$$5n + 25q = 125 \quad (2)$$

Carry out. We use the elimination method. First multiply Equation (1) by -5 and add.

$$
\begin{array}{r}
-5n - 5q = -65 \\
5n + 25q = 125 \\
\hline
20q = 60 \\
q = 3
\end{array}
$$

Substitute 3 for q in Equation (1) and solve for n.

$$n + q = 13$$
$$n + 3 = 13$$
$$n = 10$$

Check. The sum of 10 and 3 is 13. The value of the coins is $5 \cdot 10 + 3 \cdot 25$, or $50 + 75$, or 125¢, or \$1.25. These numbers check.

State. The collection contains 10 nickels and 3 quarters.

28. 7 1300-word pages, 5 1850-word pages

29. Familiarize. We let x and y represent the number of pounds of peanuts and fancy nuts in the mixture, respectively. We organize the given information in a table.

Type of nuts	Peanuts	Fancy	Mixture
Amount	x	y	10
Price per pound	\$2.50	\$7	
Value	2.5x	7y	40

Translate. We get a system of equations from the first and third rows of the table.

$$x + y = 10,$$
$$2.5x + 7y = 40$$

Clearing decimals we have

$$x + y = 10, \quad (1)$$
$$25x + 70y = 400. \quad (2)$$

Carry out. We use the elimination method. Multiply Equation (1) by -25 and add.

$$-25x - 25y = -250$$
$$\underline{25x + 70y = 400}$$
$$45y = 150$$
$$y = \frac{10}{3}, \quad \text{or } 3\frac{1}{3}$$

Substitute $\frac{10}{3}$ for y in Equation (1) and solve for x.

$$x + y = 10$$
$$x + \frac{10}{3} = 10$$
$$x = \frac{20}{3}, \quad \text{or } 6\frac{2}{3}$$

Check. The sum of $6\frac{2}{3}$ and $3\frac{1}{3}$ is 10. The value of the mixture is $2.5\left(\frac{20}{3}\right) + 7\left(\frac{10}{3}\right)$, or $\frac{50}{3} + \frac{70}{3}$, or \$40. These numbers check.

State. $6\frac{2}{3}$ lb of peanuts and $3\frac{1}{3}$ lb of fancy nuts should be used.

30. Inexpensive: \$19.408 per gallon, expensive: \$20.075

31. $25x^2 - 81 = (5x)^2 - 9^2$
$$= (5x + 9)(5x - 9)$$

32. $(6 + a)(6 - a)$

33. $4x^2 + 100 = 4(x^2 + 25)$

34. 4

35. $x^2 - 10x + 25 = 0$
$$(x - 5)(x - 5) = 0$$
$$x - 5 = 0 \quad or \quad x - 5 = 0$$
$$x = 5 \quad or \quad x = 5$$

The solution is 5.

36. $-10, 10$

37. ◈

38. ◈

39. ◈

40. ◈

41. Familiarize. In a table we arrange the information regarding the solution *after* some of the 30% solution is drained and replaced with pure antifreeze. We let x represent the amount of the original (30%) solution remaining, and we let y represent the amount of the 30% mixture that is drained and replaced with pure antifreeze.

Type of solution	Original (30%)	Pure anti-freeze	Mixture
Amount of solution	x	y	16
Percent of antifreeze	30%	100%	50%
Amount of antifreeze in solution	0.3x	$1 \cdot y$, or y	0.5(16), or 8

Translate. The table gives us two equations.

Amount of solution: $x + y = 16$

Amount of antifreeze in solution: $0.3x + y = 8$, or $3x + 10y = 80$

The resulting system is

$$x + y = 16, \quad (1)$$
$$3x + 10y = 80. \quad (2)$$

Carry out. We multiply Equation (1) by -3 and then add.

$$-3x - 3y = -48$$
$$\underline{3x + 10y = 80}$$
$$7y = 32$$
$$y = \frac{32}{7}, \text{ or } 4\frac{4}{7}$$

Then we substitute $4\frac{4}{7}$ for y in Equation (1) and solve for x.

$$x + y = 16$$
$$x + 4\frac{4}{7} = 16$$
$$x = 11\frac{3}{7}$$

Check. When $x = 11\frac{3}{7}$ L and $y = 4\frac{4}{7}$ L, the total is 16 L. The amount of antifreeze in the mixture is $0.3\left(11\frac{3}{7}\right) + 4\frac{4}{7}$, or $\frac{3}{10} \cdot \frac{80}{7} + \frac{32}{7}$, or $\frac{24}{7} + \frac{32}{7} = \frac{56}{7}$, or 8 L. This is 50% of 16 L, so the numbers check.

State. $4\frac{4}{7}$ L of the original mixture should be drained and replaced with pure antifreeze.

42. $2666\frac{2}{3}$ L

43. Familiarize. Let x represent the number of gallons of 91-octane gas to be added to the tank and let y represent the total number of gallons in the tank after the 91-octane gas is added. We organize the given information in a table.

Type of gasoline	85-octane	91-octane	Mixture
Amount of gas	5	x	y
Octane rating	85	91	87
Mixture	$85 \cdot 5$, or 425	$91x$	$87y$

Translate. We get a system of equations from the first and third rows of the table.

$$5 + x = y, \qquad (1)$$
$$425 + 91x = 87y \quad (2)$$

Carry out. Substitute $5 + x$ for y in Equation (2) and solve for x.

$$425 + 91x = 87y$$
$$425 + 91x = 87(5 + x)$$
$$425 + 91x = 435 + 87x$$
$$425 + 4x = 435$$
$$4x = 10$$
$$x = 2.5$$

Although the original problem asks us to find only x, we will find y also in order to check the answer. Substitute 2.5 for x in Equation (1) and compute y.

$$y = 5 + 2.5 = 7.5$$

Check. The mixture is $425 + 91(2.5)$, or 652.5. This is equal to $87(7.5)$, so the answer checks.

State. Kim should add 2.5 gal of 91-octane gas to her tank.

44. 43.75 L

45. Familiarize. Let x = the original number of $20 workers and y = the original number of $25 workers. Then the total cost per hour for the $20 workers is $20x$ and the total cost per hour for the $25 workers is $25y$. The original number of $20 workers increased by 50% is $x + 0.5x$, or $1.5x$. The original number of $25 workers decreased by 20% is $y - 0.2y$, or $0.8y$.

Translate.

Original total hourly cost is $325.
$$20x + 25y = 325$$

New total hourly cost is $400.
$$20(1.5x) + 25(0.8y) = 400$$

We have a system of equations.
$$20x + 25y = 325,$$
$$20(1.5x) + 25(0.8y) = 400$$

Carry out. Begin by removing parentheses in the second equation.

$$20x + 25y = 325, \quad (1)$$
$$30x + 20y = 400 \quad (2)$$

Multiply Equation (1) by 3 and Equation (2) by -2 and then add.

$$60x + 75y = 975$$
$$\underline{-60x - 40y = -800}$$
$$35y = 175$$
$$y = 5$$

Substitute 5 for y in Equation (1) and solve for x.

$$20x + 25y = 325$$
$$20x + 25 \cdot 5 = 325$$
$$20x + 125 = 325$$
$$20x = 200$$
$$x = 10$$

Check. If there are 10 $20 workers and 5 $25 workers originally, then the total hourly cost is $20 \cdot 10 + $25 \cdot 5$ or $200 + $125, or $325. If the original number of $20 workers is increased by 50%, the new number is $10 + 0.5(10)$, or $10 + 5$, or 15. If the original number of $25 workers is decreased by 20%, the new number is $5 - 0.2(5)$, or $5 - 1$, or 4. Then the new hourly cost is $20 \cdot 15 + $25 \cdot 4$, or $300 + $100, or $400. The answer checks.

State. Originally 10 workers were employed at $20 per hour and 5 were employed at $25 per hour.

46. $25,000 at 6%, $29,000 at 6.5%

47. Familiarize. Let x = the lower interest rate and y = the higher interest rate, where each expression is given in decimal notation. Then the interest earned in one year at the lower rate is $1100x$ and at the higher rate is $1800y$.

Translate.

$$\underbrace{\text{Total interest}}_{\downarrow} \quad \underbrace{\text{is}}_{\downarrow} \quad \underbrace{\$288.}_{\downarrow}$$

$$1100x + 1800y \quad = \quad 288$$

$$\underbrace{\text{Higher rate}}_{\downarrow} \quad \underbrace{\text{is}}_{\downarrow} \quad \underbrace{1.5\%}_{\downarrow} \quad \underbrace{\text{more than}}_{\downarrow} \quad \underbrace{\text{lower rate.}}_{\downarrow}$$

$$y \quad = \quad 0.015 \quad + \quad x$$

We have a system of equations.

$$1100x + 1800y = 288, \quad (1)$$
$$y = 0.015 + x \quad (2)$$

Carry out. Use the substitution method. Substitute $0.015 + x$ for y in Equation (1) and solve for x.

$$1100x + 1800y = 288 \quad (1)$$
$$1100x + 1800(0.015 + x) = 288 \quad \text{Substituting}$$
$$1100x + 27 + 1800x = 288$$
$$2900x + 27 = 288$$
$$2900x = 261$$
$$x = 0.09$$

Substitute 0.09 for x in Equation (2) and compute y.

$$y = 0.015 + 0.09 = 0.105$$

Check. The higher rate 0.105, or 10.5%, is 1.5% higher than the lower rate of 0.09, or 9%. The total interest at these rates would be $\$1100(0.09) + \$1800(0.105)$, or $\$99 + \$189$, or $\$288$.

State. The interest rates were 9% and 10.5%.

48. 54

49. *Familiarize*. Let x represent the ten's digit of the original number and y represent the one's digit. Then the original number is $10x + y$ and the number that results when the digits are reversed is $10y + x$.

Translate.

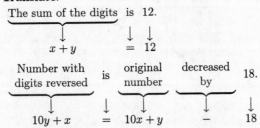

$$\underbrace{\text{The sum of the digits}}_{\downarrow} \quad \underbrace{\text{is}}_{\downarrow} \quad \underbrace{12.}_{\downarrow}$$

$$x + y \quad = \quad 12$$

$$\underbrace{\begin{array}{c}\text{Number with}\\\text{digits reversed}\end{array}}_{\downarrow} \quad \underbrace{\text{is}}_{\downarrow} \quad \underbrace{\begin{array}{c}\text{original}\\\text{number}\end{array}}_{\downarrow} \quad \underbrace{\begin{array}{c}\text{decreased}\\\text{by}\end{array}}_{\downarrow} \quad \underbrace{18.}_{\downarrow}$$

$$10y + x \quad = \quad 10x + y \quad - \quad 18$$

We have a system of equations.

$$x + y = 12,$$
$$10y + x = 10x + y - 18, \quad \text{or}$$

$$x + y = 12, \quad (1)$$
$$-9x + 9y = -18 \quad (2)$$

Carry out. We use elimination. Multiply Equation (1) by 9 and add.

$$9x + 9y = 108$$
$$\underline{-9x + 9y = -18}$$
$$18y = 90$$
$$y = 5$$

Now substitute 5 for y in Equation (1) and solve for x.

$$x + y = 12 \quad (1)$$
$$x + 5 = 12 \quad \text{Substituting}$$
$$x = 7$$

Check. The sum of the digits is $7 + 5$, or 12. If the original number is 75, then the number that results when the digits are reversed, 57, is the original number decreased by 18. The answer checks.

State. The original number is 75.

50. Bat: $\$14.50$, ball: $\$4.55$, glove: $\$79.95$

51. *Familiarize*. Let $x = $ Tweedledum's weight and $y = $ Tweedledee's weight, in pounds.

Translate.

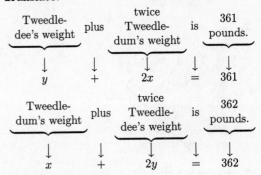

$$\underbrace{\begin{array}{c}\text{Tweedle-}\\\text{dee's weight}\end{array}}_{\downarrow} \quad \underbrace{\text{plus}}_{\downarrow} \quad \underbrace{\begin{array}{c}\text{twice}\\\text{Tweedle-}\\\text{dum's weight}\end{array}}_{\downarrow} \quad \underbrace{\text{is}}_{\downarrow} \quad \underbrace{\begin{array}{c}361\\\text{pounds.}\end{array}}_{\downarrow}$$

$$y \quad + \quad 2x \quad = \quad 361$$

$$\underbrace{\begin{array}{c}\text{Tweedle-}\\\text{dum's weight}\end{array}}_{\downarrow} \quad \underbrace{\text{plus}}_{\downarrow} \quad \underbrace{\begin{array}{c}\text{twice}\\\text{Tweedle-}\\\text{dee's weight}\end{array}}_{\downarrow} \quad \underbrace{\text{is}}_{\downarrow} \quad \underbrace{\begin{array}{c}362\\\text{pounds.}\end{array}}_{\downarrow}$$

$$x \quad + \quad 2y \quad = \quad 362$$

We have a system of equations.

$$y + 2x = 361,$$
$$x + 2y = 362, \quad \text{or}$$

$$2x + y = 361, \quad (1)$$
$$x + 2y = 362 \quad (2)$$

Carry out. We use elimination. First multiply Equation (2) by -2 and add.

$$2x + y = 361$$
$$\underline{-2x - 4y = -724}$$
$$-3y = -363$$
$$y = 121$$

Now substitute 121 for y in Equation (2) and solve for x.

$$x + 2y = 362 \quad (2)$$
$$x + 2 \cdot 121 = 362 \quad \text{Substituting}$$
$$x + 242 = 362$$
$$x = 120$$

Check. If Tweedledum weighs 120 lb and Tweedledee weighs 121 lb, then the sum of Tweedledee's weight and twice Tweedledum's is $121 + 2 \cdot 120$, or $121 + 240$, or 361 lb. The sum of Tweedledum's weight and twice Tweedledee's is $120 + 2 \cdot 121$, or $120 + 242$, or 362 lb. The answer checks.

State. Tweedledum weighs 120 lb, and Tweedledee weighs 121 lb.

Exercise Set 8.5

1. $x + y \leq 2$,
    $x - y \leq 7$

    We graph the lines $x + y = 2$ and $x - y = 7$ using solid lines. We indicate the region for each inequality by the arrows at the ends of the lines. We shade the area where the regions overlap.

    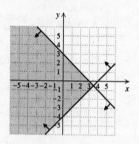

2.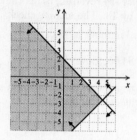

3. $y - 2x > 1$,
    $y - 2x < 3$

    We graph the lines $y - 2x = 1$ and $y - 2x = 3$ using dashed lines. We indicate the region for each inequality by the arrows at the ends of the lines. We shade the area where the regions overlap.

    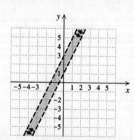

4.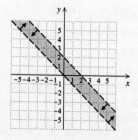

5. $y \geq -3$,
    $x > 2 + y$

    We graph the line $y = -3$ using a solid line and the line $x = 2 + y$ using a dashed line. We indicate the region for each inequality by the arrows at the ends of the lines. We shade the area where the regions overlap.

    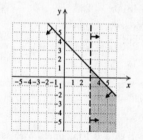

6.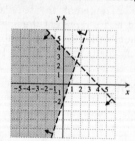

7. $y > 3x - 2$,
    $y < -x + 4$

    We graph the lines $y = 3x - 2$ and $y = -x + 4$ using dashed lines. We indicate the region for each inequality by the arrows at the ends of the lines. We shade the area where the regions overlap.

    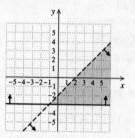

8.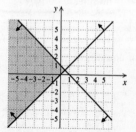

9. $x \leq 4$,
    $y \leq 5$

    We graph the lines $x = 4$ and $y = 5$ using solid lines. We indicate the region for each inequality by the

arrows at the ends of the lines. We shade the area
where the regions overlap.

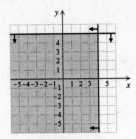

10.

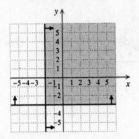

11. $x \le 0$,

$y \le 0$

We graph the lines $x = 0$ and $y = 0$ using solid
lines. We indicate the region for each inequality by
the arrows at the ends of the lines. We shade the
area where the regions overlap.

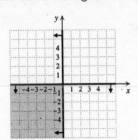

12.

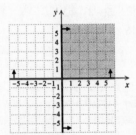

13. $2x - 3y \ge 9$,

$2y + x > 6$

We graph the line $2x - 3y = 9$ using a solid line
and the line $2y + x = 6$ using a dashed line. We
indicate the region for each inequality by the arrows
at the ends of the lines. We shade the area where
the regions overlap.

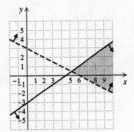

14.

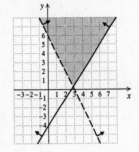

15. $y > 5x + 2$,

$y \le 1 - x$

We graph the line $y = 5x + 2$ using a dashed line and
the line $y = 1 - x$ using a solid line. We indicate the
region for each inequality by the arrows at the ends
of the lines. We shade the area where the regions
overlap.

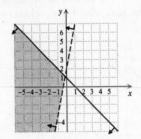

16.

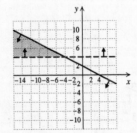

17. $x + y \le 5$,

$x \ge 0$,

$y \ge 0$,

$y \le 3$

We graph the lines $x + y = 5$, $x = 0$, $y = 0$, and
$y = 3$ using solid lines. We indicate the region for
each inequality by the arrows at the ends of the lines.
We shade the area where the regions overlap.

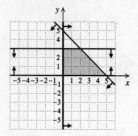

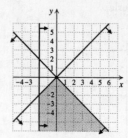

18.

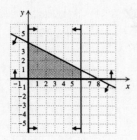

22.

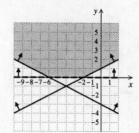

19. $y - x \geq 1,$

$y - x \leq 3,$

$x \leq 5,$

$x \geq 2$

We graph the lines $y - x = 1$, $y - x = 3$, $x = 5$, and $x = 2$ using solid lines. We indicate the region for each inequality by the arrows at the ends of the lines. We shade the area where the regions overlap.

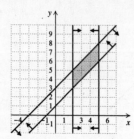

20.

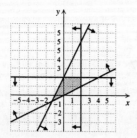

21. $y \leq x,$

$x \geq -2,$

$x \leq -y$

We graph the lines $y = x$, $x = -2$ and $x = -y$ using solid lines. We indicate the region for each inequality by the arrows at the ends of the lines. We shade the area where the regions overlap.

23.

$$\frac{3}{x^2 - 4} - \frac{2}{3x + 6}$$

$$= \frac{3}{(x+2)(x-2)} - \frac{2}{3(x+2)},$$

$$\text{LCD} = 3(x+2)(x-2)$$

$$= \frac{3}{(x+2)(x-2)} \cdot \frac{3}{3} - \frac{2}{3(x+2)} \cdot \frac{x-2}{x-2}$$

$$= \frac{9 - 2(x-2)}{3(x+2)(x-2)}$$

$$= \frac{9 - 2x + 4}{3(x+2)(x-2)}$$

$$= \frac{13 - 2x}{3(x+2)(x-2)}$$

24. $\dfrac{5a - 5 - 8a^2}{4a^2(a-3)(a-1)}$

25. $x^2 - 5x + 2 = (-3)^2 - 5(-3) + 2$

$$= 9 + 15 + 2$$

$$= 26$$

26. -153

27. $x^2 - 10x + 25 = x^2 - 2 \cdot x \cdot 5 + 5^2$

$$= (x - 5)^2$$

28. $3(a - 3)^2$

29. ◈

30. ◈

31. ◈

32. ◈

33. $5a + 3b \geq 30,$

$2a + 3b \geq 21,$

$3a + 6b \geq 36,$

$a \geq 0,$

$b \geq 0$

We graph the related equations, find the region for each inequality, and shade the area where the regions overlap.

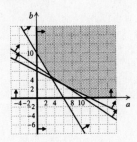

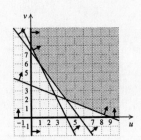

34.

Chapter 9

Radical Expressions and Equations

Exercise Set 9.1

1. The square roots of 9 are 3 and -3, since $3^2 = 9$ and $(-3)^2 = 9$.

2. 2, -2

3. The square roots of 16 are 4 and -4, because $4^2 = 16$ and $(-4)^2 = 16$.

4. 1, -1

5. The square roots of 49 are 7 and -7, because $7^2 = 49$ and $(-7)^2 = 49$.

6. 11, -11

7. The square roots of 169 are 13 and -13, because $13^2 = 169$ and $(-13)^2 = 169$.

8. 12, -12

9. $\sqrt{4} = 2$, taking the principal square root.

10. 3

11. $\sqrt{1} = 1$, so $-\sqrt{1} = -1$.

12. -5

13. $\sqrt{0} = 0$

14. -9

15. $\sqrt{121} = 11$, so $-\sqrt{121} = -11$.

16. 19

17. $\sqrt{400} = 20$

18. 21

19. $\sqrt{169} = 13$

20. 12

21. $\sqrt{625} = 25$, so $-\sqrt{625} = -25$.

22. -30

23. The radicand is the expression under the radical, $a - 4$.

24. $t + 3$

25. The radicand is the expression under the radical, $t^2 + 1$.

26. $x^2 + 5$

27. The radicand is the expression under the radical, $\dfrac{3}{x+2}$.

28. $\dfrac{a}{a-b}$

29. $\sqrt{16}$ is rational, since 16 is a perfect square.

30. Irrational

31. $\sqrt{8}$ is irrational, since 8 is not a perfect square.

32. Irrational

33. $\sqrt{32}$ is irrational, since 32 is not a perfect square.

34. Rational

35. $\sqrt{98}$ is irrational, since 98 is not a perfect square.

36. Irrational

37. $-\sqrt{4}$ is rational, since 4 is a perfect square.

38. Rational

39. $-\sqrt{12}$ is irrational, since 12 is not a perfect square.

40. Irrational

41. 2.236

42. 2.449

43. 4.123

44. 4.359

45. 9.644

46. 6.557

47. $\sqrt{t^2} = t$ Since t is assumed to be nonnegative

48. x

49. $\sqrt{25x^2} = \sqrt{(5x)^2} = 5x$ Since x is assumed to be nonnegative

50. $3a$

51. $\sqrt{(7a)^2} = 7a$ Since a is assumed to be nonnegative

52. $4x$

53. $\sqrt{(17x)^2} = 17x$ Since x is assumed to be nonnegative

54. $8ab$

55. a) We substitute 36 in the formula.
$$N = 2.5\sqrt{36} = 2.5(6) = 15$$
For an average of 36 arrivals, 15 spaces are needed.

b) We substitute 29 in the formula. We use a calculator or Table 2 to find an approximation.
$$N = 2.5\sqrt{29} \approx 2.5(5.385) \approx 13.463 \approx 14$$
For an average of 29 arrivals, 14 spaces are needed.

56. (a) 18; (b) 19

57. Substitute 36 in the formula.
$$T = 0.144\sqrt{36} = 0.144(6) = 0.864 \text{ sec}$$

58. 0.72 sec

59. $5x^3 \cdot 2x^6 = 5 \cdot 2 \cdot x^3 \cdot x^6$
$= 5 \cdot 2 \cdot x^{3+6}$
$= 10x^9$

60. $15a^4b^9$

61. $m = \dfrac{\text{change in } y}{\text{change in } x} = \dfrac{-6-4}{5-(-3)} = \dfrac{-10}{8} = -\dfrac{5}{4}$

62. $\dfrac{3}{5}$

63. Use the point-slope equation.
$$y - y_1 = m(x - x_1)$$
$$y - 4 = 2(x - (-3))$$
$$y - 4 = 2(x + 3)$$
$$y - 4 = 2x + 6$$
$$y = 2x + 10$$

64. $y = -\dfrac{5}{4}x + \dfrac{1}{4}$

65. ◈

66. ◈

67. ◈

68. ◈

69. We find the inner square root first.
$$\sqrt{\sqrt{16}} = \sqrt{4} = 2$$

70. 5

71. $-\sqrt{36} < -\sqrt{33} < -\sqrt{25}$, or $-6 < -\sqrt{33} < -5$
$-\sqrt{33}$ is between -6 and -5.

72. 64; answers may vary

73. If $\sqrt{x^2} = 8$, then $x^2 = 8^2$, or 64. Thus, $x = 8$ or $x = -8$. The solutions are 8 and -8.

74. No solution

75. If $-\sqrt{x^2} = -3$, then $\sqrt{x^2} = 3$ and $x^2 = 3^2$, or 9. Thus, $x = 3$ or $x = -3$. The solutions are 3 and -3.

76. $-7, 7$

77. $\sqrt{(9a^3b^4)^2} = 9a^3b^4$ Since all variables represent positive numbers

78. $3a$

79. $\sqrt{\dfrac{144x^8}{36y^6}} = \sqrt{\dfrac{4x^8}{y^6}} = \sqrt{\left(\dfrac{2x^4}{y^3}\right)^2} = \dfrac{2x^4}{y^3}$

80. $\dfrac{y^6}{90}$

81. $\sqrt{\dfrac{400}{m^{16}}} = \sqrt{\left(\dfrac{20}{m^8}\right)^2} = \dfrac{20}{m^8}$

82. $\dfrac{p}{60}$

83. a) Locate 3 on the x-axis, move up vertically to the graph, and then move left horizontally to the y-axis to read the approximation.
$$\sqrt{3} \approx 1.7 \quad \text{(Answers may vary.)}$$

b) Locate 5 on the x-axis, move up vertically to the graph, and then move left horizontally to the y-axis to read the approximation.
$$\sqrt{5} \approx 2.2 \quad \text{(Answers may vary.)}$$

c) Locate 7 on the x-axis, move up vertically to the graph, and then move left horizontally to the y-axis to read the approximation.
$$\sqrt{7} \approx 2.6 \quad \text{(Answers may vary.)}$$

84. 1141.6 ft/sec

85. We substitute 5 in the formula.
$$V = \dfrac{1087\sqrt{273+5}}{16.52}$$
$$V = \dfrac{1087\sqrt{278}}{16.52}$$
$$V \approx \dfrac{1087(16.673)}{16.52}$$
$$V \approx 1097.1 \text{ ft/sec}$$

86. 1067.1 ft/sec

87. We substitute 100 in the formula.
$$V = \dfrac{1087\sqrt{273+100}}{16.52}$$
$$V = \dfrac{1087\sqrt{373}}{16.52}$$
$$V \approx \dfrac{1087(19.313)}{16.52}$$
$$V \approx 1270.8 \text{ ft/sec}$$

88.
$y_1 = \sqrt{x-2};\ y_2 = \sqrt{x+7};$
$y_3 = 5 + \sqrt{x};\ y_4 = -4 + \sqrt{x}$

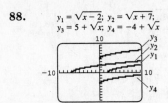

Exercise Set 9.2

1. $\sqrt{5}\sqrt{7} = \sqrt{5\cdot7} = \sqrt{35}$

2. $\sqrt{15}$

3. $\sqrt{4}\sqrt{3} = \sqrt{12}$, or
$\sqrt{4}\sqrt{3} = 2\sqrt{3}$ Taking the square root of 4

4. $\sqrt{18}$, or $3\sqrt{2}$

5. $\sqrt{\dfrac{2}{5}}\sqrt{\dfrac{3}{4}} = \sqrt{\dfrac{2\cdot3}{5\cdot4}} = \sqrt{\dfrac{3}{10}}$

6. $\sqrt{\dfrac{3}{40}}$, or $\dfrac{1}{2}\sqrt{\dfrac{3}{10}}$

7. $\sqrt{8}\sqrt{8} = \sqrt{8\cdot8} = \sqrt{64} = 8$

8. 18

9. $\sqrt{25}\sqrt{3} = \sqrt{75}$, or
$\sqrt{25}\sqrt{3} = 5\sqrt{3}$ Taking the square root of 25

10. $\sqrt{72}$, or $6\sqrt{2}$

11. $\sqrt{2}\sqrt{x} = \sqrt{2\cdot x} = \sqrt{2x}$

12. $\sqrt{3a}$

13. $\sqrt{7}\sqrt{3x} = \sqrt{7\cdot3x} = \sqrt{21x}$

14. $\sqrt{20x}$, or $2\sqrt{5x}$

15. $\sqrt{x}\sqrt{7y} = \sqrt{x\cdot7y} = \sqrt{7xy}$

16. $\sqrt{10mn}$

17. $\sqrt{3a}\sqrt{2c} = \sqrt{3a\cdot2c} = \sqrt{6ac}$

18. $\sqrt{3xyz}$

19. $\sqrt{28} = \sqrt{4\cdot7}$ 4 is a perfect square.
$\quad = \sqrt{4}\sqrt{7}$ Factoring into a product of radicals
$\quad = 2\sqrt{7}$

20. $2\sqrt{3}$

21. $\sqrt{8} = \sqrt{4\cdot2}$ 4 is a perfect square.
$\quad = \sqrt{4}\sqrt{2}$ Factoring into a product of radicals
$\quad = 2\sqrt{2}$

22. $3\sqrt{5}$

23. $\sqrt{500} = \sqrt{100\cdot5} = \sqrt{100}\sqrt{5} = 10\sqrt{5}$

24. $10\sqrt{2}$

25. $\sqrt{9x} = \sqrt{9\cdot x} = \sqrt{9}\sqrt{x} = 3\sqrt{x}$

26. $2\sqrt{y}$

27. $\sqrt{75a} = \sqrt{25\cdot3a} = \sqrt{25}\sqrt{3a} = 5\sqrt{3a}$

28. $2\sqrt{10m}$

29. $\sqrt{16a} = \sqrt{16\cdot a} = \sqrt{16}\sqrt{a} = 4\sqrt{a}$

30. $7\sqrt{b}$

31. $\sqrt{64y^2} = \sqrt{64}\sqrt{y^2} = 8y$, or
$\sqrt{64y^2} = \sqrt{(8y)^2} = 8y$

32. $3x$

33. $\sqrt{13x^2} = \sqrt{13}\sqrt{x^2} = \sqrt{13}\cdot x$, or $x\sqrt{13}$

34. $t\sqrt{29}$

35. $\sqrt{8t^2} = \sqrt{4\cdot t^2\cdot2} = \sqrt{4}\sqrt{t^2}\sqrt{2} = 2t\sqrt{2}$

36. $5a\sqrt{5}$

37. $\sqrt{80} = \sqrt{2\cdot2\cdot2\cdot2\cdot5}$ Writing the prime factorization
$\quad = \sqrt{2^2}\sqrt{2^2}\sqrt{5}$
$\quad = 2\cdot2\cdot\sqrt{5}$
$\quad = 4\sqrt{5}$

38. $7\sqrt{2}$

39. $\sqrt{288y} = \sqrt{144\cdot2y} = \sqrt{144}\sqrt{2y} = 12\sqrt{2y}$

40. $11\sqrt{3p}$

41. $\sqrt{x^{20}} = \sqrt{(x^{10})^2} = x^{10}$

42. x^{15}

43. $\sqrt{x^{12}} = \sqrt{(x^6)^2} = x^6$

44. x^8

45. $\sqrt{x^5} = \sqrt{x^4\cdot x}$ One factor is a perfect square
$\quad = \sqrt{x^4}\sqrt{x}$
$\quad = \sqrt{(x^2)^2}\sqrt{x}$
$\quad = x^2\sqrt{x}$

46. $x\sqrt{x}$

47. $\sqrt{t^{19}} = \sqrt{t^{18}\cdot t} = \sqrt{t^{18}}\sqrt{t} = \sqrt{(t^9)^2}\sqrt{t} = t^9\sqrt{t}$

48. $p^8\sqrt{p}$

49. $\sqrt{36m^3} = \sqrt{36\cdot m^2\cdot m} = \sqrt{36}\sqrt{m^2}\sqrt{m} = 6m\sqrt{m}$

50. $5y\sqrt{10y}$

51. $\sqrt{8a^5} = \sqrt{4a^4(2a)} = \sqrt{4(a^2)^2(2a)} =$
$\sqrt{4}\sqrt{(a^2)^2}\sqrt{2a} = 2a^2\sqrt{2a}$

52. $2b^3\sqrt{3b}$

53. $\sqrt{104p^{17}} = \sqrt{4p^{16}(26p)} = \sqrt{4(p^8)^2(26p)} =$
$\sqrt{4}\sqrt{(p^8)^2}\sqrt{26p} = 2p^8\sqrt{26p}$

54. $3m^{11}\sqrt{10m}$

55. $\sqrt{15} \cdot \sqrt{5} = \sqrt{15 \cdot 5}$ Multiplying
$\qquad\qquad\;\; = \sqrt{3 \cdot 5 \cdot 5}$ Writing the prime
$\qquad\qquad\qquad\qquad\qquad$ factorization
$\qquad\qquad\;\; = \sqrt{3}\sqrt{5^2}$
$\qquad\qquad\;\; = \sqrt{3} \cdot 5,$ or
$\qquad\qquad\;\;\;\; 5\sqrt{3}$

56. $3\sqrt{2}$

57. $\sqrt{3} \cdot \sqrt{27} = \sqrt{3 \cdot 27}$ Multiplying
$\qquad\qquad\;\; = \sqrt{3 \cdot 3 \cdot 3 \cdot 3}$ Writing the prime
$\qquad\qquad\qquad\qquad\qquad$ factorization
$\qquad\qquad\;\; = \sqrt{3^4}$
$\qquad\qquad\;\; = 3^2$
$\qquad\qquad\;\; = 9$

58. $7\sqrt{6}$

59. $\sqrt{3x}\sqrt{12y} = \sqrt{3x \cdot 12y}$
$\qquad\qquad\;\; = \sqrt{3 \cdot x \cdot 2 \cdot 2 \cdot 3 \cdot y}$
$\qquad\qquad\;\; = \sqrt{2^2}\sqrt{3^2}\sqrt{xy}$
$\qquad\qquad\;\; = 2 \cdot 3\sqrt{xy}$
$\qquad\qquad\;\; = 6\sqrt{xy}$

60. $10\sqrt{xy}$

61. $\sqrt{10}\sqrt{10} = \sqrt{10 \cdot 10} = \sqrt{10^2} = 10$

62. $11\sqrt{x}$

63. $\sqrt{5b}\sqrt{15b} = \sqrt{5b \cdot 15b}$
$\qquad\qquad\;\; = \sqrt{5 \cdot b \cdot 3 \cdot 5 \cdot b}$
$\qquad\qquad\;\; = \sqrt{5^2}\sqrt{b^2}\sqrt{3}$
$\qquad\qquad\;\; = 5b\sqrt{3}$

64. $6a\sqrt{3}$

65. $\sqrt{7x} \cdot \sqrt{7x} = \sqrt{7x \cdot 7x} = \sqrt{(7x)^2} = 7x$

66. $3a$

67. $\sqrt{ab}\sqrt{ac} = \sqrt{a^2bc} = \sqrt{a^2}\sqrt{bc} = a\sqrt{bc}$

68. $x\sqrt{yz}$

69. $\sqrt{2x}\sqrt{4x^5} = \sqrt{2x \cdot 4x^5}$
$\qquad\qquad\;\; = \sqrt{2 \cdot 4 \cdot x^6}$
$\qquad\qquad\;\; = \sqrt{2}\sqrt{4}\sqrt{x^6}$
$\qquad\qquad\;\; = \sqrt{2} \cdot 2 \cdot x^3,$ or
$\qquad\qquad\;\;\;\; 2x^3\sqrt{2}$

70. $5m^4\sqrt{3}$

71. $\sqrt{x^2y^3}\sqrt{xy^4}$
$\;\; = \sqrt{x^2y^3}\sqrt{x(y^2)^2}$ x^2 and $(y^2)^2$ are
$\qquad\qquad\qquad\qquad\qquad$ perfect squares
$\;\; = \sqrt{x^2} \cdot \sqrt{y^3} \cdot \sqrt{x} \cdot \sqrt{(y^2)^2}$
$\;\; = x \cdot \sqrt{y^3} \cdot \sqrt{x} \cdot y^2$
$\;\; = xy^2\sqrt{y^3 \cdot x}$
$\;\; = xy^2\sqrt{x \cdot y^2 \cdot y}$
$\;\; = xy^2\sqrt{y^2}\sqrt{xy}$
$\;\; = xy^2 \cdot y \cdot \sqrt{xy}$
$\;\; = xy^3\sqrt{xy}$

72. $x^2y\sqrt{y}$

73. $\sqrt{50ab}\sqrt{10a^2b^4} = \sqrt{50ab} \cdot \sqrt{10} \cdot \sqrt{a^2} \cdot \sqrt{(b^2)^2}$
$\qquad\qquad\qquad\;\; = ab^2\sqrt{50ab}\sqrt{10}$
$\qquad\qquad\qquad\;\; = ab^2\sqrt{50ab \cdot 10}$
$\qquad\qquad\qquad\;\; = ab^2\sqrt{2 \cdot 5 \cdot 5 \cdot a \cdot b \cdot 2 \cdot 5}$
$\qquad\qquad\qquad\;\; = ab^2\sqrt{2^2 \cdot 5^2 \cdot 5 \cdot a \cdot b}$
$\qquad\qquad\qquad\;\; = ab^2\sqrt{2^2}\sqrt{5^2}\sqrt{5ab}$
$\qquad\qquad\qquad\;\; = ab^2 \cdot 2 \cdot 5\sqrt{5ab}$
$\qquad\qquad\qquad\;\; = 10ab^2\sqrt{5ab}$

74. $5xy^2\sqrt{2xy}$

75. First we substitute 20 for L in the formula:
$r = 2\sqrt{5L} = 2\sqrt{5 \cdot 20} = 2\sqrt{100} = 2 \cdot 10 = 20$ mph
Then we substitute 150 for L:
$r = 2\sqrt{5 \cdot 150} = 2\sqrt{750} = 2\sqrt{25 \cdot 30} = 2\sqrt{25}\sqrt{30} =$
$2 \cdot 5\sqrt{30} = 10\sqrt{30} \approx 10(5.477) \approx 54.77$ mph, or
54.8 mph (rounded to the nearest tenth)

76. 24.5 mph, 37.4 mph

77. **Familiarize.** We present the information in a table.

	d	$=$	r	$\cdot$	t
	Distance	Speed	Time
First car	d	56	t
Second car	d	84	$t - 1$

Translate. From the rows of the table we get two equations:
$$d = 56t,$$
$$d = 84(t - 1).$$

Carry out. We use the substitution method.

$$56t = 84(t - 1) \quad \text{Substituting } 56t \text{ for } d$$
$$56t = 84t - 84$$
$$-28t = -84$$
$$t = 3$$

The problem asks how far from Hereford the second car will overtake the first, so we need to find d. Substitute 3 for t in the first equation.

$$d = 56t$$
$$d = 56 \cdot 3$$
$$d = 168$$

Check. If $t = 3$, then the first car travels $56 \cdot 3$, or 168 km, and the second car travels $84(3 - 1)$, or $84 \cdot 2$, or 168 km. Since the distances are the same, the answer checks.

State. The second car overtakes the first 168 km from Hereford.

78. $a^2 - 25b^2$

79. $\dfrac{15a^3b^7}{5ab^2} = \dfrac{15}{5} \cdot a^{3-1}b^{7-2} = 3a^2b^5$

80. $4x^6y^4$

81. ◈

82. ◈

83. ◈

84. ◈

85. $\sqrt{0.01} = \sqrt{(0.1)^2} = 0.1$

86. 0.5

87. $\sqrt{0.0625} = \sqrt{(0.25)^2} = 0.25$

88. 0.001

89. $15 = \sqrt{225}$ and $4\sqrt{14} = \sqrt{16 \cdot 14} = \sqrt{224}$, so $15 > 4\sqrt{14}$.

90. $\sqrt{450}$

91. $16 = \sqrt{256}$ and $\sqrt{15}\sqrt{17} = \sqrt{255}$, so $16 > \sqrt{15}\sqrt{17}$.

92. $3\sqrt{11} > 7\sqrt{2}$

93. $5\sqrt{7} = \sqrt{25 \cdot 7} = \sqrt{175}$ and $4\sqrt{11} = \sqrt{16 \cdot 11} = \sqrt{176}$, so $5\sqrt{7} < 4\sqrt{11}$.

94. $8 > \sqrt{15} + \sqrt{17}$

95. $\sqrt{27(x + 1)}\,\sqrt{12y(x + 1)^2} =$
$\sqrt{27(x + 1) \cdot 12y(x + 1)^2} =$
$\sqrt{9 \cdot 3 \cdot (x + 1) \cdot 4 \cdot 3 \cdot y(x + 1)^2} =$
$\sqrt{9 \cdot 3 \cdot 3 \cdot 4 \cdot (x + 1)^2 \cdot (x + 1)y} =$
$\sqrt{9}\,\sqrt{3 \cdot 3}\,\sqrt{4}\,\sqrt{(x + 1)^2}\,\sqrt{(x + 1)y} =$
$3 \cdot 3 \cdot 2(x + 1)\sqrt{(x + 1)y} = 18(x + 1)\sqrt{(x + 1)y}$

96. $6(x - 2)^2\sqrt{10}$

97. $\sqrt{x}\,\sqrt{2x}\,\sqrt{10x^5} = \sqrt{x \cdot 2x \cdot 10x^5} =$
$\sqrt{x \cdot 2 \cdot x \cdot 2 \cdot 5 \cdot x^4 \cdot x} = \sqrt{x \cdot x \cdot 2 \cdot 2 \cdot x^4 \cdot 5 \cdot x} =$
$\sqrt{x \cdot x}\,\sqrt{2 \cdot 2}\,\sqrt{x^4}\,\sqrt{5x} = x \cdot 2 \cdot x^2\sqrt{5x} = 2x^3\sqrt{5x}$

98. $2^{54}x^{158}\sqrt{2x}$

99. $\sqrt{x^{8n}} = \sqrt{(x^{4n})^2} = x^{4n}$

100. $0.2x^{2n}$

101. If n is an odd whole number greater than or equal to 3, then $n = 2k + 1$, where k is a natural number.
$\sqrt{y^n} = \sqrt{y^{2k+1}} = \sqrt{y^{2k} \cdot y} = y^k\sqrt{y}$, where $k = \dfrac{n - 1}{2}$.

Exercise Set 9.3

1. $\dfrac{\sqrt{12}}{\sqrt{3}} = \sqrt{\dfrac{12}{3}} = \sqrt{4} = 2$

2. 2

3. $\dfrac{\sqrt{60}}{\sqrt{15}} = \sqrt{\dfrac{60}{15}} = \sqrt{4} = 2$

4. 6

5. $\dfrac{\sqrt{75}}{\sqrt{15}} = \sqrt{\dfrac{75}{15}} = \sqrt{5}$

6. $\sqrt{6}$

7. $\dfrac{\sqrt{5}}{\sqrt{80}} = \sqrt{\dfrac{5}{80}} = \sqrt{\dfrac{1}{16}} = \dfrac{1}{4}$

8. $\dfrac{1}{4}$

9. $\dfrac{\sqrt{12}}{\sqrt{75}} = \sqrt{\dfrac{12}{75}} = \sqrt{\dfrac{4}{25}} = \dfrac{2}{5}$

10. $\dfrac{3}{4}$

11. $\dfrac{\sqrt{8x}}{\sqrt{2x}} = \sqrt{\dfrac{8x}{2x}} = \sqrt{4} = 2$

12. 3

13. $\dfrac{\sqrt{63y^3}}{\sqrt{7y}} = \sqrt{\dfrac{63y^3}{7y}} = \sqrt{9y^2} = 3y$

14. $4x$

15. $\dfrac{\sqrt{27x^5}}{\sqrt{3x}} = \sqrt{\dfrac{27x^5}{3x}} = \sqrt{9x^4} = 3x^2$

16. $2a^3$

17. $\dfrac{\sqrt{75x}}{\sqrt{25x^9}} = \sqrt{\dfrac{75x}{25x^9}} = \sqrt{\dfrac{3}{x^8}} = \dfrac{\sqrt{3}}{x^4}$

18. $x^3\sqrt{3}$

19. $\sqrt{\dfrac{36}{25}} = \dfrac{\sqrt{36}}{\sqrt{25}} = \dfrac{6}{5}$

20. $\dfrac{3}{7}$

21. $\sqrt{\dfrac{49}{16}} = \dfrac{\sqrt{49}}{\sqrt{16}} = \dfrac{7}{4}$

22. $\dfrac{1}{2}$

23. $-\sqrt{\dfrac{1}{81}} = -\dfrac{\sqrt{1}}{\sqrt{81}} = -\dfrac{1}{9}$

24. $-\dfrac{5}{7}$

25. $\sqrt{\dfrac{64}{144}} = \sqrt{\dfrac{4}{9}} = \dfrac{\sqrt{4}}{\sqrt{9}} = \dfrac{2}{3}$

26. $\dfrac{9}{11}$

27. $\sqrt{\dfrac{36}{a^2}} = \dfrac{\sqrt{36}}{\sqrt{a^2}} = \dfrac{6}{a}$

28. $\dfrac{5}{x}$

29. $\sqrt{\dfrac{9a^2}{625}} = \dfrac{\sqrt{9a^2}}{\sqrt{625}} = \dfrac{3a}{25}$

30. $\dfrac{xy}{12}$

31. $\sqrt{\dfrac{5}{3}} = \dfrac{\sqrt{5}}{\sqrt{3}} = \dfrac{\sqrt{5}}{\sqrt{3}} \cdot \dfrac{\sqrt{3}}{\sqrt{3}} = \dfrac{\sqrt{15}}{3}$

32. $\dfrac{\sqrt{14}}{7}$

33. $\sqrt{\dfrac{7}{20}} = \dfrac{\sqrt{7}}{\sqrt{20}} = \dfrac{\sqrt{7}}{\sqrt{4}\sqrt{5}} = \dfrac{\sqrt{7}}{2\sqrt{5}} = \dfrac{\sqrt{7}}{2\sqrt{5}} \cdot \dfrac{\sqrt{5}}{\sqrt{5}} =$
$\dfrac{\sqrt{35}}{2\cdot 5} = \dfrac{\sqrt{35}}{10}$

34. $\dfrac{\sqrt{3}}{6}$

35. $\sqrt{\dfrac{1}{45}} = \dfrac{\sqrt{1}}{\sqrt{45}} = \dfrac{1}{\sqrt{9}\sqrt{5}} = \dfrac{1}{3\sqrt{5}} = \dfrac{1}{3\sqrt{5}} \cdot \dfrac{\sqrt{5}}{\sqrt{5}} =$
$\dfrac{\sqrt{5}}{3\cdot 5} = \dfrac{\sqrt{5}}{15}$

36. $\dfrac{\sqrt{14}}{6}$

37. $\dfrac{3}{\sqrt{5}} = \dfrac{3}{\sqrt{5}} \cdot \dfrac{\sqrt{5}}{\sqrt{5}} = \dfrac{3\sqrt{5}}{5}$

38. $\dfrac{4\sqrt{3}}{3}$

39. $\sqrt{\dfrac{8}{3}} = \dfrac{\sqrt{8}}{\sqrt{3}} = \dfrac{\sqrt{4}\sqrt{2}}{\sqrt{3}} = \dfrac{2\sqrt{2}}{\sqrt{3}} = \dfrac{2\sqrt{2}}{\sqrt{3}} \cdot \dfrac{\sqrt{3}}{\sqrt{3}} = \dfrac{2\sqrt{6}}{3}$

40. $\dfrac{2\sqrt{15}}{5}$

41. $\sqrt{\dfrac{3}{x}} = \dfrac{\sqrt{3}}{\sqrt{x}} = \dfrac{\sqrt{3}}{\sqrt{x}} \cdot \dfrac{\sqrt{x}}{\sqrt{x}} = \dfrac{\sqrt{3x}}{x}$

42. $\dfrac{\sqrt{2x}}{x}$

43. $\dfrac{\sqrt{7}}{\sqrt{3}} = \dfrac{\sqrt{7}}{\sqrt{3}} \cdot \dfrac{\sqrt{3}}{\sqrt{3}} = \dfrac{\sqrt{21}}{3}$

44. $\dfrac{\sqrt{77}}{7}$

45. $\dfrac{\sqrt{9}}{\sqrt{8}} = \dfrac{3}{\sqrt{4}\sqrt{2}} = \dfrac{3}{2\sqrt{2}} = \dfrac{3}{2\sqrt{2}} \cdot \dfrac{\sqrt{2}}{\sqrt{2}} = \dfrac{3\sqrt{2}}{2\cdot 2} = \dfrac{3\sqrt{2}}{4}$

46. $\dfrac{2\sqrt{3}}{9}$

47. $\dfrac{\sqrt{3}}{\sqrt{14}} = \dfrac{\sqrt{3}}{\sqrt{14}} \cdot \dfrac{\sqrt{14}}{\sqrt{14}} = \dfrac{\sqrt{42}}{14}$

48. $\dfrac{\sqrt{6}}{2}$

49. $\dfrac{\sqrt{7}}{\sqrt{12}} = \dfrac{\sqrt{7}}{\sqrt{4}\sqrt{3}} = \dfrac{\sqrt{7}}{2\sqrt{3}} = \dfrac{\sqrt{7}}{2\sqrt{3}} \cdot \dfrac{\sqrt{3}}{\sqrt{3}} = \dfrac{\sqrt{21}}{2\cdot 3} = \dfrac{\sqrt{21}}{6}$

50. $\dfrac{\sqrt{10}}{6}$

51. $\dfrac{\sqrt{x}}{\sqrt{32}} = \dfrac{\sqrt{x}}{\sqrt{16}\sqrt{2}} = \dfrac{\sqrt{x}}{4\sqrt{2}} = \dfrac{\sqrt{x}}{4\sqrt{2}} \cdot \dfrac{\sqrt{2}}{\sqrt{2}} =$
$\dfrac{\sqrt{2x}}{4\cdot 2} = \dfrac{\sqrt{2x}}{8}$

52. $\dfrac{\sqrt{10a}}{20}$

53. $\dfrac{4y}{\sqrt{3}} = \dfrac{4y}{\sqrt{3}} \cdot \dfrac{\sqrt{3}}{\sqrt{3}} = \dfrac{4y\sqrt{3}}{3}$

54. $\dfrac{8x\sqrt{5}}{5}$

55. $\dfrac{\sqrt{6a}}{\sqrt{8}} = \sqrt{\dfrac{6a}{8}} = \sqrt{\dfrac{3a \cdot 2}{4 \cdot 2}} = \sqrt{\dfrac{3a \cdot \cancel{2}}{4 \cdot \cancel{2}}} = \dfrac{\sqrt{3a}}{\sqrt{4}} = \dfrac{\sqrt{3a}}{2}$

56. $\dfrac{\sqrt{x}}{3}$

57. $\dfrac{\sqrt{72}}{\sqrt{20x}} = \dfrac{\sqrt{36}\sqrt{2}}{\sqrt{4}\sqrt{5x}} = \dfrac{6\sqrt{2}}{2\sqrt{5x}} = \dfrac{3\sqrt{2}}{\sqrt{5x}} =$

$\dfrac{3\sqrt{2}}{\sqrt{5x}} \cdot \dfrac{\sqrt{5x}}{\sqrt{5x}} = \dfrac{3\sqrt{10x}}{5x}$

58. $\dfrac{3\sqrt{10a}}{4a}$

59. $\dfrac{\sqrt{27c}}{\sqrt{32c^3}} = \sqrt{\dfrac{27c}{32c^3}} = \sqrt{\dfrac{27}{32c^2}} = \dfrac{\sqrt{9}\sqrt{3}}{\sqrt{16}\sqrt{c^2}\sqrt{2}} =$

$\dfrac{3\sqrt{3}}{4c\sqrt{2}} = \dfrac{3\sqrt{3}}{4c\sqrt{2}} \cdot \dfrac{\sqrt{2}}{\sqrt{2}} = \dfrac{3\sqrt{6}}{4c \cdot 2} = \dfrac{3\sqrt{6}}{8c}$

60. $\dfrac{x\sqrt{21}}{6}$

61. 2 ft: $T \approx 2(3.14)\sqrt{\dfrac{2}{32}} \approx 6.28\sqrt{\dfrac{1}{16}} \approx 6.28\left(\dfrac{1}{4}\right) \approx$

1.57 sec

8 ft: $T \approx 2(3.14)\sqrt{\dfrac{8}{32}} \approx 6.28\sqrt{\dfrac{1}{4}} \approx 6.28\left(\dfrac{1}{2}\right) \approx$

3.14 sec

64 ft: $T \approx 2(3.14)\sqrt{\dfrac{64}{32}} \approx 6.28\sqrt{2} \approx$

$(6.28)(1.414) \approx 8.88$ sec

100 ft: $T \approx 2(3.14)\sqrt{\dfrac{100}{32}} \approx 6.28\sqrt{\dfrac{50}{16}} \approx$

$\dfrac{6.28\sqrt{50}}{4} \approx \dfrac{6.28(7.071)}{4} \approx 11.10$ sec

62. 0.26 sec

63. Substitute $\dfrac{45}{\pi^2}$ for L in the formula.

$T = 2\pi\sqrt{\dfrac{L}{32}} = 2\pi\sqrt{\dfrac{\frac{45}{\pi^2}}{32}} = 2\pi\sqrt{\dfrac{45}{\pi^2} \cdot \dfrac{1}{32}} =$

$2\pi\sqrt{\dfrac{45}{32\pi^2}} = 2\pi \cdot \dfrac{\sqrt{45}}{\sqrt{32\pi^2}} = 2\pi \cdot \dfrac{\sqrt{9}\sqrt{5}}{\sqrt{16}\sqrt{\pi^2}\sqrt{2}} =$

$2\pi \cdot \dfrac{3\sqrt{5}}{4\pi\sqrt{2}} = \dfrac{3\sqrt{5}}{2\sqrt{2}} = \dfrac{3\sqrt{5}}{2\sqrt{2}} \cdot \dfrac{\sqrt{2}}{\sqrt{2}} = \dfrac{3\sqrt{10}}{4}$ sec

It takes $\dfrac{3\sqrt{10}}{4}$ sec to move from one side to the other

and back. Thus it takes $\dfrac{1}{2} \cdot \dfrac{3\sqrt{10}}{4} = \dfrac{3\sqrt{10}}{8}$ sec, or

approximately 1.19 sec, to swing from one side to the other.

64. 1 sec

65. $x = y + 2$,

$x + y = 6$

We first write both equations in the form $Ax + By = C$.

$$\begin{array}{rl} x - y = 2 & (1) \\ \underline{x + y = 6} & (2) \\ 2x = 8 & \text{Adding} \\ x = 4 & \end{array}$$

Substitute 4 for x in one of the original equations and solve for y.

$4 + y = 6$ (2)

$y = 2$

The ordered pair $(4, 2)$ checks in both equations. It is the solution.

66. $\left(4, \dfrac{1}{3}\right)$

67. $(3x - 7)(3x + 7) = (3x)^2 - 7^2 = 9x^2 - 49$

68. $16a^2 - 25b^2$

69. $9x - 5y + 12x - 4y$

$= (9x + 12x) + (-5y - 4y)$

$= 21x - 9y$

70. $14a - 6b$

71. ◈

72. ◈

73. ◈

74. ◈

75. $\sqrt{\dfrac{7}{800}} = \dfrac{\sqrt{7}}{\sqrt{800}} = \dfrac{\sqrt{7}}{\sqrt{400}\sqrt{2}} = \dfrac{\sqrt{7}}{20\sqrt{2}} =$

$\dfrac{\sqrt{7}}{20\sqrt{2}} \cdot \dfrac{\sqrt{2}}{\sqrt{2}} = \dfrac{\sqrt{14}}{20 \cdot 2} = \dfrac{\sqrt{14}}{40}$

76. $\dfrac{\sqrt{30}}{100}$

77. $\sqrt{\dfrac{5}{8x^7}} = \dfrac{\sqrt{5}}{\sqrt{8x^7}} = \dfrac{\sqrt{5}}{\sqrt{4x^6}\sqrt{2x}} = \dfrac{\sqrt{5}}{2x^3\sqrt{2x}} =$

$\dfrac{\sqrt{5}}{2x^3\sqrt{2x}} \cdot \dfrac{\sqrt{2x}}{\sqrt{2x}} = \dfrac{\sqrt{10x}}{2x^3 \cdot 2x} = \dfrac{\sqrt{10x}}{4x^4}$

78. $\dfrac{\sqrt{3xy}}{ax^2}$

79. $\sqrt{\dfrac{2a}{5b^3c^9}} = \dfrac{\sqrt{2a}}{\sqrt{5b^3c^9}} = \dfrac{\sqrt{2a}}{\sqrt{b^2c^8}\sqrt{5bc}} = \dfrac{\sqrt{2a}}{bc^4\sqrt{5bc}} =$

$\dfrac{\sqrt{2a}}{bc^4\sqrt{5bc}} \cdot \dfrac{\sqrt{5bc}}{\sqrt{5bc}} = \dfrac{\sqrt{10abc}}{bc^4 \cdot 5bc} = \dfrac{\sqrt{10abc}}{5b^2c^5}$

80. $\dfrac{\sqrt{5z}}{5zw}$

81. $\sqrt{\dfrac{1}{x^2} - \dfrac{2}{xy} + \dfrac{1}{y^2}}$, LCD is x^2y^2

$$= \sqrt{\dfrac{1}{x^2} \cdot \dfrac{y^2}{y^2} - \dfrac{2}{xy} \cdot \dfrac{xy}{xy} + \dfrac{1}{y^2} \cdot \dfrac{x^2}{x^2}}$$

$$= \sqrt{\dfrac{y^2 - 2xy + x^2}{x^2y^2}}$$

$$= \sqrt{\dfrac{(y - x)^2}{x^2y^2}}$$

$$= \dfrac{\sqrt{(y - x)^2}}{\sqrt{x^2y^2}}$$

$$= \dfrac{y - x}{xy}$$

An alternate method of simplifying this expression is shown below.

$$\sqrt{\dfrac{1}{x^2} - \dfrac{2}{xy} + \dfrac{1}{y^2}} = \sqrt{\left(\dfrac{1}{x} - \dfrac{1}{y}\right)^2}$$

$$= \dfrac{1}{x} - \dfrac{1}{y}$$

The two answers are equivalent.

82. $\dfrac{\sqrt{2}(z^2 - 1)}{z^2}$, or $\left(1 - \dfrac{1}{z^2}\right)\sqrt{2}$

Exercise Set 9.4

1. $7\sqrt{2} + 4\sqrt{2}$

$= (7 + 4)\sqrt{2}$ Using the distributive law

$= 11\sqrt{2}$

2. $7\sqrt{3}$

3. $9\sqrt{5} - 6\sqrt{5}$

$= (9 - 6)\sqrt{5}$ Using the distributive law

$= 3\sqrt{5}$

4. $3\sqrt{2}$

5. $6\sqrt{x} + 7\sqrt{x} = (6 + 7)\sqrt{x}$

$= 13\sqrt{x}$

6. $12\sqrt{y}$

7. $9\sqrt{x} - 11\sqrt{x} = (9 - 11)\sqrt{x} = -2\sqrt{x}$

8. $-8\sqrt{a}$

9. $5\sqrt{2a} + 3\sqrt{2a} = (5 + 3)\sqrt{2a} = 8\sqrt{2a}$

10. $9\sqrt{6x}$

11. $9\sqrt{10y} - \sqrt{10y} = (9 - 1)\sqrt{10y} = 8\sqrt{10y}$

12. $11\sqrt{14y}$

13. $5\sqrt{7} + 2\sqrt{7} + 4\sqrt{7} = (5 + 2 + 4)\sqrt{7} = 11\sqrt{7}$

14. $15\sqrt{5}$

15. $7\sqrt{2} - 9\sqrt{2} + 4\sqrt{2} = (7 - 9 + 4)\sqrt{2} = 2\sqrt{2}$

16. 0

17. $5\sqrt{3} + \sqrt{8} = 5\sqrt{3} + \sqrt{4 \cdot 2}$ Factoring 8

$= 5\sqrt{3} + \sqrt{4}\sqrt{2}$

$= 5\sqrt{3} + 2\sqrt{2}$

$5\sqrt{3} + \sqrt{8}$, or $5\sqrt{3} + 2\sqrt{2}$, cannot be simplified further.

18. $5\sqrt{5}$

19. $\sqrt{x} - \sqrt{9x} = \sqrt{x} - \sqrt{9}\sqrt{x}$

$= \sqrt{x} - 3\sqrt{x}$

$= (1 - 3)\sqrt{x}$

$= -2\sqrt{x}$

20. $4\sqrt{a}$

21. $5\sqrt{8} + 15\sqrt{2} = 5\sqrt{4 \cdot 2} + 15\sqrt{2}$

$= 5 \cdot 2\sqrt{2} + 15\sqrt{2}$

$= 10\sqrt{2} + 15\sqrt{2}$

$= 25\sqrt{2}$

22. $26\sqrt{3}$

23. $\sqrt{27} - 2\sqrt{3} = \sqrt{9 \cdot 3} - 2\sqrt{3}$

$= 3\sqrt{3} - 2\sqrt{3}$

$= (3 - 2)\sqrt{3}$

$= 1\sqrt{3}$

$= \sqrt{3}$

24. $32\sqrt{2}$

25. $\sqrt{72} + \sqrt{98} = \sqrt{36 \cdot 2} + \sqrt{49 \cdot 2}$

$= 6\sqrt{2} + 7\sqrt{2}$

$= (6 + 7)\sqrt{2}$

$= 13\sqrt{2}$

26. $7\sqrt{5}$

27. $4\sqrt{12} + \sqrt{27} - \sqrt{8}$

$= 4\sqrt{4 \cdot 3} + \sqrt{9 \cdot 3} - \sqrt{4 \cdot 2}$

$= 4 \cdot 2\sqrt{3} + 3\sqrt{3} - 2\sqrt{2}$

$= 8\sqrt{3} + 3\sqrt{3} - 2\sqrt{2}$

$= (8 + 3)\sqrt{3} - 2\sqrt{2}$

$= 11\sqrt{3} - 2\sqrt{2}$

28. $19\sqrt{2}$

29.
$$5\sqrt{18} - 2\sqrt{32} - \sqrt{50}$$
$$= 5\sqrt{9\cdot 2} - 2\sqrt{16\cdot 2} - \sqrt{25\cdot 2}$$
$$= 5\cdot 3\sqrt{2} - 2\cdot 4\sqrt{2} - 5\sqrt{2}$$
$$= 15\sqrt{2} - 8\sqrt{2} - 5\sqrt{2}$$
$$= (15 - 8 - 5)\sqrt{2}$$
$$= 2\sqrt{2}$$

30. $-3\sqrt{2} + 5\sqrt{3}$

31. $\sqrt{9x} + \sqrt{49x} - 9\sqrt{x} = 3\sqrt{x} + 7\sqrt{x} - 9\sqrt{x}$
$$= (3 + 7 - 9)\sqrt{x}$$
$$= 1\sqrt{x}$$
$$= \sqrt{x}$$

32. $5\sqrt{a}$

33.
$$\sqrt{2}(\sqrt{5} + \sqrt{7})$$
$$= \sqrt{2}\sqrt{5} + \sqrt{2}\sqrt{7} \quad \text{Using the distributive law}$$
$$= \sqrt{10} + \sqrt{14}$$

34. $\sqrt{10} + \sqrt{55}$

35. $\sqrt{5}(\sqrt{6} - \sqrt{10}) = \sqrt{5}\sqrt{6} - \sqrt{5}\sqrt{10}$
$$= \sqrt{30} - \sqrt{50}$$
$$= \sqrt{30} - \sqrt{25\cdot 2}$$
$$= \sqrt{30} - 5\sqrt{2}$$

36. $3\sqrt{10} - \sqrt{42}$

37.
$$(4 + \sqrt{2})(5 + \sqrt{2})$$
$$= 4\cdot 5 + 4\cdot\sqrt{2} + \sqrt{2}\cdot 5 + \sqrt{2}\cdot\sqrt{2} \quad \text{Using FOIL}$$
$$= 20 + 4\sqrt{2} + 5\sqrt{2} + 2$$
$$= 22 + 9\sqrt{2}$$

38. $26 + 8\sqrt{11}$

39.
$$(\sqrt{6} - 2)(\sqrt{6} - 5)$$
$$= \sqrt{6}\cdot\sqrt{6} - \sqrt{6}\cdot 5 - 2\cdot\sqrt{6} + 2(5) \quad \text{Using FOIL}$$
$$= 6 - 5\sqrt{6} - 2\sqrt{6} + 10$$
$$= 16 - 7\sqrt{6}$$

40. $-18 - 3\sqrt{10}$

41.
$$(\sqrt{5} + 7)(\sqrt{5} - 7)$$
$$= (\sqrt{5})^2 - 7^2 \quad \text{Using } (A + B)(A - B) = A^2 - B^2$$
$$= 5 - 49$$
$$= -44$$

42. -4

43.
$$(\sqrt{6} - \sqrt{3})(\sqrt{6} + \sqrt{3})$$
$$= (\sqrt{6})^2 - (\sqrt{3})^2 \quad \text{Using } (A - B)(A + B) = A^2 - B^2$$
$$= 6 - 3$$
$$= 3$$

44. -4

45.
$$(5 + 3\sqrt{2})(1 - \sqrt{2})$$
$$= 5\cdot 1 - 5\cdot\sqrt{2} + 3\sqrt{2}\cdot 1 - 3\sqrt{2}\cdot\sqrt{2} \quad \text{Using FOIL}$$
$$= 5 - 5\sqrt{2} + 3\sqrt{2} - 3\cdot 2$$
$$= 5 - 2\sqrt{2} - 6$$
$$= -1 - 2\sqrt{2}$$

46. $10 + 13\sqrt{7}$

47.
$$(7 + \sqrt{3})^2$$
$$= 7^2 + 2\cdot 7\cdot\sqrt{3} + (\sqrt{3})^2 \quad \text{Using } (A + B)^2 = A^2 + 2AB + B^2$$
$$= 49 + 14\sqrt{3} + 3$$
$$= 52 + 14\sqrt{3}$$

48. $9 + 4\sqrt{5}$

49.
$$(1 - 2\sqrt{3})^2$$
$$= 1^2 - 2\cdot 1\cdot 2\sqrt{3} + (2\sqrt{3})^2 \quad \text{Using } (A - B)^2 = A^2 - 2AB + B^2$$
$$= 1 - 4\sqrt{3} + 4\cdot 3$$
$$= 1 - 4\sqrt{3} + 12$$
$$= 13 - 4\sqrt{3}$$

50. $81 - 36\sqrt{5}$

51. $(\sqrt{x} - \sqrt{10})^2 = (\sqrt{x})^2 - 2\sqrt{x}\sqrt{10} + (\sqrt{10})^2$
$$= x - 2\sqrt{10x} + 10$$

52. $a - 2\sqrt{6a} + 6$

53.
$$\frac{9}{5 + \sqrt{2}}$$
$$= \frac{9}{5 + \sqrt{2}}\cdot\frac{5 - \sqrt{2}}{5 - \sqrt{2}} \quad \text{Multiplying by 1}$$
$$= \frac{9(5 - \sqrt{2})}{(5 + \sqrt{2})(5 - \sqrt{2})}$$
$$= \frac{45 - 9\sqrt{2}}{5^2 - (\sqrt{2})^2}$$
$$= \frac{45 - 9\sqrt{2}}{25 - 2}$$
$$= \frac{45 - 9\sqrt{2}}{23}$$

54. $\dfrac{3 - \sqrt{5}}{2}$

55. $\dfrac{6}{2-\sqrt{7}}$

$= \dfrac{6}{2-\sqrt{7}} \cdot \dfrac{2+\sqrt{7}}{2+\sqrt{7}}$

$= \dfrac{6(2+\sqrt{7})}{(2-\sqrt{7})(2+\sqrt{7})}$

$= \dfrac{12+6\sqrt{7}}{2^2-(\sqrt{7})^2}$

$= \dfrac{12+6\sqrt{7}}{4-7}$

$= \dfrac{12+6\sqrt{7}}{-3}$ Since 3 is a common factor, we simplify.

$= \dfrac{\cancel{3}(4+2\sqrt{7})}{\cancel{3}(-1)}$ Factoring and removing a factor equal to 1

$= \dfrac{4+2\sqrt{7}}{-1}$

$= -4-2\sqrt{7}$

56. $\dfrac{21+3\sqrt{2}}{47}$

57. $\dfrac{2}{\sqrt{7}+3}$

$= \dfrac{2}{\sqrt{7}+3} \cdot \dfrac{\sqrt{7}-3}{\sqrt{7}-3}$

$= \dfrac{2(\sqrt{7}-3)}{(\sqrt{7})^2-3^2}$

$= \dfrac{2\sqrt{7}-6}{7-9}$

$= \dfrac{2\sqrt{7}-6}{-2}$ Since 2 is a common factor, we simplify.

$= \dfrac{\cancel{2}(\sqrt{7}-3)}{\cancel{2}(-1)}$ Factoring and removing a factor equal to 1

$= \dfrac{\sqrt{7}-3}{-1}$

$= \dfrac{-(\sqrt{7}-3)}{1}$

$= -\sqrt{7}+3,$ or $3-\sqrt{7}$

58. $-\dfrac{2\sqrt{10}-10}{5}$

59. $\dfrac{\sqrt{6}}{\sqrt{6}-5} = \dfrac{\sqrt{6}}{\sqrt{6}-5} \cdot \dfrac{\sqrt{6}+5}{\sqrt{6}+5} = \dfrac{\sqrt{6}\sqrt{6}+\sqrt{6}\cdot5}{(\sqrt{6})^2-5^2} =$

$\dfrac{6+5\sqrt{6}}{6-25} = \dfrac{6+5\sqrt{6}}{-19} = -\dfrac{6+5\sqrt{6}}{19}$

60. $-\dfrac{10+7\sqrt{10}}{39}$

61. $\dfrac{\sqrt{5}}{\sqrt{5}-\sqrt{3}} = \dfrac{\sqrt{5}}{\sqrt{5}-\sqrt{3}} \cdot \dfrac{\sqrt{5}+\sqrt{3}}{\sqrt{5}+\sqrt{3}} =$

$\dfrac{\sqrt{5}\sqrt{5}+\sqrt{5}\sqrt{3}}{(\sqrt{5})^2-(\sqrt{3})^2} = \dfrac{5+\sqrt{15}}{5-3} = \dfrac{5+\sqrt{15}}{2}$

62. $\dfrac{7+\sqrt{35}}{2}$

63. $\dfrac{\sqrt{3}}{\sqrt{5}+\sqrt{3}} = \dfrac{\sqrt{3}}{\sqrt{5}+\sqrt{3}} \cdot \dfrac{\sqrt{5}-\sqrt{3}}{\sqrt{5}-\sqrt{3}} =$

$\dfrac{\sqrt{3}\sqrt{5}-\sqrt{3}\sqrt{3}}{(\sqrt{5})^2-(\sqrt{3})^2} = \dfrac{\sqrt{15}-3}{5-3} = \dfrac{\sqrt{15}-3}{2}$

64. $\sqrt{42}+6$

65. $\dfrac{2}{\sqrt{7}-\sqrt{2}} = \dfrac{2}{\sqrt{7}-\sqrt{2}} \cdot \dfrac{\sqrt{7}+\sqrt{2}}{\sqrt{7}+\sqrt{2}} =$

$\dfrac{2\sqrt{7}+2\sqrt{2}}{(\sqrt{7})^2-(\sqrt{2})^2} = \dfrac{2\sqrt{7}+2\sqrt{2}}{7-2} = \dfrac{2\sqrt{7}+2\sqrt{2}}{5}$

66. $3\sqrt{5}+3\sqrt{3}$

67. $\dfrac{\sqrt{6}+\sqrt{5}}{\sqrt{6}-\sqrt{5}} = \dfrac{\sqrt{6}+\sqrt{5}}{\sqrt{6}-\sqrt{5}} \cdot \dfrac{\sqrt{6}+\sqrt{5}}{\sqrt{6}+\sqrt{5}}$

$= \dfrac{(\sqrt{6}+\sqrt{5})^2}{(\sqrt{6}-\sqrt{5})(\sqrt{6}+\sqrt{5})}$

$= \dfrac{(\sqrt{6})^2+2\sqrt{6}\sqrt{5}+(\sqrt{5})^2}{(\sqrt{6})^2-(\sqrt{5})^2}$

$= \dfrac{6+2\sqrt{30}+5}{6-5} = \dfrac{11+2\sqrt{30}}{1}$

$= 11+2\sqrt{30}$

68. $\dfrac{17-2\sqrt{70}}{3}$

69. $3x+5+2(x-3) = 4-6x$

$3x+5+2x-6 = 4-6x$

$5x-1 = 4-6x$

$11x-1 = 4$

$11x = 5$

$x = \dfrac{5}{11}$

The solution is $\dfrac{5}{11}$.

70. $-\dfrac{38}{13}$

71. $\quad x^2-5x = 6$

$x^2-5x-6 = 0$

$(x+1)(x-6) = 0$

$x+1 = 0 \quad or \quad x-6 = 0$

$x = -1 \quad or \quad\quad x = 6$

The solutions are -1 and 6.

72. 2, 5

73. *Familiarize.* Let x = the number of liters of Jolly Juice and y = the number of liters of Real Squeeze in the mixture. We organize the given information in a table.

	Jolly Juice	Real Squeeze	Mixture
Amount	x	y	8
Percent real fruit juice	3%	6%	5.4%
Amount of real fruit juice	$0.03x$	$0.06y$	$0.054(8)$, or 0.432

Translate. We get two equation from the first and third rows of the table.

$$x + y = 8,$$
$$0.03x + 0.06y = 0.432$$

Clearing decimals gives

$$x + y = 8, \quad (1)$$
$$30x + 60y = 432. \quad (2)$$

Carry out. We use elimination. Multiply Equation (1) by -30 and add.

$$\begin{array}{r} -30x - 30y = -240 \\ 30x + 60y = 432 \\ \hline 30y = 192 \\ y = 6.4 \end{array}$$

Now substitute 6.4 for y in Equation (1) and solve for x.

$$x + y = 8$$
$$x + 6.4 = 8$$
$$x = 1.6$$

Check. The sum of 1.6 and 6.4 is 8. The amount of real fruit juice in this mixture is $0.03(1.6) + 0.06(6.4)$, or $0.048 + 0.384$, or 0.432 L. The answer checks.

State. 1.6 L of Jolly Juice and 6.4 L of Real Squeeze should be used.

74. At least 3 hr

75. ◈

76. ◈

77. ◈

78. ◈

79.
$$7\sqrt{\frac{1}{2}} + \frac{5}{2}\sqrt{18} + \sqrt{98}$$
$$= 7 \cdot \frac{1}{\sqrt{2}} + \frac{5}{2}\sqrt{9 \cdot 2} + \sqrt{49 \cdot 2}$$
$$= \frac{7}{\sqrt{2}} + \frac{5}{2} \cdot 3\sqrt{2} + 7\sqrt{2}$$
$$= \frac{7}{\sqrt{2}} + \frac{15\sqrt{2}}{2} + 7\sqrt{2}$$
$$= \frac{7}{\sqrt{2}} \cdot \frac{\sqrt{2}}{2} + \frac{15\sqrt{2}}{2} + 7\sqrt{2}$$
$$= \frac{7\sqrt{2}}{2} + \frac{15\sqrt{2}}{2} + 7\sqrt{2}$$
$$= \left(\frac{7}{2} + \frac{15}{2} + 7\right)\sqrt{2}$$
$$= 18\sqrt{2}$$

80. $(ab^3 - a^2b + a^4)\sqrt{a}$

81.
$$\sqrt{\frac{9}{x}} + \frac{\sqrt{x}}{2x}$$
$$= \frac{3}{\sqrt{x}} + \frac{\sqrt{x}}{2x}$$
$$= \frac{3}{\sqrt{x}} \cdot \frac{\sqrt{x}}{x} + \frac{\sqrt{x}}{2x}$$
$$= \frac{3\sqrt{x}}{x} + \frac{\sqrt{x}}{2x}$$
$$= \frac{3\sqrt{x}}{x} \cdot \frac{2}{2} + \frac{\sqrt{x}}{2x}$$
$$= \frac{6\sqrt{x}}{2x} + \frac{\sqrt{x}}{2x}$$
$$= \frac{7\sqrt{x}}{2x}$$

82. $(b^3 + ab + a)\sqrt{a}$

83.
$$x\sqrt{2y} - \sqrt{8x^2y} + \frac{x}{3}\sqrt{18y}$$
$$= x\sqrt{2y} - \sqrt{4x^2 \cdot 2y} + \frac{x}{3}\sqrt{9 \cdot 2y}$$
$$= x\sqrt{2y} - 2x\sqrt{2y} + \frac{x}{3} \cdot 3\sqrt{2y}$$
$$= (x - 2x + x)\sqrt{2y}$$
$$= 0$$

84. $37xy\sqrt{3x}$

85. Any pair of numbers a, b such that $a = 0$ or $b = 0$ will do.

86. All of them

Exercise Set 9.5

1. $\sqrt{x} = 8$

$(\sqrt{x})^2 = 8^2$ Squaring both sides

$x = 64$ Simplifying

Check: $\dfrac{\sqrt{x} = 8}{\sqrt{64} \ ? \ 8}$

$8 \ | \ 8$ TRUE

The solution is 64.

2. 49

3. $\sqrt{x + 3} = 4$

$(\sqrt{x + 3})^2 = 4^2$ Squaring both sides

$x + 3 = 16$

$x = 13$

Check: $\dfrac{\sqrt{x + 3} = 4}{\sqrt{13 + 3} \ ? \ 4}$

$\sqrt{16}$

$4 \ | \ 4$ TRUE

The solution is 13.

4. 117

5. $\sqrt{2x + 4} = 9$

$(\sqrt{2x + 4})^2 = 9^2$

$2x + 4 = 81$

$2x = 77$

$x = \dfrac{77}{2}$

Check: $\dfrac{\sqrt{2x + 4} = 9}{\sqrt{2 \cdot \dfrac{77}{2} + 4} \ ? \ 9}$

$\sqrt{77 + 4}$

$\sqrt{81}$

$9 \ | \ 9$ TRUE

6. 84

7. $3 + \sqrt{x - 1} = 5$

$\sqrt{x - 1} = 2$ Subtracting 3

$(\sqrt{x - 1})^2 = 2^2$ Squaring both sides

$x - 1 = 4$

$x = 5$

Check: $\dfrac{3 + \sqrt{x - 1} = 5}{3 + \sqrt{5 - 1} \ ? \ 5}$

$3 + \sqrt{4}$

$3 + 2$

$5 \ | \ 5$ TRUE

The solution is 5.

8. 52

9. $6 - 2\sqrt{3n} = 0$

$6 = 2\sqrt{3n}$ Adding $2\sqrt{3n}$

$6^2 = (2\sqrt{3n})^2$ Squaring both sides

$36 = 4 \cdot 3n$

$36 = 12n$

$3 = n$

Check: $\dfrac{6 - 2\sqrt{3n} = 0}{6 - 2\sqrt{3 \cdot 3} \ ? \ 0}$

$6 - 2 \cdot 3$

$6 - 6$

$0 \ | \ 0$ TRUE

The solution is 3.

10. $\dfrac{4}{5}$

11. $\sqrt{4x + 7} = \sqrt{2x + 13}$

$(\sqrt{4x + 7})^2 = (\sqrt{2x + 13})^2$ Squaring both sides

$4x + 7 = 2x + 13$

$2x + 7 = 13$

$2x = 6$

$x = 3$

Check: $\dfrac{\sqrt{4x + 7} = \sqrt{2x + 13}}{\sqrt{4 \cdot 3 + 7} \ ? \ \sqrt{2 \cdot 3 + 13}}$

$\sqrt{12 + 7} \ | \ \sqrt{6 + 13}$

$\sqrt{19} \ | \ \sqrt{19}$ TRUE

The solution is 3.

12. $\dfrac{14}{3}$

13. $\sqrt{x} = -2$

The principal square root of x cannot be negative. There is no solution.

14. No solution

15. $\sqrt{2y + 6} = \sqrt{2y - 5}$

$(\sqrt{2y + 6})^2 = (\sqrt{2y - 5})^2$

$2y + 6 = 2y - 5$

$6 = -5$

The equation $6 = -5$ is false; there is no solution.

16. 6

17.
$$x - 7 = \sqrt{x - 5}$$
$$(x - 7)^2 = (\sqrt{x - 5})^2$$
$$x^2 - 14x + 49 = x - 5$$
$$x^2 - 15 + 54 = 0$$
$$(x - 9)(x - 6) = 0$$
$$x - 9 = 0 \quad or \quad x - 6 = 0$$
$$x = 9 \quad or \qquad x = 6$$

Check:

$x - 7 = \sqrt{x - 5}$	
$9 - 7$? $\sqrt{9 - 5}$	
2	$\sqrt{4}$
2	2 TRUE

$x - 7 = \sqrt{x - 5}$	
$6 - 7$? $\sqrt{6 - 5}$	
-1	$\sqrt{1}$
-1	1 FALSE

The number 9 checks, but 6 does not. The solution is 9.

18. 9

19.
$$\sqrt{x + 18} = x - 2$$
$$(\sqrt{x + 18})^2 = (x - 2)^2$$
$$x + 18 = x^2 - 4x + 4$$
$$0 = x^2 - 5x - 14$$
$$0 = (x - 7)(x + 2)$$
$$x - 7 = 0 \quad or \quad x + 2 = 0$$
$$x = 7 \quad or \qquad x = -2$$

Check:

$\sqrt{x + 18} = x - 2$	
$\sqrt{7 + 18}$? $7 - 2$	
$\sqrt{25}$	5
5	5 TRUE

$\sqrt{x + 18} = x - 2$	
$\sqrt{-2 + 18}$? $-2 - 2$	
$\sqrt{16}$	-4
4	-4 FALSE

The number 7 checks, but −2 does not. The solution is 7.

20. 12

21.
$$x - 5 = \sqrt{15 - 3x}$$
$$(x - 5)^2 = (\sqrt{15 - 3x})^2$$
$$x^2 - 10x + 25 = 15 - 3x$$
$$x^2 - 7x + 10 = 0$$
$$(x - 2)(x - 5) = 0$$
$$x - 2 = 0 \quad or \quad x - 5 = 0$$
$$x = 2 \quad or \qquad x = 5$$

Check:

$x - 5 = \sqrt{15 - 3x}$	
$2 - 5$? $\sqrt{15 - 3 \cdot 2}$	
-3	$\sqrt{15 - 6}$
	$\sqrt{9}$
-3	3 FALSE

$x - 5 = \sqrt{15 - 3x}$	
$5 - 5$? $\sqrt{15 - 3 \cdot 5}$	
0	$\sqrt{15 - 15}$
	$\sqrt{0}$
0	0 TRUE

The number 5 checks, but 2 does not. The solution is 5.

22. 13, 25

23.
$$\sqrt{5x + 21} = x + 3$$
$$(\sqrt{5x + 21})^2 = (x + 3)^2$$
$$5x + 21 = x^2 + 6x + 9$$
$$0 = x^2 + x - 12$$
$$0 = (x + 4)(x - 3)$$
$$x + 4 = 0 \quad or \quad x - 3 = 0$$
$$x = -4 \quad or \qquad x = 3$$

Check:

$\sqrt{5x + 21} = x + 3$	
$\sqrt{5(-4) + 21}$? $-4 + 3$	
$\sqrt{1}$	-1
1	-1 FALSE

$\sqrt{5x + 21} = x + 3$	
$\sqrt{5 \cdot 3 + 21}$	$3 + 3$
$\sqrt{36}$	6
6	6 TRUE

The number 3 checks, but −4 does not. The solution is 3.

24. 6

25.
$$x + 4 = 4\sqrt{x + 1}$$
$$(x + 4)^2 = (4\sqrt{x + 1})^2$$
$$x^2 + 8x + 16 = 16(x + 1)$$
$$x^2 + 8x + 16 = 16x + 16$$
$$x^2 - 8x = 0$$
$$x(x - 8) = 0$$
$$x = 0 \quad or \quad x - 8 = 0$$
$$x = 0 \quad or \qquad x = 8$$

Check: $\dfrac{x + 4 = 4\sqrt{x+1}}{\begin{array}{c|c} 0 + 4 \ ? \ 4\sqrt{0+1} \\ 4 & 4\sqrt{1} \\ & 4 \cdot 1 \\ 4 & 4 \end{array}}$ TRUE

$\dfrac{x + 4 = 4\sqrt{x+1}}{\begin{array}{c|c} 8 + 4 \ ? \ 4\sqrt{8+1} \\ 12 & 4\sqrt{9} \\ & 4 \cdot 3 \\ 12 & 12 \end{array}}$ TRUE

The solutions are 0 and 8.

26. 1, 5

27. $\sqrt{x^2 + 6} - x + 3 = 0$

$\qquad \sqrt{x^2 + 6} = x - 3 \quad$ Isolating the radical

$\qquad (\sqrt{x^2 + 6})^2 = (x-3)^2$

$\qquad x^2 + 6 = x^2 - 6x + 9$

$\qquad -3 = -6x \quad$ Adding $-x^2$ and -9

$\qquad \dfrac{1}{2} = x$

Check: $\dfrac{\sqrt{x^2 + 6} - x + 3 = 0}{\begin{array}{c|c} \sqrt{\left(\dfrac{1}{2}\right)^2 + 6} - \dfrac{1}{2} + 3 & 0 \\ \sqrt{\dfrac{25}{4}} - \dfrac{1}{2} + 3 & \\ \dfrac{5}{2} - \dfrac{1}{2} + 3 & \\ 5 & 0 \end{array}}$ FALSE

The number $\dfrac{1}{2}$ does not check. There is no solution.

28. No solution

29. $\sqrt{(p+6)(p+1)} - 2 = p + 1$

$\qquad \sqrt{(p+6)(p+1)} = p + 3 \quad$ Isolating the radical

$\qquad \left(\sqrt{(p+6)(p+1)}\right)^2 = (p+3)^2$

$\qquad (p+6)(p+1) = p^2 + 6p + 9$

$\qquad p^2 + 7p + 6 = p^2 + 6p + 9$

$\qquad p = 3$

The number 3 checks. It is the solution.

30. 5

31. $\sqrt{x - 2} = \sqrt{5 - 2x}$

$\qquad (\sqrt{x-2})^2 = (\sqrt{5-2x})^2$

$\qquad x - 2 = 5 - 2x$

$\qquad 3x - 2 = 5$

$\qquad 3x = 7$

$\qquad x = \dfrac{7}{3}$

The number $\dfrac{7}{3}$ checks. It is the solution.

32. $\dfrac{9}{4}$

33. $x - 1 = \sqrt{(x+1)(x-2)}$

$\qquad (x-1)^2 = (\sqrt{(x+1)(x-2)})^2$

$\qquad x^2 - 2x + 1 = (x+1)(x-2)$

$\qquad x^2 - 2x + 1 = x^2 - x - 2$

$\qquad -2x + 1 = -x - 2 \quad$ Adding $-x^2$ on both sides

$\qquad -x + 1 = -2$

$\qquad -x = -3$

$\qquad x = 3$

The number 3 checks. It is the solution.

34. 1

35. *Familiarize and Translate.* We substitute 21 for V in equation $V = 3.5\sqrt{h}$.

$\qquad 21 = 3.5\sqrt{h}$

Carry out. We solve the equation.

$\qquad 21 = 3.5\sqrt{h}$

$\qquad 6 = \sqrt{h} \quad$ Dividing by 3.5

$\qquad 6^2 = (\sqrt{h})^2$

$\qquad 36 = h$

Check. We go over the computation.

State. The altitude of the steeplejack's eyes is 36 m.

36. 576 m

37. *Familiarize and Translate.* We substitute 378 for V in the equation $V = 3.5\sqrt{h}$.

$\qquad 378 = 3.5\sqrt{h}$

Carry out.

$\qquad 378 = 3.5\sqrt{h}$

$\qquad 108 = \sqrt{h} \quad$ Dividing by 3.5

$\qquad 108^2 = (\sqrt{h})^2$

$\qquad 11,664 = h$

Check. We go over the computation.

State. The airplane is 11,664 m high.

38. 806.56 m

39.
$$T = 2\pi\sqrt{\frac{L}{32}}$$

$$1.6 = 2(3.14)\sqrt{\frac{L}{32}} \quad \begin{array}{l}\text{Substituting 1.6 for } T \\ \text{and 3.14 for } \pi\end{array}$$

$$1.6 = 6.28\sqrt{\frac{L}{32}}$$

$$\frac{1.6}{6.28} = \sqrt{\frac{L}{32}}$$

$$\left(\frac{1.6}{6.18}\right)^2 = \left(\sqrt{\frac{L}{32}}\right)^2$$

$$0.065 \approx \frac{L}{32}$$

$$2.08 \approx L$$

The pendulum is about 2.08 ft long.

40. 7.30 ft

41.
$$r = 2\sqrt{5L}$$

$$40 = 2\sqrt{5L} \quad \text{Substituting 40 for } r$$

$$20 = \sqrt{5L}$$

$$20^2 = (\sqrt{5L})^2$$

$$400 = 5L$$

$$80 = L$$

The car will skid 80 ft at 40 mph.

$$80 = 2\sqrt{5L} \quad \text{Substituting 80 for } r$$

$$40 = \sqrt{5L}$$

$$40^2 = (\sqrt{5L})^2$$

$$1600 = 5L$$

$$320 = L$$

The car will skid 320 ft at 80 mph.

42. 115.2 ft, 180 ft

43. *Familiarize.* Let x represent the number. Then the opposite of three times its square root is $-3\sqrt{x}$.

Translate.

$$\underbrace{\text{The opposite of three times the square root of a number}}_{-3\sqrt{x}} \quad \underset{=}{\text{is}} \quad \underset{-33}{-33.}$$

Carry out.
$$-3\sqrt{x} = -33$$
$$\sqrt{x} = 11$$
$$(\sqrt{x})^2 = 11^2$$
$$x = 121$$

Check. The opposite of $3\sqrt{121}$ is $-3 \cdot 11$, or -33. The answer checks.

State. The number is 121.

44. 32

45. $(-2)^5 = (-2)(-2)(-2)(-2)(-2) = -32$

46. -125

47. $\dfrac{7x^9}{27} \cdot \dfrac{9}{7x^3} = \dfrac{63x^9}{189x^3} = \dfrac{63}{189}x^{9-3} = \dfrac{1}{3}x^6$, or $\dfrac{x^6}{3}$

48. $\dfrac{x-3}{4(x+3)}$

49.
$$\frac{x}{x+5} + \frac{x^2-20}{x+5}$$

$$= \frac{x+x^2-20}{x+5}$$

$$= \frac{x^2+x-20}{x+5} \quad \text{Rearranging in the numerator}$$

$$= \frac{(x+5)(x-4)}{x+5}$$

$$= \frac{(x\!\!\!/+5)(x-4)}{(x\!\!\!/+5)\cdot 1}$$

$$= x-4$$

50. $\dfrac{13}{a-1}$

51. ◈

52. ◈

53. ◈

54. ◈

55.
$$5 - \sqrt{x} = \sqrt{x-5}$$
$$(5-\sqrt{x})^2 = (\sqrt{x-5})^2$$
$$25 - 10\sqrt{x} + x = x - 5$$
$$25 - 10\sqrt{x} = -5 \quad \text{Adding } -x$$
$$-10\sqrt{x} = -30$$
$$\sqrt{x} = 3 \quad \text{Dividing by } -10$$
$$(\sqrt{x})^2 = 3^2$$
$$x = 9$$

The number 9 checks. It is the solution.

56. 16

57.
$$\sqrt{3x+1} = 1 - \sqrt{x+4}$$
$$(\sqrt{3x+1})^2 = (1-\sqrt{x+4})^2$$
$$3x+1 = 1 - 2\cdot 1\cdot\sqrt{x+4} + (\sqrt{x+4})^2$$
$$3x+1 = 1 - 2\sqrt{x+4} + x + 4$$
$$3x+1 = 5 - 2\sqrt{x+4} + x$$
$$2x-4 = -2\sqrt{x+4} \quad \text{Isolating the radical}$$
$$2(x-2) = -2\sqrt{x+4}$$
$$x-2 = -\sqrt{x+4} \quad \text{Multiplying by } \frac{1}{2}$$
$$(x-2)^2 = (-\sqrt{x+4})^2$$
$$x^2-4x+4 = x+4$$
$$x^2-5x = 0$$
$$x(x-5) = 0$$

$x = 0$ *or* $x - 5 = 0$

$x = 0$ *or* $x = 5$

Check: $\sqrt{3x+1} = 1 - \sqrt{x+4}$

$$\begin{array}{c|c} \sqrt{3\cdot 0+1} \overset{?}{} 1-\sqrt{0+4} & \\ \sqrt{1} & 1-2 \\ 1 & -1 \qquad \text{FALSE} \end{array}$$

$\sqrt{3x+1} = 1 - \sqrt{x+4}$

$$\begin{array}{c|c} \sqrt{3\cdot 5+1} \overset{?}{} 1-\sqrt{5+4} & \\ \sqrt{16} & 1-3 \\ 4 & -2 \qquad \text{FALSE} \end{array}$$

Neither number checks. There is no solution.

58. 1

59. $4 + \sqrt{19-x} = 6 + \sqrt{4-x}$

$\sqrt{19-x} = 2 + \sqrt{4-x}$ Isolating one radical

$(\sqrt{19-x})^2 = (2 + \sqrt{4-x})^2$

$19 - x = 4 + 4\sqrt{4-x} + (4-x)$

$19 - x = 4\sqrt{4-x} + 8 - x$

$11 = 4\sqrt{4-x}$

$11^2 = (4\sqrt{4-x})^2$

$121 = 16(4-x)$

$121 = 64 - 16x$

$57 = -16x$

$-\dfrac{57}{16} = x$

$-\dfrac{57}{16}$ checks, so it is the solution.

60. 3

61.

$2\sqrt{x-1} - \sqrt{3x-5} = \sqrt{x-9}$

$(2\sqrt{x-1} - \sqrt{3x-5})^2 = (\sqrt{x-9})^2$

$4(x-1) - 4\sqrt{(x-1)(3x-5)} + (3x-5) = x-9$

$4x - 4 - 4\sqrt{3x^2 - 8x + 5} + 3x - 5 = x-9$

$7x - 9 - 4\sqrt{3x^2 - 8x + 5} = x-9$

$-4\sqrt{3x^2 - 8x + 5} = -6x$

$2\sqrt{3x^2 - 8x + 5} = 3x$

$(2\sqrt{3x^2 - 8x + 5})^2 = (3x)^2$

$4(3x^2 - 8x + 5) = 9x^2$

$12x^2 - 32x + 20 = 9x^2$

$3x^2 - 32x + 20 = 0$

$(3x-2)(x-10) = 0$

$3x - 2 = 0$ *or* $x - 10 = 0$

$3x = 2$ *or* $x = 10$

$x = \dfrac{2}{3}$ *or* $x = 10$

The number 10 checks, but $\dfrac{2}{3}$ does not. The solution is 10.

62. 0, 4

63. *Familiarize.* We will use the formula $V = 3.5\sqrt{h}$. We present the information in a table.

	Height	Distance to the horizon
First sighting	h	V
Second sighting	$h + 100$	$V + 20$

Translate. The rows of the table give us two equations.

$V = 3.5\sqrt{h},$ (1)

$V + 20 = 3.5\sqrt{h + 100}$ (2)

Carry out. We substitute $3.5\sqrt{h}$ for V in Equation (2) and solve for h.

$3.5\sqrt{h} + 20 = 3.5\sqrt{h+100}$

$(3.5\sqrt{h} + 20)^2 = (3.5\sqrt{h+100})^2$

$12.25h + 140\sqrt{h} + 400 = 12.25(h+100)$

$12.25h + 140\sqrt{h} + 400 = 12.25h + 1225$

$140\sqrt{h} = 825$

$28\sqrt{h} = 165$ Multiplying by $\dfrac{1}{5}$

$(28\sqrt{h})^2 = (165)^2$

$784h = 27,225$

$h \approx 34.726$

Check. When $h \approx 34.726$, then $V \approx 3.5\sqrt{34.726} \approx 20.625$ km. When $h \approx 100 + 34.726$, or 134.726, then $V \approx 3.5\sqrt{134.726} \approx 40.625$ km. This is 20 km more than 20.625. The answer checks.

State. The climber was at a height of about 34.726 m when the first computation was made.

64. $b = \dfrac{a}{A^4 - 2A^2 + 1}$

65. Graph $y = \sqrt{x}$.

We make a table of values. Note that we must choose nonnegative values of x in order to have a nonnegative radicand.

x	y
0	0
1	1
2	1.414
4	2
5	2.236

We plot these points and connect them with a smooth curve.

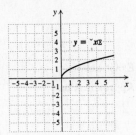

66.

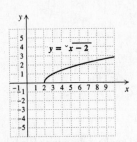

67. Graph $y = \sqrt{x - 3}$.

We make a table of values. Note that we must choose values for x that are greater than or equal to 3 in order to have a nonnegative radicand.

x	y
3	0
4	1
5	1.414
6	1.732
7	2

We plot these points and connect them with a smooth curve.

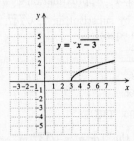

68.

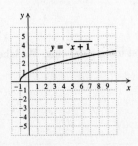

69. We can graph $y = x - 7$ using the intercepts, $(0, -7)$ and $(7, 0)$.

We make a table of values for $y = \sqrt{x - 5}$.

x	y
5	0
6	1
7	1.414
8	1.732
9	2

We plot these points and connect them with a smooth curve.

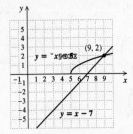

The graphs intersect at $(9, 2)$, so the solution of $x - 7 = \sqrt{x - 5}$ is 9.

70.

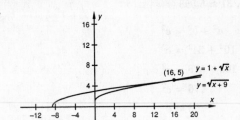

The solution is 16.

71. Graph $y_1 = \sqrt{x + 3}$ and $y_2 = 2x - 1$ and then find the first coordinate(s) of the point(s) of intersection. The solution is about 1.57.

72. -0.32

Exercise Set 9.6

1. $a^2 + b^2 = c^2$
    $8^2 + 15^2 = c^2$ Substituting
    $64 + 225 = c^2$
    $289 = c^2$
    $\sqrt{289} = c$
    $17 = c$

2. $\sqrt{34} \approx 5.831$

3. $a^2 + b^2 = c^2$

$6^2 + 6^2 = c^2$ Substituting

$36 + 36 = c^2$

$72 = c^2$

$\sqrt{72} = c$ Exact answer

$8.485 \approx c$ Approximation

4. $\sqrt{98} \approx 9.899$

5. $a^2 + b^2 = c^2$

$5^2 + b^2 = 13^2$

$25 + b^2 = 169$

$b^2 = 144$

$b = 12$

6. 5

7. $a^2 + b^2 = c^2$

$(6\sqrt{3})^2 + b^2 = 12^2$

$36 \cdot 3 + b^2 = 144$

$108 + b^2 = 144$

$b^2 = 36$

$b = 6$

8. $\sqrt{31} \approx 5.568$

9. $a^2 + b^2 = c^2$

$10^2 + 24^2 = c^2$

$100 + 576 = c^2$

$676 = c^2$

$26 = c$

10. 13

11. $a^2 + b^2 = c^2$

$9^2 + b^2 = 15^2$

$81 + b^2 = 225$

$b^2 = 144$

$b = 12$

12. 24

13. $a^2 + b^2 = c^2$

$a^2 + 1^2 = (\sqrt{5})^2$

$a^2 + 1 = 5$

$a^2 = 4$

$a = 2$

14. 1

15. $a^2 + b^2 = c^2$

$1^2 + b^2 = (\sqrt{3})^2$

$1 + b^2 = 3$

$b^2 = 2$

$b = \sqrt{2}$ Exact answer

$b \approx 1.414$ Approximation

16. $\sqrt{8} \approx 2.828$

17. $a^2 + b^2 = c^2$

$a^2 + (5\sqrt{3})^2 = 10^2$

$a^2 + 25 \cdot 3 = 100$

$a^2 + 75 = 100$

$a^2 = 25$

$a = 5$

18. $\sqrt{50} \approx 7.071$

19. *Familiarize.* We first make a drawing. We label the unknown height h.

Translate. We use the Pythagorean theorem, substituting 7 for a, h for b, and 14 for c.

$7^2 + h^2 = 14^2$

Carry out. We solve the equation.

$7^2 + h^2 = 14^2$

$49 + h^2 = 196$

$h^2 = 147$

$h = \sqrt{147}$ Exact answer

$h \approx 12.124$ Approximation

Check. We check by substituting 7, $\sqrt{147}$, and 14 in the Pythagorean equation.

$$a^2 + b^2 = c^2$$

$$\begin{array}{c|c} 7^2 + (\sqrt{147})^2 \ ? \ 14^2 & \\ 49 + 147 & 196 \\ 196 & 196 \quad \text{TRUE} \end{array}$$

State. The top of the ladder is $\sqrt{147}$ m, or about 12.124 m high.

20. $\sqrt{32}$ cm ≈ 5.657 cm

21. *Familiarize.* We first make a drawing. We label the unknown length w.

Translate. We use the Pythagorean theorem, substituting 8 for a, 12 for b, and w for c.

$$8^2 + 12^2 = w^2$$

Carry out. We solve the equation.

$$8^2 + 12^2 = w^2$$
$$64 + 144 = w^2$$
$$208 = w^2$$
$$\sqrt{208} = w \quad \text{Exact answer}$$
$$14.422 \approx w \quad \text{Approximation}$$

Check. We check by substituting 8, 12, and $\sqrt{208}$ into the Pythagorean equation:

$$a^2 + b^2 = c^2$$

$8^2 + 12^2$?	$(\sqrt{208})^2$
$64 + 144$	208
208	208 TRUE

State. The pipe should be $\sqrt{208}$ feet or about 14.422 feet long.

22. $\sqrt{250}$ m ≈ 15.811 m

23. *Familiarize*. We first make a drawing. We label the diagonal d.

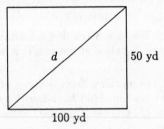

Translate. We use the Pythagorean theorem, substituting 50 for a, 100 for b, and d for c.

$$a^2 + b^2 = c^2$$
$$50^2 + 100^2 = d^2$$

Carry out. We solve the equation.

$$50^2 + 100^2 = d^2$$
$$2500 + 10,000 = d^2$$
$$12,500 = d^2$$
$$\sqrt{12,500} = d \quad \text{Exact answer}$$
$$111.803 \approx d \quad \text{Approximation}$$

Check. We check by substituting 50, 100, and $\sqrt{12,500}$ in the Pythagorean equation:

$$a^2 + b^2 = c^2$$

$50^2 + 100^2$?	$(\sqrt{12,500})^2$
$2500 + 10,000$	$12,500$
$12,500$	$12,500$ TRUE

State. The length of a diagonal is $\sqrt{12,500}$ yd or about 111.803 yd.

24. $\sqrt{26,900}$ yd ≈ 164.012 yd

25. *Familiarize*. We first make a drawing. Let d represent the distance Becky can move away from the building while using the telephone.

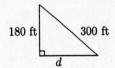

Translate. We use the Pythagorean theorem, substituting 180 for a, d for b, and 300 for c.

$$180^2 + d^2 = 300^2$$

Carry out. We solve the equation.

$$180^2 + d^2 = 300^2$$
$$32,400 + d^2 = 90,000$$
$$d^2 = 57,600$$
$$d = 240$$

Check. We check by substituting 180, 240, and 300 in the Pythagorean equation.

$$a^2 + b^2 = c^2$$

$180^2 + 240^2$?	300^2
$32,400 + 57,600$	$90,000$
$90,000$	$90,000$ TRUE

State. Becky can use her telephone 240 ft into her backyard.

26. $\sqrt{16,200}$ ft ≈ 127.279 ft

27. *Familiarize*. Referring to the drawing in the text, we let d represent the distance the plane will travel. The vertical length of the descent is $30,000 - 20,000$ or 10,000 ft.

Translate. We use the Pythagorean theorem, substituting 10,000 for a, 50,000 for b, and d for c.

$$10,000^2 + 50,000^2 = d^2$$

Carry out. We solve the equation.

$$10,000^2 + 50,000^2 = d^2$$
$$100,000,000 + 2,500,000,000 = d^2$$
$$2,600,000,000 = d^2$$
$$\sqrt{2,600,000,000} = d$$
$$50,990.195 \approx d$$

Check. We check by substituting 10,000, 50,000, and $\sqrt{2,600,000,000}$ in the Pythagorean equation.

$$a^2 + b^2 = c^2$$

$10,000^2 + 50,000^2$?	$(\sqrt{2,600,000,000})^2$
$100,000,000+$	
$2,500,000,000$	$2,600,000,000$
$2,600,000,000$	$2,600,000,000$ TRUE

State. The plane will travel $\sqrt{2,600,000,000}$ ft, or about 50,990.195 ft.

28. $\sqrt{74}$ km ≈ 8.602 km

29. Write the equation in the slope-intercept form.

$$4 - x = 3y$$
$$\frac{1}{3}(4 - x) = y$$
$$\frac{4}{3} - \frac{1}{3}x = y, \text{ or}$$
$$y = -\frac{1}{3}x + \frac{4}{3}$$

The slope is $-\frac{1}{3}$

30. $\frac{5}{8}$

31. $-\frac{3}{5}x < 15$

$$x > -\frac{5}{3} \cdot 15 \quad \text{Multiplying by } -\frac{5}{3} \text{ and}$$
$$\text{reversing the inequality}$$
$$x > -25$$

The solution set is $\{x | x > -25\}$.

32. $\{x | x \le 1\}$

33. $(3x)^4 = 3^4 \cdot x^4 = 81x^4$

34. $-\frac{8}{125}$

35. ◈

36. ◈

37. ◈

38. ◈

39. *Familiarize*. We make a drawing.

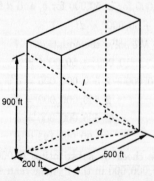

First we will find d, the diagonal distance, in feet, across the building. This is the length of the hypotenuse of a right triangle with legs of 200 ft and 500 ft. Then we will find l, the distance from Vance's office to the restaurant. This is the length, in feet, of the hypotenuse of a right triangle with legs of 900 ft and d.

Note that $\frac{1}{4}$ mi $= \frac{1}{4} \cdot 5280$ ft $= 1320$ ft. Then if $l \le 1320$ ft, Vance can use the handset at the restaurant.

Translate. To find d we use the Pythagorean theorem, substituting 200 for a, 500 for b, and d for c.

$$200^2 + 500^2 = d^2$$

After we find d we will use the Pythagorean theorem again, this time substituting 900 for a, d for b, and l for c. (Since we will use d^2 in this equation we need only solve the equation above for d^2.)

$$900^2 + d^2 = l^2$$

Carry out. First find d^2.

$$200^2 + 500^2 = d^2$$
$$40,000 + 250,000 = d^2$$
$$290,000 = d^2$$

Now we substitute 290,000 for d^2 in the second equation in the Translate step.

$$900^2 + d^2 = l^2$$
$$900^2 + 290,000 = l^2$$
$$810,000 + 290,000 = l^2$$
$$1,100,000 = l^2$$
$$\sqrt{1,100,000} = l$$
$$1049 \approx l$$

Check. We can check by substituting in the Pythagorean equation as before. This is left to the student.

State. The distance from Vance's office to the restaurant is about 1049 ft. Since this is less than 1320 ft, or 1/4 mi, Vance can use the handset at the restaurant.

40. 8 ft

41. *Familiarize*. Let $s =$ the length of a side of the square. Recall that the area A of a square with side s is given by $A = s^2$.

Translate. We substitute 7 for A in the formula.

$$7 = s^2$$

Carry out. We solve the equation.

$$7 = s^2$$
$$\sqrt{7} = s$$
$$2.646 \approx s$$

Check. If the length of a side of a square is $\sqrt{7}$ m, then the area of the square is $(\sqrt{7}$ m$)^2$, or 7 m^2. The result checks.

State. The length of a side of the square is $\sqrt{7}$ m or about 2.646 m.

42. 3, 4, 5

43.

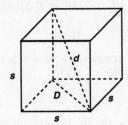

From the drawing we see that the diagonal d of the cube is the hypotenuse of a right triangle with one leg of length s, where s is the length of a side of the cube, and the other leg of length D, where D is the length of the diagonal of the base of the cube. First we find D:

Using the Pythagorean theorem we have:

$$s^2 + s^2 = D^2$$
$$2s^2 = D^2$$
$$\sqrt{2s^2} = D$$
$$s\sqrt{2} = D$$

Then we find d:

Using the Pythagorean theorem again we have:

$$s^2 + (s\sqrt{2})^2 = d^2$$
$$s^2 + 2s^2 = d^2$$
$$3s^2 = d^2$$
$$\sqrt{3s^2} = d$$
$$s\sqrt{3} = d$$

44. $\dfrac{2}{3}$

45. Using the Pythagorean theorem we get:

$$h^2 + \left(\frac{a}{2}\right)^2 = a^2$$
$$h^2 + \frac{a^2}{4} = a^2$$
$$h^2 = a^2 - \frac{a^2}{4}$$
$$h^2 = \frac{3a^2}{4}$$
$$h = \sqrt{\frac{3a^2}{4}}$$
$$h = \frac{a\sqrt{3}}{2}$$

46. $\sqrt{2400}$ ft ≈ 48.990 ft

47. *Familiarize.* After one-half hour, the car traveling east has gone $\frac{1}{2} \cdot 50$, or 25 mi, and the car traveling south has gone $\frac{1}{2} \cdot 60$, or 30 mi. We make a drawing. We label the distance between the cars, d, where d is in miles.

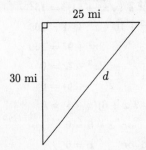

Translate. We use the Pythagorean theorem, substituting 25 for a, 30 for b, and d for c.

$$25^2 + 30^2 = d^2$$

Carry out. We solve the equation.

$$30^2 + 25^2 = d^2$$
$$900 + 625 = d^2$$
$$1525 = d^2$$
$$\sqrt{1525} \text{ mi} = d \quad \text{Exact answer}$$
$$39.051 \text{ mi} \approx d \quad \text{Approximation}$$

Check. We check by substituting 25, 30, and $\sqrt{1525}$ in the Pythagorean equation.

$$\frac{a^2 + b^2 = c^2}{}$$

$25^2 + 30^2$?	$(\sqrt{1525})^2$	
$625 + 900$	1525	
1525	1525	TRUE

State. After one half hour, the cars are 1525 mi $\approx$ 39.051 mi apart.

48. $12 - 2\sqrt{6} \approx 7.101$

49. Using the Pythagorean equation we can label the figure with additional information.

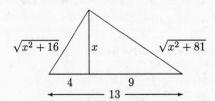

Next we use the Pythagorean equation with the largest right triangle and solve for x.

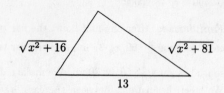

$$(\sqrt{x^2+16})^2 + (\sqrt{x^2+81})^2 = 13^2$$
$$x^2 + 16 + x^2 + 81 = 169$$
$$2x^2 + 97 = 169$$
$$2x^2 - 72 = 0$$
$$2(x^2 - 36) = 0$$
$$2(x+6)(x-6) = 0$$
$$x+6 = 0 \quad or \quad x-6 = 0$$
$$x = -6 \quad or \qquad x = 6$$

Since a length cannot be negative, only 6 has meaning in this problem Thus, $x = 6$.

50. 50 ft^2

51. The perimeter of the smaller square plot is 2 mi, so each side is $\frac{1}{4} \cdot 2$, or $\frac{1}{2}$ mi, and the area is $\left(\frac{1}{2}\text{ mi}\right)^2$, or $\frac{1}{4}$ mi^2. This tells us that $\frac{1}{4}$ mi^2 is equivalent to 160 acres. The perimeter of the larger square plot is 4 mi, so each side is $\frac{1}{4} \cdot 4$, or 1 mi, and the area is $(1 \text{ mi})^2$, or 1 mi^2. Since 1 mi$^2 = 4 \cdot \frac{1}{4}$ mi^2, then 1 mi^2 is equivalent to $4 \cdot 160$, or 640 acres. Thus, 4 mi of fencing will enclose a square whose area is 640 acres.

Exercise Set 9.7

1. $\sqrt[3]{-64} = -4 \qquad (-4)^3 = (-4)(-4)(-4) = -64$

2. -2

3. $\sqrt[3]{1000} = 10 \qquad 10^3 = 10 \cdot 10 \cdot 10 = 1000$

4. -3

5. $-\sqrt[3]{125} = -5 \qquad \sqrt[3]{125} = 5$, so $-\sqrt[3]{125} = -5$

6. -2

7. $\sqrt[3]{216} = 6 \qquad 6^3 = 6 \cdot 6 \cdot 6 = 216$

8. -7

9. $\sqrt[4]{625} = 5 \qquad 5^4 = 5 \cdot 5 \cdot 5 \cdot 5 = 625$

10. 3

11. $\sqrt[5]{0} = 0 \qquad 0^5 = 0 \cdot 0 \cdot 0 \cdot 0 \cdot 0 = 0$

12. 1

13. $\sqrt[5]{-1} = -1 \quad (-1)^5 = (-1)(-1)(-1)(-1)(-1) = -1$

14. -3

15. $\sqrt[4]{-81}$ is not a real number, because it is an even root of a negative number.

16. Not a real number

17. $\sqrt[4]{10,000} = 10 \quad 10^4 = 10 \cdot 10 \cdot 10 \cdot 10 = 10,000$

18. 10

19. $\sqrt[3]{5^3} = 5 \qquad 5^3 = 5 \cdot 5 \cdot 5$

20. 7

21. $\sqrt[6]{64} = 2 \qquad 2^6 = 2 \cdot 2 \cdot 2 \cdot 2 \cdot 2 \cdot 2 = 64$

22. 1

23. $\sqrt[9]{a^9} = a \qquad a^9 = a \cdot a \cdot a \cdot a \cdot a \cdot a \cdot a \cdot a \cdot a$

24. n

25. $\sqrt[3]{32} = \sqrt[3]{8 \cdot 4} = \sqrt[3]{8}\sqrt[3]{4} = 2\sqrt[3]{4}$

26. $3\sqrt[3]{2}$

27. $\sqrt[4]{48} = \sqrt[4]{16 \cdot 3} = \sqrt[4]{16}\sqrt[4]{3} = 2\sqrt[4]{3}$

28. $2\sqrt[5]{5}$

29. $\sqrt[3]{\dfrac{64}{125}} = \dfrac{\sqrt[3]{64}}{\sqrt[3]{125}} = \dfrac{4}{5}$

30. $\dfrac{5}{3}$

31. $\sqrt[5]{\dfrac{32}{243}} = \dfrac{\sqrt[5]{32}}{\sqrt[5]{243}} = \dfrac{2}{3}$

32. $\dfrac{5}{4}$

33. $\sqrt[3]{\dfrac{17}{8}} = \dfrac{\sqrt[3]{17}}{\sqrt[3]{8}} = \dfrac{\sqrt[3]{17}}{2}$

34. $\dfrac{\sqrt[5]{11}}{2}$

35. $\sqrt[4]{\dfrac{13}{81}} = \dfrac{\sqrt[4]{13}}{\sqrt[4]{81}} = \dfrac{\sqrt[4]{13}}{3}$

36. $\dfrac{\sqrt[3]{10}}{3}$

37. $25^{1/2} = \sqrt{25} = 5$

38. 3

39. $1000^{1/3} = \sqrt[3]{1000} = 10$

40. 5

41. $32^{1/5} = \sqrt[5]{32} = 2$

42. 2

43. $16^{3/4} = (16^{1/4})^3 = (\sqrt[4]{16})^3 = 2^3 = 8$

44. 16

45. $4^{5/2} = (4^{1/2})^5 = (\sqrt{4})^5 = 2^5 = 32$

46. 27

47. $64^{2/3} = (64^{1/3})^2 = (\sqrt[3]{64})^2 = 4^2 = 16$

48. 4

49. $8^{5/3} = (8^{1/3})^5 = (\sqrt[3]{8})^5 = 2^5 = 32$

50. 32

51. $25^{5/2} = (25^{1/2})^5 = (\sqrt{25})^5 = 5^5 = 3125$

52. $\dfrac{1}{2}$

53. $36^{-1/2} = \dfrac{1}{36^{1/2}} = \dfrac{1}{\sqrt{36}} = \dfrac{1}{6}$

54. $\dfrac{1}{2}$

55. $256^{-1/4} = \dfrac{1}{256^{1/4}} = \dfrac{1}{\sqrt[4]{256}} = \dfrac{1}{4}$

56. $\dfrac{1}{1000}$

57. $81^{-3/4} = \dfrac{1}{81^{3/4}} = \dfrac{1}{(\sqrt[4]{81})^3} = \dfrac{1}{3^3} = \dfrac{1}{27}$

58. $\dfrac{1}{8}$

59. $81^{-5/4} = \dfrac{1}{81^{5/4}} = \dfrac{1}{(\sqrt[4]{81})^5} = \dfrac{1}{3^5} = \dfrac{1}{243}$

60. $\dfrac{1}{4}$

61. $8^{-2/3} = \dfrac{1}{8^{2/3}} = \dfrac{1}{(\sqrt[3]{8})^2} = \dfrac{1}{2^2} = \dfrac{1}{4}$

62. $\dfrac{1}{125}$

63. $x^2 - 10x + 25 = x^2 - 2 \cdot x \cdot 5 + 5^2 = (x - 5)^2$

64. $(a - 6)^2$

65. $t^2 + 18t + 81 = t^2 + 2 \cdot t \cdot 9 + 9^2 = (t + 9)^2$

66. $y = 0.625x$

67.
$$\begin{aligned}
\frac{a+5}{a^2-25} - \frac{3a+15}{a^2-25} &= \frac{a+5-(3a+15)}{a^2-25} \\
&= \frac{a+5-3a-15}{a^2-25} \\
&= \frac{-2a-10}{a^2-25} \\
&= \frac{-2(a+5)}{(a+5)(a-5)} \\
&= \frac{-2}{a-5}
\end{aligned}$$

68. $\dfrac{x-1}{3x^3(x+3)}$

69. ◈

70. ◈

71. ◈

72. ◈

73. Enter 10, press the power key, enter 0.8 (or $(4 \div 5)$), and then press $\boxed{=}$.
$$10^{4/5} \approx 6.310$$
(Some calculators have a 10^x key which might have to be accessed using the $\boxed{\text{SHIFT}}$ key.)

74. 2.213

75. Enter 10, press the power key, enter 1.5 (or $(3 \div 2)$), and then press $\boxed{=}$.
$$10^{3/2} \approx 31.623$$
(Some calculators have a 10^x key which might have to be accessed using the $\boxed{\text{SHIFT}}$ key.)

76. 9.391

77. $(x^{2/3})^{7/3} = x^{\frac{2}{3} \cdot \frac{7}{3}} = x^{14/9}$

78. $a^{7/4}$

79. $\dfrac{p^{4/5}}{p^{2/3}} = p^{4/5 - 2/3} = p^{12/15 - 10/15} = p^{2/15}$

80. $m^{7/12}$

81. Graph $y = \sqrt[3]{x}$

We make a table of values.

x	y
-8	-2
-1	-1
0	0
1	1
8	2

We plot these points and connect them with a smooth curve.

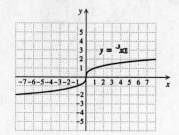

82.

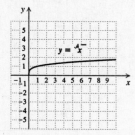

83. $y_1 = x^{2/3},\ y_2 = x^1,$
$y_3 = x^{5/4},\ y_4 = x^{3/2}$

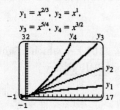

Chapter 10

Quadratic Equations

Exercise Set 10.1

1.
$$x^2 = 100$$
$x = \sqrt{100} \;\; or \;\; x = -\sqrt{100}$ Using the princi-
$x = 10 \;\;\;\; or \;\; x = -10$ ple of square roots

We can check mentally that $10^2 = 10$ and $(-10)^2 = 100$. The solutions are 10 and -10.

2. 5, -5

3.
$$a^2 = 36$$
$a = \sqrt{36} \;\; or \;\; a = -\sqrt{36}$ Using the principle
$a = 6 \;\;\;\; or \;\; a = -6$ of square roots

We can check mentally that $6^2 = 36$ and $(-6)^2 = 36$. The solutions are 6 and -6.

4. 7, -7

5.
$$t^2 \;= 23$$
$t = \sqrt{23} \;\; or \;\; t \;= -\sqrt{23}$ Using the principle
of square roots

Check: For $\sqrt{23}$:   For $-\sqrt{23}$:

$t^2 = 23$	$t^2 = 23$
$(\sqrt{23})^2$? 23	$(-\sqrt{23})^2$? 23
23 | 23 TRUE	23 | 23 TRUE

The solutions are $\sqrt{23}$ and $-\sqrt{23}$.

6. $\sqrt{13}$, $-\sqrt{13}$

7.
$$3x^2 \;= 30$$
$$x^2 \;= 10 \;\;\;\; \text{Dividing by 3}$$
$x = \sqrt{10} \;\; or \;\; x = -\sqrt{10}$ Using the principle
of square roots

Both numbers check. The solutions are $\sqrt{10}$ and $-\sqrt{10}$.

8. $\sqrt{14}$, $-\sqrt{14}$

9.
$$9t^2 = 72$$
$$t^2 = 8 \;\;\;\; \text{Dividing by 9}$$
$t = \;\;\;\; \sqrt{8} \; or \; t = -\sqrt{8}$
$t = \sqrt{4 \cdot 2} \; or \; t = -\sqrt{4 \cdot 2}$
$t = \;\; 2\sqrt{2} \; or \; t = -2\sqrt{2}$

Both numbers check. The solutions are $2\sqrt{2}$ and $-2\sqrt{2}$.

10. $2\sqrt{5}$, $-2\sqrt{5}$

11.
$$4 - 9x^2 = 0$$
$$4 = 9x^2$$
$$\frac{4}{9} = x^2$$
$x = \sqrt{\dfrac{4}{9}} \;\; or \;\; x = -\sqrt{\dfrac{4}{9}}$
$x = \dfrac{2}{3} \;\;\; or \;\; x = -\dfrac{2}{3}$

Both numbers check. The solutions are $\dfrac{2}{3}$ and $-\dfrac{2}{3}$.

12. $\dfrac{5}{2}$, $-\dfrac{5}{2}$

13.
$$49y^2 - 5 = 15$$
$$49y^2 = 20$$
$$y^2 = \frac{20}{49}$$
$y = \sqrt{\dfrac{20}{49}} \;\; or \;\; y = -\sqrt{\dfrac{20}{49}}$
$y = \dfrac{\sqrt{20}}{7} \;\; or \;\; y = -\dfrac{\sqrt{20}}{7}$
$y = \dfrac{\sqrt{4 \cdot 5}}{7} \;\; or \;\; y = -\dfrac{\sqrt{4 \cdot 5}}{7}$
$y = \dfrac{2\sqrt{5}}{7} \;\; or \;\; y = -\dfrac{2\sqrt{5}}{7}$

The solutions are $\dfrac{2\sqrt{5}}{7}$ and $-\dfrac{2\sqrt{5}}{7}$.

14. $\sqrt{3}$, $-\sqrt{3}$

15.
$$5x^2 - 120 = 0$$
$$5x^2 = 120$$
$$x^2 = 24$$
$x = \sqrt{24} \;\;\; or \;\; x = -\sqrt{24}$
$x = \sqrt{4 \cdot 6} \;\; or \;\; x = -\sqrt{4 \cdot 6}$
$x = 2\sqrt{6} \;\;\; or \;\; x = -2\sqrt{6}$

The solutions are $2\sqrt{6}$ and $-2\sqrt{6}$.

16. $\sqrt{\dfrac{7}{5}}$, $-\sqrt{\dfrac{7}{5}}$, or $\dfrac{\sqrt{35}}{5}$, $-\dfrac{\sqrt{35}}{5}$

17.
$$(x + 1)^2 = 25$$
$x + 1 = 5 \;\; or \;\; x + 1 = -5$ Using the principle
of square roots
$x = 4 \;\; or \;\;\;\; x \;\; = -6$

The solutions are 4 and -6.

18. 9, -5

19. $(x-4)^2 = 81$

$x - 4 = 9 \quad or \quad x - 4 = -9$ Using the principle
of square roots

$x = 13 \ or \qquad x = -5$

The solutions are 13 and -5.

20. 3, -9

21. $(m+3)^2 = 6$

$m + 3 = \sqrt{6} \qquad or \ m + 3 = -\sqrt{6}$

$m = -3 + \sqrt{6} \ or \qquad m = -3 - \sqrt{6}$

The solutions are $-3+\sqrt{6}$ and $-3-\sqrt{6}$, or $-3\pm\sqrt{6}$.

22. $4 \pm \sqrt{21}$

23. $(a-13)^2 = 64$

$a - 13 = 8 \quad or \ a - 13 = -8$

$a = 21 \ or \qquad a = 5$

The solutions are 21 and 5.

24. $-13 \pm 2\sqrt{2}$

25. $(x-5)^2 = 14$

$x - 5 = \sqrt{14} \qquad or \ x - 5 = -\sqrt{14}$

$x = 5 + \sqrt{14} \ or \qquad x = 5 - \sqrt{14}$

The solutions are $5+\sqrt{14}$ and $5-\sqrt{14}$, or $5\pm\sqrt{14}$.

26. $7 \pm 2\sqrt{3}$

27. $(t+2)^2 = 25$

$t + 2 = 5 \ or \ t + 2 = -5$

$t = 3 \ or \qquad t = -7$

The solutions are 3 and -7.

28. $-9 \pm \sqrt{34}$

29. $\left(y - \dfrac{3}{4}\right)^2 = \dfrac{17}{16}$

$y - \dfrac{3}{4} = \sqrt{\dfrac{17}{16}} \qquad or \ y - \dfrac{3}{4} = -\sqrt{\dfrac{17}{16}}$

$y - \dfrac{3}{4} = \dfrac{\sqrt{17}}{4} \qquad or \ y - \dfrac{3}{4} = -\dfrac{\sqrt{17}}{4}$

$y = \dfrac{3}{4} + \dfrac{\sqrt{17}}{4} \ or \qquad y = \dfrac{3}{4} - \dfrac{\sqrt{17}}{4}$

$y = \dfrac{3+\sqrt{17}}{4} \quad or \qquad y = \dfrac{3-\sqrt{17}}{4}$

The solutions are $\dfrac{3+\sqrt{17}}{4}$ and $\dfrac{3-\sqrt{17}}{4}$, or $\dfrac{3\pm\sqrt{17}}{4}$.

30. $\dfrac{-3 \pm \sqrt{14}}{2}$

31. $x^2 + 10x + 25 = 100$

$(x+5)^2 = 100$

$x + 5 = 10 \ or \ x + 5 = -10$

$x = 5 \ or \qquad x = -15$

The solutions are 5 and -15.

32. 11, -5

33. $p^2 - 8p + 16 = 1$

$(p-4)^2 = 1$

$p - 4 = 1 \ or \ p - 4 = -1$

$p = 5 \ or \qquad p = 3$

The solutions are 5 and 3.

34. -5, -9

35. $t^2 + 6t + 9 = 13$

$(t+3)^2 = 13$

$t + 3 = \sqrt{13} \qquad or \ t + 3 = -\sqrt{13}$

$t = -3 + \sqrt{13} \ or \qquad t = -3 - \sqrt{13}$

The solutions are $-3 + \sqrt{13}$ and $-3 - \sqrt{13}$, or $-3 \pm \sqrt{13}$.

36. $1 \pm \sqrt{5}$

37. $x^2 - 12x + 36 = 18$

$(x-6)^2 = 18$

$x - 6 = \sqrt{18} \qquad or \ x - 6 = -\sqrt{18}$

$x - 6 = 3\sqrt{2} \qquad or \ x - 6 = -3\sqrt{2}$

$x = 6 + 3\sqrt{2} \ or \qquad x = 6 - 3\sqrt{2}$

The solutions are $6 + 3\sqrt{2}$ and $6 - 3\sqrt{2}$, or $6 \pm 3\sqrt{2}$.

38. $-2 \pm 2\sqrt{3}$

39. Write each equation in slope-intercept form.

$y + 5 = 2x \qquad y - 2x = 7$

$y = 2x - 5 \qquad y = 2x + 7$

Since the slopes are the same ($m = 2$) and the y-intercepts are different, the equations represent parallel lines.

40. No

41. Write each equation in slope-intercept form.

$3x - 2y = 9 \qquad\qquad 4y - 6x = 8$

$-2y = -3x + 9 \qquad\qquad 4y = 6x + 8$

$y = \dfrac{3}{2}x - \dfrac{9}{2} \qquad\qquad y = \dfrac{3}{2}x + 2$

Since the slopes are the same $\left(m = \dfrac{3}{2}\right)$ and the y-intercepts are different, the equations represent parallel lines.

42. Yes

43. $\left(x - \dfrac{3}{8}\right)^2 = x^2 - 2 \cdot x \cdot \dfrac{3}{8} + \left(\dfrac{3}{8}\right)^2$

$\qquad\qquad = x^2 - \dfrac{3}{4}x + \dfrac{9}{16}$

44. $t^2 + \dfrac{5}{2}t + \dfrac{25}{16}$

45. ◈

46. ◈

47. ◈

48. ◈

49. $x^2 + \dfrac{7}{3}x + \dfrac{49}{36} = \dfrac{7}{36}$

$\qquad \left(x + \dfrac{7}{6}\right)^2 = \dfrac{7}{36}$

$\quad x + \dfrac{7}{6} = \dfrac{\sqrt{7}}{6} \quad or \quad x + \dfrac{7}{6} = -\dfrac{\sqrt{7}}{6}$

$\qquad x = -\dfrac{7}{6} + \dfrac{\sqrt{7}}{6} \quad or \qquad x = -\dfrac{7}{6} - \dfrac{\sqrt{7}}{6}$

The solutions are $\dfrac{-7 + \sqrt{7}}{6}$ and $\dfrac{-7 - \sqrt{7}}{6}$, or $\dfrac{-7 \pm \sqrt{7}}{6}$.

50. $\dfrac{5 \pm \sqrt{13}}{2}$

51. $m^2 - \dfrac{3}{2}m + \dfrac{9}{16} = \dfrac{17}{16}$

$\qquad \left(m - \dfrac{3}{4}\right)^2 = \dfrac{17}{16}$

$\quad m - \dfrac{3}{4} = \dfrac{\sqrt{17}}{4} \quad or \quad m - \dfrac{3}{4} = -\dfrac{\sqrt{17}}{4}$

$\qquad m = \dfrac{3 + \sqrt{17}}{4} \quad or \qquad m = \dfrac{3 - \sqrt{17}}{4}$

The solutions are $\dfrac{3 \pm \sqrt{17}}{4}$.

52. $2, -5$

53. $x^2 + 2.5x + 1.5625 = 9.61$

$\qquad (x + 1.25)^2 = 9.61$

$\quad x + 1.25 = 3.1 \quad or \quad x + 1.25 = -3.1$

$\qquad x = 1.85 \quad or \qquad\quad x = -4.35$

The solutions are 1.85 and -4.35.

54. $7.1, -3.3$

55. From the graph we see that when $y = 4$, then $x = -5$ or $x = -1$. Thus, the solution of $(x + 3)^2 = 4$ are -5 and -1.

56. $-6, 0$

57. $f = \dfrac{kMm}{d^2}$

$d^2 f = kMm$ Multiplying by d^2

$d^2 = \dfrac{kMm}{f}$ Dividing by f

$d = \sqrt{\dfrac{kMm}{f}}$ Taking the principal square root

Exercise Set 10.2

1. To complete the square for $x^2 + 8x$, we take half the coefficient of x and square it:

$$\left(\dfrac{8}{2}\right)^2 = 4^2 = 16$$

The trinomial $x^2 + 8x + 16$ is the square of $x + 4$. That is, $x^2 + 8x + 16 = (x + 4)^2$.

2. $x^2 + 4x + 4$

3. To complete the square for $x^2 - 14x$, we take half the coefficient of x and square it:

$$\left(\dfrac{-14}{2}\right)^2 = (-7)^2 = 49$$

The trinomial $x^2 - 14x + 49$ is the square of $x - 7$. That is, $x^2 - 14x + 49 = (x - 7)^2$.

4. $x^2 - 12x + 36$

5. To complete the square for $x^2 - 3x$, we take half the coefficient of x and square it:

$$\left(\dfrac{-3}{2}\right)^2 = \dfrac{9}{4}$$

The trinomial $x^2 - 3x + \dfrac{9}{4}$ is the square of $x - \dfrac{3}{2}$. That is, $x^2 - 3x + \dfrac{9}{4} = \left(x - \dfrac{3}{2}\right)^2$.

6. $x^2 - x + \dfrac{1}{4}$

7. To complete the square for $t^2 + t$, we take half the coefficient of t and square it:

$$\left(\dfrac{1}{2}\right)^2 = \dfrac{1}{4}$$

The trinomial $t^2 + t + \dfrac{1}{4}$ is the square of $t + \dfrac{1}{2}$. That is, $t^2 + t + \dfrac{1}{4} = \left(t + \dfrac{1}{2}\right)^2$.

8. $y^2 - 9y + \dfrac{81}{4}$

9. To complete the square for $x^2 + \dfrac{5}{4}x$, we take half the coefficient of x and square it:

$$\left(\dfrac{1}{2} \cdot \dfrac{5}{4}\right)^2 = \left(\dfrac{5}{8}\right)^2 = \dfrac{25}{64}$$

The trinomial $x^2 + \frac{5}{4}x + \frac{25}{64}$ is the square of $x + \frac{5}{8}$.

That is, $x^2 + \frac{5}{4}x + \frac{25}{64} = \left(x + \frac{5}{8}\right)^2$.

10. $x^2 + \frac{4}{3}x + \frac{4}{9}$

11. To complete the square for $m^2 - \frac{9}{2}m$, we take half the coefficient of m and square it:

$$\left[\frac{1}{2}\left(-\frac{9}{2}\right)\right]^2 = \left(-\frac{9}{4}\right)^2 = \frac{81}{16}$$

The trinomial $m^2 - \frac{9}{2}m + \frac{81}{16}$ is the square of $m - \frac{9}{4}$.

That is, $m^2 - \frac{9}{2}m + \frac{81}{16} = \left(x - \frac{9}{4}\right)^2$.

12. $r^2 - \frac{2}{5}r + \frac{1}{25}$

13. $x^2 - 8x + 15 = 0$

$x^2 - 8x \quad\quad = -15$ Subtracting 15

$x^2 - 8x + 16 = -15 + 16$ Adding 16:

$$\left(\frac{-8}{2}\right)^2 = (-4)^2 = 16$$

$(x - 4)^2 = 1$

$x - 4 = 1 \;\; or \;\; x - 4 = -1$ Principle of square roots

$x = 5 \;\; or \;\;\;\;\; x = 3$

The solutions are 5 and 3.

14. 7, -1

15. $x^2 + 22x + 21 = 0$

$x^2 + 22x \quad\quad = -21$ Subtracting 21

$x^2 + 22x + 121 = -21 + 121$ Adding 121:

$$\left(\frac{22}{2}\right)^2 = 11^2 = 121$$

$(x + 11)^2 = 100$

$x + 11 = 10 \;\; or \;\; x + 11 = -10$ Principle of square roots

$x = -1 \;\; or \;\;\;\;\; x = -21$

The solutions are -1 and -21.

16. -4, -10

17. $\quad\quad 3x^2 - 6x - 15 = 0$

$\frac{1}{3}(3x^2 - 6x - 15) = \frac{1}{3} \cdot 0$

$x^2 - 2x - 5 = 0$

$x^2 - 2x \quad\quad = 5$

$x^2 - 2x + 1 = 5 + 1$ Adding 1: $\left(\frac{-2}{2}\right)^2 =$

$\quad\quad\quad\quad\quad\quad\quad (-1)^2 = 1$

$(x - 1)^2 = 6$

$x - 1 = \sqrt{6} \quad\quad or \quad x - 1 = -\sqrt{6}$

$x = 1 + \sqrt{6} \;\; or \;\;\;\;\; x = 1 - \sqrt{6}$

The solutions are $1 \pm \sqrt{6}$.

18. $2 \pm \sqrt{15}$

19. $x^2 - 22x + 102 = 0$

$x^2 - 22x \quad\quad = -102$

$x^2 - 22x + 121 = -102 + 121$ Adding 121:

$$\left(\frac{-22}{2}\right)^2 = (-11)^2 = 121$$

$(x - 11)^2 = 19$

$x - 11 = \sqrt{19} \quad\quad or \;\; x - 11 = -\sqrt{19}$

$x = 11 + \sqrt{19} \;\; or \;\;\;\;\; x = 11 - \sqrt{19}$

The solutions are $11 \pm \sqrt{19}$.

20. $9 \pm \sqrt{7}$

21. $x^2 + 10x - 4 = 0$

$x^2 + 10x \quad\quad = 4$

$x^2 + 10x + 25 = 4 + 25$ Adding 25:

$$\left(\frac{10}{2}\right)^2 = 5^2 = 25$$

$(x + 5)^2 = 29$

$x + 5 = \sqrt{29} \quad\quad or \;\; x + 5 = -\sqrt{29}$

$x = -5 + \sqrt{29} \;\; or \;\;\;\;\; x = -5 - \sqrt{29}$

The solutions are $-5 \pm \sqrt{29}$.

22. $\dfrac{7 \pm \sqrt{57}}{2}$

23. $x^2 + 5x - 2 = 0$

$x^2 + 5x \quad\quad = 2$

$x^2 + 5x + \frac{25}{4} = 2 + \frac{25}{4}$ Adding $\frac{25}{4}$:

$$\left(\frac{5}{2}\right)^2 = \frac{25}{4}$$

$$\left(x + \frac{5}{2}\right)^2 = \frac{8}{4} + \frac{25}{4} = \frac{33}{4}$$

$x + \frac{5}{2} = \frac{\sqrt{33}}{2} \quad\quad or \;\; x + \frac{5}{2} = -\frac{\sqrt{33}}{2}$

$x = -\frac{5}{2} + \frac{\sqrt{33}}{2} \;\; or \;\;\;\;\; x = -\frac{5}{2} - \frac{\sqrt{33}}{2}$

$x = \frac{-5 + \sqrt{33}}{2} \;\; or \;\;\;\;\; x = \frac{-5 - \sqrt{33}}{2}$

The solutions are $\dfrac{-5 \pm \sqrt{33}}{2}$.

24. 4, -7

25.
$$x^2 + \frac{3}{2}x - 2 = 0$$
$$x^2 + \frac{3}{2}x = 2$$
$$x^2 + \frac{3}{2}x + \frac{9}{16} = 2 + \frac{9}{16} \quad \text{Adding } \frac{9}{16}:$$
$$\left(\frac{1}{2} \cdot \frac{3}{2}\right)^2 = \left(\frac{3}{4}\right)^2 = \frac{9}{16}$$
$$\left(x + \frac{3}{4}\right)^2 = \frac{32}{16} + \frac{9}{16} = \frac{41}{16}$$
$$x + \frac{3}{4} = \frac{\sqrt{41}}{4} \quad \text{or} \quad x + \frac{3}{4} = -\frac{\sqrt{41}}{4}$$
$$x = -\frac{3}{4} + \frac{\sqrt{41}}{4} \quad \text{or} \quad x = -\frac{3}{4} - \frac{\sqrt{41}}{4}$$
$$x = \frac{-3 + \sqrt{41}}{4} \quad \text{or} \quad x = \frac{-3 - \sqrt{41}}{4}$$
The solutions are $\dfrac{-3 \pm \sqrt{41}}{4}$.

26. $\dfrac{3 \pm \sqrt{41}}{4}$

27.
$$2x^2 + 3x - 16 = 0$$
$$\frac{1}{2}(2x^2 + 3x - 16) = \frac{1}{2} \cdot 0$$
$$x^2 + \frac{3}{2}x - 8 = 0$$
$$x^2 + \frac{3}{2}x = 8$$
$$x^2 + \frac{3}{2}x + \frac{9}{16} = 8 + \frac{9}{16} \quad \text{Adding } \frac{9}{16}:$$
$$\left(\frac{1}{2} \cdot \frac{3}{2}\right)^2 = \left(\frac{3}{4}\right)^2 = \frac{9}{16}$$
$$\left(x + \frac{3}{4}\right)^2 = \frac{128}{16} + \frac{9}{16} = \frac{137}{16}$$
$$x + \frac{3}{4} = \frac{\sqrt{137}}{4} \quad \text{or} \quad x + \frac{3}{4} = -\frac{\sqrt{137}}{4}$$
$$x = -\frac{3}{4} + \frac{\sqrt{137}}{4} \quad \text{or} \quad x = -\frac{3}{4} - \frac{\sqrt{137}}{4}$$
The solutions are $\dfrac{-3 \pm \sqrt{137}}{4}$.

28. $\dfrac{3 \pm \sqrt{73}}{4}$

29.
$$3x^2 + 4x - 1 = 0$$
$$\frac{1}{3}(3x^2 + 4x - 1) = \frac{1}{3} \cdot 0$$
$$x^2 + \frac{4}{3}x - \frac{1}{3} = 0$$
$$x^2 + \frac{4}{3}x = \frac{1}{3}$$
$$x^2 + \frac{4}{3}x + \frac{4}{9} = \frac{1}{3} + \frac{4}{9}$$
$$\left(x + \frac{2}{3}\right)^2 = \frac{7}{9}$$

$$x + \frac{2}{3} = \frac{\sqrt{7}}{3} \quad \text{or} \quad x + \frac{2}{3} = -\frac{\sqrt{7}}{3}$$
$$x = \frac{-2 + \sqrt{7}}{3} \quad \text{or} \quad x = -\frac{-2 - \sqrt{7}}{3}$$
The solutions are $\dfrac{-2 \pm \sqrt{7}}{3}$.

30. $\dfrac{2 \pm \sqrt{13}}{3}$

31.
$$2x^2 = 9 + 5x$$
$$2x^2 - 5x - 9 = 0$$
$$\frac{1}{2}(2x^2 - 5x - 9) = \frac{1}{2} \cdot 0$$
$$x^2 - \frac{5}{2}x - \frac{9}{2} = 0$$
$$x^2 - \frac{5}{2}x = \frac{9}{2}$$
$$x^2 - \frac{5}{2}x + \frac{25}{16} = \frac{9}{2} + \frac{25}{16}$$
$$\left(x - \frac{5}{4}\right)^2 = \frac{72 + 25}{16} = \frac{97}{16}$$
$$x - \frac{5}{4} = \frac{\sqrt{97}}{4} \quad \text{or} \quad x - \frac{5}{4} = -\frac{\sqrt{97}}{4}$$
$$x = \frac{5}{4} + \frac{\sqrt{97}}{4} \quad \text{or} \quad x = \frac{5}{4} - \frac{\sqrt{97}}{4}$$
The solutions are $\dfrac{5 \pm \sqrt{97}}{4}$.

32. $5, -\dfrac{1}{2}$

33.
$$4x^2 + 12x = 7$$
$$x^2 + 3x = \frac{7}{4}$$
$$x^2 + 3x + \frac{9}{4} = \frac{7}{4} + \frac{9}{4}$$
$$\left(x + \frac{3}{2}\right)^2 = 4$$
$$x + \frac{3}{2} = 2 \quad \text{or} \quad x + \frac{3}{2} = -2$$
$$x = \frac{1}{2} \quad \text{or} \quad x = -\frac{7}{2}$$
The solutions are $\dfrac{1}{2}$ and $-\dfrac{7}{2}$.

34. $\dfrac{2}{3}, -\dfrac{5}{2}$

35.
$$y - x = 5, \quad (1)$$
$$y + 2x = 7 \quad (2)$$
We solve Equation (1) for y.
$$y - x = 5 \qquad (1)$$
$$y = x + 5 \quad (3)$$

Then we substitute $x + 5$ for y in Equation (2) and solve for x.

$$y + 2x = 7 \quad (2)$$
$$x + 5 + 2x = 7$$
$$3x + 5 = 7$$
$$3x = 2$$
$$x = \frac{2}{3}$$

Substitute $\frac{2}{3}$ for x in Equation (3).

$$y = x + 5 = \frac{2}{3} + 5 = \frac{17}{3}.$$

The solution is $\left(\dfrac{2}{3}, \dfrac{17}{3}\right)$.

36. $(-2, 4)$

37. Graph $y = \dfrac{3}{5}x - 1$.

First we plot the y-intercept $(0, -1)$. The slope is $\dfrac{3}{5}$. Starting at $(0, -1)$, we move up 3 units and right 5 units to another point on the graph, $(5, 2)$. We can also think of the slope as $\dfrac{-3}{-5}$. Starting at $(0, -1)$ again, we move down 3 units and left 5 units to a third point on the graph, $(-5, -4)$. We draw the line through these points.

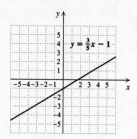

38.

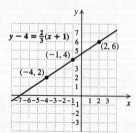

39. $\sqrt{84} = \sqrt{4 \cdot 21} = \sqrt{4}\sqrt{21} = 2\sqrt{21}$

40. $3\sqrt{10}$

41. ◈

42. ◈

43. ◈

44. ◈

45. $x^2 + bx + 49$

The trinomial is a square if the square of one-half the x-coefficient is equal to 49. Thus, we have:

$$\left(\frac{b}{2}\right)^2 = 49$$
$$\frac{b^2}{4} = 49$$
$$b^2 = 196$$
$$b = 14 \ \ or \ \ b = -14$$

46. $12, \ -12$

47. $x^2 + bx + 50$

The trinomial is a square if the square of one-half the x-coefficient is equal to 50. Thus, we have:

$$\left(\frac{b}{2}\right)^2 = 50$$
$$\frac{b^2}{4} = 50$$
$$b^2 = 200$$
$$b = \sqrt{200} \ \ or \ \ b = -\sqrt{200}$$
$$b = \sqrt{100 \cdot 2} \ \ or \ \ b = -\sqrt{100 \cdot 2}$$
$$b = 10\sqrt{2} \ \ or \ \ b = -10\sqrt{2}$$

48. $6\sqrt{5}, \ -6\sqrt{5}$

49. $x^2 - bx + 48$

The trinomial is a square if the square of one-half the x-coefficient is equal to 48. Thus, we have:

$$\left(\frac{-b}{2}\right)^2 = 48$$
$$\frac{b^2}{4} = 48$$
$$b^2 = 192$$
$$b = \sqrt{192} \ \ or \ \ b = -\sqrt{192}$$
$$b = \sqrt{64 \cdot 3} \ \ or \ \ b = -\sqrt{64 \cdot 3}$$
$$b = 8\sqrt{3} \ \ or \ \ b = -8\sqrt{3}$$

50. $16, \ -16$

51. $-0.39, \ -7.61$

52. $7.41, \ 4.59$

53. $0.37, \ -5.37$

54. $7.27, \ -0.27$

55. $3.71, \ -1.21$

56. $5, \ -0.5$

Exercise Set 10.3

1.
$$x^2 - 7x = 18$$
$$x^2 - 7x - 18 = 0 \qquad \text{Standard form}$$

We can factor.
$$x^2 - 7x - 18 = 0$$
$$(x - 9)(x + 2) = 0$$

$$x - 9 = 0 \quad or \quad x + 2 = 0$$
$$x = 9 \quad or \qquad x = -2$$

The solutions are 9 and -2.

2. $-7, 3$

3.
$$x^2 = 6x - 9$$
$$x^2 - 6x + 9 = 0 \qquad \text{Standard form}$$

We can factor.
$$x^2 - 6x + 9 = 0$$
$$(x - 3)(x - 3) = 0$$

$$x - 3 = 0 \quad or \quad x - 3 = 0$$
$$x = 3 \quad or \qquad x = 3$$

The solution is 3.

4. 4

5. $3y^2 + 7y + 4 = 0$

We can factor.
$$3y^2 + 7y + 4 = 0$$
$$(3y + 4)(y + 1) = 0$$

$$3y + 4 = 0 \quad or \quad y + 1 = 0$$
$$3y = -4 \quad or \qquad y = -1$$
$$y = -\frac{4}{3} \quad or \qquad y = -1$$

The solutions are $-\frac{4}{3}$ and -1.

6. $\frac{4}{3}, -2$

7.
$$4x^2 - 12x = 7$$
$$4x^2 - 12x - 7 = 0$$

We can factor.
$$4x^2 - 12x - 7 = 0$$
$$(2x + 1)(2x - 7) = 0$$

$$2x + 1 = 0 \quad or \quad 2x - 7 = 0$$
$$2x = -1 \quad or \qquad 2x = 7$$
$$x = -\frac{1}{2} \quad or \qquad x = \frac{7}{2}$$

The solutions are $-\frac{1}{2}$ and $\frac{7}{2}$.

8. $\frac{3}{2}, -\frac{5}{2}$

9.
$$x^2 - 64 = 0$$
$$(x + 8)(x - 8) = 0$$

$$x + 8 = 0 \quad or \quad x - 8 = 0$$
$$x = -8 \quad or \qquad x = 8$$

The solutions are -8 and 8.

10. $-2, 2$

11. $x^2 + 4x - 7 = 0$

We use the quadratic formula.

$a = 1, b = 4, c = -7$

$$x = \frac{-b \pm \sqrt{b^2 - 4ac}}{2a}$$

$$x = \frac{-4 \pm \sqrt{4^2 - 4 \cdot 1 \cdot (-7)}}{2 \cdot 1}$$

$$x = \frac{-4 \pm \sqrt{16 + 28}}{2}$$

$$x = \frac{-4 \pm \sqrt{44}}{2} = \frac{-4 \pm \sqrt{4 \cdot 11}}{2}$$

$$x = \frac{-4 \pm 2\sqrt{11}}{2} = \frac{2(-2 \pm \sqrt{11})}{2 \cdot 1}$$

$$x = -2 \pm \sqrt{11}$$

The solutions are $-2 + \sqrt{11}$ and $-2 - \sqrt{11}$, or $-2 \pm \sqrt{11}$.

12. $-1 \pm \sqrt{3}$

13. $y^2 - 10y + 22 = 0$

We use the quadratic formula.

$a = 1, b = -10, c = 22$

$$y = \frac{-b \pm \sqrt{b^2 - 4ac}}{2a}$$

$$y = \frac{-(-10) \pm \sqrt{(-10)^2 - 4 \cdot 1 \cdot 22}}{2 \cdot 1}$$

$$y = \frac{10 \pm \sqrt{100 - 88}}{2}$$

$$y = \frac{10 \pm \sqrt{12}}{2} = \frac{10 \pm \sqrt{4 \cdot 3}}{2}$$

$$y = \frac{10 \pm 2\sqrt{3}}{2} = \frac{2(5 \pm \sqrt{3})}{2 \cdot 1}$$

$$y = 5 \pm \sqrt{3}$$

The solutions are $5 + \sqrt{3}$ and $5 - \sqrt{3}$, or $5 \pm \sqrt{3}$.

14. $-3 \pm \sqrt{10}$

15.
$$x^2 + 2x + 1 = 7$$
$$x^2 + 2x - 6 = 0 \quad \text{Standard form}$$

We use the quadratic formula.

$a = 1, b = 2, c = -6$

$$x = \frac{-b \pm \sqrt{b^2 - 4ac}}{2a}$$

$$x = \frac{-2 \pm \sqrt{2^2 - 4 \cdot 1 \cdot (-6)}}{2 \cdot 1}$$

$$x = \frac{-2 \pm \sqrt{4 + 24}}{2} = \frac{-2 \pm \sqrt{28}}{2}$$

$$x = \frac{-2 \pm \sqrt{4 \cdot 7}}{2} = \frac{-2 \pm 2\sqrt{7}}{2}$$

$$x = \frac{2(-1 \pm \sqrt{7})}{2 \cdot 1} = -1 \pm \sqrt{7}$$

The solutions are $-1 + \sqrt{7}$ and $-1 - \sqrt{7}$, or $-1 \pm \sqrt{7}$.

16. $2 \pm \sqrt{5}$

17. $3x^2 + 4x - 2 = 0$

We use the quadratic formula.

$a = 3, b = 4, c = -2$

$$x = \frac{-b \pm \sqrt{b^2 - 4ac}}{2a}$$

$$x = \frac{-4 \pm \sqrt{4^2 - 4 \cdot 3 \cdot (-2)}}{2 \cdot 3}$$

$$x = \frac{-4 \pm \sqrt{16 + 24}}{6} = \frac{-4 \pm \sqrt{40}}{6}$$

$$x = \frac{-4 \pm \sqrt{4 \cdot 10}}{6} = \frac{-4 \pm 2\sqrt{10}}{6}$$

$$x = \frac{2(-2 \pm \sqrt{10})}{2 \cdot 3} = \frac{-2 \pm \sqrt{10}}{3}$$

The solutions are $\dfrac{-2 + \sqrt{10}}{3}$ and $\dfrac{-2 - \sqrt{10}}{3}$, or $\dfrac{-2 \pm \sqrt{10}}{3}$.

18. $\dfrac{4 \pm \sqrt{10}}{3}$

19. $2x^2 - 5x = 1$

$2x^2 - 5x - 1 = 0$ Standard form

We use the quadratic formula.

$a = 2, b = -5, c = -1$

$$x = \frac{-b \pm \sqrt{b^2 - 4ac}}{2a}$$

$$x = \frac{-(-5) \pm \sqrt{(-5)^2 - 4 \cdot 2 \cdot (-1)}}{2 \cdot 2}$$

$$x = \frac{5 \pm \sqrt{25 + 8}}{4} = \frac{5 \pm \sqrt{33}}{4}$$

The solutions are $\dfrac{5 + \sqrt{33}}{4}$ and $\dfrac{5 - \sqrt{33}}{4}$, or $\dfrac{5 \pm \sqrt{33}}{4}$.

20. $\dfrac{-1 \pm \sqrt{7}}{2}$

21. $4y^2 + 4y - 1 = 0$

We use the quadratic formula.

$a = 4, b = 4, c = -1$

$$y = \frac{-b \pm \sqrt{b^2 - 4ac}}{2a}$$

$$y = \frac{-4 \pm \sqrt{4^2 - 4 \cdot 4 \cdot (-1)}}{2 \cdot 4}$$

$$y = \frac{-4 \pm \sqrt{16 + 16}}{8} = \frac{-4 \pm \sqrt{32}}{8}$$

$$y = \frac{-4 \pm \sqrt{16 \cdot 2}}{8} = \frac{-4 \pm 4\sqrt{2}}{8}$$

$$y = \frac{4(-1 \pm \sqrt{2})}{4 \cdot 2} = \frac{-1 \pm \sqrt{2}}{2}$$

The solutions are $\dfrac{-1 + \sqrt{2}}{2}$ and $\dfrac{-1 - \sqrt{2}}{2}$, or $\dfrac{-1 \pm \sqrt{2}}{2}$.

22. $\dfrac{1 \pm \sqrt{2}}{2}$

23. $2t^2 - 3t + 2 = 0$

We use the quadratic formula.

$a = 2, b = -3, c = 2$

$$t = \frac{-b \pm \sqrt{b^2 - 4ac}}{2a}$$

$$t = \frac{-(-3) \pm \sqrt{(-3)^2 - 4 \cdot 2 \cdot 2}}{2 \cdot 2}$$

$$t = \frac{3 \pm \sqrt{9 - 16}}{4} = \frac{3 \pm \sqrt{-7}}{4}$$

Since the radicand, -7, is negative, there are no real-number solutions.

24. No real-number solutions

25. $3x^2 = 5x + 4$

$3x^2 - 5x - 4 = 0$

We use the quadratic formula.

$a = 3, b = -5, c = -4$

$$x = \frac{-b \pm \sqrt{b^2 - 4ac}}{2a}$$

$$x = \frac{-(-5) \pm \sqrt{(-5)^2 - 4 \cdot 3 \cdot (-4)}}{2 \cdot 3}$$

$$x = \frac{5 \pm \sqrt{25 + 48}}{6} = \frac{5 \pm \sqrt{73}}{6}$$

The solutions are $\dfrac{5 + \sqrt{73}}{6}$ and $\dfrac{5 - \sqrt{73}}{6}$, or $\dfrac{5 \pm \sqrt{73}}{6}$.

26. $\dfrac{-3 \pm \sqrt{17}}{4}$

27. $2y^2 - 6y = 10$

$2y^2 - 6y - 10 = 0$

$y^2 - 3y - 5 = 0$ Multiplying by $\dfrac{1}{2}$

We use the quadratic formula.

$a = 1, b = -3, c = -5$

$$y = \frac{-b \pm \sqrt{b^2 - 4ac}}{2a}$$

$$y = \frac{-(-3) \pm \sqrt{(-3)^2 - 4 \cdot 1 \cdot (-5)}}{2 \cdot 1}$$

$$y = \frac{3 \pm \sqrt{9 + 20}}{2} = \frac{3 \pm \sqrt{29}}{2}$$

The solutions are $\dfrac{3 + \sqrt{29}}{2}$ and $\dfrac{3 - \sqrt{29}}{2}$, or $\dfrac{3 \pm \sqrt{29}}{2}$.

28. $\dfrac{11 \pm \sqrt{181}}{10}$

29. $10x^2 - 15x = 0$

We can factor.

$10x^2 - 15x = 0$

$5x(2x - 3) = 0$

$5x = 0 \quad or \quad 2x - 3 = 0$

$x = 0 \quad or \quad \qquad 2x = 3$

$x = 0 \quad or \quad \qquad x = \dfrac{3}{2}$

The solutions are 0 and $\dfrac{3}{2}$.

30. No real-number solutions

31. $\qquad 5t^2 - 7t = -4$

$5t^2 - 7t + 4 = 0 \quad$ Standard form

We use the quadratic formula.

$a = 5, \; b = -7, \; c = 4$

$t = \dfrac{-b \pm \sqrt{b^2 - 4ac}}{2a}$

$t = \dfrac{-(-7) \pm \sqrt{(-7)^2 - 4 \cdot 5 \cdot 4}}{2 \cdot 5}$

$t = \dfrac{7 \pm \sqrt{49 - 80}}{10} = \dfrac{7 \pm \sqrt{-31}}{10}$

Since the radicand, -31, is negative, there are no real-number solutions.

32. $0, \; -\dfrac{2}{3}$

33. $9y^2 = 162$

$y^2 = 18 \quad$ Dividing by 9

$y = \sqrt{18} \; or \; y = -\sqrt{18} \quad$ Principle of square roots

$y = 3\sqrt{2} \; or \; y = -3\sqrt{2}$

The solutions are $3\sqrt{2}$ and $-3\sqrt{2}$, or $\pm 3\sqrt{2}$.

34. $2\sqrt{5}, \; -2\sqrt{5}$

35. $x^2 - 4x - 7 = 0$

$a = 1, \; b = -4, \; c = -7$

$x = \dfrac{-(-4) \pm \sqrt{(-4)^2 - 4 \cdot 1 \cdot (-7)}}{2 \cdot 1}$

$= \dfrac{4 \pm \sqrt{16 + 28}}{2} = \dfrac{4 \pm \sqrt{44}}{2}$

$= \dfrac{4 \pm \sqrt{4 \cdot 11}}{2} = \dfrac{4 \pm 2\sqrt{11}}{2}$

$= \dfrac{2(2 \pm \sqrt{11})}{2} = 2 \pm \sqrt{11}$

Using a calculator or Table 2, we see that $\sqrt{11} \approx 3.317$:

$2 + \sqrt{11} \approx 2 + 3.317 \; or \; 2 - \sqrt{11} \approx 2 - 3.317$

$\approx 5.317 \qquad or \qquad\qquad \approx -1.317$

The approximate solutions, to the nearest thousandth, are 5.317 and -1.317.

36. $0.732, \; -2.732$

37. $y^2 - 6y - 1 = 0$

$a = 1, \; b = -6, \; c = -1$

$y = \dfrac{-(-6) \pm \sqrt{(-6)^2 - 4 \cdot 1 \cdot (-1)}}{2 \cdot 1}$

$= \dfrac{6 \pm \sqrt{36 + 4}}{2} = \dfrac{6 \pm \sqrt{40}}{2}$

$= \dfrac{6 \pm \sqrt{4 \cdot 10}}{2} = \dfrac{6 \pm 2\sqrt{10}}{2}$

$= \dfrac{2(3 \pm \sqrt{10})}{2} = 3 \pm \sqrt{10}$

Using a calculator or Table 2, we see that $\sqrt{10} \approx 3.162$:

$3 + \sqrt{10} \approx 3 + 3.162 \; or \; 3 - \sqrt{10} \approx 3 - 3.162$

$\approx 6.162 \qquad or \qquad\qquad \approx -0.162$

The approximate solutions, to the nearest thousandth, are 6.162 and -0.162.

38. $-3.268, \; -6.732$

39. $\qquad 4x^2 + 4x = 1$

$4x^2 + 4x - 1 = 0 \quad$ Standard form

$a = 4, \; b = 4, \; c = -1$

$x = \dfrac{-4 \pm \sqrt{4^2 - 4 \cdot 4 \cdot (-1)}}{2 \cdot 4}$

$= \dfrac{-4 \pm \sqrt{16 + 16}}{8} = \dfrac{-4 \pm \sqrt{32}}{8}$

$= \dfrac{-4 \pm \sqrt{16 \cdot 2}}{8} = \dfrac{-4 \pm 4\sqrt{2}}{8}$

$= \dfrac{4(-1 \pm \sqrt{2})}{4 \cdot 2} = \dfrac{-1 \pm \sqrt{2}}{2}$

Using a calculator or Table 2, we see that $\sqrt{2} \approx 1.414$:

$\dfrac{-1 + \sqrt{2}}{2} \approx \dfrac{-1 + 1.414}{2} \; or \; \dfrac{-1 - \sqrt{2}}{2} \approx \dfrac{-1 - 1.414}{2}$

$\approx \dfrac{0.414}{2} \qquad or \qquad\qquad \approx \dfrac{-2.414}{2}$

$\approx 0.207 \qquad or \qquad\qquad \approx -1.207$

The approximate solutions, to the nearest thousandth, are 0.207 and -1.207.

40. $1.207, \; -0.207$

41. *Familiarize*. We will use the formula

$d = \dfrac{n^2 - 3n}{2},$

where d is the number of diagonals and n is the number of sides.

Translate. We substitute 35 for d.

$35 = \dfrac{n^2 - 3n}{2}$

Carry out. We solve the equation.

$$\frac{n^2 - 3n}{2} = 35$$
$$n^2 - 3n = 70 \quad \text{Multiplying by 2}$$
$$n^2 - 3n - 70 = 0$$
$$(n - 10)(n + 7) = 0$$
$$n - 10 = 0 \quad or \quad n + 7 = 0$$
$$n = 10 \quad or \qquad n = -7$$

Check. Since the number of sides cannot be negative, -7 cannot be a solution. To check 10, we substitute 10 for n in the original formula and determine if this yields $d = 35$. This is left to the student.

State. The polygon has 10 sides.

42. 8

43. Familiarize. We will use the formula $s = 16t^2$.

Translate. We substitute 1012 for s.

$$1012 = 16t^2$$

Carry out. We solve the equation.

$$1012 = 16t^2$$
$$\frac{1012}{16} = t^2$$
$$\sqrt{\frac{1012}{16}} = t \quad or \quad -\sqrt{\frac{1012}{16}} = t \quad \text{Principle of square roots}$$
$$7.95 \approx t \quad or \qquad -7.95 \approx t$$

Check. The number -7.95 cannot be a solution, because time cannot be negative in this situation. We substitute 7.95 in the original equation:

$$s = 16(7.95)^2 = 16(63.2025) = 1011.24.$$

This is close to 1012. Remember that we approximated the solution. Thus, we have a check.

State. It would take about 7.95 sec for an object to fall to the ground from the top of Library Square Tower.

44. 8.43 sec

45. Familiarize. We will use the formula $s = 16t^2$.

Translate. We substitute 700 for s.

$$700 = 16t^2$$

Carry out. We solve the equation.

$$700 = 16t^2$$
$$\frac{700}{16} = t^2$$
$$\sqrt{\frac{700}{16}} = t \quad or \quad -\sqrt{\frac{700}{16}} = t \quad \text{Principle of square roots}$$
$$6.61 \approx t \quad or \qquad -6.61 \approx t$$

Check. The number -6.61 cannot be a solution, because time cannot be negative in this situation. We substitute 6.61 in the original equation:

$$s = 16(6.61)^2 = 16(43.6921) = 699.0736.$$

This is close to 700. Remember that we approximated the solution. Thus, we have a check.

State. The free-fall portion of the jump lasted about 6.61 sec.

46. 3.31 sec

47. Familiarize. From the drawing in the text we have $s =$ the length of the shorter leg and $s + 17 =$ the length of the longer leg, in feet.

Translate. We use the Pythagorean theorem.

$$x^2 + (x + 17)^2 = 25^2$$

Carry out. We solve the equation.

$$x^2 + x^2 + 34x + 289 = 625$$
$$2x^2 + 34x - 336 = 0$$
$$x^2 + 17x - 168 = 0 \quad \text{Multiplying by } \frac{1}{2}$$
$$(x - 7)(x + 24) = 0$$
$$x - 7 = 0 \quad or \quad x + 24 = 0$$
$$x = 7 \quad or \qquad x = -24$$

Check. Since the length of a leg cannot be negative, -24 does not check. But 7 does check. If the smaller leg is 7, the other leg is $7+17$, or 24. Then, $7^2 + 24^2 = 49 + 576 = 625$, and $\sqrt{625} = 25$, the length of the hypotenuse.

State. The legs measure 7 ft and 24 ft.

48. 10 yd, 24 yd

49. Familiarize. From the drawing in the text, we see that w represents the width of the rectangle and $w+4$ represents the length, in centimeters.

Translate. The area is length × width. Thus, we have two expressions for the area of the rectangle: $(w + 4)w$ and 60. This gives us a translation.

$$(w + 4)w = 60$$

Carry out. We solve the equation.

$$w^2 + 4w = 60$$
$$w^2 + 4w - 60 = 0$$
$$(w + 10)(w - 6) = 0$$
$$w + 10 = 0 \quad or \quad w - 6 = 0$$
$$w = -10 \quad or \qquad w = 6$$

Check. Since the length of a side cannot be negative, -10 does not check. But 6 does check. If the width is 6, then the length is $6 + 4$, or 10. The area is 10×6, or 60. This checks.

State. The length is 10 cm, and the width is 6 cm.

50. 10 m, 7 m

51. Familiarize. We make a drawing. We let $w =$ the width of the yard. Then $w + 5 =$ the length, in meters.

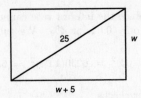

Translate. We use the Pythagorean theorem.

$$w^2 + (w+5)^2 = 25^2$$

Carry out. We solve the equation.

$$w^2 + w^2 + 10w + 25 = 625$$
$$2w^2 + 10w - 600 = 0$$
$$w^2 + 5w - 300 = 0$$
$$(w+20)(w-15) = 0$$
$$w + 20 = 0 \quad or \quad w - 15 = 0$$
$$w = -20 \quad or \quad w = 15$$

Check. Since the width cannot be negative, -20 does not check. But 15 does check. If the width is 15, then the length is $15 + 5$, or 20, and $15^2 + 20^2 = 225 + 400 = 625 = 25^2$.

State. The yard is 15 m by 20 m.

52. 24 ft

53. Familiarize. We make a drawing. Let $x =$ the length of the shorter leg of the right triangle. Then $x + 2.5 =$ the length of the longer leg, in meters.

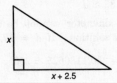

Translate. Using the formula $A = \frac{1}{2}bh$, we substitute 13 for A, $x + 2.5$ for b, and x for h.

$$13 = \frac{1}{2}(x + 2.5)(x)$$

Carry out. We solve the equation.

$$13 = \frac{1}{2}(x + 2.5)(x)$$
$$26 = (x + 2.5)(x) \qquad \text{Multiplying by 2}$$
$$26 = x^2 + 2.5x$$
$$0 = x^2 + 2.5x - 26$$
$$0 = 10x^2 + 25x - 260 \qquad \text{Multiplying by 10 to}$$
$$\text{clear the decimal}$$
$$0 = 5(2x^2 + 5x - 52)$$
$$0 = 5(2x + 13)(x - 4)$$
$$2x + 13 = 0 \quad or \quad x - 4 = 0$$
$$2x = -13 \quad or \quad x = 4$$
$$x = -6.5 \quad or \quad x = 4$$

Check. Since the length cannot be negative, -6.5 does not check. But 4 does check. If the shorter leg is 4, then the longer leg is $4 + 2.5$, or 6.5, and $A = \frac{1}{2}(6.5)(4) = 13$.

State. The legs are 4 m and 6.5 m.

54. 5 cm, 6.2 cm

55. Familiarize. We first make a drawing. We let x represent the width and $x + 2$ the length, in inches.

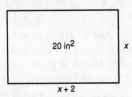

Translate. The area is length $\times$ width. We have two expressions for the area of the rectangle: $(x+2)x$ and 20. This gives us a translation.

$$(x + 2)x = 20$$

Carry out. We solve the equation.

$$x^2 + 2x = 20$$
$$x^2 + 2x - 20 = 0$$
$$a = 1, b = 2, c = -20$$
$$x = \frac{-2 \pm \sqrt{2^2 - 4 \cdot 1 \cdot (-20)}}{2 \cdot 1}$$
$$x = \frac{-2 \pm \sqrt{4 + 80}}{2} = \frac{-2 \pm \sqrt{84}}{2}$$
$$x = \frac{-2 \pm \sqrt{4 \cdot 21}}{2} = \frac{-2 \pm 2\sqrt{21}}{2}$$
$$x = \frac{2(-1 \pm \sqrt{21})}{2} = -1 \pm \sqrt{21}$$

Using a calculator or Table 2 we find that $\sqrt{21} \approx 4.58$:

$$-1 + \sqrt{21} \approx -1 + 4.58 \quad or \quad -1 - \sqrt{21} \approx -1 - 4.58$$
$$\approx 3.58 \qquad or \qquad \approx -5.58$$

Check. Since the length of a side cannot be negative, -5.58 does not check. But 3.58 does check. If the width is 3.58, then the length is $3.58 + 2$, or 5.58. The area is $5.58(3.58)$, or $19.9764 \approx 20$.

State. The length is about 5.58 in., and the width is about 3.58 in.

56. 5.65 ft, 2.65 ft

57. Familiarize. We first make a drawing. We let x represent the width and $2x$ the length, in meters.

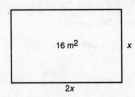

Translate. The area is length $\times$ width. We have two expressions for the area of the rectangle: $2x \cdot x$ and 16. This gives us a translation.

$$2x \cdot x = 16$$

Carry out. We solve the equation.
$$2x^2 = 16$$
$$x^2 = 8$$
$$x = \sqrt{8} \quad or \quad x = -\sqrt{8}$$
$$x = 2.83 \quad or \quad x \approx -2.83 \quad \text{Using a calculator}$$
$$\text{or Table 2}$$

Check. Since the length cannot be negative, -2.83 does not check. But 2.83 does check. If the width is $\sqrt{8}$ m, then the length is $(2\sqrt{8})$ or 5.66 m. The area is $(5.66)(2.83)$, or $16.0178 \approx 16$.

State. The length is about 5.66 m, and the width is about 2.83 m.

58. 6.32 cm, 3.16 cm

59. Familiarize. We will use the formula $A = P(1+r)^t$.

Translate. We substitute 1000 for P, 1440 for A, and 2 for t.
$$1440 = 1000(1+r)^2$$

Carry out. We solve the equation.
$$1440 = 1000(1+r)^2$$
$$\frac{144}{100} = (1+r)^2$$
$$\sqrt{\frac{144}{100}} = 1+r \quad or \quad -\sqrt{\frac{144}{100}} = 1+r \quad \text{Principle}$$
$$\text{of square roots}$$
$$\frac{12}{10} = 1+r \quad or \quad -\frac{12}{10} = 1+r$$
$$1.2 = 1+r \quad or \quad -1.2 = 1+r$$
$$0.2 = r \quad or \quad -2.2 = r$$

Check. Since the interest rate cannot be negative, we check only 0.2, or 20%. We substitute in the formula:
$$1000(1+0.2)^2 = 1000(1.2)^2 = 1000(1.44) = 1440.$$
The answer checks.

State. The interest rate is 20%.

60. 18.75%

61. Familiarize. We will use the formula $A = P(1+r)^t$.

Translate. We substitute 6250 for P, 6760 for A, and 2 for t.
$$6760 = 6250(1+r)^2$$

Carry out. We solve the equation.
$$6760 = 6250(1+r)^2$$
$$\frac{676}{625} = (1+r)^2$$
$$\sqrt{\frac{676}{625}} = 1+r \quad or \quad -\sqrt{\frac{676}{625}} = 1+r \quad \text{Principle}$$
$$\text{of square roots}$$
$$\frac{26}{25} = 1+r \quad or \quad -\frac{26}{25} = 1+r$$
$$1.04 = 1+r \quad or \quad -1.04 = 1+r$$
$$0.04 = r \quad or \quad -2.04 = r$$

Check. Since the interest rate cannot be negative, we check only 0.04, or 4%. We substitute in the formula:
$$6250(1+0.04)^2 = 6250(1.04)^2 = 6250(1.0816) = 6760.$$
The answer checks.

State. The interest rate is 4%.

62. 8%

63. Familiarize. Let d = the diameter (or width) of the flower garden, in feet. Then $\frac{d}{2}$ = the radius. We will use the formula for the area of the circle, $A = \pi r^2$.

Translate. We substitute 250 for A, 3.14 for π, and $\frac{d}{2}$ for r in the formula.
$$250 = 3.14\left(\frac{d}{2}\right)^2$$
$$250 = 3.14\left(\frac{d^2}{4}\right)$$

Carry out. We solve the equation.
$$250 = 3.14\left(\frac{d^2}{4}\right)$$
$$250 = 0.785d^2 \qquad \left(\frac{3.14}{4} = 0.785\right)$$
$$\frac{250}{0.785} = d^2$$
$$17.85 \approx d \quad or \quad -17.85 \approx d$$

Check. Since the diameter cannot be negative, -17.85 cannot be a solution. If $d = 17.85$, then $A = 3.14\left(\frac{17.85}{2}\right)^2 = 250.1186625 \approx 250$. The answer checks.

State. The width of the largest circular garden Laura can cover with the mulch is about 17.85 ft. (If we had used the π key on a calculator, the result would have been 17.84 ft.)

64. About 160 m

65.
$$5(2x-3) + 4x = 9 - 6x$$
$$10x - 15 + 4x = 9 - 6x \quad \text{Removing parentheses}$$
$$14x - 15 = 9 - 6x$$
$$20x = 24 \quad \text{Adding 15 and } 6x$$
$$x = \frac{24}{20}$$
$$x = \frac{6}{5} \quad \text{Simplifying}$$

66. -20

67.
$$\sqrt{40} - 2\sqrt{10} + \sqrt{90} = \sqrt{4\cdot10} - 2\sqrt{10} + \sqrt{9\cdot10}$$
$$= \sqrt{4}\sqrt{10} - 2\sqrt{10} + \sqrt{9}\sqrt{10}$$
$$= 2\sqrt{10} - 2\sqrt{10} + 3\sqrt{10}$$
$$= (2-2+3)\sqrt{10}$$
$$= 3\sqrt{10}$$

68. $30x^5\sqrt{10}$

69. $(-1)^7 = (-1)(-1)(-1)(-1)(-1)(-1)(-1)$
$= 1 \cdot 1 \cdot 1 \cdot (-1)$
$= 1 \cdot (-1)$
$= -1$

70. 1

71. ◈

72. ◈

73. ◈

74. ◈

75. $x(3x + 7) - 3x = 0$
$3x^2 + 7x - 3x = 0$
$3x^2 + 4x = 0$
$x(3x + 4) = 0$
$x = 0 \quad or \quad 3x + 4 = 0$
$x = 0 \quad or \quad 3x = -4$
$x = 0 \quad or \quad x = -\dfrac{4}{3}$

The solutions are 0 and $-\dfrac{4}{3}$.

76. 0, 2

77. $x(5x - 7) = 1$
$5x^2 - 7x = 1$
$5x^2 - 7x - 1 = 0$
$a = 5, \, b = -7, \, c = -1$
$$x = \frac{-(-7) \pm \sqrt{(-7)^2 - 4 \cdot 5 \cdot (-1)}}{2 \cdot 5}$$
$$x = \frac{7 \pm \sqrt{69}}{10}$$

The solutions are $\dfrac{7 + \sqrt{69}}{10}$ and $\dfrac{7 - \sqrt{69}}{10}$, or $\dfrac{7 \pm \sqrt{69}}{10}$.

78. $\dfrac{3 \pm \sqrt{5}}{2}$

79. $x^2 + (x + 2)^2 = 7$
$x^2 + x^2 + 4x + 4 = 7$
$2x^2 + 4x + 4 = 7$
$2x^2 + 4x - 3 = 0$
$a = 2, \, b = 4, \, c = -3$
$$x = \frac{-4 \pm \sqrt{4^2 - 4 \cdot 2 \cdot (-3)}}{2 \cdot 2}$$
$$x = \frac{-4 \pm \sqrt{40}}{2 \cdot 2} = \frac{-4 \pm \sqrt{4 \cdot 10}}{4}$$
$$x = \frac{-4 \pm 2\sqrt{10}}{4} = \frac{2(-2 \pm \sqrt{10})}{2 \cdot 2}$$

$$x = \frac{-2 \pm \sqrt{10}}{2}$$

The solutions are $\dfrac{-2 + \sqrt{10}}{2}$ and $\dfrac{-2 - \sqrt{10}}{2}$, or $\dfrac{-2 \pm \sqrt{10}}{2}$.

80. $\dfrac{-7 \pm \sqrt{61}}{2}$

81.
$$\frac{x^2}{x + 5} - \frac{7}{x + 5} = 0, \quad \text{LCM is } x + 5$$
$$(x + 5)\left(\frac{x^2}{x + 5} - \frac{7}{x + 5}\right) = (x + 5) \cdot 0$$
$$x^2 - 7 = 0$$
$$x^2 = 7$$

$x = \sqrt{7} \text{ or } x = -\sqrt{7}$ Principle of square roots

Both numbers check. The solutions are $\sqrt{7}$ and $-\sqrt{7}$, or $\pm\sqrt{7}$.

82. $\sqrt{5}, \, -\sqrt{5}$

83.
$$\frac{1}{x} + \frac{1}{x + 6} = \frac{1}{5}, \quad \text{LCM is } 5x(x + 6)$$
$$5x(x + 6)\left(\frac{1}{x} + \frac{1}{x + 6}\right) = 5x(x + 6) \cdot \frac{1}{5}$$
$$5(x + 6) + 5x = x(x + 6)$$
$$5x + 30 + 5x = x^2 + 6x$$
$$10x + 30 = x^2 + 6x$$
$$0 = x^2 - 4x - 30$$
$a = 1, \, b = -4, \, c = -30$
$$x = \frac{-(-4) \pm \sqrt{(-4)^2 - 4 \cdot 1 \cdot (-30)}}{2 \cdot 1}$$
$$x = \frac{4 \pm \sqrt{136}}{2} = \frac{4 \pm 2\sqrt{34}}{2}$$
$$x = 2 \pm \sqrt{34}$$

Both numbers check. The solutions are $2 + \sqrt{34}$ and $2 - \sqrt{34}$, or $2 \pm \sqrt{34}$.

84. $\dfrac{5 \pm \sqrt{37}}{2}$

85. *Familiarize.* From the drawing in the text, we see that we have a right triangle where $r =$ the length of each leg and $r + 2 =$ the length of the hypotenuse, in centimeters.

Translate. We use the Pythagorean theorem.
$$r^2 + r^2 = (r + 2)^2.$$

Carry out. We solve the equation.
$$2r^2 = r^2 + 4r + 4$$
$$r^2 - 4r - 4 = 0$$
$a = 1, \, b = -4, \, c = -4$

$$r = \frac{-(-4) \pm \sqrt{(-4)^2 - 4 \cdot 1 \cdot (-4)}}{2 \cdot 1}$$

$$r = \frac{4 \pm \sqrt{16 + 16}}{2} = \frac{4 \pm \sqrt{32}}{2}$$

$$r = \frac{4 \pm \sqrt{16 \cdot 2}}{2} = \frac{4 \pm 4\sqrt{2}}{2}$$

$$r = \frac{2(2 \pm 2\sqrt{2})}{2 \cdot 1} = 2 \pm 2\sqrt{2}$$

$x = 2 - 2\sqrt{2} \quad or \quad x = 2 + 2\sqrt{2}$

$x \approx 2 - 2.828 \quad or \quad x \approx 2 + 2.828$

$x \approx -0.828 \quad or \quad x \approx 4.828$

$x \approx -0.83 \quad or \quad x \approx 4.83$ Rounding to the nearest hundredth

Check. Since the length of a leg cannot be negative, -0.83 cannot be a solution of the original equation. When $x \approx 4.83$, then $x + 2 \approx 6.83$ and $(4.83)^2 + (4.83)^2 = 23.3289 + 23.3289 = 46.6578 \approx (5.83)^2$. This checks.

State. In the figure, $r = 2 + 2\sqrt{2}$ cm ≈ 4.83 cm.

86. $3 + 2\sqrt{2} \approx 5.828$ square units

87. Familiarize. We make a drawing. Let x represent the distance above the ground at which the pole broke, in feet. Then $20 - x$ represents the length of the pole that fell.

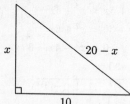

Translate. We use the Pythagorean theorem.
$$x^2 + 10^2 = (20 - x)^2$$

Carry out. We solve the equation.
$$x^2 + 10^2 = (20 - x)^2$$
$$x^2 + 100 = 400 - 40x + x^2$$
$$100 = 400 - 40x$$
$$-300 = -40x$$
$$7.5 = x$$

Check. When $x = 7.5$, then $20 - x = 20 - 7.5$, or 12.5.
$$(7.5)^2 + 10^2 = 56.25 + 100 = 156.25 = (12.5)^2$$
The answer checks.

State. The pole broke 7.5 ft above the ground.

88. $3 + 3\sqrt{2}$ cm ≈ 7.2 cm

89. Familiarize. We will use the formula $A = P(1 + r)^t$.

Translate. Substitute 8000 for P, 8904.20 for A, and 2 for t.
$$8904.20 = 8000(1 + r)^2$$

Carry out. We solve the equation.
$$8904.20 = 8000(1 + r)^2$$
$$\frac{8904.20}{8000} = (1 + r)^2$$
$$\sqrt{\frac{8904.20}{8000}} = 1 + r \quad or \quad -\sqrt{\frac{8904.20}{8000}} = 1 + r$$
$$1.055 = 1 + r \quad or \quad -1.055 = 1 + r$$
$$0.055 = r \quad or \quad -2.055 = r$$

Check. Since the interest rate cannot be negative, we check only 0.055, or 5.5%. We substitute in the formula:
$$8000(1 + 0.055)^2 = 8000(1.055)^2 = 8904.2$$
The answer checks.

State. The interest rate is 5.5%

90. $2682.63

91. Familiarize. The area of the actual strike zone is $15(40)$, so the area of the enlarged zone is $15(40) + 0.4(15)(40)$, or $1.4(15)(40)$. From the drawing in the text we see that the dimensions of the enlarged strike zone are $15 + 2x$ by $40 + 2x$.

Translate. Using the formula $A = lw$, we write an equation for the area of the enlarged strike zone.
$$1.4(15)(40) = (15 + 2x)(40 + 2x)$$

Carry out. We solve the equation.
$$1.4(15)(40) = (15 + 2x)(40 + 2x)$$
$$840 = 600 + 110x + 4x^2 \quad \text{Multiplying on both sides}$$
$$0 = 4x^2 + 110x - 240$$
$$0 = 2x^2 + 55x - 120 \quad \text{Dividing by 2}$$
$$a = 2, b = 55, c = -120$$
$$x = \frac{-55 \pm \sqrt{55^2 - 4 \cdot 2 \cdot (-120)}}{2 \cdot 2}$$
$$x = \frac{-55 \pm \sqrt{3985}}{4}$$
$$x \approx 2.0 \quad or \quad x \approx -29.5$$

Check. Since the measurement cannot be negative, -29.5 cannot be a solution. If $x = 2$, then the dimensions of the enlarged strike zone are $15 + 2 \cdot 2$, or 19, by $40 + 2 \cdot 2$, or 44, and the area is $19 \cdot 44 = 836 \approx 840$. The answer checks.

State. The dimensions of the enlarged strike zone are 19 in. by 44 in.

92. ▮

Exercise Set 10.4

1. $\sqrt{-1} = i$

2. $6i$

3. $\sqrt{-9} = \sqrt{-1 \cdot 9} = \sqrt{-1} \cdot \sqrt{9} = i \cdot 3 = 3i$

4. $9i$

5. $\sqrt{-50} = \sqrt{-1 \cdot 25 \cdot 2} = \sqrt{-1} \cdot \sqrt{25}\sqrt{2} =$
$i \cdot 5\sqrt{2} = 5i\sqrt{2}$, or $5\sqrt{2}i$

6. $2\sqrt{11}i$

7. $-\sqrt{-20} = -\sqrt{-1 \cdot 4 \cdot 5} = -\sqrt{-1} \cdot \sqrt{4} \cdot \sqrt{5} =$
$-i \cdot 2\sqrt{5} = -2i\sqrt{5}$, or $-2\sqrt{5}i$

8. $-3\sqrt{5}i$

9. $-\sqrt{-18} = -\sqrt{-1 \cdot 9 \cdot 2} = -\sqrt{-1} \cdot \sqrt{9} \cdot \sqrt{2} =$
$-i \cdot 3\sqrt{2} = -3i\sqrt{2}$, or $-3\sqrt{2}i$

10. $-2\sqrt{7}i$

11. $4 + \sqrt{-49} = 4 + \sqrt{-1 \cdot 49} = 4 + \sqrt{-1} \cdot \sqrt{49} =$
$4 + i \cdot 7 = 4 + 7i$

12. $7 + 2i$

13. $7 + \sqrt{-16} = 7 + \sqrt{-1 \cdot 16} = 7 + \sqrt{-1} \cdot \sqrt{16} =$
$7 + i \cdot 4 = 7 + 4i$

14. $-8 - 6i$

15. $3 - \sqrt{-98} = 3 - \sqrt{-1 \cdot 98} = 3 - \sqrt{-1} \cdot \sqrt{98} =$
$3 - i \cdot 7\sqrt{2} = 3 - 7i\sqrt{2}$

16. $-2 + 5i\sqrt{5}$

17. $x^2 + 9 = 0$
$x^2 = -9$
$x = \sqrt{-9} \quad$ or $\quad x = -\sqrt{-9}$
$x = \sqrt{-1}\sqrt{9} \quad$ or $\quad x = -\sqrt{-1}\sqrt{9}$
$x = 3i \qquad$ or $\quad x = -3i \qquad$ Principle of
square roots

The solutions are $3i$ and $-3i$, or $\pm 3i$.

18. $\pm 2i$

19. $x^2 = -28$
$x = \sqrt{-28} \qquad$ or $\quad x = -\sqrt{-28} \quad$ Principle
of square roots
$x = \sqrt{-1 \cdot 4 \cdot 7} \quad$ or $\quad x = -\sqrt{-1 \cdot 4 \cdot 7}$
$x = i \cdot 2\sqrt{7} \qquad$ or $\quad x = -i \cdot 2\sqrt{7}$
$x = 2i\sqrt{7} \qquad$ or $\quad x = -2i\sqrt{7}$

The solutions are $2i\sqrt{7}$ and $-2i\sqrt{7}$, or $\pm 2i\sqrt{7}$.

20. $\pm 4i\sqrt{3}$

21. $x^2 - 4x + 6 = 0$
$a = 1, \, b = -4, \, c = 6$
$$x = \frac{-b \pm \sqrt{b^2 - 4ac}}{2a}$$
$$x = \frac{-(-4) \pm \sqrt{(-4)^2 - 4 \cdot 1 \cdot 6}}{2 \cdot 1}$$
$$x = \frac{4 \pm \sqrt{-8}}{2}$$
$$x = \frac{4 \pm \sqrt{-1}\sqrt{8}}{2} = \frac{4 \pm i \cdot 2\sqrt{2}}{2}$$
$$x = \frac{4}{2} \pm \frac{2\sqrt{2}}{2}i \qquad \text{Writing in the form } a + bi$$
$$x = 2 \pm \sqrt{2}i$$

The solutions are $2 \pm \sqrt{2}i$.

22. $-2 \pm i$

23. $(x - 3)^2 = -25$
$x - 3 = \sqrt{-25} \quad$ or $\quad x - 3 = -\sqrt{-25} \quad$ Principle
of square roots
$x - 3 = 5i \qquad$ or $\quad x - 3 = -5i$
$x = 3 + 5i \quad$ or $\qquad x = 3 - 5i$

The solutions are $3 + 5i$ and $3 - 5i$, or $3 \pm 5i$.

24. $-1 \pm 4i$

25. $x^2 + 2x + 2 = 0$
$a = 1, \, b = 2, \, c = 2$
$$x = \frac{-b \pm \sqrt{b^2 - 4ac}}{2a}$$
$$x = \frac{-2 \pm \sqrt{2^2 - 4 \cdot 1 \cdot 2}}{2 \cdot 1}$$
$$x = \frac{-2 \pm \sqrt{-4}}{2} = \frac{-2 \pm i\sqrt{4}}{2} = \frac{-2 \pm 2i}{2}$$
$$x = \frac{-2}{2} \pm \frac{2i}{2} = -1 \pm i$$

The solutions are $-1 \pm i$.

26. $1 \pm 2i$

27. $\qquad x^2 + 7 = 4x$
$x^2 - 4x + 7 = 0 \quad$ Standard form
$a = 1, \, b = -4, \, c = 7$
$$x = \frac{-b \pm \sqrt{b^2 - 4ac}}{2a}$$
$$x = \frac{-(-4) \pm \sqrt{(-4)^2 - 4 \cdot 1 \cdot 7}}{2 \cdot 1}$$
$$x = \frac{4 \pm \sqrt{-12}}{2} = \frac{4 \pm i\sqrt{12}}{2}$$
$$x = \frac{4 \pm 2i\sqrt{3}}{2} = \frac{4}{2} \pm \frac{2\sqrt{3}}{2}i = 2 \pm \sqrt{3}i$$

The solutions are $2 \pm \sqrt{3}i$.

28. $-2 \pm \sqrt{3}i$

29. $2t^2 + 6t + 5 = 0$

$a = 2,\ b = 6,\ t = 5$

$$t = \frac{-b \pm \sqrt{b^2 - 4ac}}{2a}$$

$$t = \frac{-6 \pm \sqrt{6^2 - 4 \cdot 2 \cdot 5}}{2 \cdot 2}$$

$$t = \frac{-6 \pm \sqrt{-4}}{4} = \frac{-6 \pm i\sqrt{4}}{4} = \frac{-6 \pm 2i}{4}$$

$$t = \frac{-6}{4} \pm \frac{2}{4}i = -\frac{3}{2} \pm \frac{1}{2}i$$

The solutions are $-\dfrac{3}{2} \pm \dfrac{1}{2}i$.

30. $-\dfrac{3}{8} \pm \dfrac{\sqrt{23}}{8}i$

31. $1 + 2m + 3m^2 = 0$

$3m^2 + 2m + 1 = 0$ Standard form

$a = 3,\ b = 2,\ c = 1$

$$m = \frac{-b \pm \sqrt{b^2 - 4ac}}{2a}$$

$$m = \frac{-2 \pm \sqrt{2^2 - 4 \cdot 3 \cdot 1}}{2 \cdot 3}$$

$$m = \frac{-2 \pm \sqrt{-8}}{6} = \frac{-2 \pm i\sqrt{8}}{6} = \frac{-2 \pm 2i\sqrt{2}}{6}$$

$$m = \frac{-2}{6} \pm \frac{2\sqrt{2}}{6}i = -\frac{1}{3} \pm \frac{\sqrt{2}}{3}i$$

The solutions are $-\dfrac{1}{3} \pm \dfrac{\sqrt{2}}{3}i$.

32. $\dfrac{3}{4} \pm \dfrac{\sqrt{3}}{4}i$

33. Graph $y = \dfrac{3}{5}x - 2$.

x	y	(x, y)
-5	-5	$(-5, -5)$
0	-2	$(0, -2)$
5	1	$(5, 1)$

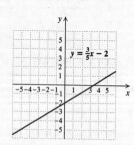

34.

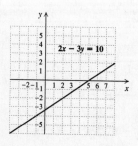

35. Graph $y = -4$.

The graph is a horizontal line with y-intercept $(0, -4)$.

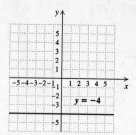

36.

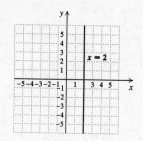

37. $-1 - 4^2 = -1 - 16 = -17$

38. -16

39. ◈

40. ◈

41. ◈

42. ◈

43. $(x + 1)^2 + (x + 3)^2 = 0$

$x^2 + 2x + 1 + x^2 + 6x + 9 = 0$

$2x^2 + 8x + 10 = 0$

$x^2 + 4x + 5 = 0$ Dividing by 2

$a = 1,\ b = 4,\ c = 5$

$$x = \frac{-b \pm \sqrt{b^2 - 4ac}}{2a}$$

$$x = \frac{-4 \pm \sqrt{4^2 - 4 \cdot 1 \cdot 5}}{2 \cdot 1}$$

$$x = \frac{-4 \pm \sqrt{16 - 20}}{2} = \frac{-4 \pm \sqrt{-4}}{2}$$

$$x = \frac{-4 \pm 2i}{2} = \frac{2(-2 \pm i)}{2 \cdot 1}$$

$$x = -2 \pm i$$

The solutions are $-2 \pm i$.

44. $-3 \pm 2i$

45. $\dfrac{2x-1}{5} - \dfrac{2}{x} = \dfrac{x}{2}$

We multiply by $10x$, the LCD.

$$10x\left(\frac{2x-1}{5} - \frac{2}{x}\right) = 10x \cdot \frac{x}{2}$$

$$2x(2x-1) - 10 \cdot 2 = 5x \cdot x$$

$$4x^2 - 2x - 20 = 5x^2$$

$$0 = x^2 + 2x + 20$$

$a = 1,\ b = 2,\ c = 20$

$$x = \frac{-b \pm \sqrt{b^2 - 4ac}}{2a}$$

$$x = \frac{-2 \pm \sqrt{2^2 - 4 \cdot 1 \cdot 20}}{2 \cdot 1}$$

$$x = \frac{-2 \pm \sqrt{-76}}{2} = \frac{-2 \pm i\sqrt{76}}{2} = \frac{-2 \pm 2i\sqrt{19}}{2}$$

$$x = \frac{-2}{2} \pm \frac{2\sqrt{19}}{2}i = -1 \pm \sqrt{19}i$$

The solutions are $-1 \pm \sqrt{19}i$.

46. $\dfrac{1}{2} \pm \dfrac{\sqrt{3}}{6}i$

47. Example 2(a):

Graph $y = x^2 + 3x + 4$.

There are no x-intercepts, so the equation $x^2 + 3x + 4 = 0$ has no real-number solutions.

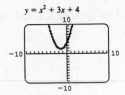

Example 2(b):

Graph $y_1 = x^2 + 2$ and $y_2 = 2x$.

The graphs do not intersect, so the equation $x^2 + 2 = 2x$ has no real-number solutions.

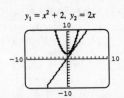

Exercise Set 10.5

1. $y = x^2 - 2$

We first find the vertex. The x-coordinate is

$$-\frac{b}{2a} = -\frac{0}{2 \cdot 1} = 0.$$

We substitute into the equation to find the second coordinate of the vertex.

$$x^2 - 2 = 0^2 - 2 = -2$$

The vertex is $(0, -2)$. The line of symmetry is $x = 0$, the y-axis.

We choose some x-values on both sides of the vertex and graph the parabola.

When $x = 1$, $y = 1^2 - 2 = 1 - 2 = -1$.

When $x = -1$, $y = (-1)^2 - 2 = 1 - 2 = -1$.

When $x = 2$, $y = 2^2 - 2 = 4 - 2 = 2$.

When $x = -2$, $y = (-2)^2 - 2 = 4 - 2 = 2$.

x	y
0	-2
1	-1
-1	-1
2	2
-2	2

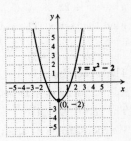

2.

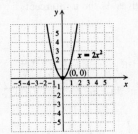

3. $y = -1 \cdot x^2$

Find the vertex. The x-coordinate is

$$-\frac{b}{2a} = -\frac{0}{2(-1)} = 0.$$

The y-coordinate is

$$-1 \cdot x^2 = -1 \cdot 0^2 = 0.$$

The vertex is $(0, 0)$. The line of symmetry is $x = 0$, the y-axis.

Choose some x-values on both sides of the vertex and graph the parabola.

When $x = -2$, $y = -1 \cdot (-2)^2 = -1 \cdot 4 = -4$.

When $x = -1$, $y = -1 \cdot (-1)^2 = -1 \cdot 1 = -1$.

When $x = 1$, $y = -1 \cdot 1^2 = -1 \cdot 1 = -1$.

When $x = 2$, $y = -1 \cdot 2^2 = -1 \cdot 4 = -4$.

x	y
0	0
-2	-4
-1	-1
1	-1
2	-4

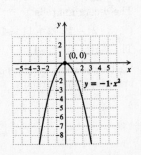

4.

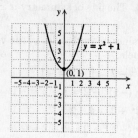

x	y
0	11
1	2
3	2
4	11

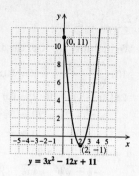

$y = 3x^2 - 12x + 11$

5. $y = -x^2 + 2x$

Find the vertex. The x-coordinate is
$$-\frac{b}{2a} = -\frac{2}{2(-1)} = -\frac{2}{-2} = 1.$$
The y-coordinate is
$$-x^2 + 2x = -(1)^2 + 2 \cdot 1 = -1 + 2 = 1.$$
The vertex is $(1, 1)$.

We choose some x-values on both sides of the vertex and graph the parabola. We make sure we find y when $x = 0$. This gives us the y-intercept.

x	y
1	1
0	0
−1	−3
2	0
3	−3

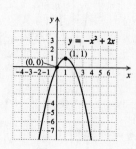

6.

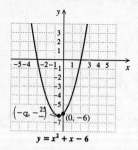

$y = x^2 + x - 6$

7. $y = 3x^2 - 12x + 11$

Find the vertex. The x-coordinate is
$$-\frac{b}{2a} = -\frac{-12}{2 \cdot 3} = 2.$$
The y-coordinate is
$$3 \cdot 2^2 - 12 \cdot 2 + 11 = 12 - 24 + 11 = -1.$$
The vertex is $(2, -1)$.

We choose some x-values on both sides of the vertex and graph the parabola.

8.

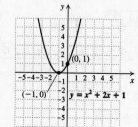

$y = x^2 + 2x + 1$

9. $y = x^2 - 2x - 3$

Find the vertex. The x-coordinate is
$$-\frac{b}{2a} = -\frac{-2}{2 \cdot 1} = 1.$$
The y-coordinate is
$$1^2 - 2 \cdot 1 - 3 = 1 - 2 - 3 = -4.$$
The vertex is $(1, -4)$.

We choose some x-values on both sides of the vertex and graph the parabola.

x	y
−1	0
0	−3
2	−3
3	0

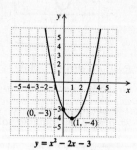

$y = x^2 - 2x - 3$

10.

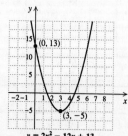

$y = 2x^2 - 12x + 13$

11. $y = -2x^2 - 4x + 1$

Find the vertex. The x-coordinate is
$$-\frac{b}{2a} = -\frac{-4}{2(-2)} = -1.$$

The y-coordinate is

$$-2(-1)^2 - 4(-1) + 1 = -2 + 4 + 1 = 3.$$

The vertex is $(-1, 3)$.

We choose some x-values on both sides of the vertex and graph the parabola.

x	y
-3	-5
-2	1
0	1
1	-5

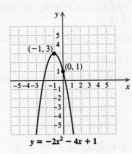

$y = -2x^2 - 4x + 1$

12.

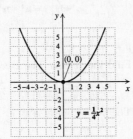

$y = -3x^2 - 2x + 8$

13. $y = \dfrac{1}{4}x^2$

Find the vertex. The x-coordinate is

$$-\frac{b}{2a} = -\frac{0}{2 \cdot \frac{1}{4}} = 0.$$

The y-coordinate is

$$\frac{1}{4} \cdot 0^2 = 0.$$

The vertex is $(0, 0)$.

We choose some x-values on both sides of the vertex and graph the parabola.

x	y
-4	4
-2	1
2	1
4	4

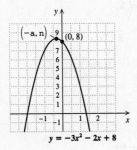

$y = \frac{1}{4}x^2$

14.

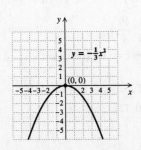

$y = -\frac{1}{3}x^2$

15. $y = -\dfrac{1}{2}x^2 + 5$

Find the vertex. The x-coordinate is

$$-\frac{b}{2a} = -\frac{0}{2\left(-\dfrac{1}{2}\right)} = 0.$$

The y-coordinate is

$$-\frac{1}{2} \cdot 0^2 + 5 = 0 + 5 = 5.$$

The vertex is $(0, 5)$.

We choose some x-values on both sides of the vertex and graph the parabola.

x	y
-4	-3
-2	3
2	3
4	-3

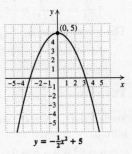

$y = -\frac{1}{2}x^2 + 5$

16.

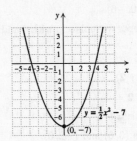

$y = \frac{1}{2}x^2 - 7$

17. $y = x^2 - 3x$

Find the vertex. The x-coordinate is

$$-\frac{b}{2a} = -\frac{-3}{2 \cdot 1} = \frac{3}{2}.$$

The y-coordinate is

$$\left(\frac{3}{2}\right)^2 - 3 \cdot \frac{3}{2} = \frac{9}{4} - \frac{9}{2} = -\frac{9}{4}.$$

The vertex is $\left(\dfrac{3}{2}, -\dfrac{9}{4}\right)$.

We choose some x-values on both sides of the vertex and graph the parabola.

x	y
0	0
1	-2
2	-2
3	0

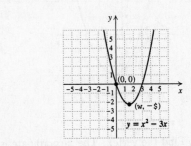

$y = x^2 - 3x$

18.

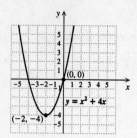

19. $y = x^2 - x - 6$

Find the vertex. The x-coordinate is
$$-\frac{b}{2a} = -\frac{-1}{2 \cdot 1} = \frac{1}{2}.$$
The y-coordinate is
$$\left(\frac{1}{2}\right)^2 - \frac{1}{2} - 6 = \frac{1}{4} - \frac{1}{2} - 6 = -\frac{25}{4}.$$
The vertex is $\left(\frac{1}{2}, -\frac{25}{4}\right)$.

To find the y-intercept we replace x with 0 and compute y:
$$y = 0^2 - 0 - 6 = 0 - 0 - 6 = -6.$$
The y-intercept is $(0, -6)$.

To find the x-intercepts we replace y with 0 and solve for x.
$$0 = x^2 - x - 6$$
$$0 = (x - 3)(x + 2)$$
$$x - 3 = 0 \quad or \quad x + 2 = 0$$
$$x = 3 \quad or \qquad x = -2$$
The x-intercepts are $(3, 0)$ and $(-2, 0)$.

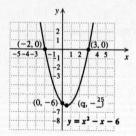

20.

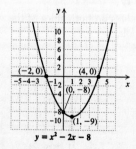

21. $y = 2x^2 - 5x$

Find the vertex. The x-coordinate is
$$-\frac{b}{2a} = -\frac{-5}{2 \cdot 2} = \frac{5}{4}.$$

The y-coordinate is
$$2\left(\frac{5}{4}\right)^2 - 5 \cdot \frac{5}{4} = \frac{25}{8} - \frac{25}{4} = -\frac{25}{8}.$$
The vertex is $\left(\frac{5}{4}, -\frac{25}{8}\right)$.

To find the y-intercept we replace x with 0 and compute y:
$$y = 2 \cdot 0^2 - 5 \cdot 0 = 0 - 0 = 0.$$
The y-intercept is $(0, 0)$.

To find the x-intercepts we replace y with 0 and solve for x.
$$0 = 2x^2 - 5x$$
$$0 = x(2x - 5)$$
$$x = 0 \quad or \quad 2x - 5 = 0$$
$$x = 0 \quad or \qquad 2x = 5$$
$$x = 0 \quad or \qquad x = \frac{5}{2}$$
The x-intercepts are $(0, 0)$ and $\left(\frac{5}{2}, 0\right)$.

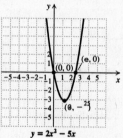

22.

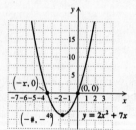

23. $y = -x^2 - x + 12$

Find the vertex. The x-coordinate is
$$-\frac{b}{2a} = -\frac{-1}{2(-1)} = -\frac{1}{2}.$$
The y-coordinate is
$$-\left(-\frac{1}{2}\right)^2 - \left(-\frac{1}{2}\right) + 12 = -\frac{1}{4} + \frac{1}{2} + 12 = \frac{49}{4}.$$
The vertex is $\left(-\frac{1}{2}, \frac{49}{4}\right)$.

To find the y-intercept we replace x with 0 and compute y:
$$y = -0^2 - 0 + 12 = -0 - 0 + 12 = 12.$$
The y-intercept is $(0, 12)$.

To find the x-intercepts we replace y with 0 and solve for x.

$$0 = -x^2 - x + 12$$
$$0 = x^2 + x - 12 \qquad \text{Multiplying by } -1$$
$$0 = (x+4)(x-3)$$
$$x + 4 = 0 \quad or \quad x - 3 = 0$$
$$x = -4 \quad or \qquad x = 3$$

The x-intercepts are $(-4, 0)$ and $(3, 0)$.

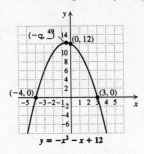

$$y = -x^2 - x + 12$$

24.

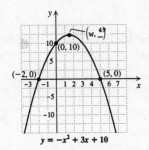

$$y = -x^2 + 3x + 10$$

25. $y = -3x^2 + 6x - 1$

Find the vertex. The x-coordinate is
$$-\frac{b}{2a} = -\frac{6}{2(-3)} = 1.$$
The y-coordinate is
$$-3 \cdot 1^2 + 6 \cdot 1 - 1 = -3 + 6 - 1 = 2.$$
The vertex is $(1, 2)$.

To find the y-intercept we replace x with 0 and compute y:
$$y = -3 \cdot 0^2 + 6 \cdot 0 - 1 = 0 + 0 - 1 = -1.$$
The y-intercept is $(0, -1)$.

To find the x-intercepts we replace y with 0 and solve for x.

$$0 = -3x^2 + 6x - 1$$
$$0 = 3x^2 - 6x + 1 \qquad \text{Multiplying by } -1$$
$$x = \frac{-b \pm \sqrt{b^2 - 4ac}}{2a}$$
$$x = \frac{-(-6) \pm \sqrt{(-6)^2 - 4 \cdot 3 \cdot 1}}{2 \cdot 3}$$
$$x = \frac{6 \pm \sqrt{36 - 12}}{6} = \frac{6 \pm \sqrt{24}}{6}$$
$$x = \frac{6 \pm 2\sqrt{6}}{6} = \frac{2(3 \pm \sqrt{6})}{2 \cdot 3}$$
$$x = \frac{3 \pm \sqrt{6}}{3}$$

The x-intercepts are $\left(\frac{3 - \sqrt{6}}{3}, 0\right)$ and $\left(\frac{3 + \sqrt{6}}{3}, 0\right)$, or about $(-0.184, 0)$ and $(1.816, 0)$.

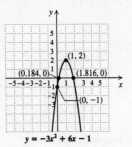

$$y = -3x^2 + 6x - 1$$

26.

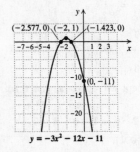

$$y = -3x^2 - 12x - 11$$

27. $y = x^2 - 2x + 3$

Find the vertex. The x-coordinate is
$$-\frac{b}{2a} = -\frac{-2}{2 \cdot 1} = 1.$$
The y-coordinate is
$$1^2 - 2 \cdot 1 + 3 = 1 - 2 + 3 = 2.$$
The vertex is $(1, 2)$.

To find the y-intercept we replace x with 0 and compute y:
$$y = 0^2 - 2 \cdot 0 + 3 = 0 - 0 + 3 = 3.$$
The y-intercept is $(0, 3)$.

To find the x-intercepts we replace y with 0 and solve for x.

$$0 = x^2 - 2x + 3$$
$$x = \frac{-(-2) \pm \sqrt{(-2)^2 - 4 \cdot 1 \cdot 3}}{2 \cdot 1}$$
$$x = \frac{2 \pm \sqrt{4 - 12}}{2} = \frac{2 \pm \sqrt{-8}}{2}$$

Since the radicand, -8, is negative the equation has no real-number solutions and hence the graph has no x-intercepts.

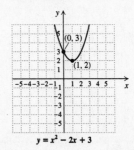

$$y = x^2 - 2x + 3$$

28.

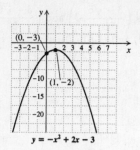

$$y = -x^2 + 2x - 3$$

29. $y = 1 - 4x - 2x^2$

Find the vertex. The x-coordinate is

$$-\frac{b}{2a} = -\frac{-4}{2(-2)} = -1.$$

The y-coordinate is

$$1 - 4(-1) - 2(-1)^2 = 1 + 4 - 2 = 3.$$

The vertex is $(-1, 3)$.

To find the y-intercept we replace x with 0 and compute y:

$$y = 1 - 4 \cdot 0 - 2 \cdot 0^2 = 1 - 0 - 0 = 1.$$

The y-intercept is $(0, 1)$.

To find the x-intercepts we replace y with 0 and solve for x.

$$0 = 1 - 4x - 2x^2$$
$$0 = 2x^2 + 4x - 1 \qquad \text{Standard form}$$

$$x = \frac{-4 \pm \sqrt{4^2 - 4 \cdot 2 \cdot (-1)}}{2 \cdot 2}$$

$$x = \frac{-4 \pm \sqrt{16 + 8}}{4} = \frac{-4 \pm \sqrt{24}}{4}$$

$$x = \frac{-4 \pm 2\sqrt{6}}{4} = \frac{2(-2 \pm \sqrt{6})}{2 \cdot 2}$$

$$x = \frac{-2 \pm \sqrt{6}}{2}$$

The x-intercepts are $\left(\dfrac{-2 - \sqrt{6}}{2}, 0\right)$ and

$\left(\dfrac{-2 + \sqrt{6}}{2}, 0\right)$, or about $(-2.225, 0)$ and $(0.225, 0)$.

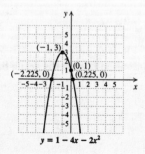

$$y = 1 - 4x - 2x^2$$

30.

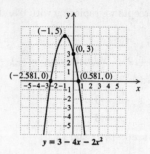

$$y = 3 - 4x - 2x^2$$

31. *Familiarize*. We make a drawing. Let $h =$ the height of the top of the pipe, in feet.

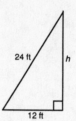

***Translate*.** We use the Pythagorean theorem.

$$12^2 + h^2 = 24^2$$

***Carry out*.** We solve the equation.

$$12^2 + h^2 = 24^2$$
$$144 + h^2 = 576$$
$$h^2 = 432$$

$$h = \sqrt{432} \quad \text{or} \quad h = -\sqrt{432}$$

***Check*.** Since the height of the top of the pipe cannot be negative, $-\sqrt{432}$ is not a solution. If $h = \sqrt{432}$, then $12^2 + (\sqrt{432})^2 = 144 + 432 = 576 = 24^2$ and $\sqrt{432}$ is the solution.

***State*.** The height of the top of the pipe is $\sqrt{432} \approx$ 20.78 ft.

32. $\sqrt{11,336} \approx 106.47$ ft

33. First we find the slope.

$$m = \frac{\text{change in } y}{\text{change in } x} = \frac{-3 - 7}{4 - (-2)} = \frac{-10}{6} = -\frac{5}{3}$$

Now we use the two-point equation and solve for y.

$$y - y_1 = m(x - x_1)$$

$$y - 7 = -\frac{5}{3}(x - (-2)) \qquad \text{Substituting } -\frac{5}{3} \text{ for } m, -2 \text{ for } x_1, \text{ and } 7 \text{ for } y_1$$

$$y - 7 = -\frac{5}{3}(x + 2)$$

$$y - 7 = -\frac{5}{3}x - \frac{10}{3}$$

$$y = -\frac{5}{3}x + \frac{11}{3} \qquad \text{Slope-intercept equation}$$

34. -3

35. $5a^3 - 2a = 5(-1)^3 - 2(-1)$
$$= 5(-1) - 2(-1)$$
$$= -5 + 2$$
$$= -3$$

36. 68

37.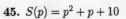

38. ◈

39. ◈

40. ◈

41. See the answer section in the text.

42. (a) After 2 sec, after 4 sec; (b) after 3 sec; (c) after 6 sec

43. a) For $r = 25$, $d = 25 + 0.05(25)^2 = 56.25$ ft
For $r = 40$, $d = 40 + 0.05(40)^2 = 120$ ft
For $r = 55$, $d = 55 + 0.05(55)^2 = 206.25$ ft
For $r = 65$, $d = 65 + 0.05(65)^2 = 276.25$ ft
For $r = 75$, $d = 75 + 0.05(75)^2 = 356.25$ ft
For $r = 100$, $d = 100 + 0.05(100)^2 = 600$ ft

b)

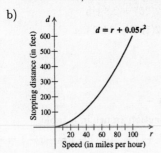

44.

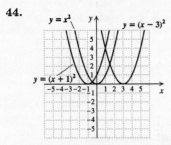

We can move the graph of $y = x^2$ to the right h units if $h \geq 0$ or to the left $|h|$ units if $h < 0$ to obtain the graph of $y = (x - h)^2$.

45. $S(p) = p^2 + p + 10$

p	S
0	10
1	12
2	16
3	22
4	30
5	40
6	52

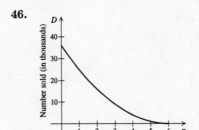

46.

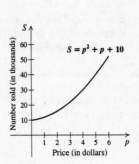

47.
$$D = S$$
$$(p - 6)^2 = p^2 + p + 10$$
$$p^2 - 12p + 36 = p^2 + p + 10$$
$$26 = 13p$$
$$2 = p$$

For $p = \$2$, $D = S$.

When $p = \$2$, $D(p) = (p - 6)^2 = (2 - 6)^2 = 16$, so 16,000 will be sold at that price.

48.

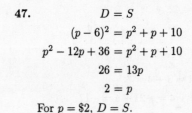

We can move the graph of $y = x^2$ up k units if $k \geq 0$ or down $|k|$ units if $k < 0$ to obtain the graph of $y = x^2 + k$.

49. Graph $y = x^2 - 5$ and find the first coordinate of the right-hand x-intercept. We find that $\sqrt{5} \approx 2.2361$.

Exercise Set 10.6

1. Yes; each member of the domain is matched to only one member of the range.

2. Yes

3. Yes; each member of the domain is matched to only one member of the range.

4. No

5. No; a member of the domain is matched to more than one member of the range. In fact, each member of the domain is matched to 3 members of the range.

6. Yes

7. Yes; each member of the domain is matched to only one member of the range.

8. Yes

9. $f(x) = x + 5$

$f(4) = 4 + 5 = 9$

$f(7) = 7 + 5 = 12$

$f(-3) = -3 + 5 = 2$

10. $-6, \ 0, \ 7$

11. $h(p) = 3p$

$h(-7) = 3(-7) = -21$

$h(5) = 3 \cdot 5 = 15$

$h(14) = 3 \cdot 14 = 42$

12. $-24, \ 2, \ -80$

13. $g(s) = 3s + 4$

$g(1) = 3 \cdot 1 + 4 = 3 + 4 = 7$

$g(-7) = 3(-7) + 4 = -21 + 4 = -17$

$g(6.7) = 3(6.7) + 4 = 20.1 + 4 = 24.1$

14. $19, \ 19, \ 19$

15. $F(x) = 2x^2 - 3x$

$F(0) = 2 \cdot 0^2 - 3 \cdot 0 = 0 - 0 = 0$

$F(-1) = 2(-1)^2 - 3(-1) = 2 + 3 = 5$

$F(2) = 2 \cdot 2^2 - 3 \cdot 2 = 8 - 6 = 2$

16. $0, \ 16, \ 21$

17. $f(t) = |t| + 1$

$f(-5) = |-5| + 1 = 5 + 1 = 6$

$f(0) = |0| + 1 = 0 + 1 = 1$

$f\left(-\dfrac{9}{4}\right) = \left|-\dfrac{9}{4}\right| + 1 = \dfrac{9}{4} + 1 = \dfrac{13}{4}$

18. $1, \ 91, \ 98$

19. $g(t) = t^3 + 3$

$g(1) = 1^3 + 3 = 1 + 3 = 4$

$g(-5) = (-5)^3 + 3 = -125 + 3 = -122$

$g(0) = 0^3 + 3 = 0 + 3 = 3$

20. $-3, \ -2, \ 78$

21. $F(x) = 2.75x + 71.48$

a) $F(32) = 2.75(32) + 71.48$

$\qquad = 88 + 71.48$

$\qquad = 159.48$ cm

b) $F(30) = 2.75(30) + 71.48$

$\qquad = 82.5 + 71.48$

$\qquad = 153.98$ cm

22. (a) 157.34 cm; (b) 171.79 cm

23. $P(d) = 1 + \dfrac{d}{33}$

$P(20) = 1 + \dfrac{20}{33} = 1\dfrac{20}{33}$

$P(30) = 1 + \dfrac{30}{33} = 1\dfrac{10}{11}$

$P(100) = 1 + \dfrac{100}{33} = 1 + 3\dfrac{1}{33} = 4\dfrac{1}{33}$

24. $70°$ C, $220°$ C, $10{,}020°$ C

25. $W(d) = 0.112d$

$W(16) = 0.112(16) = 1.792$ cm

$W(25) = 0.112(25) = 2.8$ cm

$W(100) = 0.112(100) = 11.2$ cm

26. $16\dfrac{2}{3}°, \ 25°, \ -5°$

27. Graph $f(x) = 3x - 1$

Make a list of function values in a table.

When $x = -1$, $f(-1) = 3(-1) - 1 = -3 - 1 = -4$.

When $x = 0$, $f(0) = 3 \cdot 0 - 1 = 0 - 1 = -1$.

When $x = 2$, $f(2) = 3 \cdot 2 - 1 = 6 - 1 = 5$.

x	$f(x)$
-1	-4
0	-1
2	5

Plot these points and connect them.

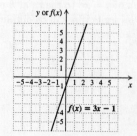

28.

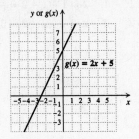

x	$f(x)$
-2	0
0	1
4	3

Plot these points and connect them.

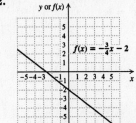

29. Graph $g(x) = -2x + 3$

Make a list of function values in a table.

When $x = -1$, $g(-1) = -2(-1) + 3 = 2 + 3 = 5$.

When $x = 0$, $g(0) = -2 \cdot 0 + 3 = 0 + 3 = 3$.

When $x = 3$, $g(3) = -2 \cdot 3 + 3 = -6 + 3 = -3$.

x	$g(x)$
-1	5
0	3
3	-3

Plot these points and connect them.

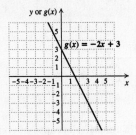

30.

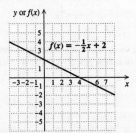

31. Graph $f(x) = \dfrac{1}{2}x + 1$.

Make a list of function values in a table.

When $x = -2$, $f(-2) = \dfrac{1}{2}(-2) + 1 = -1 + 1 = 0$.

When $x = 0$, $f(0) = \dfrac{1}{2} \cdot 0 + 1 = 0 + 1 = 1$.

When $x = 4$, $f(4) = \dfrac{1}{2} \cdot 4 + 1 = 2 + 1 = 3$.

32.

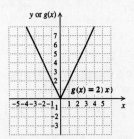

33. Graph $g(x) = 2|x|$.

Make a list of function values in a table.

When $x = -3$, $g(-3) = 2|-3| = 2 \cdot 3 = 6$.

When $x = -1$, $g(-1) = 2|-1| = 2 \cdot 1 = 2$.

When $x = 0$, $g(0) = 2|0| = 2 \cdot 0 = 0$.

When $x = 1$, $g(1) = 2|1| = 2 \cdot 1 = 2$.

When $x = 3$, $g(3) = 2|3| = 2 \cdot 3 = 6$.

x	$g(x)$
-3	6
-1	2
0	0
1	2
3	6

Plot these points and connect them.

34.

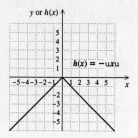

35. Graph $g(x) = x^2$.

Recall from Section 10.5 that the graph is a parabola. Make a list of function values in a table.

When $x = -2$, $g(-2) = (-2)^2 = 4$.

When $x = -1$, $g(-1) = (-1)^2 = 1$.

When $x = 0$, $g(0) = 0^2 = 0$.

When $x = 1$, $g(1) = 1^2 = 1$.

When $x = 2$, $g(2) = 2^2 = 4$.

x	$g(x)$
-2	4
-1	1
0	0
1	1
2	4

Plot these points and connect them.

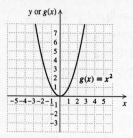

36.

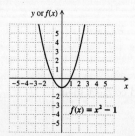

37. Graph $f(x) = x^2 - x - 2$.

Recall from Section 10.5 that the graph is a parabola. Make a list of function values in a table.

When $x = -1$, $f(-1) = (-1)^2 - (-1) - 2 = 1 + 1 - 2 = 0$.

When $x = 0$, $f(0) = 0^2 - 0 - 2 = -2$.

When $x = 1$, $f(1) = 1^2 - 1 - 2 = 1 - 1 - 2 = -2$.

When $x = 2$, $f(2) = 2^2 - 2 - 2 = 4 - 2 - 2 = 0$.

x	$f(x)$
-1	0
0	-2
1	-2
2	0

Plot these points and connect them.

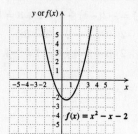

38.

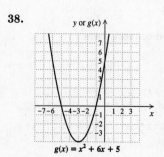

39. The graph is that of a function because no vertical line can cross the graph at more than one point.

40. No

41. The graph is not that of a function because a vertical line, say $x = -2$, crosses the graph at more than one point.

42. No

43. The graph is that of a function because no vertical line can cross the graph at more than one point.

44. Yes

45. The first equation is in slope-intercept form:

$$y = \frac{3}{4}x - 7, \ m = \frac{3}{4}$$

We write the second equation in slope-intercept form.

$$3x + 4y = 7$$
$$4y = -3x + 7$$
$$y = -\frac{3}{4}x + \frac{7}{4}, \ m = -\frac{3}{4}$$

Since the slopes are different, the equations do not represent parallel lines.

46. Yes

47. $2x - y = 6$, (1)

$4x - 2y = 5$ (2)

We solve Equation (1) for y.

$2x - y = 6$ (1)

$2x - 6 = y$ Adding y and -6

Substitute $2x - 6$ for y in Equation (2) and solve for x.

$$4x - 2y = 5 \quad (2)$$
$$4x - 2(2x - 6) = 5$$
$$4x - 4x + 12 = 5$$
$$12 = 5$$

We get a false equation, so the system has no solution.

48. Infinite number of solutions

49.

50. ◈

51. ◈

52. ◈

53. Graph $g(x) = x^3$.

Make a list of function values in a table. Then plot the points and connect them.

x	$g(x)$
-2	-8
-1	-1
0	0
1	1
2	8

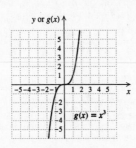

54.

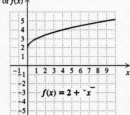

55. Graph $g(x) = |x| + x$.

Make a list of function values in a table. Then plot the points and connect them.

x	$f(x)$
-3	0
-2	0
-1	0
0	0
1	2
2	4
3	6

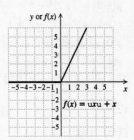

56.

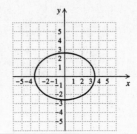

57. Answers may vary.

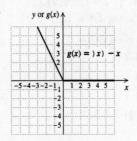

58. $y = \dfrac{15}{4}x - \dfrac{13}{4}$

59. $g(x) = ax^2 + bx + c$

$-4 = a \cdot 0^2 + b \cdot 0 + c$, or $-4 = c$ (1)

$0 = a(-2)^2 + b(-2) + c$, or $0 = 4a - 2b + c$ (2)

$0 = a \cdot 2^2 + b \cdot 2 + c$, or $0 = 4a + 2b + c$ (3)

Substitute -4 for c in Equations (2) and (3).

$0 = 4a - 2b - 4$, or $4 = 4a - 2b$ (5)

$0 = 4a + 2b - 4$, or $4 = 4a + 2b$ (6)

Add Equations (5) and (6).

$8 = 8a$

$1 = a$

Substitute 1 for a in Equation (6).

$4 = 4 \cdot 1 + 2b$

$0 = 2b$

$0 = b$

We have $a = 1$, $b = 0$, $c = -4$, so $g(x) = x^2 - 4$.

60. $\{5, 8, 11, 14\}$

61. $g(t) = t^2 - 5$

The domain is the set $\{-3, -2, -1, 0, 1\}$.

$g(-3) = (-3)^2 - 5 = 9 - 5 = 4$

$g(-2) = (-2)^2 - 5 = 4 - 5 = -1$

$g(-1) = (-1)^2 - 5 = 1 - 5 = -4$

$g(0) = 0^2 - 5 = 0 - 5 = -5$

$g(1) = 1^2 - 5 = 1 - 5 = -4$

The range is the set $\{-5, -4, -1, 4\}$.

62. $\{0, 2\}$

63. $f(m) = m^3 + 1$

The domain is the set $\{-2, -1, 0, 1, 2\}$.

$f(-2) = (-2)^3 + 1 = -8 + 1 = -7$

$f(-1) = (-1)^3 + 1 = -1 + 1 = 0$

$f(0) = 0^3 + 1 = 0 + 1 = 1$

$f(1) = 1^3 + 1 = 1 + 1 = 2$

$f(2) = 2^3 + 1 = 8 + 1 = 9$

The range is the set $\{-7, 0, 1, 2, 9\}$.

64.

Answers for Exercises in the Appendixes

Exercise Set A

1. $\{3, 4, 5, 6, 7, 8\}$ 2. $\{101, 102, 103, 104, 105, 106, 107\}$

3. $\{41, 43, 45, 47, 49\}$ 4. $\{15, 20, 25, 30, 35\}$

5. $\{-3, 3\}$ 6. $\{0.008\}$ 7. False 8. True 9. True

10. True 11. True 12. True 13. True 14. True

15. True 16. False 17. False 18. True

19. $\{c, d, e\}$ 20. $\{u, i\}$ 21. $\{1, 10\}$ 22. $\{0, 1\}$

23. $\emptyset$ 24. $\emptyset$ 25. $\{a, e, i, o, u, q, c, k\}$

26. $\{a, b, c, d, e, f, g\}$ 27. $\{0, 1, 2, 5, 7, 10\}$

28. $\{1, 2, 5, 10, 0, 7\}$ 29. $\{a, e, i, o, u, m, n, f, g, h\}$

30. $\{1, 2, 5, 10, a, b\}$ 31. Set-builder notation allows us to name a very large set compactly. 32. Roster notation allows us to see all the members of a set. It is also useful when a set cannot be named using a general rule or statement. 33. The set of integers 34. $\emptyset$ 35. The set of real numbers 36. The set of positive even integers 37. $\emptyset$ 38. The set of integers

39. (a) A; (b) A; (c) A; (d) $\emptyset$ 40. (a) Yes; (b) no; (c) no; (d) yes; (e) yes; (f) no 41. True

Exercise Set B

1. $(t + 3)(t^2 - 3t + 9)$ 2. $(p + 2)(p^2 - 2p + 4)$

3. $(a - 1)(a^2 + a + 1)$ 4. $(w - 4)(w^2 + 4w + 16)$

5. $(z + 5)(z^2 - 5z + 25)$ 6. $(x + 1)(x^2 - x + 1)$

7. $(2a - 1)(4a^2 + 2a + 1)$ 8. $(3x - 1)(9x^2 + 3x + 1)$

9. $(y - 3)(y^2 + 3y + 9)$ 10. $(p - 2)(p^2 + 2p + 4)$

11. $(4 + 5x)(16 - 20x + 25x^2)$

12. $(2 + 3b)(4 - 6b + 9b^2)$ 13. $(5p - 1)(25p^2 + 5p + 1)$

14. $(4w - 1)(16w^2 + 4w + 1)$

15. $(3m + 4)(9m^2 - 12m + 16)$

16. $(2t + 3)(4t^2 - 6t + 9)$ 17. $(p - q)(p^2 + pq + q^2)$

18. $(a + b)(a^2 - ab + b^2)$ 19. $\left(x + \dfrac{1}{2}\right)\left(x^2 - \dfrac{1}{2}x + \dfrac{1}{4}\right)$

20. $\left(y + \dfrac{1}{3}\right)\left(y^2 - \dfrac{1}{3}y + \dfrac{1}{9}\right)$ 21. $2(y - 4)(y^2 + 4y + 16)$

22. $3(z - 1)(z^2 + z + 1)$ 23. $3(2a + 1)(4a^2 - 2a + 1)$

24. $2(3x + 1)(9x^2 - 3x + 1)$ 25. $r(s + 4)(s^2 - 4s + 16)$

26. $a(b + 5)(b^2 - 5b + 25)$

27. $5(x - 2z)(x^2 + 2xz + 4z^2)$

28. $2(y - 3z)(y^2 + 3yz + 9z^2)$

29. $(x + 0.1)(x^2 - 0.1x + 0.01)$

30. $(y + 0.5)(y^2 - 0.5y + 0.25)$ 31. Observe that the product $(a - b)(a^2 + b^2) \neq a^3 - b^3$. 32. The number c is not a perfect cube. Otherwise, $x^3 + c$ could be factored as $(x + \sqrt[3]{c})(x^2 - \sqrt[3]{c}x + (\sqrt[3]{c})^2)$.

33. $(5c^2 - 2d^2)(25c^4 + 10c^2d^2 + 4d^4)$

34. $8(2x^2 - t^2)(4x^4 + 2x^2t^2 + t^4)$

35. $3(x^a + 2y^b)(x^{2a} - 2x^ay^b + 4y^{2b})$

36. $\left(\dfrac{2}{3}x + \dfrac{1}{4}y\right)\left(\dfrac{4}{9}x^2 - \dfrac{1}{6}xy + \dfrac{1}{16}y^2\right)$

37. $\dfrac{1}{3}\left(\dfrac{1}{2}xy + z\right)\left(\dfrac{1}{4}x^2y^2 - \dfrac{1}{2}xyz + z^2\right)$

38. $\dfrac{1}{2}\left(\dfrac{1}{2}x^a + y^{2a}z^{3b}\right)\left(\dfrac{1}{4}x^{2a} - \dfrac{1}{2}x^ay^{2a}z^{3b} + y^{4a}z^{6b}\right)$